PEASANTS, FARMERS AND SCIENTISTS

PEASANTS, FARMERS AND SCIENTISTS

A Chronicle of Tropical Agricultural Science in the Twentieth Century

H.J.W. Mutsaers

Springer

A C.I.P. Catalogue record for this book is available from the Library of Congress.

ISBN 978-1-4020-6165-3 (HB)
ISBN 978-1-4020-6166-0 (e-book)

Published by Springer,
P.O. Box 17, 3300 AA Dordrecht, The Netherlands.

www.springer.com

Front cover legend: One part of the cover shows a former 'pioneer' from one of the Dutch polders (Noordoostpolder), Henk te Raa, ploughing with two 'Belgian' horses on the enactment of a traditional farm called 'Op d'n Akker' (www.opdnakker.nl), owned by the van Oorschot family. The other half shows a farmer in southwest Nigeria in a field where he is testing a new crop, soybeans, among traditional crops in the area - maize, cocoyamsand in the background cassava.

Printed on acid-free paper

Acronyms

ADP	Agricultural Development Programme
ANOVA	Analysis of Variance
BACROS	Basic Crop Simulator
CGR	Crop Growth Rate
CIMMYT	Centro Internacional de Mejoramiento de Maiz y Trigo
CIRAD	Centre de Coopération Internationale en Recherche Agronomique pour le Développement
CMDT	Compagnie Malienne pour le Développement des Textiles
COR	Client-Oriented Research
CPM	Cotton Production Model
DSSAT	Decision Support for Agrotechnology Transfer
ELCROS	Elementary Crop growth Simulator
FACU	Federal Agricultural Coordination Unit
FAO	Food and Agriculture Organisation
FSP	Farming Systems Program
FSR	Farming Systems Research
FSSP	Farming Systems Support Program
GDD	Growing Degree Days
GIS	Geographic Information System
GMO	Genetically Modified Organism
GTZ	German Organisation for Technical Cooperation
HYV	High Yielding Variety
IAR	Institute for Agricultural Research
ICRISAT	International Crops Research Institute for the Semi-Arid Tropics
IDRC	International Development Research Centre (Canada)
IFSA	International Farming Systems Association
IITA	International Institute of Tropical Agriculture
ILACO	International Land Development Consultants
ILCA	International Livestock Centre for Africa
IMF	International Monetary Fund
IPM	Integrated Pest Management
IPNM	Integrated Plant Nutrient Management

IRRI   International Rice Research Institute
ISFM   Integrated Soil Fertility Management
ISNAR   International Service for National Agricultural
                Research
KIT   Royal Tropical Institute
LAI   Leaf Area Index
LEI   Agro-Economic Institute
LEXSYS   Legume Expert System
M&E   Monitoring and Evaluation
NAR   Net Assimilation Rate
NGO   Non-Government Organisation
NPT   New Plant Type
OFR   On-Farm Research
PAR   Photosynthetically Active Radiation
PRA   Participatory Rural Appraisal
RCMP   Resource and Crop Management Program (IITA)
RGR   Relative Growth Rate
RRA   Rapid Rural Appraisal
RUE   Radiation Use Efficiency
RYT   Relative Yield Total
SARP   Systems Analysis for Rice Production
SLA   Specific Leaf Area
SPARC   Support Group for Adaptive Research
                Cooperation
SUCROS   Simple and Universal Crop Simulator
SWAp   Sector-Wide Approach
SWOT   Strengths, Weaknesses, Opportunities and Threats
T&V   Training and Visit
UN   United Nations
USAID   United States Agency for International
                Development
VHYV   Very High Yielding Variety
ZOPP   Objective-Oriented Programme Planning

Contents

Preface

This book is a critical account of tropical agricultural science in the twentieth century, with its successes as well as its fads and failures. Its coverage reaches back into the beginning of the twentieth century, but the second half of the century takes place of choice. Which means that the story is closely linked with that of post-colonial development and scrutinises the role agricultural science has played (or failed to play) in the advancement of the tropical smallholder, in particular in Africa.

The book was written from the perspective of a practitioner, with the inevitable result that the treatment is skewed by this practitioner's own experience. That is an obvious drawback of a personalised account, but not a serious one I think, provided the author manages to make up for that disadvantage by getting across to the reader the thrill of his direct involvement.

A wide range of topics are covered, including interesting early science from the beginning of the last century (sugarcane breeding and shifting cultivation in Indonesia, anthropological studies in German West Africa), fascinating indigenous farming practices in Africa, the history of Farming Systems Research and its offshoots, computer modelling of crop growth, and the role of development projects, donors and consultants. Technical topics (such as soil fertility, plant breeding and crop modelling) are treated in an elementary but non-trivial way, such that intelligent lay persons will be able to understand them and form an opinion about their importance or otherwise. A considerable part of the book deals with such fashionable topics as Farming Systems Research and 'participatory' development approaches, which have dominated the field for 30 years, and analyses their sense and nonsense. The technical subjects are always treated in the context of the real African peasant's conditions and the few successes and numerous failures of agricultural research and development are thoroughly analysed.

The primary target readership are the agricultural science and development community and university students. The latter, particularly those specialising in agriculture in developing countries, will benefit from a critical historical review of their field of study. Since the book touches on a very wide range of topics related to tropical agriculture it provides a broad *tour d'horizon*, particularly useful for graduate students who have had extensive exposure to theory and perhaps some practical experience as well.

Finally, I am confident that the book will also appeal to the general educated public, who are becoming increasingly concerned about the obvious lack of impact of four or five decades of development aid. They will be interested to find out, at a non-trivial level, what tropical agricultural science has been about and understand what have been its challenges and promises, and what have been the factors affecting its achievements and failures, especially in Africa.

Acknowledgements

This book would never have seen the light of day without the help of numerous people, some of whom I want to acknowledge explicitly here.

First and foremost I would like to thank my former IITA colleague Peter Walker, statistician and agriculturist as well as gifted linguist, who went through all the chapters with a fine comb and pointed out many flaws – linguistic, stylistic, historical and agronomic. He has been my major guardian on the bumpy road to the completion of this work. It does not mean that he is in any way responsible for the remaining flaws. I have been stubbornly attached to some of my idiosyncrasies.

The other person who has contributed significantly to the concept and set-up of the book is Dirk Zoebl, tropical agronomist, philosopher of science and avid debater. His many comments on some of the key chapters helped me to keep the book readable for more than just a few initiated.

I have also been happy to find several recognised experts prepared to go through some of the more technical chapters and appendixes. In particular I wish to acknowledge the help of Jan Goudriaan and Jan Vos of Wageningen University for going through the modelling chapter and Jim Jones of the University of Florida and Basil Acock, formerly with the USDA-ARS, for commenting on the sections dealing with the American crop models. The comments by Vinus Zachariasse, Director of the Institute for Agricultural Economics (LEI) on the sections on farmer skills and factor analysis were much appreciated. Another former IITA colleague, B.T. Kang, 'father of alley cropping' as well as accomplished soil scientist, commented on my modelling acrobatics with alley cropping, sometimes disapprovingly, while accepting the overall conclusions.

In addition to the avoidance of conceptual and factual errors, a major benefit derived from the comments of all my reviewers was the (implicit or explicit) assurance that there was nothing seriously wrong with what I was saying, because it generally stood up to their scrutiny. I am therefore confident that the book's contents, while perhaps provocative and controversial at times, passes the test of scientific rigour.

List of Figures, Tables and Boxes

Figures, Photos

Tables

Boxes

Chapter 1. What Is Tropical Agronomy?

This book is a chronicle as well as a critical account of tropical agricultural science in the twentieth century, its successes as well as its fads and failures. Its ambition is to cover most of what has been significant in tropical agricultural science and agricultural development, although that qualification becomes less certain as one draws closer to one's own time. The book was written by a practitioner who has been directly involved in tropical agriculture for 45 years in different capacities. But this author is not a genius who can speak with equal authority about the very diverse disciplines which together form the agricultural sciences, or about all three continents where tropical climates are found, so I must make a few important restrictions at once. First, the emphasis in this book will be on the areas I am most familiar with, that is tropical West, Central and East Africa and Indonesia. Furthermore, I will draw most extensively, some will say excessively, on Dutch sources, both historical and more recent. That is because I am a Dutchman. Finally, and most importantly, the perspective from which the chronicle was written is that of an agronomist whose discipline, although referred to unquestioningly by its practitioners as 'agronomy', is rather ill-defined. So let us see first whether we can decide what tropical agronomy and agronomists are.

At best, agronomists are generalists with the necessary broadness of coverage as well as depth of understanding to separate the significant from the trivial in a wide range of agriculture-related disciplines. At worst they are jacks of all trades and masters of none. In which category this author belongs it will be the reader's privilege to judge. That is not very helpful, I am afraid; so I will try a little harder. Agronomy is not really something. It has to do with agriculture of course, but otherwise it is easier to explain what it is not than what it is. For example, it does not include plant breeding. Plant breeders make crosses between plants of the same or related species and then search among the progeny for individuals which are better than the ones they started from. If they are lucky and skilful, or rather both, they will come up with a better variety. Today, they may even transfer genes from an entirely unrelated species if the traits they are looking for are not available within the species. Plant breeding has glamour. It has always had, but especially now that genetically modified crops (or genetically manipulated if you are a pessimist) are coming inexorably to your table and wardrobe. Agronomists are also not soil scientists. Soil scientists deal with soils and they come in various breeds, for example soil chemists, soil physicists and soil taxonomists. We will meet some of them repeatedly in due course. Then there are several kinds of crop protectionists. They deal with insects and fungi, viruses and nematodes which attack crops, and with the things you can do to keep them out or get rid of them once they are there.

Maybe agronomists are general practitioners. They know a few things and, with time, perhaps even a lot, about most of the other disciplines, which is useful when they are extension agronomists whose job is to advise farmers. They may be research agronomists who do research about things which the specialists are not interested in. You may start wondering if there is much left for them, but there is. For example, how a crop reacts when it is planted at different dates and densities or fertiliser rates or combinations of those. Or what happens when you grow different kinds of crops in sequence or in a mixture and which of those sequences or mixtures give the best results. Some research agronomists chart out a sub-area for themselves and make that into their specialty, like crop physiology or even computer modelling. That does not mean that by so specialising one necessarily becomes successful. You may still end up being an unsuccessful researcher in some narrow sub-discipline, but the chances are probably a little better than when you flit from one area to another, which agronomy is eminently suitable for.

Agronomists may also specialise in a particular crop, such as cotton or bananas, get to know almost everything about it and call themselves cotton or banana agronomists. Or they may work in tropical countries, like I did, which makes them tropical agronomists. But that does not tell you much about what tropical agronomy is either.

So let us try something else and look at the kind of papers which are being published in the *Agronomy Journal*. In Box 1-1, I have made a small selection from the year 2000 which I think fairly represents the variety of topics that that year's issue of the journal dealt with.

This list is not of much help in defining agronomy either, apart from confirming that it is more difficult to say what it is than what

> *Box 1-1.* Representative titles from the 2000 issue of the *Agronomy Journal*
>
> 1. Simplifying daily evapotranspiration estimates over short full-canopy crops
> 2. Genetically modified crops and the environment
> 3. Simulating inbred-maize yields with CERES-IM
> 4. Agronomic changes from 53 years of genetic improvement of short-season soybean cultivars in Canada
> 5. Symptoms and growth of potato leafhopper-tolerant alfalfa in response to potato leafhopper feeding
> 6. Reversal of rice yield decline in a long-term continuous cropping experiment
> 7. Corn production with Kura clover as living mulch
> 8. Cropping system effects on weed emergence and densities in corn
> 9. Agronomic and economic analyses of cotton starter fertilisers

it is not. And it shows that the range of topics which qualify as agronomy is even wider than you may have thought; so perhaps agronomy is a little of everything after all. Most of the titles in the box sound rather unsexy, but that is normal for scientific papers. If you look at the titles of Einstein's papers you would not suspect that there is something exciting hidden (nor would most of us by reading the papers themselves). If you are an agronomist, however, you will appreciate that this list reflected some of the concerns of the profession at the turn of the century and in some cases, like number 2, of the population at large.

Apart from being an agronomist, what is particular about a tropical agronomist? For a European before 1960 it usually meant that you went to work in a colonial country. Colonial agriculture was mainly about growing crops like tea or rubber or oil palm for the European market in plantations managed by Europeans and using the natives to do most of the work. A lot of research was done on these crops in excellent research institutes, also run by Europeans. The links between plantation agriculture and research were tight, in some cases the industry itself funded the research institutes, which ensured that research addressed real practical problems put forward by the plantations. As a result, dissociation of research and practical farming, which is probably one of today's most important problems in tropical agriculture, was hardly an issue.

There was also an ethical side to colonialism. In the Dutch East Indies, for example, research institutes were set up early in the twentieth century to study indigenous, mainly food crops and extension services were created to bring the research results to the indigenous farmers. In other colonies something similar happened, although in Africa mostly a few decades later. The research institutes dealing with indigenous farming were also organised in a western way and, although the crops were different, the research itself was quite similar to that in the agricultural research institutes in Europe.

Things really started changing after most colonies had gained independence in the 1960s and 1970s. Although some of my classmates still ended up in foreign-owned sugar, rubber or oil palm plantations or in research institutes which continued to be run and financed by the former colonial powers, many found employment in development projects in Africa and Asia, which were meant to help indigenous farmers become modern producers as quickly as possible. That turned out to be easier said than done. Transferring so-called modern technology to traditional husbandmen did not meet with the expected success and after some time people started questioning the western model for agricultural research and development. Soon it was decided that the entire process was in need of overhaul and so were we, tropical agronomists. In fact, much of the story I have to tell is about that transformation process and what it has led to.

So there are two threads running through this book. One is about the agricultural sciences, in particular agronomy and its tropical variant, and how it has evolved as a scientific discipline. The other is about international development in which tropical agronomists have played a major role. My professional life spans an entire era. It started when I went to university, right at the end of the colonial period, and ended (well, almost) 45 years later. The scientific story is fascinating and the one about development, although sad in many respects, is not devoid of hope and often hilarious, so I think there is enough to enjoy. And if in the end we have disappointed those who put so much hope in us, we can always tell our successors what went wrong and how they should try and do better next time. That message runs like a red thread through this book.

Chapter 2. A Tropical Agronomist's Education

2.1 Wageningen and tropical agronomy around 1960

What I like are clear problems with clear answers, or at least the promise that such answers can somehow be found. Agriculture in developing countries mostly does not belong in that category. In Africa in particular, peasant agriculture is not easily amenable to exact study. The reason is that, as in much of Europe a few hundred years ago, agriculture is not simply a repertoire of techniques to produce agricultural goods. It is a way of life, steeped in folklore and tradition if you look at it with a romantic eye, and one which is marred by outdated beliefs and superstition if you are a sceptic. Whereas in industrialised countries farming is the chosen profession of a few skilled farmers, in most tropical countries it is an inherited way of life and one which many parents, especially in Africa, would rather like to see their children escape from if they can. So, when you deal with tropical agriculture there are many things which will distract you from the strictly technical or even hide it from view entirely.

Yet, many traditional farming practices, when looked at close enough, turn out to make a lot of sense. They will be found to be not just anthropologically interesting, but technically sound, although their rationale may have been long forgotten. We will come across many examples of highly ingenious 'traditional practices' in this book. Common sense, I suppose, would tell you that it has to be so, otherwise peasants would not have survived the harsh conditions under which they have to produce their food. What makes working in peasant agriculture a sometimes enervating experience, however, is its lack of the purposefulness, which is the trademark of today's western agriculture, and which make farmers highly receptive to the results of the agricultural sciences. No wonder then that many of my colleagues who started out as tropical agronomists in the end became some sort of sociologists, because the challenges of advancing tropical agriculture appeared to be more related with the human than the technical.

At the time I went to Wageningen Agricultural College in the Netherlands in 1959, tropical agriculture still had a romantic ring about it of pioneering in strange lands or standing at the top of a hill overseeing a colonial plantation with beautiful women picking tea leaves. I think that is what attracted me, although I would not have admitted it at the time. But when I arrived in Wageningen it was already too late for that. In the 1950s and 1960s colonial countries became independent in quick succession and many of the plantations were nationalised. Not all of them and not immediately, but a career in the plantations did not seem such a good idea anymore. A new era was dawning and the small tropical farmer was moving or rather was pushed to centre stage as the main actor of people-centred development. Plantations were 'out', except perhaps as a necessary evil to earn foreign exchange, with which to pay for more important things, like building infrastructure, setting up industries and pulling peasant agriculture out of its stagnation.

After Indonesia had gained independence from Dutch rule in 1948,[1] teaching about tropical agriculture in Wageningen continued to be dominated for some time by the Dutch East Indies, as Indonesia was called before independence. That is where practically all graduates from the tropical departments used to go. There were several tropical departments for agriculture, forestry, animal science, sociology and economics. The Department of Tropical Agriculture, which had been called Department of Colonial Agriculture before the war, was to become my academic home base. It was really a Department of Tropical Crop Husbandry and its name was changed into that in the late 1950s, just before I arrived. Although the department had always devoted a lot of time to teaching plantation crops, it had not neglected 'indigenous agriculture', in particular the intricate Javanese sawah rice system. The reason was that the colonial government had become increasingly concerned with peasant agriculture since the mid-nineteenth century and tried to stimulate its development through research and extension. And although many pre-war graduates from Wageningen continued to work in the plantation sector or in the specialised research stations

---

[1] Indonesia declared its independence in 1945 but had to fight a war of independence for 3 years before 'sovereignty was transferred'.

which worked on plantation crops, many also found employment in the colonial extension service and in the research institutions which were involved with peasant agriculture. The Tropical Crops Department kept a large collection of pre-war glass-mounted transparencies with scenes from the East Indies. When I was a student some of them were still shown at lectures by the associate professor who had made his career there. Most of the transparencies were beginning to fade, like the departmental memory of the colonial world itself.

If anything, tropical agronomy, like its temperate counterpart, is an interdisciplinary science. It cannot stand on its own but needs a host of other disciplines. The bulk of the undergraduate curriculum in Wageningen, as probably everywhere else, consisted therefore of introductory courses on a wide variety of agronomy-related subjects, ranging from the plant sciences, chemistry and physics through geology and soil science to statistics and economics. In later years, when peasant or smallholder agriculture had become the main focus, sociology and extension methodology were added, but in the early 1960s that was not the case. The basic sciences were taught quite thoroughly, including in most cases effective drill in the basic handiwork. With an old-fashioned shaving blade and a piece of elder pith I will still cut thin microscopic slices from a piece of alcohol-impregnated plant root. Then there were lectures on general tropical agronomy, which was something like production ecology, and on individual crops and groups of crops, their botany, nutrition, pests and diseases and methods of cultivation: interesting in some cases, but dull and eminently forgettable in others. Practical crop science is experiential and it is very difficult to teach it adequately in classroom except by the occasional exceptionally gifted teacher, who could bring practical and broad-ranging experience to the task. Only one of our teachers of tropical agronomy in the 1960s satisfied both these conditions.

I must say a few words, however, in favour of the very thorough botanical practicals run by the department, all the way up to the department's demise in the 1990s. The practicals were inherited from the colonial times, along with the person who delivered them, Dr. Frahm-Lelyveld, and were for most of us our first encounter with tropical crops. They were of course updated regularly over the years, but in my own student days they still had a

definite colonial flavour. We had to make detailed drawings of
sugarcane cuttings with precise rendering of the details of the
stem segment and its root primordia and buds. Those kinds of
drawings were still being made of new cane selections in the
research station in eastern Java when I visited there in 1970. And
we had to pin up rice panicles and describe their branching pattern and spikelets in detail and with precision. These sugarcane
and rice drawings had been used in the research institutes in the
East Indies to characterise sugarcane and rice varieties and they
became anchored in my mind as images from the colonial heritage
which attracted me to the profession.

Teaching tropical crop husbandry would perhaps have been more
effective if the students had had previous exposure at least to the
tropical environment, like the sons and daughters of the colonials
in the pre-war days, but most of our generation had no such exposure. Our first experience would come from working as trainees in a
tropical country for 6 months, sandwiched between undergraduate
and graduate training.

2.2 The story of sugarcane

Although the time for a career in a tropical plantation had
passed, I did spend almost a year as a trainee in a large sugar
estate in the Dominican Republic in 1963/64, and that also
somewhat cooled my enthusiasm for a life in the plantations. I
was 23 years old and the rum and the local ladies were good, but
every morning, including Saturdays, we had to go back beneath
the sugarcane canopy and spend the entire day piercing holes in
cane stalks to measure their sugar content (Figure 2-1). That was
because I was a sugarcane breeding assistant and we had to
examine endless numbers of mature cane seedlings during endless numbers of days in the intense heat and humidity of the
cane field. But we had a nice field team, two Dominicans and
one Haitian. Their favourite expression was: *tamos jodío*, we are
screwed here, followed by much merriment. They were very poor
people living in what I thought were disgraceful sheds owned
by the company. But much better off than the Haitian cane
cutters in the commercial plantation, many of whom were illegal
immigrants.

Figure 2-1. Sugarcane seedling selection at Central Romana, Dominican Republic, 1964

Plant breeding was not really my field of study, so that trainee-ship was not entirely fitting, but the important thing for us youngsters was to work in the tropics in some agriculture-related area, it did not really matter what exactly. I did like the theoretical part of sugarcane breeding. I will explain in a moment. I will first say something about my boss. He was born in a colonial family in the Dutch East Indies, but too late for a career of his own there. So, after graduating from a technical college in the Netherlands he ended up in the American-owned La Romana sugar estate. He judged everything against the background of his memories of the East Indies, like many of his generation, and nothing could match that. Colonial society had been closely knit and provided a strong identity if you fitted in. That identity was lost with the loss of the colonies and could not be regained by joining an American sugar estate in the Caribbean. His life and that of his family in La Romana was a poor shadow of the past. To make things worse, he did not have a university education, which meant that he was supervised by an elderly American,

Dr. Arceneaux, who came twice a year to design the breeding programme and left detailed instructions for him to carry out. He had to do the hard work of making crosses, growing seedlings and, the sweaty part, stamping through the cane fields looking for promising offspring. That is, until we came, another student and I, and took over the sweating. I do not remember what the boss did meanwhile but I do not think it was much. I even wondered whether he had more than a cursory understanding of sugarcane breeding, but the stories he told, over tea in his house, about the glories of the sugar industry in the East were interesting nevertheless. In particular the story about the famous Javanese sugarcane variety POJ 2878 is fascinating. I delved into that a little further on my own.

Between the early 1930s and the mid-1950s POJ 2878 was the most important sugarcane variety from Argentina to Louisiana and from Cuba to Indonesia. What was so special about POJ 2878? In order to explain that I will have to go far back in history and also into some technical details, but it is very interesting and quite simple really. In the early twentieth century sugarcane production in Java was threatened by a number of serious diseases to which all the cultivated varieties were susceptible, in particular the sereh and mosaic diseases, both caused by viruses. A group of breeders at the Proefstation Oost Java (the East Java Experimental Station), most prominent among them, J. Jeswiet, in a very short time created a new high-yielding variety which was insensitive to mosaic, by crossing the species of commerce, *Saccharum officinarum*, or noble cane, with a wild cane which occurred in Java and which was free of mosaic. I could not resist the temptation to tell the POJ 2878 story in detail, it is one of my favourites. I have put it in Box 2-1 which you may skip if you wish, or read later.

The sugar research stations in Java, early in the twentieth century, must have been a fertile environment for science. Although the stations were funded entirely by the sugar industry, their scientists were allowed to do what they thought was right, even though they produced no commercial varieties for several years before they hit on POJ 2878. Several of its scientists later on were called to university chairs in Holland, including Jeswiet (in Wageningen), F.A.F.C. Went and V.J. Koningsberger (in Utrecht), the latter two plant physiologists of international repute.

Box 2-1. Creating sugarcane variety POJ 2878

While reading this colonial breeding saga you have to keep an eye on the pedigree diagram shown below (Figure 2-2).

Sugarcane belongs to the genus *Saccharum*, of the grasses family. From an agricultural point of view the most interesting species in that genus was *Saccharum officinarum*, which has long thick stalks with a soft rind and a high sugar content, making it ideal for sugar extraction. It was called 'noble' cane in the Dutch East Indies and that name is still in use internationally today. All the cane varieties grown in the sugar plantations of the late nineteenth century belonged to this species. Many varieties were collected all over the archipelago and they were often named after the place where they were found. The important ones for our story are 'Black Cheribon' from West Java, a commercial variety planted extensively during the latter part of the nineteenth century, and 'Bandjarmasin hitam', a good variety from Borneo but very susceptible to diseases. When you look at those canes with their thick beautifully coloured stalks and luxurious leaf canopy the word noble is appropriate. Another noble cane, 'Loethers', had been imported from Mauritius. It did poorly as a production crop in Java, but turned out to be useful for breeding. Sugarcane is propagated vegetatively by stem cuttings, so every generation is identical to the previous one, except when there is some local 'sport' mutation.[2] So when you find an interesting individual you just go on cutting it up and planting the cuttings until you have a big area. This is straightforward and simple, not something messy like grain crops where every next generation starts from seed. Most sugarcane varieties would eventually flower and produce seed or you could manipulate them to do so; so it was only natural for breeders to start crossing them, plant the seed and see what the offspring would look like when they matured. And if they found a good one they would multiply it and if it stood the test of time they would have a new variety. That is what happened and in 1893 J.H. Wakker, a scientist at the East Java Experimental Station obtained an interesting plant from a cross between the noble varieties 'Bandjarmasin hitam' and 'Loethers'. He called this POJ 100. POJ stands for 'Proefstation Oost Java', which is Dutch for Experimental Station East Java.

Since its parents were noble (*Saccharum officinarum*), so is POJ 100 (see Figure 2-2). Remember this name because it comes back as the grandmother of POJ 2878. POJ 100 was grown for a while as a commercial variety but it was actually quite susceptible to various diseases. When in the early twentieth century sugarcane production in Java was threatened by the sereh and mosaic diseases, the breeders started looking outside the species *S. officinarum* for resistance. There was a sugarcane-like plant occurring in the wild, with thin, very hard stalk and little sugar, which was free of all common diseases and very vigorous. It was locally known under the name 'Kassoer'. In 1916 G. Bremer proved that this was actually a natural hybrid between a noble cane, probably 'Black Cheribon', and a bushy grass-like species with thin stalks, locally called

(continued)

---

[2] Sport mutations occur with a low frequency in vegetative buds.

Box 2-1. (continued)

'Glagah'. The scientific name of 'Glagah' is *Saccharum spontaneum*. At the research station 'Kassoer' had been crossed with noble canes as early as 1911, but its offspring was unsuitable for production because of its low sugar content. Then, in 1916, J. Jeswiet picked up a good looking seedling with reasonable sugar content from a cross between 'Kassoer' and POJ 100, which he named POJ 2364. He crossed it again with another noble cane, EK 28, which at the time was the most widely grown commercial variety on Java. Its pedigree was uncertain. The reason why Jeswiet used POJ 2364 as a parent was that it had inherited the resistance to diseases and the vigour of 'Kassoer' and, when crossed with other varieties, produced large numbers of good-looking individuals. That of course is what makes a variety suitable for breeding. One individual from the offspring of a 1921 cross was POJ 2878. The variety was tested in the field for a few years and officially released in 1925.

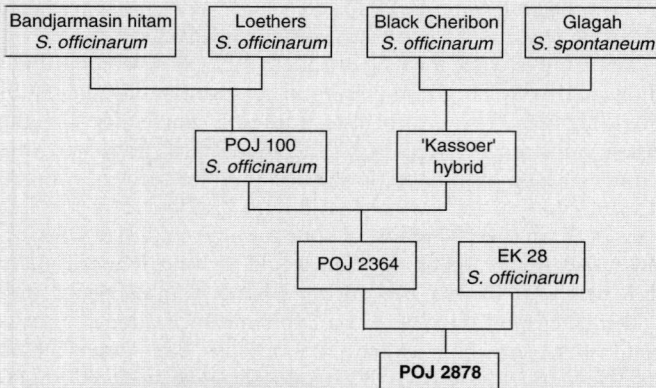

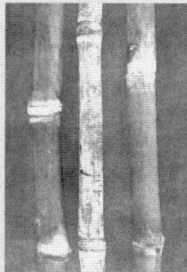

POJ 100 x Kassoer = POJ 2364   POJ 2364 x EK 28 = POJ 2878
(Bremer, 1928)   (Bremer, 1928)

Figure 2-2. The pedigree of Java's most famous sugarcane variety: POJ 2878 and photographs of its parents and grandparents

Four years after its release POJ 2878 occupied 90% of the total sugarcane area of Java, which was close to 200,000 ha at the time. As early as 1924 the variety was introduced in the western hemisphere and it saved the Caribbean and Louisiana sugar industries from extinction by the rapidly spreading mosaic virus. By 1928, 85% of the Louisiana sugar crop was POJ 2878.

By the time we arrived at the plantation in the Dominican Republic, all this was history. At the La Romana estate POJ 2878 had been replaced almost completely by two varieties from Barbados, B 41227 and B 4362. So was that the end of POJ 2878 then? Not quite. POJ 2878 was one of the parents of both these new Barbados varieties and conferred its high degree of mosaic resistance and wide adaptability to them, which was the reason why they had become important varieties in many countries. In Cuba POJ 2878 itself remained the number one variety up to the late 1960s. And even as late as 1987, POJ 2878 contributed 25% of the genomes of some of the world's top varieties.[3]

And what happened to Jeswiet himself? That is a sad story told by van der Haar (1993). Jeswiet must have become afflicted with the Dutch variety of the national socialist doctrine while still in Indonesia, where it had many sympathisers in colonial society. After returning to Holland to become professor of plant taxonomy and dendrology at Wageningen University, he joined the NSB[4] party. He quit the party in 1934 because its membership became illegal for civil servants, but joined again in 1941, after Holland had been occupied by Germany. He attained a high position in the party during the wartime and was sentenced to 3.5 years of internment and loss of all his civil rights after the war. His greatness as a scientist was completely eclipsed by his moral failure.

What I like about the sugarcane story is that it seems almost inevitable, like the discovery of Newton's laws, or the short straw varieties of wheat and rice which triggered the green revolution in Asia and Latin America many decades later. Perhaps I should have become a breeder then, but that would probably not have worked out, because I missed the single-minded determination needed for that profession.

---

[3] CP 70–321, NCo 310 and NCo 376 (Heinz, 1987).
[4] Nationaal Socialistische Beweging (National Socialist Movement).

2.3 Good and not so good science

After coming back from traineeship in 1964 there were another few years to go until graduation. Studying was a rather leisurely affair in the Netherlands in those years. First you had to put together a graduate programme. I majored in tropical crop husbandry, of course, but I had to choose three more subjects. Together they should constitute a kind of professional profile. Since tropical agronomy was so ill-defined, almost anything would go, but I am still surprised that my choice was accepted. The subjects were: general genetics, mathematical statistics and plant nutrition. That combination of course did not make much sense from a career point of view, nor was there an obvious logic behind it, but otherwise I think it was quite nice. There was something precise and reliable about genetics and statistics. With hard work you could actually master those subjects. Not so with plant nutrition, which consisted of a lot of empirical knowledge, useful no doubt, but quite dull, and some unconvincing theory about how plants procure and use nutrients. But it was necessary to study something practical in relation to crops if one wanted to find a job, that was the main reason why I picked it, otherwise I might have chosen organic chemistry or something like that (that would not have been accepted, though). I must say a few things about these four subjects, since they continued to play a role in my life later on, and also because they give a bird's-eye view of what these disciplines looked like in the mid-1960s.

2.3.1 Tropical crop husbandry

I majored in tropical crop husbandry because the romantic ring of colonial plantations had attracted me as a freshman. In 1959 the professor of tropical crops was Coolhaas, a typical colonial agronomist, from the Java coffee industry. The second man, G.G. Bolhuis, also came from the East Indies where he had worked in research and extension on indigenous agriculture and food crops. Almost everything I learned at the university about tropical crops, which was not all that much, came from him. He combined considerable practical knowledge with a benign attitude to the tropical peasant farmer. He once said that the most reliable sign that a new crop variety was any good was that farmers came to steal it from the station's multiplication plots, a surprisingly modern point of view.

Coolhaas retired in 1960 and was succeeded by J.D. Ferwerda, an oil palm researcher from the Belgian Congo, the most colonial setting imaginable at the time. That was a surprising choice and it suggests that the university did not quite understand how the times were moving. When he arrived his technical knowledge about tropical crops was mostly limited to the Congolese oil palm. He had to teach the fundamentals of tropical crop husbandry and tried to explain what he had read the previous night about the work of some muddle-head in Yangambi, also in the Congo, about the relationship between climate and crop growth. He also mentioned the work by Dr. C.T. de Wit who had recently published a brilliant theory on plant competition, which I doubt he had understood, because he never explained what that brilliant theory was. I will come back to that later. Furthermore, he had to lecture on all sorts of crops which he had no apparent knowledge of. So he had a real hard time to find his bearings, in fact he never did. I only started sympathising with his ordeal when I became a lecturer in his department some years later.

My graduate research work in the department is not really worth mentioning, except that I started wondering vaguely whether there was not some more basic way to approach tropical crop husbandry.

2.3.2 Genetics

Genetics was good basic stuff. The genetics professor, R. Prakken, had worked his way up from primary school teacher to university professor. He taught a charming and up-to-date undergraduate course, but his graduate classes were a disappointment. He had spent a lifetime crossing beans and analysing the inheritance of their seed coat colours. That was typical classical genetics and pretty boring as lecture material, but what did get across was that there was something concrete and knowable there which could be discovered if you were smart enough. Other staff members gave mostly forgettable lectures on a host of other topics. We also had to participate in research projects by the staff, do some literature research and study the excellent textbook by Srb, Owen and Edgar, on general genetics.

I must tell a little anecdote about my experimental work in genetics, because it shows how one may pick up scientific culture. I was involved in a project on the inheritance of biochemical deficiencies in *Arabidopsis thaliana* (Figure 2-3). That is a tiny little plant from

Figure 2-3. Arabidopsis thaliana. (Reproduced by permission of the Arabidopsis Information Resource (TAIR))

the *Cruciferae* (now *Brassicaceae*) family, whose complete genome has been sequenced some years ago. There were two genes (or cistrons really) on the same chromosome coding for the synthesis of pyrimidine. If either of these was damaged on both chromosomes the plantlets' leaves remained white and the plants died an early death, unless pyrimidine was added to the growing medium. I had to carry out a lot of crosses to see whether the chromosomes could break and recombine right between those two cistrons. Whenever that had happened it would show up as a green individual among all white progeny, never mind the details. So I got up early in the morning and spent a lot of time doing dull work crossing many plants until I would have enough seed to have a reasonable chance to find a recombinant, if they occurred at all. In one of the petri dishes with young plantlets from these crosses there was a small sector with green leaves close to the edge and a transition with pale green ones. I was looking at that, thinking that there must have been a trace of pyrimidine in the dish, when Professor Prakken came in. He asked me what I was looking at and I told him what I thought. He said that might be true, but that I should always remember what somebody (I do not remember who) said about such things: 'always be careful with your exceptions'. So I had to analyse the progeny of a number of those green plantlets to see what they were. They all turned out to be white, so my assumption was right. This may sound insignificant, but a little remark like that, made at the right time, becomes part of one's awareness of the essence of science.

Genetics played practically no role in my professional life. But at least it gave me a pretty good understanding of what was happening in that field, partly through the pages of the once-unsurpassed *Scientific American*, until that journal declined to the state of triviality it had reached at the time of writing this book.

2.3.3 Statistics

Then there was statistics. If there is one thing all agronomists have in common it is their use of, and in many cases their problems with, statistics. In agronomy, statistics is mainly about drawing conclusions from experiments. The kind of questions it allows you to answer is like this. Suppose you want to know which of two crop varieties A and B is the best and you decide to grow them in several pairs of small plots in the same field to find out. You cannot just grow one or two pairs because the difference between the varieties in such a small trial may be due to chance alone. Now suppose you have planted six pairs and the average yield of the A's is 20% higher than that of the B's, does that mean that variety A is indeed better than B? Not necessarily. If A did much better than B in only one or two of the pairs and about the same in the others, the higher average yield of A could be entirely due to chance, for instance because A happened to be planted in one or two favourable plots. But if A did better than B in most of the pairs that becomes less likely. So you need a mathematical test which calculates how big the chance is that you get that kind of result *if the varieties were not really different*. If that chance is smaller than, say, one in a hundred ($P<0.01$), you would feel confident that A is really better than B. The measured yield difference between the varieties is then said to be significant at the 1% (probability) level. The argument may not be entirely clear at first reading but if you ponder it for a while it will take shape, like those 3D images embedded in colour patterns which were popular a decade ago. The mathematical test involved here is called Analysis of Variance (ANOVA). It is the kind of analysis agronomists most often deal with, the rest is mainly about designing experiments from which you can draw such conclusions about more complicated comparisons. For instance, about several varieties planted at different densities in different locations and at different levels of fertiliser.

There are two possible ways to teach statistics. The hard way is to start from the underlying mathematics, which is probability theory and linear algebra, derive mathematical tests and apply those to the results of the experiments. The other approach is to present it as a box of tricks about how to choose treatment combinations and lay them out in the field, how to add up the yields or whatever you measure, what to divide by what and where to look in a table to see whether there are significant differences. I am exaggerating now, but that is the way most agronomists do statistics in practice. In my university they opted for the first approach. That was all right for me because I like basics and, with considerable effort, I mastered it pretty well in the end. There is nothing like learning things from basics, if you can handle it. Later on I studied the excellent practical textbook on statistics by G.W. Snedecor and W.G. Cochran, as a kind of refresher course, and went through it in a breeze, precisely because of the way I had been taught basic statistics first. But most people in Wageningen entirely missed the point of the mathematically rigorous treatment favoured by the Statistics Department. As a result, several generations of students graduated without a clue about statistics. I know this for a fact because as a graduate student I had to teach practicals on statistics to undergraduates (oh horror!) and help correct their exams.

2.3.4 Soil fertility

Plant nutrition or rather soil fertility was my third minor subject and that was a disappointment. I will try to explain and in order to do that I need to go into some details about soils and plants. Plant nutrition is obviously very important for tropical agronomy and it will show up many times in this book, so this will also serve as an introduction for things to come.

Plants take up nutrients from a watery solution around the roots which is replenished from the nutrient stocks in the soil. That sounds simple, but it is not. Nutrients are bound to the soil in various ways, as part of undecomposed or partly decomposed minerals, adsorbed[5] to clay and humus particles and bound to organic matter. Soils are very different in nutrient content and in the way they hold and release nutrients. There are different types of minerals, some of them rich in plant nutrients, some poor like quartzite, the main constituent of sand. There are also different types of clay particles and humus, each with their own adsorption and release properties. Also, soil processes, like decomposition of minerals and changes in humus content depend on temperature and soil moisture during the year. Simply knowing the total nutrient content, therefore, does not tell you much about the availability of nutrients to the plants. Furthermore, nutrient uptake is an active process. Something happens at the root surfaces which allows them to take up the nutrients across the root cells' membranes and against a concentration gradient, because the solute concentration inside the roots is higher than that of the soil solution. Nutrient uptake must therefore involve some kind of 'carrier' which takes the nutrients inside the roots through the cell membranes against the concentration gradient with expenditure of energy. Plant species are quite different in their ability to take up different nutrients and that ability in turn depends strongly on the composition of the soil solution, like the absolute and relative nutrient concentration, the acidity of the solution and things like that. That is why a particular plant species may grow well in one type of soil and poorly in another and it also explains differences in fertiliser response.

---

[5] 'adsorbed', not absorbed. The former means bound to the surface, while the latter means sucked in.

So, when you want to understand the processes involved in plant nutrition you have to look at both sides of the root surface: the soil side and the plant side. On the soil side there are chemical and physical processes which govern the weathering of soil minerals, the formation and decomposition of organic material, the exchange of ions between the soil particles and the soil solution and the movement of water in the soil. If these processes were completely understood it should be possible in principle to predict the composition of the soil solution from the physical and chemical properties of the soil. But of course they are not and furthermore the plants themselves and other living organisms influence the release of nutrients by secreting compounds into their immediate surroundings which may dissolve minerals, releasing nutrients contained in them. The plants in turn take up nutrients selectively from the soil solution immediately adjacent to the roots through some kind of carrier mechanism whereby different ions compete for entrance, and the composition of the solution therefore also affects uptake. Ion uptake must be electrically neutral, hence roots have to secrete hydrogen ions into the solution to compensate for excess uptake of positive ions. It is clear that the situation becomes very complicated and it was impossible to bring it all together in a predictive soil fertility theory in the early 1960s; nor is a comprehensive theory available today.

Practical agriculturists did not wait until all the processes involved were understood. They went ahead and devised pragmatic approaches to formulate useful recommendations to farmers in spite of their incomplete understanding of the processes. They wanted to measure the nutrient content of different soils in the laboratory and then grow crops in the field to see whether the laboratory results could help to predict their yield. If that worked you would only have to analyse soil samples in the laboratory to advise farmers what to apply. But what kind of laboratory analysis can actually predict the availability of nutrients to the plants? The total nutrient content extracted by some powerful chemicals means little or nothing because much of what is in the soil is so tightly bound that the plants have no access to it. So a measure was needed for 'available nutrients'. But wasn't that precisely the problem we started from, the inability to predict nutrient availability from the knowledge of the soil's and the plants' properties?

Even today that remains an elusive goal. Nevertheless, a number of laboratory extraction methods were developed which at least resemble the way plants extract nutrients from the soil, for example, the positively charged ions (cations) potassium, magnesium and calcium in the soil solution. They are called 'exchangeable' cations, that is, exchangeable between the adsorption sites of the clay and humus particles and the soil solution. Ions in the soil solution are in equilibrium with those adsorbed and when the plant takes them up from the soil solution they are replenished from this adsorption complex. The usual soil tests mimic that process, albeit crudely, by shaking the soil sample with a solution of ammonium acetate which pushes most of them off the adsorption complex into the solutions where their concentration can be measured.

Measuring available phosphorus is much more problematic. P is not bound to the adsorption complex but it is part of the organic matter and also occurs as inorganic P in various forms. A whole range of extractants is in use, including water, and different ones are recommended for different climates and soils. When applied to a particular soil they all give different results, so the statement that a soil contains 10 ppm P means very little unless you know what extractant was used. Even then P-tests remain notoriously unreliable.

What about N, the most crucial element of all? Well, 'most crucial' is nonsense really, because all nutrients are equally crucial, but N is almost invariably the most limiting nutrient in the soil, especially in the tropics. N is a major constituent of soil organic matter or humus and in the absence of N-fertiliser the N released by humus decomposition is the main source for plant growth. Humus-N content is fairly easy to measure, and it gives you some idea about the yield you can expect without fertiliser. Water-soluble N is sometimes also measured but it is so variable with season and rainfall that it has little predictive value.

Once you have these laboratory measurements, what next? The idea was that they would allow you to say something about the crop yields that can be expected and the likely response to fertiliser. There is no way you can predict that from laboratory data and soil information alone. You need to carry out a lot of field trials as well to find the relationship between the outcome of the soil tests and the response to fertiliser for each type of soil and

each crop. That is what practical soil fertility research has been
doing mostly. In the Netherlands, as in most other western
countries, there is a whole network of testing sites covering the
entire spectrum of soil types where the relationships between soil
tests and crop yields and fertiliser responses are continuously
monitored. That allows the extension services to give site-specific
and up-to-date fertiliser recommendations. In developing countries there is usually no such thing, which makes fertiliser recommendations on the basis of laboratory analyses very unreliable.
Nevertheless, soil scientists have established 'thresholds' for
P and K and other nutrients: if the amount measured in the
laboratory is lower than the threshold you can expect that plant
growth will be reduced. It is not very precise, but at least it is
better than nothing. And precision counts less when the overall
level of productivity is low anyway. I will come back to that
later on.

So the knowledge about physical and chemical processes in the
soil, although insufficient to base predictive theory on, has been
useful to develop laboratory tests which mimic to some extent
what happens at the root surface of crop plants. There are many
other research results as well, which have helped to establish the
broad pattern of plant response to nutrient supply. They can
reduce the number of trials you need to carry out, because they
allow you to construct response curves from just a small number
of tests. Take, for example, the law of diminishing returns which
is illustrated in Box 2-2. It says that the yield increment due to the
application of a unit of nutrient diminishes as more of it is
applied. So, if you have measured a few points on the response
curve you can fairly well predict what will happen at other values,
even those outside the range of measured values, because you
roughly know the overall shape of the response curve. Different
versions of this law have been formulated starting with J. von
Liebig in 1855, until de Wit gave a satisfactory synthesis of all of
them in 1992. Another law states that excess K inhibits uptake of
Mg. That may cause grass tetany in cattle feeding on grass which
have been manured with K-rich cattle urine in spring. There
are many more empirical laws like that, which have made soil
fertility and plant nutrition a fairly successful although not very
exciting discipline.

> *Box 2-2.* The law of dimirishing returns applied to plant response to fertiliser
>
> The first figure shows the schematic relationship between *absorbed* N and yield for two varieties of, say, wheat. Yield increase per unit N at low yield levels is the same for both and diminishes rapidly for variety 2 as maximum yield is approached.
>
>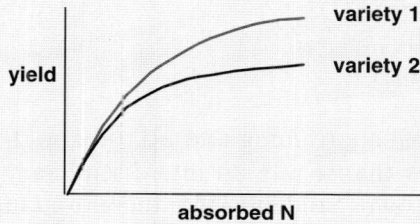
>
> The same for *applied* N. At zero application the soil still supplies N from its own store. The efficiency of applied N is lower for variety 2 throughout.
>
>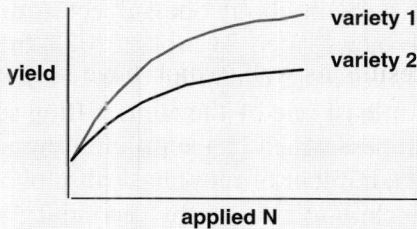
>
> Maximum yields may be higher if other limiting factors (e.g. another nutrient) are corrected. The graphs were adapted from a paper by de Wit (1992).

The main reason why the Wageningen graduate course on soil fertility was a disappointment is that it was all done very poorly. A good course should first of all teach chemical soil analysis adequately, explain what the results mean for the availability of nutrients to the plants, and how you correlate the results of the laboratory analysis with the outcome of field trails. But for those of us, who did not major in plant nutrition, training in chemical soil analysis was very poor and the mechanics of routine fertility research was not explained at all. Furthermore, the empirical laws should be taught

very well, which they were not. Our classroom lectures were a chaotic
collection of experimental data and a lot of half-baked theories about
nutrient carriers and organic matter. So, I learned very little about
plant nutrition, which is bad because plant nutrition is obviously a
very important subject for an agronomist. What I did find out is that
learning a little may be almost as bad as learning nothing at all and it
takes a lot of effort to correct that later on, if it is even possible.

2.4 Summing Up

What did this graduate training add up to? The trouble with tropical agronomy was that it was an interdisciplinary science without
much body of its own. Students were therefore taught the basics of
a wide variety of scientific disciplines, which in practice resulted in
learning too little about too many subjects. The Dutch biologist and
writer Dick Hillenius, when asked a long time ago how he managed
to be so knowledgeable in so many areas, answered: 'That is because
I know really very much about one subject.'(quoted in the *NRC-
Handelsblad* daily newspaper, 23 April 2002). Interdisciplinarity is
something you cannot learn in a more than trivial way unless you are
firmly rooted in at least one of the contributing sciences. Perhaps it
is the lack of firmness which has kept haunting me and caused me
to try my hand at a number of more basic disciplines related to tropical agronomy, including two of my secondary graduate subjects,
statistics and genetics, and later on crop modelling.

Apart from the romanticism surrounding the tropics at the time
I started my studies, I do not quite know why I chose tropical
agronomy and it does not really matter now. Most people make
choices early in life based on some vague notions about what they
want to be and then start groping their way forward with more or
less success. If it is in their nature and they are lucky they will find
their bearings and chart out their career course. I am sure that many
agronomists have felt at some time during this process that they
were jacks of all trades and masters of none, as I did. There are
many routes leading out of this dilemma, though, and choosing one
in good time is essential for a successful career. Perhaps that is
where I failed. In spite of some useful things I have done over the
years, I did not build a consistent career, perhaps because I tend to
get fed up with what I do every 5 or 6 years. I was told that is

because I was born under the constellation of Gemini. On the other hand, by sampling quite a diverse area within and outside the realm of tropical agronomy, and doing it rather thoroughly, I claim competence to review the entire tropical agronomy scene in a more than trivial manner.

While I was working for my last exams early in 1967, I was visited by a senior staff member of a Dutch consultancy firm, the International Land Development Consultants (ILACO), who wanted to know if I was interested in a job with them in a cotton project in Indonesia. I was; and after some tests and interviews I was hired and flown to Indonesia by KLM DC8, which took almost 24 h at the time because the plane stopped in practically every country it met on the way. It was January 1968. After working for 2 years with the company I was given continuing appointment status. ILACO was a daughter of the *Nederlandsche Heidemaatschappij*, which means something like Netherlands Wasteland Development Company, a company with a strong *esprit de corps*. They used to present new recruits with a sort of ceremonial dress, a forestry officer's uniform with hat. I narrowly missed that, the tradition was abolished just before I was elevated to permanent status, another sign that the times were changing.

Chapter 3. Old and New: The 1960s and 1970s

3.1 New times, old reflexes: A cotton project on the island of Lombok

Sending a Dutchman without any relevant experience to Indonesia as a development worker in 1968 was perhaps more than a little arrogant. But development projects were the craze of the time, the need for personnel was large and available expertise scarce. The new development industry therefore recruited young graduates who were shoved into the projects from which it was hoped they would emerge a few years later as new experts. Their seniors were men (mostly) of the old school, many of them retired, not always voluntarily, from the colonial services.

The idea of the project I was attached to was to grow cotton on the island of Lombok. Today Lombok is a major tourist spot but in 1968 it was completely off the beaten track. Cotton growing had been tried in Indonesia before the Second World War with little success because of insect problems. Insects are always the biggest problem with cotton, today as much as 40 years ago. With the rapid expansion of insecticide use after the war that problem seemed to be under control and in the early 1960s the Indonesians had started trying cotton again in the drier parts of the archipelago, in East Java and the eastern islands of Lombok, Sumbawa and Flores. On Lombok the crop was grown in the dry parts of the island in the shadow of the volcano Rinjani, but we were also going to try it in the lowlands, as a second crop after irrigated rice. That was a bad idea.

I do not think the Indonesians really needed or wanted us, except perhaps for the money and equipment, which invariably came with a batch of experts attached. The Dutch project manager, my immediate boss, was just a few years older than I and he had some project experience, although not with cotton and not in Indonesia. But he was the son of a pre-war general of the Dutch marines, born in Surabaya. Our supervisor who came to inspect us several times a year was also from an old East Indies colonial family. He had worked as a cotton agronomist in the Belgian Congo, but I do not think he understood a great deal about the crop. That did not matter.

There was enough money and he simply recruited a variety of short-term experts who came to tell us what to do. Actually that did a lot of good for my professional education. You might think that our presence there was rather unnecessary and perhaps even a little embarrassing, but that was not how our company saw it. We were supposed to bring discipline, scientific rigour and devotion to the task at hand, things the Indonesians were thought to master in insufficient measure, possibly because independence had come prematurely. It was our task to teach them these values.

We went about the cotton business in a highly technical way. That was how things were done at the time. Participatory approaches and environmental friendliness were not yet part of our vocabulary. We wanted farmers to grow cotton as a cash crop after rice because somebody somewhere, I suppose our supervisor, had decided that was a good idea. If there were problems with that we were going to solve them one way or the other. And there were many problems. The first was that a year has only 12 months. Cotton took between 5 and 6 months to mature and the local rice varieties also took up to 5 months. After the rice harvest the land had to be ploughed and ridged and when that was done there was barely enough time left for the cotton to mature before the onset of the next rains. And you can imagine what happens when open cotton bolls are hit by a good shower. So maybe the farmers should grow an earlier maturing rice variety to allow more time for cotton. Farmers are understandably conservative about the crop which ensures their livelihood and convincing them to tamper with it in favour of an unknown and unproven new crop is like telling a frugal family man to risk his savings at the races. But we were helped by a contemporary development, the advent of the new 'green revolution' rice varieties which were coming from the International Rice Research Institute (IRRI) in the Philippines. They matured earlier than the local varieties and produced much more grain, especially when given a lot of fertiliser. The Indonesian government mounted an aggressive campaign for the adoption of these varieties and we joined in with that effort of course, so that problem got more or less solved in the end.

The next and most serious problem was the insect pests. It is clear, in retrospect of course, that the way we tried to solve it became part of the problem, i.e. it made things worse. Before the Second World War the insect which prevented cotton growing in the East Indies was the spiny bollworm (Figure 3-1). That is an interesting animal which starts tunnelling in the tops of young plants when there is nothing else

Figure 3-1. The spiny bollworm (*Earias insulana*) and the damage it causes. (From Bayer Pflanzenschutz Compendium II, reproduced by permission of Bayer Crop Science A.G.)

to eat and moves to the flower buds and bolls as soon as they show up. That was also our most devastating pest but by the time I left Lombok there were at least five other major insects which could do serious damage: two leaf eaters, two leaf suckers and another bollworm. The reason was that we created the conditions for their explosion by our excessive use of insecticides, which destroyed their normal predators but not necessarily the pests themselves. Routine spraying such as we did can unleash pests which normally have a low incidence, by killing their natural enemies. Cotton entomologists in the USA were already warning against that kind of insect control, but our advisers apparently did not do their homework very well. At least I do not remember that the visiting entomologist was particularly worried about the approach we were using. He merrily went about testing new insecticides one after the other, most of which have now been banned. Of course it is easy to be wise now that every schoolkid knows about the nasty effects of insecticide abuse. In 1968 we were perhaps a little uneasy about exposing farmers to extremely toxic pesticides, but we thought that was all part of progress. And we were not really aware that we were disturbing the ecological balance and causing perhaps more problems than we solved. It was not for lack of signs, though. I once observed an interesting phenomenon in a farmer's cassava crop next to a cotton field. A strip of about 3 m of cassava was heavily attacked by spider mites while the rest of the field was clean. That of course was because the pesticides applied to the cotton drifted into the cassava and killed the spider mites' enemies while the mites themselves were not affected. Another example was a very heavy rice crop at our experimental station which suddenly started to show large areas with severe leaf wilting. That was due to the brown leafhopper being unchained by our chemical control of stem borers. One of the great advances of the last 30 years has been the much more careful use of pesticides, especially in rice, based on a concept called Integrated Pest Management (IPM). I will come back to that later.

So what was the verdict about our cotton project? In spite of our close watch average farmer yields kept hovering around 1 t/ha, not enough to pay for all the inputs and make the crop profitable compared with other crops, which were also much easier to grow. The final report of the project (which I wrote mainly) admitted as much, but cheerfully claimed that the 20–30% higher yields needed for profitable cotton growing could easily be obtained. How? Well, better pest control, of course, better land preparation, fertiliser, the usual stuff agronomists can think of. Now, after more than 35 years, I know it could never have worked. First, it was all far too complicated, especially the pest control by teams going around doing the 'calendar spraying' equipped with fragile petrol-engine sprayers. And the risk involved was unacceptable. If a farmer missed a spraying date by even a few days he might lose most of the crop. Second, it was too much work for too little gain. We fooled ourselves by paying little or no attention to the amount of labour needed, so our economic calculations were flawed, if we did any which I think we did not. Towards the end of the project a team from the *Faroka* tobacco company landed on Lombok and in less than 2 years managed to get an area of tobacco planted by farmers which was larger than the total cotton area we ever attained. Tobacco was a really profitable crop, eminently suited to the small-scale rice farmers' conditions and it showed that it was not because of lack of entrepreneurship that farmers would not go for cotton.

Though the whole project made no sense, I learned a lot of things. ILACO wanted to maintain a high level of professionalism and supervised their new brood closely. That did not always work out well, but my agronomy supervisor, Gerard Kerkhoven, taught me several important things about field experiments and his way with figures and rules of thumb was a revelation, for example, how to estimate the water flow through an irrigation ditch without any instruments. On the other hand there was a certain rigidity. We once started using a hand-drawn wooden roll with equally spaced disks across a puddled rice field to mark out planting lines, which was much faster than using planting ropes. The lines became a bit wobbly, of course, but that would not harm in any way. When the project supervisor came he was quite upset, because the wobbly lines of rice were a disgrace and unworthy of employees of his company. We said we would prove that it did not matter and that the yields would be exactly the same, but that was not the point: things had to be done the 'correct' way.

I have come to realise the importance of one's first job as an essential part of one's training, especially for a profession where most of the skills can only be acquired by doing rather than from books. During the colonial times a standing joke was that a new recruit in the plantations was not supposed to open his mouth, until he came back from his first home leave (which was after 5 years). ILACO of course was more modern than that, yet the company did manage to transfer some of its institutional culture and professional skills to its new recruits. In that respect I had more luck than many of my contemporaries and the following waves of graduates who went to work as so-called associate experts on overseas projects run by Food and Agriculture Organisation (FAO) and other international development organisations, and often came out just about as wise as they went in. In particular, agricultural know-how is not to be picked up in the short contracts which were becoming increasingly the norm.

3.2 Old and new in crop science: Growth analysis and modelling

After Indonesia I left the company and worked for 3 years as a lecturer in my old Department of Tropical Crops at the Agricultural University in Wageningen. I think that department

never recovered from the disorientation caused by the loss of the
colonies. During the colonial era everything was clear. Students
were prepared for a career in the East Indies and the department
staff having worked there themselves knew exactly what their
students would be expected to do after graduation. So the department taught courses about plantation crops and about the major
peasant crops like rice, and how they were grown in the East
Indies, from their own experience. That became all different after
the Second World War when Indonesia gained independence.
Graduates could now end up in any of a large number of tropical countries and work on any of the very large number of crops
grown there. The department had no clear concept about the kind
of teaching that was called for by this new situation and chose the
least imaginative approach: trying to teach about as many crops
as possible, grouped in clusters like cereals, fruits, oil crops, etc.
Since I had worked in cotton, my allocated group was the fibre
crops. Now the only similarity between cotton, sisal and kapok is
that their end product is some kind of useful fibre, otherwise they
are as different as wheat, cabbage and apple trees. And I knew
nothing about sisal and kapok, or about jute, kenaf, coir, ramieh
or manilla hemp, nor does one get to know much more by teaching about them. I think, in retrospect, that the department's thorough botanical practicals inherited from the colonial times, of
which all former students have fond memories, did more good
than any of the classroom lectures about crops.

Apart from teaching and some administrative chores I did a few
useful things, like organising student excursions to Dutch fibre mills
and to the south of Spain, which is as close as you can get to subtropical conditions in Europe. And then there was research, of
course, which all of us were expected to spend part of our time on
and which students would participate in as part of their degree
work. I had nurtured some vague feeling that because of the lack of
adequate theory many agronomists were endlessly carrying out trials without getting any nearer to a fundamental explanation of
their results. When interviewing for the job at the tropical crops
department I had actually tried to explain that I wanted to look at
some of those series of agronomic trials and see whether one could
not come up with some theory that would make most of them
redundant because the results could be predicted. I do not think
it sounded very convincing, nor did I have the faintest idea how

I would do that, but I was hired anyway. Fortunately, I was not the first with the ambition to replace the endless agronomic trials by process-based models supported by precise trials to find some essential model parameters, and for model verification. A new school of agricultural scientists was emerging, in Wageningen and elsewhere, which was doing just that, using computers as a tool to model crop growth. As soon as I had found out about that, things started to fall into place, but before I found my way into crop modelling I made a kind of false start, by indulging for a while in what can be seen as a precursor of crop modelling: Growth Analysis. Apart from a few concepts and growth parameters which it contributed to crop science, Growth Analysis is dead now, but at that time it was a creative attempt to bring quantitative methods of analysis into crop physiology which until then was a largely empirical science. I think Growth Analysis deserves some space in a history of twentieth century agronomy.

3.2.1 Growth analysis

If you want to economise on agronomic trials you need a theory of crop growth processes that makes it possible to predict growth under different conditions. One such theory, or rather research method, was 'Growth Analysis', developed first in Great Britain in the 1920s and 1930s. It is an interesting example of a fashionable method which for some time attracts a lot of adherents but eventually fades away for lack of results. There were some intriguing ideas in Growth Analysis. The first was the basic concept which was borrowed from a completely different branch of science – economics. That concept was the compound interest law, which is not really a law but rather the mathematical expression of the growth of a capital which earns a fixed interest rate. The simple interest rate is a fraction of the capital while compound interest results when each year's interest is added to the capital. V.H. Blackman saw in 1919 that the compound interest law would also apply to the growth of young plants: each increment starts participating immediately in the plant's growth. The equivalent in Growth Analysis of the interest rate was called Relative Growth Rate (RGR) which is the increase in plant weight as a fraction of the weight already there, in the same way as interest is a fraction of the capital which is also already there. It is expressed by the equations:

$$\frac{dw}{dt} = RGR \cdot w \text{ or } \frac{1}{w}\frac{dw}{dt} = RGR, \text{where } w \text{ is the weight of the plant.}$$

Contrary to capital, however, plant growth is continuous while the bank calculates the growth of the capital in discrete time steps, usually once a year. And another important difference is that as a plant ages some of its tissue no longer takes part in the growth and the compound interest law no longer applies, whereas with money every increment earns the same interest as the previous ones.

There are two other important quantities in Growth Analysis:

Net Assimilation Rate (NAR), defined by another differential equation:

$$\frac{dw}{dt} = NAR \cdot A, \text{ where } A \text{ is leaf area}$$

which states that a plant's growth rate is the product of its leaf area and the NAR (average biomass produced per unit leaf area).

When dealing with a crop instead of individual plants, leaf area is represented by:

Leaf Area Index (LAI), that is the leaf area per unit ground area. Crop growth is also expressed relative to ground area, so that the above equations become[1]:

$$\frac{1}{W}\frac{dW}{dt} = RGR \text{ and } \frac{dW}{dt} = NAR \cdot LAI$$

These are the basic equations of Growth Analysis which define the three key quantities RGR, NAR and LAI.

I would now like to go into some further technical detail, but if you think this will do, just skip Box 3-1.

If Growth Analysis has not met the ambitious goals I thought it had set for itself, it has nevertheless contributed a number of very useful quantities to characterise plant growth, like RGR, LAI and NAR, which continue to be used in plant and crop physiology. They are useful shorthand for certain growth phenomena, but do not lead to a predictive theory of crop growth. Such theory emerged from the pioneering work by de Wit in Wageningen.

---

[1] I use lower case w for an individual plant and capital W for plant weight per unit ground area.

Box 3-1. Growth analysis, some technical details

Consider a crop which is in an early-growth stage where the compound interest law still holds for the entire plants. The mathematical expression for the increase in crop weight, taken per unit area is then a differential equation:

$$\frac{1}{W}\frac{dW}{dt} = RGR \qquad (1)$$

and when RGR is constant, integration results in the equation for exponential growth: $W = W_0 e^{RGR \cdot t}$. W_0 is the weight at time zero, which can be chosen arbitrarily, it is simply the point where you start measuring.

Very young seedlings consist almost entirely of meristematic tissue, all of which partakes in growth, so they grow exponentially, feeding on the reserves stored in the seed. Once these are exhausted, the young leaves take over to supply the necessary growth substrates. The amount of growth per unit leaf area and per day is called Net Assimilation Rate (NAR), so the Crop Growth Rate (CGR) also equals the product of leaf area per unit land area times NAR:

$$\frac{dW}{dt} (=CGR) = NAR \cdot LAI \qquad (2)$$

The expression only states that the crop's growth rate is directly proportional to the leaf area index and to the NAR. It does not say whether growth is exponential or linear or something else. We may now divide both sides by W and get:

$$\frac{1}{W}\frac{dW}{dt} (=RGR) = NAR\frac{LAI}{W} \qquad (3)$$

The left-hand side, by the definition given in expression (1), is the relative growth rate. Growth will be exponential (RGR is constant) if $NAR\frac{LAI}{W}$ is constant. Now suppose that NAR is constant, reasonable assumption for young plants as long as all the leaves are exposed to direct sunlight and there is no significant change in solar radiation. Growth will then be exponential provided $\frac{LAI}{W}$ is constant, which means that plants invest a constant proportion of new growth in leaf tissue. Very young plants may therefore continue to grow exponentially for a while even after the reserves in the seed have been depleted. As the plants get bigger the canopy gets denser and the leaves start shading each other. The average amount of light absorbed per unit leaf area will therefore go down and so will NAR. The plants will also start converting more assimilates into stem tissue and perhaps in storage organs like tubers and eventually in fruits and seeds. Leaf area growth will then be slower than that of total plant weight and plant growth is no longer exponential. Finally, as the plants mature, growth will slow down further until it stops altogether at full maturity.

(continued)

Box 3-1. (continued)

Somewhere between the juvenile, exponential phase, and maturity growth must necessarily pass through a linear phase, however short that may be. Let us see how the transition takes place from exponential to linear growth and from linear to zero growth. In a young crop where growth is no longer exponential, the leaf area will still increase more rapidly than NAR decreases and the product of NAR and LAI, that is, the CGR continues to increase. As the canopy becomes denser a point is reached where the product of NAR and LAI becomes constant, because the increase in LAI is offset by a similar decrease in NAR, and consequently crop growth becomes linear. There is a maximum to LAI, however, when the canopy becomes very dense the lowest leaves do not receive enough light to survive. An equilibrium will then be attained between the loss of shaded leaves and the growth of new ones, which results in an LAI that remains constant for a while. If solar radiation does not change during that period NAR will also be constant and so is the product of LAI and NAR, so growth is linear, until the plants approach maturity, when fewer leaves are formed and the ageing ones assimilate less. Growth will then slow down and eventually stops.

Summing up, plant growth passes through an exponential phase, followed by a linear phase and then gradually grinds to a halt. That is what causes the 'sigmoid' shape of a plant's or crop's growth curve. Two such curves are shown in Figure 3-2. Apart from the fact that plant growth is sigmoid what does this tell us? There is an infinite number of such curves which all show the sequence of exponential-linear-declining growth and a good predictive theory should be able to do more than just saying that this sequence will occur. It should be able to predict the precise shape of the curve for a particular species, growing under a particular set of conditions. Can Growth Analysis do that? Expression (2) looks deceptively simple, but its two factors, NAR and LAI, are actually very complex, mutually dependent quantities. There is nothing in the expression itself that gives any clue about their values and how they change. A lot of experimental research has been done to find relationships between *W* and *LAI*, in other words how leaf area changes with plant weight and what is the effect of growing conditions on this relationship. That may be interesting, but it is not what we were looking for. We would like to *predict* growth, not fitting functions to measured data.

One could of course argue that prediction was not what Growth Analysis was meant for, as its name indicates. But that is nonsense. The analysis was meant to clarify the processes under one set of conditions in order to predict them under another. A simple expression like (2) could not do that because its variables were too complex, so Growth Analysts broke them down into less complex one like Leaf Area Ratio (LAR), Leaf Weight Ratio (LWR), Specific Leaf Area (SLA), Leaf Area Duration (LAD).[2] Hundreds of papers and

---

[2] Leaf Area Ratio (LAR) is the leaf area per unit of plant weight, Leaf Weight Ratio (LWR), the leaf weight per unit plant weight, Specific Leaf Area (SLA), the ratio between leaf area and leaf weight, Leaf Area Duration (LAD), the leaf area summed over the crop's lifetime.

several books have been published about how all these quantities relate to each other and to the crop's growth rate and what is the effect of growing conditions on them. They have clarified many things about the physiology of organ growth and dry matter distribution, but they have not much advanced the prediction of growth. The reason is that the growth analytical parameters are not really parameters but rather complex quantities which are themselves the outcome of the complex processes they wish to explain. The parameters are almost as complex as growth itself and it is therefore impossible to relate them in an unambiguous way to environmental factors and to one another. The bible of Growth Analysis published by G.C. Evans in 1972 tried to do that but after going through its 734 pages it becomes clear that it will never work. The Growth Analysis quantities themselves are in need of explanation in terms of underlying, more basic processes. P.J. Radford in a paper published in 1967, while talking about future research needs, says that Growth Analysis should be based on *measured* growth curves of plant weight and leaf area from which the other quantities should then be derived. That sounds like the ultimate reversal of objectives using measured growth curves to derive explanatory parameters instead of the other way around. Or perhaps it is not. It is from my own bias that I made demands on the technique beyond what is was meant for, although I keep thinking that the conceivers of Growth Analysis did have a predictive theory in mind as the final result of their efforts.

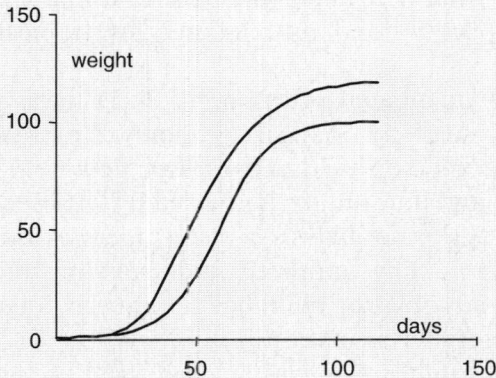

Figure 3-2. Typical whole-plant sigmoid growth curves

3.2.2 Novel approaches: Crop modelling

In 1968, the year after I graduated, de Wit had been appointed professor of theoretical production ecology in Wageningen, a chair created especially for him. I had heard about his theoretical work but
I barely knew what it was all about, until I started digging into it in
search of predictive theories for crop growth. In the late 1950s and
early 1960s de Wit, who was then an agronomist in one of the
Wageningen research institutes, had started to take a fresh look at
some half-understood crop growth processes related to plant production: nutrient uptake, water use, interplant competition and
photosynthesis of crop canopies. De Wit had a brilliant mind and
over a period of 5 or 6 years he managed to give a satisfactory
quantitative treatment of each of those processes, practically without doing any experiments himself but using results from those carried out by others. When I joined the tropical crops department in
1971 his fame had spread across the world, especially to the USA,
where several scientists had also started using computer-aided
research methods in biology and plant production. I will give a brief
overview of his work here and come back to it in a separate chapter on crop modelling and its relevance for tropical agronomy
(Chapter 9).

De Wit's work on mineral nutrition, his Ph.D. thesis published in
1953, was still essentially classical agronomy. He broke down the
effect of fertiliser on crop yield in two parts: how yield is related to
nutrient *uptake*, and how uptake is affected by fertiliser *application*,
and showed that separate analysis of these processes and the factors
influencing them greatly improved understanding of what happened. That sounds obvious, but it had never been done in that way
before, at least not as systematically.

The next major publication was on crop water use and appeared
in 1958. First, de Wit argued that both transpiration and photosynthesis depended on solar radiation and that water use and dry
matter yield of a crop must therefore be closely connected. He
derived two simple expressions for this relationship, one which was
valid for temperate climates and one for arid climates. The expression for temperate climates is $P = k \cdot T$, which states that the ratio
of Production (P) to Transpiration (T) is constant. That was the old
'Transpiration Coefficient', a very convenient quantity, popular
among irrigation agronomists, which had been rejected as useless by

the greatest authority in the field, H.L. Penman. For arid climates de Wit showed that the transpiration coefficient indeed did not apply, but with a simple but crucial modification an equally simple expression was found: $P = k \dfrac{T}{E_0}$. E_0 is the evaporation from an open-water surface. A massive amount of experimental data from the literature was presented which supported this formulation.

Next, in an essay published in 1960, de Wit developed a complete and very successful theory of plant competition from a small number of basic principles. The title of the essay was 'On Competition', not a modest title, but one which was justified by the richness of the essay's contents. The approach was that of a physicist and the work fittingly started with a quotation from de Wit's physics professor who had been his Ph.D. thesis supervisor. To get an idea of his way of thinking the competition theory is summarised in Appendix 1. It is rather technical, though, so if you feel it is too much algebra, you may just skip it. Anyway, the competition theory has turned out to be particularly relevant for tropical agronomy because growing crops in mixtures is such an important practice in the tropics.

De Wit's work on plant nutrition, crop water use and competition was not modelling in the modern sense, which usually means the step-by-step simulation of growth processes using dynamic computer-modelling techniques. The early models were static in that they dealt with the outcome of growth processes, rather than the processes themselves. In that respect they resembled Growth Analysis, but contrary to that almost forgotten approach, de Wit's models went right into the heart of the growth processes. In 'On Competition' there were hints, however, that he was not quite satisfied with his static models and felt that one should analyse the competition processes themselves instead of just looking at the final outcome. He made some unconvincing attempts to make his competition formulas dynamic but when they did not really work he gave up and turned to the basic growth processes.

The first one, crop photosynthesis, was tackled as an essentially geometrical problem. Photosynthesis by the canopy was found by summing the contribution of a large number of individual leaves, whose exposure to radiation depended on orientation and location in the canopy. There had been important precursors. In 1953, Monsi and Saeki in Japan published an epochal paper in the Japanese Journal of Botany on light interception and assimilation by plant communities. Their work formed already a sharp contrast with Growth Analysis, in that they calculated carbon assimilation by a canopy from the physics of light interception at different depths in a canopy. The paper was written in rather awful German, apparently German was still considered in Japan as a suitable language to reach the international scientific community. The paper's wide citation lends support to de Wit's opinion that it matters little where something is published and whether the style satisfies literary standards, what matters is that the content is good. And the fact that de Wit's own much-cited early papers were mostly written in rather awful English provides further evidence.

In de Wit's analysis a crop canopy is a collection of leaves with different inclinations and orientations. When there is no stress, photosynthesis by an individual leaf depends on the amount of radiation it receives, which in turn depends on the leaf's habit and its depth inside the canopy as well as on the position of the sun in the sky throughout the day. Calculating the distribution of light intensities in the canopy with time of day is a computationally complex but conceptually straightforward geometric problem and since the photosynthetic rates of leaves as a function of light intensity were know for many species, daily canopy photosynthesis in principle could be calculated. In an earlier paper de Wit had proposed an approximate analytical formula, but in 1965 he carried out the exact calculation using the first computer installed at Wageningen University, an IBM 1620. That launched the use of computer modelling for the quantification of crop production processes. The canopy photosynthesis paper was a real breakthrough because it was for the first time that it became possible to calculate the maximum daily assimilation by a crop for any date at any place on earth. The model was still static because the canopy was treated as a fixed entity and growth of the canopy itself was not looked at, only the amount of photosynthate a canopy with a given structure would produce.

The canopy assimilation model was still several steps removed from what had clearly become de Wit's goal: a truly predictive model for crop growth. He therefore now tackled plant growth itself, whereby the assimilates are built into plant biomass and each increment in biomass depends on what is already there, in other words, modelling became dynamic. That was made possible by the rapid increase in computing power and the development of programming techniques suitable for process simulation which de Wit himself made significant contributions to. Dynamic crop-modelling was one of the defining features of twentieth century agronomy and I will return to the subject in much more detail later, in Chapter 9.

So a lot of things had been happening which I was totally unaware of, until I joined the university in the early 1970s. They held promise to find the kind of clear answers to clear problems which I was looking for. Obviously I wanted to be part of that and I had all the freedom I needed in the tropical crops department where I worked. The crop I was most familiar with was cotton, which looked ideal to develop a growth model for. It has all these branches coming out of the mainstem in a very regular way and producing flowers and bolls in neat succession. I went to see de Wit with some rather vague notions about how cotton growth could be modelled. He was a little sceptical at first but thought that I had some reasonable ideas, so he agreed to supervise the work and I started thinking about how to model a cotton plant and did some greenhouse research on the growth of cotton bolls. But my job at the tropical crops department was temporary and after 3 years I had to find something else again. My cotton model was not even remotely ready, but I had to put it on ice for sometime. In the end I did complete it, but that was several years later. I will come back to that also in Chapter 9.

3.3 Academic exercises in Africa

It so happened that around the time I had to find a new job, our tropical crops department was asked to help develop a university course on crop production in Cameroun. One of my colleagues, Egbert Westphal and I were going to do that. In January 1975 we started work at the Ecole National Supérieure Agronomique at N'Kolbisson, near Yaoundé. That was my first encounter with Africa.

In the mid-1970s the new African countries were still busy groping their way out of the colonial era. Most of them had become independent around 1960 and in many respects little had changed since then. Whereas the key positions in government and in the armed forces had of course been taken over rapidly by the new elites who had participated in the struggle for independence, the more technical positions continued to be occupied for some years by former colonial officers, especially in the former French colonies.

Even if those officers thought things had gone too fast, many of them sincerely tried to adjust to the new situation and help conserve what they saw as the achievements of the colonial period. They were especially numerous in agriculture and forestry, and in the few institutions for higher learning which had been set up in the pre- and early post-independence years.

In Cameroun there were still a good many Frenchmen working in all kinds of positions in agricultural research and development, now on the payroll of the French Ministry of Cooperation or of one of the many tropical research institutes based in France which amounted to the same thing. Meanwhile, the international aid bureaucracies had also been building up their forces and sent a variety of development workers around the globe, most prominent among them, at least in numbers, the United Nations organisations. I think these newcomers were looked upon by the old hands, not entirely unjustifiably, as dilettantes. The agricultural college where we worked had been set up in the late 1960s and was mainly populated by this new generation of experts, supplied by FAO and by Belgian, Dutch and British bilateral projects. There were also some young Camerounian graduates who were expected to take over once they had learned enough from us.

Our teaching job was rather dull and the way we went about it was rather unimaginative, I am afraid. We were supposed to teach about tropical food crops, not just a few but all of them, in the style of our department in Holland, but this time between the two of us, assisted by our Cameroonian colleagues. That of course was impossible, but we did it nevertheless. I suppose I only had myself to blame for the insignificance of my courses, since we had quite some freedom to decide on how to do it. Maybe it is simply because I do not really like teaching, at least about things I do not know too much about myself. Anyway, it is not something I am particularly proud of.

The research was a little more interesting. Our two Cameroonian colleagues and I decided to study the local cropping system in the vicinity of the capital Yaoundé. Yaoundé has a humid tropical climate with two rainy seasons, from March to June and from September to November and about 1,600 mm mean annual rainfall. That means that the natural climax vegetation would be high forest but around Yaoundé it was long gone. What remained was a lot of secondary bush at different stages of maturity, some older

'sub-climax' forest[3] and crop fields with cocoa or food crops scattered all around. We went around the farms and saw mainly two
kinds of food crop fields which at first sight looked as if they
belonged to two different cropping systems. The most common type
was the groundnut field, called 'afub owondo' in the local language
(Figure 3-3). It contained many more crops than just groundnuts,

Figure 3-3. Woman collecting cassava leaves in the latter stage of an *afub owondo*

---

[3] Secondary bush is what you get after leaving the land fallow for a sufficiently long
time, say 5–6 years, provided the soil has not become too exhausted. If it is left alone
for another 20 years or so it becomes sub-climax. Whether the real climax will eventually come back is a much debated issue. I think nobody really knows.

but that was how it was called. It belonged exclusively to women. That may sound strange but it is very common in Africa for men and women to have their own separate fields. The other, much rarer type was the *ngôn* field, or *eseb*. It was planted with *ngôn*, a kind of pumpkin belonging to the *Cucurbitaceae* family, and usually also plantains.

When starting a groundnut field a plot would be cleared from rather light secondary vegetation, which was not too difficult to slash and burn, leaving a fairly clean field. The groundnuts were then planted and cassava cuttings were stuck in at much wider spacing and often also plantain bananas at still wider spacing. A variety of other crops were planted in the same field as well, such as maize and vegetables, scattered through the field at low density. In fact, the *afub owondo* contained practically everything a family needed to feed itself, and in more or less the right proportions. The interesting thing from an agronomic point of view is that the three main crops grown together in the association have very different growth cycles. Groundnuts grow fast and mature in about 100 days. The cassava has a much slower start. It grows little while the groundnuts are there but when they are gone cassava takes over the field, to be harvested about 18 months after planting. Meanwhile the plantains have produced their first bunch and once the cassava has been harvested the field becomes a plantain field, or rather a fallow plot with plantains, because the new fallow had already started taking over by the time the cassava had been harvested. The nice thing about this system was that the field had to be prepared only once, all crops were planted at the same time and the late maturing ones took over from the earlier ones when their time had come. That is a system that is hard to beat for environmental friendliness. At least until the human population density becomes too high, the resting period between the cropping cycles becomes too short, the weeds become troublesome, plantains are no longer planted because the fertility gets too low, in short the system starts degenerating. That process was in full progress around Yaoundé in the late 1970s.

In order to start a *ngôn* field a forest plot was needed, where the undergrowth was slashed and burned and the trees were cut down and left scattered around. If you were to grow a grain crop like maize or groundnuts in a field like that you would have to do a lot of work cutting up and moving the tree branches and stems to make space. A better solution is to first plant something at wide spacing between the scattered trees and then after a few years remove the trees and rubble when it is easier to cut them up and burn them. That is what the farmer, mostly the man this time, would do. He planted *ngôn*, a rapidly growing species with vines which wound around the scattered tree stems and branches and formed a nice canopy. Usually he would also plant plantain bananas (Figure 3-4). That is a good choice too. Plantains do best on pristine soils. They grew slowly in the beginning and eventually took over the field once the *ngôn* was gone. The plantains would stay for a few years and produce two or three bunches. By then the forest trees had partly decomposed and they were easier

Figure 3-4. Eseb field, just planted with *ngôn* and plantains

to get rid of. The field was then converted into a groundnut field or left under fallow for another few years first.

In 1975 there was not much good forest left and the *ngôn* field was vanishing. Today it may be hard to find any in the area at all. We figured that in the original system the first thing after clearing a forest plot would always have been to start with the *ngôn* field, followed after a few years by its conversion into a groundnut field, after which the field would return to long fallow again.

We were not the first to study indigenous farming in that part of the country. At the beginning of the twentieth century a German anthropologist, Günter Tessmann, had done the same thing. When we started we were not aware of that, but Hans van de Belt, a sociologist who also worked at the school in an FAO project, had found the book Tessmann had written about his work. It is interesting to hear what he had to say about the cropping system from which the one we saw had evolved.

The only fields which are pleasing to the eye of a European are the groundnut fields. When, after crossing the dense cassava and sugarcane plantings, and the *ngôn* fields, whereby one has to climb over all the scattered tree stems, or the dense bush or forest, then one is relieved to alight upon a groundnut field, lying open in front of the wanderer, bordered by the dark frame of the forest . . . which reminds him so much of the cleanliness and clear structure of a European vegetable field.

One does not read that kind of language much these days in scientific texts.
Of the *ngôn* field Tessmann remarks:

The bush is felled and burned, but the land is not cleared. Everything is left lying as it falls, except the largest rubble along the direction of planting, which makes crossing a *ngôn* field into something like a gymnastic exercise.

Although Tessmann preferred the neatness of the groundnut farm, he was well aware of the great importance of the *ngôn*, a truly indigenous crop species, and expressed surprise that it was not even mentioned in the textbooks on tropical crops of his day. Every community planted a large *ngôn* field every year and the crop was surrounded by much ritual and taboos, for example: "the *ngôn* prohibition . . . which forbids women, who have planted *ngôn* to have sexual intercourse for three months", otherwise the crop would not grow.

Surprisingly, Tessmann did not make a link between the *ngôn* and the groundnut field. In fact he treated all field types as if they were unrelated. Surely, after all the work of cutting down and moving around the trees, the field would be used again once the crops from the *ngôn* field were harvested. I think he simply missed the point that the groundnut field and the *ngôn* field were different phases of the same cropping pattern, which started with clearing the forest and planting *ngôn*, followed by groundnuts, and finally letting it return to long fallow again. Tessmann must have been confused by what looked like a rather chaotic collection of seemingly unrelated field types.

In Tessmann's days cassava was grown in separate fields together with sugarcane, if we may trust his observations here. He also mentions separate plantain plots and a host of minor species grouped as 'horticultural' which apparently were grown close to the homesteads. Today all these crops are part of the groundnut field. So, in addition to the decline of *ngôn* a major change would have been that practically all the food crops have moved inside the groundnut field.

Two other novelties have occurred since Tessmann's time. One was the rapid expansion of cocoa as a small farmers' cash crop shortly after Tessmann's study. That was quite a case of small African farmers' capacity for change, with little technical support and no development projects. The other was the invasion of the fallow by eupatorium, today's major fallow species. Just a few words about that, I will come back to it in the next chapter. The scientific name of eupatorium (or Siam weed) used to be *Eupatorium odoratum*, but it was changed to *Chromolaena odorata* in the 1980s, as a result of the taxonomists' tiresome habit of renaming species. That happens when someone discovers that a different name was published earlier, but I think a more important reason is that there has not been much else to do since Linnaeus' monumental work was completed in the eighteenth century (I know this is unfair, it is how agronomists talk about taxonomists). Eupatorium reputedly was introduced by chance from southern Asia into West Africa in the 1940s by returning African soldiers who had fought in European armies and carried the seed in their boots. It spread very rapidly across the humid belt of western and Central Africa. Eupatorium is considered as a noxious weed by some and as an effective fallow species by others, the latter I think with more justice. When the fallow becomes too short for secondary forest to reappear, eupatorium is the best alternative you can get. When left alone it grows

very rapidly, shades out noxious weeds and grasses and is easy to
remove when you want to replant the field.[4]

But let us return to our field studies now and answer a question
which may have occurred to you: what was the point of doing them?
I could say for instance that it was useful for training the young
Cameroonian faculty in doing research (although I had never done
anything of the kind myself) or gathering material for our agronomy courses based on local information, both of them reasonable
justifications. Another reason could have been that we wanted to
develop ideas on how to improve the system, if indeed it needed
improvement of course, and therefore we had to understand it first.
For an agronomist that would be the natural argument, but we did
not have any plans to improve the system. It was mostly out of academic curiosity that we started these studies and we also intended
to further our academic careers by publishing the results. In academic circles that is usually sufficient justification for undertaking
a research project and the information so collected may even turn
out to be useful one day. It would have been difficult in the 1970s to
get purely descriptive studies like Tessmann's published, but we
were agronomists so we also measured things like the planting densities of the crops in a large number of fields. Such figures mean little by themselves, they must be converted into an objective figure
which characterises the fields in an agronomically meaningful way.
It should tell you for example, whether the combined density of the
crops in a field is high or low or something in between. And that,
you would hope, may tell you something about the fertility of a particular field, or perhaps about land scarcity, or some other factor
which might cause farmers to choose a high or low density. In other
words you can start to frame some hypothesis about it. We therefore
devised what I think was a rather clever way to represent the combined crop density in a field by a single figure which we called 'total
population density'. That figure would be unity if there was full
crop coverage, never mind the details, read the papers if you do. In
the fields we visited the total population density on average
remained close to unity throughout the cropping cycle, up to the
cassava harvest. And even after that, because the eupatorium filled

---

[4] It was Dick Lowe, a forester and statistician working at ENSA, who first drew
our attention to the favourable properties of eupatorium.

the gaps between the plantains as the field gradually reverted to fallow. Except for a short while after planting, the soil remained practically completely covered throughout the entire cycle. So one of the conclusions was that the farmers' system managed to mimic the conditions of full vegetation coverage of the forest reasonably well, which is very nice in a high rainfall area like this.

If the world were static you would conclude that the farmers' system is hard to beat for agronomists and that they should leave it alone. But what if it starts to break down because farmers can no longer wait until nature has run its course and restored the soil fertility before planting the next crop? That was clearly happening around Yaoundé and even more so in some other villages not far from Yaoundé with a higher population density than the villages we worked in. We did some arithmetic about fertility export by the harvested products which showed that there was not enough in the soil to continue like this indefinitely. As long as the farmers left the land under fallow for 15 or 20 years, enough nutrients could be pumped up by the trees from down below and put back into the topsoil within reach of the next crop. But now that the fallow was much shorter and dominated by fairly shallow-rooted eupatorium that was no longer the case. As a result the plantains were dropping out of the system, which was particularly bad because it was a highly valuable crop. So, what did agronomy have in store to fix that? Fertiliser, of course. Yes, but what kind of fertiliser and when or to which crop should it be applied? Those are interesting questions. You would want to apply the fertiliser when it had most effect and to the crop which suffered most from the decline in soil fertility. Those undoubtedly were the plantains; so if fertiliser were applied to the plantains perhaps they could still be grown successfully. The nutrients which the plantains had not been able to take up would be captured by the eupatorium growing up around them and later be released again when the land was cleared for a new cropping cycle. It would have been interesting to carry out some on-farm trials about this which we could easily have done together with the school's extension department. Because that department took the students to another village once a week to analyse farmers' problems, see how they themselves dealt with them and then, after consulting some technical experts, propose possible solutions. It was quite similar to what would be called Farming Systems Research (FSR) and Participatory Rural Appraisal some years later.

One problem they had diagnosed was plantain decline and the solution proposed by the department was to grow the plantains in intensively managed plantain orchards. Farmers had to dig big holes, fill them with topsoil mixed with manure or fertiliser, and plant carefully prepared and disinfected plantain suckers in them. That worked quite well, but it was a lot of work, manure, fertiliser and pesticides were not readily available and on the whole it was very different from the way farmers used to grow plantains. I wonder if farmers continued doing it once the extension team stopped coming. I would be surprised if they did. It would perhaps have been a good system for someone who wanted to invest some money in commercial farming, like a retired teacher or civil servant. If the purpose was to improve plantain growing by traditional farmers, however, I think it would have been better to start from the cropping system which they were actually practising. That means, leaving the plantains where they were, in the groundnut–cassava–plantain association, and treating them with the same care as in the proposed orchard system. That is what we agronomists should perhaps have done.

On the other hand, this raises the more basic question whether it actually made sense to try and improve the peasants' system and whether it would not be better to target a new kind of more progressive farmer, perhaps someone coming from outside agriculture and seeing new opportunities in a different way of farming. During the many years I have observed African peasant farming since leaving Cameroun I have had little doubt that the key to the development of African farming was in fact the gradual conversion of peasant farmers into rational and wholly or partially market-oriented producers. Today I am not so sure anymore. The issue will be returned to in later chapters, but I want to mention the grain of doubt seeded in my mind even while in Cameroun by an experienced Briton who taught statistics at the agricultural college, but who was really a forester, Richard G. Lowe. Lowe had worked all his professional life in Africa, starting as a colonial forestry officer in the 1950s and continuing with the Nigerian government for several years after independence. He was a keen and sympathetic observer of what was happening in Africa, culturally and agriculturally. Lowe was convinced that the future African farmer would not be a direct descendant of the current peasant, whose strongest motivation seemed to be to extract himself or at least his children at the earliest opportunity from the drudgery of rural life. In a book he published 10 years later, entitled *Agricultural Revolution in Africa?*, he argued for an entirely different approach to farming based on mixed farms with crops and animals and rooted in the findings of modern agronomic research. Surprisingly, the book does not explicitly raise the issue of who would be the modern African farmer of the future, but the author's opinion does show through that future farming is unlikely to evolve from today's peasant farming, as he tried to impress on me at many occasions.

3.4 Is station research in Africa useful?

As I said earlier, we should have done some on-farm experiments to test some new things for the traditional-crop production system we were studying, together with the extension department. That would have constituted a nice integrated programme of on-farm studies, technology testing and student training. Instead, we decided to do something entirely different and much easier to organise: mixed cropping trials at the school's experimental farm.[5] I have since become very critical of that kind of station research and I will explain why, with my own work as an example.

There are two possible valid reasons to carry out on-station experiments in Africa. The first is to test new ideas which you do not want to expose farmers to, until you have some confidence that they will work. The second argument, which I mention with hesitation, is to study basic processes, either for the advancement of science, or in order to better understand the results of the first kind of experiments, or both. The reason for my hesitation will become clear later. I will look at our own station trials and see whether they fell in either of these categories.

The trials were about intercropping of maize and groundnuts. Endless numbers of intercropping trials have been and probably continue to be carried out all over the world. Like the good old planting date and planting density trials they have become part of tropical agronomists' standard repertoire, and not infrequently they

---

[5] In fact it was I who decided to do that, because I did the experiments mostly on my own for reasons I do not remember, but which were probably not very good reasons.

have degenerated into the same mindless repetition of over-researched themes. Our trial had treatments with sole maize and sole groundnuts and four intercrop combinations, each at two levels of fertiliser application. In order not to confuse the reader with unnecessary detail I will only look at the sole crops and two of the intercrop combinations, without fertiliser. The sole crops of maize and groundnuts, were planted at their usual density and formed a 'replacement series' with the two mixtures, which means that a proportion of the maize plants were replaced by an equivalent number of groundnuts.[6] All treatments therefore had the same total population density. Figure 3-5 shows the relative maize densities on the abscissa (the groundnut densities are of course the complements of these with unity) and the relative maize and groundnut yields (i.e. relative to the sole crop yield) obtained in one of the trial seasons. The interesting thing here is the Relative Yield Total (RYT), which is the sum of the relative yields of both species. It was significantly higher than one for the intercrops, which means, in de Wit's terminology (see Appendix 1), that the species were not competing for the same space. In other words, the two crops together appeared to have more space available than when both were planted separately. That is probably because the maize can explore the entire area for nitrogen, the groundnuts obtaining theirs at least partially from their own N fixation. It means that if you want to harvest a certain quantity of maize and groundnuts you need less land when growing them together than as sole crops.

Was this useful on-station research? Remember that a valid reason could be that you have a new idea, which may or may not work when used by farmers in their own fields, and which should therefore be tested at the research station first before farmers are exposed to it. Obviously that was not the case here. Mixed cropping is the rule in peasant farming in most areas except the driest parts of the savannah and maize–groundnut intercropping is also quite common, so the idea of intercropping of maize and groundnuts is not something that needs to be tested on-station before farmers are exposed to it. Perhaps there was some other novel idea included in the test which farmers were not familiar with? Well, we applied fertiliser to half of each plot to see what that would do and we oriented half the replicates in a north-south and half in an east-west direction to see

---

[6] 'equivalent' meaning in proportion to their sole crop density.

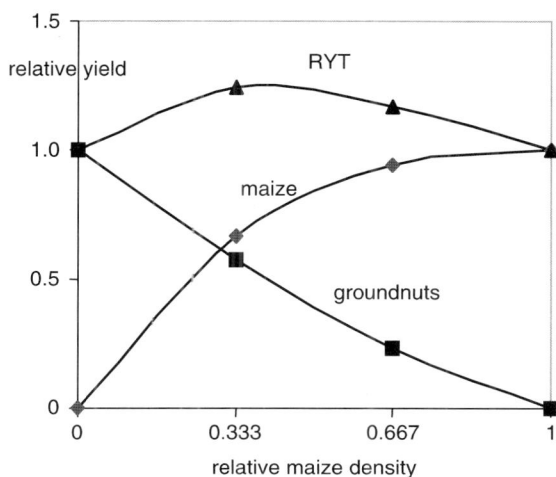

Figure 3-5. Relative Yield Total diagram for a maize + groundnut trial

which orientation would benefit the low-growing groundnuts more because of better light penetration. I will not go into the details of what came out of that, but the question is whether it is necessary to do those kinds of straightforward things on-station. If the results are supposed to be useful for farmers and they do not involve any risk, why not do it directly on-farm? So according to the first criterion there was no convincing case for doing this work at the station.

Could the work be expected to contribute something to the advancement of science then? By the mid-1970s the biological advantages of mixed cropping were quite well established and a lot of research had been done on all kinds of mixtures all over Africa. Mixtures of cereals and grain legumes had been shown repeatedly to exhibit 'over-yielding' of the kind noted in our experiments, unless factors other than soil fertility were limiting, in particular water. So the experiments mainly confirmed what had been found over and over in many experiments and they therefore also failed the second criterion. There was just one small redeeming element in the trials which is explained in Box 3-2. You may skip that if it is too much detail for your liking.

Was it all worth the effort? The trials did add a little to the already large and quite convincing body of evidence that mixed cropping was a rational thing to do. But knowing what I know now, I think it would have been much better if we had joined the

---

Box 3-2. Further analysis of maize + groundnut mixtures

The redeeming element in this work came from an additional trial, carried out
under conditions which were more similar to those in farmers' fields than the
others. In order to explain that, I must briefly describe the school's experimental farm. Its location, like that of many such farms and experimental stations
in Africa, was exceptionally favourable. Most of it was on deep, fertile alluvial
soil in a fairly wide valley, bordered by undulating land. The soils were quite
uniform and therefore ideal for controlled experiments, but completely unrep-
resentative of the agricultural soils in the area. Valid or not, uniformity is the
reason why agronomists choose that kind of place to locate their experiments
on. In the second year, however, I decided to carry out a maize–groundnut trial
in the sloping area adjacent to the experimental farm, because the soil conditions there were much more similar to those of normal farmers' fields (the land
was actually used by farm labourers to grow food crops for their own use). The
trial was also different in another respect, it was an 'addition series': groundnuts were planted at full density with different densities of maize added onto
that. The treatments in this trial were as follows:

1. Full density groundnuts only (without maize)
2. Full density groundnut with one-third of the full maize density added to it
3. Full density groundnut with two-thirds of the full maize density
4. Full density maize only

Contrary to the replacement series, the treatments now have different 'total
population densities'; treatment 1: 1.0, treatment 2: 1.33, treatment 3: 1.67,
treatment 4: 1.0. The only reason why the sole maize treatment is there is to
have something to compare the maize yields in the mixtures against.

The soil fertility in the experimental field was very patchy. That is
considered very undesirable by agronomists because a lot of variability
reduces the chance that they will be able to draw precise conclusions about
the effect of their treatments. Avoidance of variability will come back in later
chapters as a factor which has hampered the agronomists' acceptance of on-
farm experimentation. But it was precisely because of this microvariability
that an interesting effect showed up in this trial. In order to appreciate that
I must say something about the maize plant and how it reacts to poor growing conditions, which the trial conditions definitely represented.

The maize plant is very different from other cereals in that its growing point
terminates in a male inflorescence, while the female inflorescences, which produce the maize cobs, are carried in the leaf axils. Actually, the female inflorescence itself is also a terminal organ, borne at the end of a modified branch in
a leaf axil of the mainstem. That is shown schematically in the drawing above.
It is from Cobley's *The Botany of Tropical Crops.* When the maize plant is given
a lot of space and good growing conditions it produces more than one cob, at
least some varieties will, the so-called prolific ones. When conditions are bad or
the maize is planted at a high density, or both, some plants may produce no
cobs at all, remaining barren. Barren stalk is a phenomenon which is feared by
maize producing farmers in the USA, who push the density close to the limit.
In three out of our trial's four replicates the soil fertility turned out to be very

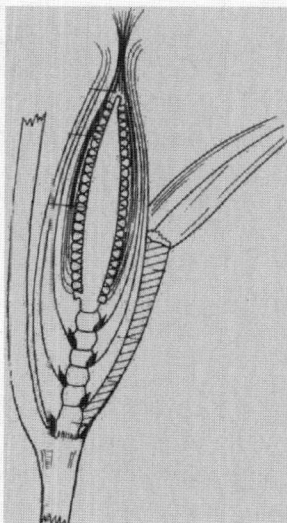

Figure 3-6. Diagram of a female inflorescence of maize. (From Cobley, 1957; reproduced by permission of Pearson, London.)

Table 3-1. Average number of cobs per plant in a trial at three maize and total population densisties

Treatment	2	3	4
Maize density	0.33	0.67	1.00
Total population density	1.33	1.67	1.00
Cobs per plant	0.91	0.75	0.69

low and the sole maize treatment (treatment 4) had a high percentage of barren plants as Table 3-1 shows: the average number of cobs per plant was 0.69, which means that the maize density of almost 42,000 plants/ha was way too high under these conditions.

In the mixed plots the total population density was of course much higher but the percentage of barren maize plants was lower, especially in treatment 2. The maize can now scavenge a larger area for N and hardly feels the presence of the groundnuts, which fix their own. So the mixed crop with one-third maize was well buffered against local variation in fertility: even where fertility was low, each maize plant still managed to grow one cob. If the two crops had been grown separately at their 'normal' sole crop density, the maize yield would have been severely depressed in the poor fertility spots, without any compensation. It was not a spectacular result, but at least the trial gave support to the rationale of growing mixed species under the highly heterogeneous soil conditions which are common in low external input agriculture in the tropics.

extension department and carried out some more interesting trials
in real farmer fields. Anyway, the station trials allowed us (or rather
me) to play around a little with mixed cropping experiments which
I had never done before. That was OK, but it would have been even
better if the students had been involved, which they were not
because of the school's quite rigorous curriculum and because we
did not make the effort, or not strongly enough.

3.5 And what about teaching tropical agronomy in the Netherlands?

After 3 years in Cameroon I went back to my old department in
Wageningen, surely not the most progressive step in my career. It
was like exchanging one backwater for another. Tropical agricultural science in the francophone world was very insular and had few
contacts with the anglophone world at that time. And although we
were not really part of the francophone culture, our own contacts
with the outside world were also very limited. There were occasional visits from Wageningen, by ageing men from the tropical-
crops department, itself a remnant of a glorious past, which had
failed to re-equip itself for modern times. We did catch some voices
from the other world, though, like Jane Guyer's, an anthropologist
of great reputation, I believe from Harvard University, who did
interesting studies about women's agriculture in the Yaoundé area.
But we were ignorant about the things which were happening in
Tanzania, northern Nigeria and Latin America and which eventually would set in motion the Farming Systems Research movement.

As I said, the tropical crops department at Wageningen
University to which I returned was a backwater in many respects.
Its staff was a group of highly individualistic people, each with
their own little area of expertise. There were specialists on rice,
cocoa, tropical fruits and fibre crops (me), and two professors who
dealt with their own range of crops as well as with tropical crop
husbandry in general. Then there was a plant physiologist who did
theoretical studies on flower induction, a botanist who had a small
tissue culture laboratory, all on his own, and a taxonomist who
wrote books on farming systems in Africa and ran the department's
botanical practicals. We all met mainly in the coffee room and in
staff meetings which dealt largely with administrative matters nor

did we have a shared vision about where tropical crop science teaching should be going. I am good at hindsight as you will have noticed and I now think we should have completely overhauled our courses. As a final farewell to the department, to which I owed a lot inspite of my criticism, I will have a look at the way its curriculum could have been reformed, whereby it might have extended its useful existence into the twenty-first century.

In the old days specialists used to teach courses on crops they were most familiar with. That was fine then, in that it served the dual purpose of familiarising the students with the crops themselves as well as with the environment they were grown in – in that context, this meant the former Dutch East Indies. After Indonesia's independence, however, the tropical crops department had had to broaden its scope to cover tropical agriculture worldwide. This blurred the image of the production systems in which the crops were grown, because there was no longer a dominant one on which the courses focussed. As a result, the courses gradually became disjointed collections of miscellaneous facts which no longer yielded a view of the underlying system. The old way of teaching tropical crops had become obsolete with the demise of the colonies, but it continued practically unchanged into the 1960s, lending the department's teaching a peculiar colonial flavour which only vanished when the last of the old guard retired.

Once the glue provided by the colonial setting had dissolved, a new organising principle was needed to restore the coherence of agronomy teaching. There were two possible ways to do that. One of them was to take cropping or farming systems as entry point and treat the individual crops as components of those systems. During the early 1980s the department made a half-hearted attempt to reorganise its courses on that basis. I think that was a mistake. It is very difficult to get a more than trivial understanding of 'systems' unless you have a thorough knowledge of the systems' contents, in the case of farming systems that is the crops grown inside them. In the colonial days knowledge about crops and cropping systems emerged simultaneously, because the same system fabric bound them all together.[7] Once that was no longer the case the concept of

---

[7] That is not entirely true. In Indonesia there were three major systems: the intensive paddy rice, the shifting cultivation and the plantation systems, but there were strong interfaces between them as we shall see later on.

'cropping systems' or 'farming systems' embraced a bewildering range of conditions in Asia, Africa and South America and the peasant and market-oriented production which cut across them. Teaching these systems at a more than trivial level in my opinion requires that the students first get thoroughly familiar with the crops, the way they grow and how their growth can be manipulated to the benefit of the grower. So, was there and alternative? When there is no longer one or a small number of common farming systems underlying all crop production, a new organising principle must be found to bring some order in the bewildering variety of crops and the way they are grown. The only alternative is to look for agronomically meaningful properties of the crops themselves, that is grouping them on the basis of their morphological and physiological similarities, which have implications for the way they are grown.

What kind of groups would you get from applying that principle? Surely, it would never bring the 'fibre crops' together into one group, because their main similarity is in the nature of their produce, not in their agronomy. An obvious group would consist of the crops belonging to the grass family, such as rice, maize, sorghum, millets, wheat, barley and even sugarcane. Other examples are grain legumes, which would include groundnuts, soybeans, cowpeas, beans, peas and a host of others. These two groups, the grasses and the legumes, are at the same time taxonomic groups, which is not surprising, because many agronomically important properties are closely associated with taxonomic ones. Another interesting taxonomy-based group would be the *Musaceae*, with bananas, plantains, Manila hemp and Enset as representatives, interesting and economically important crops all of them, with fascinating similarities and differences in growth habits and the way they are produced in three continents. The palms family would also be nice, with oil palm, date palm, coconuts, *Borassus*, betel nut, aren palm. In other cases, the grouping would not be based on taxonomic, but rather on morphological similarity, like fruit and nut bearing tree species, including fruit trees and coffee. And leafy vegetables and may be a few others. The course would look for similarities and differences between the crops within a group and the way they are grown in different climates and different cropping systems. That was not the way the crops were handled at all. Take the cereals, for example.

Rice was taught by one person, maize and sorghum by another and sugarcane not at all while other cereals were taught by the temperate crops department, like wheat, barley, oats and rye and maize, grown in the Netherlands for silage. That was a waste of effort, I think. Collaboration should have been sought with the temperate crop departments to work out and teach common trunk courses on cereal crops and fruit trees for example, perhaps with a second cycle for more typically tropical or temperate aspects. The trunk courses, supplemented with thorough morphological and physiological practicals, would have laid the foundation on which tropical cereals and tree crops and their role in cropping systems could be taught effectively and, perhaps as important, in an interesting way.

Professor Ferwerda and I actually worked out a sketch for these courses but I think we simply did not put enough energy into it. Marius Wessel, the department's cocoa specialist, did develop a course on tree crops a few years later. My priorities were certainly elsewhere, with modelling cotton and completing my Ph.D. thesis. That had to be finished by the time my temporary contract ran out again, in early 1982. A few years later it was all overtaken by events. By the late 1980s the department had joined the systems movement along with almost everybody else. It had given up its crop focus in favour of the fashion of the time, which was looking at agricultural production as part of the wider ecological system with micro, meso and macro levels and got lost somewhere among them. In the end the department stopped preparing its students for anything at all, except to take part in fashionable holistic discourse with like-minded intellectuals. Tropical agriculture as an academic discipline had now degenerated into a philosophy of agriculture with little or no relevance for a practical career in tropical agronomy. And perhaps that was just as well, because the demand for tropical crop scientists was in steep decline. The department's choice for the systems approach marked a significant step towards its ultimate demise in the late 1990s, along with that of the temperate crop departments.

But I am running ahead of my story. We are still in the early 1980s. Apart from participating in departmental organisation and supervision of students' M.Sc. thesis work I did my part of crop teaching and ran a practical course on research methods in tropical agronomy (which should also have been done jointly with the other crops departments). But my priority was completing my research on cotton modelling, which some of the students' M.Sc. research contributed to, and writing up my Ph.D. dissertation. The dissertation was finished by early 1982 just after my temporary contract with the University had run out and I had to look for a new job again, which I found at the International Institute of Tropical Agriculture (IITA) in Ibadan, Nigeria.

Chapter 4. Farmers Are Smarter Than You Think

4.1 In search of a new development vision

Paradoxically, working at the International Institute for Tropical Agriculture (IITA) was my first real encounter with African agriculture. Paradoxically, because I had already spent 3 years in Cameroon working at a genuine national institution, whereas the international research centres were often regarded as luxurious anomalies, where overpaid scientists carried out their leisurely research in splendid isolation from a destitute host country. That of course was not entirely fair. International research has had some spectacular successes for the benefit of peasant farmers, especially in crop improvement and in biological control of crop pests. The centres also encouraged their scientists to maintain intensive contacts with development organisations and projects in the host country and in other countries of the region. If you were interested in getting to know real African farming, working for an international centre was not at all a bad idea. Of course, some of the scientists made a mockery of the ideals of the centres' founding fathers by hardly ever leaving the campus, except to the airport to go on their annual leave or attend the annual meetings of the American Society of Agronomy. When I worked at IITA they were a minority, although not a negligible one, and most of the scientists were keenly interested in African farming. My job was to spend a considerable part of my time outside the campus, going around national research institutes and development projects in West and Central Africa to preach the FSR gospel. That is how I really learned what African agriculture was about.

When I arrived in Nigeria in 1982 a small revolution was going on in agricultural research and development, especially in the anglophone countries. In the early post-independence years of the 1960s and 1970s, development models had been borrowed from the west, assuming that African agriculture could be quickly pulled out of subsistence and transformed into modern inputs-based market production by copying western models. Traditional agriculture was looked at as unsuitable for stepwise improvement. In francophone countries like Ivory Coast I often heard peasant farming being referred to as Stone

Age agriculture, even in the late 1980s, by indigenous scientists. The foreign meanwhile had become too politically correct to express that kind of opinion. A quantum leap into the future was thought to be needed, to be brought about by the introduction of entirely new production concepts. The early-development projects which promoted western production models met with very little success, however, in spite or perhaps even because of considerable donor investment and large numbers of expatriate experts. Only the introduction of animal traction and the expansion of cotton in savannah areas had considerable and lasting impact, but the foundations for those successes had been laid well before independence (Figure 4-1).

In the 1980s things were becoming different. The peasant farmer was rediscovered and the way he/she[1] farmed turned out to be

Figure 4-1. Animal traction in maize, northern Nigeria

---

[1] I will use 'he/she' just once here. From now on when I say 'he' it is shorthand for 'he/she', whenever applicable. Some authors use 'she' instead of 'he' as a generic term for person, but the initial hilarious effect wears off quickly and then becomes slightly ludicrous, or so I think.

much more rational than had been thought. So it was just a matter of finding out what kind of help farmers needed and then one had to tap into their own considerable skills, rather than try to push inappropriate western technology down their throats. By doing all this, we thought, success would be inevitable, because change would now be driven by the peasants' own motivation. This change in philosophy did not happen overnight. There had been precursors, even from long before the age of decolonisation, when keen and thoughtful observers had watched with awe how seemingly primitive African husbandmen were able to live in harmony with their environment while extracting from it what they needed. Tessmann, whom I introduced in Chapter 3 was one of them. And some agriculturists had known all along that indigenous farming was actually pretty smart. But then, so was using village drums for telecommunication and that did not necessarily make them suitable for the twentieth century. After independence there had been little patience among the new elites in the developing countries and the development workers alike with what they looked upon as hopelessly outdated. Take mixed cropping, which is almost universally practised by tropical peasant farmers and whereby several crops are crammed into the same field. Mixed cropping had gained some respectability because agronomic research had shown it to be a surprisingly efficient use of space and time. For the development projects of the 1960s and 1970s, however, it was just a nuisance, hampering the use of modern technology, like fertiliser and mechanised tillage. Indeed, sole cropping makes mechanical tillage easier, because there are just straight lines of one crop with enough space in between for the oxen or the tractor wheels, but it is less obvious why it is necessary when everything is done manually. Fertiliser, for example, was usually applied by hand, even in technically advanced projects, and it is not evident that it required sole cropping and straight lines. But all the fertiliser recommendations had been developed at the research station in sole crop trials and furthermore I think sole cropping became linked up in people's mind with modern agriculture because it is the way crops are grown in the West. That has done a lot of harm to the credibility of the extension workers who were still promoting sole cropping for its own sake, even after the development projects and their inputs were long gone, and in spite of the fact that they would rarely use it themselves in their own family plots. I know, because I made it a habit to ask the extension worker, in a somewhat devious way to avert suspicion, to show me his own plot.

The failure of the early-development projects to get western methods widely adopted forced everyone to rethink their assumptions. That meant taking the farmers' own production practices seriously, in spite of their limitations, because after all they had obviously been successful, otherwise the farmers would no longer be there. This time around we were going to look at those indigenous farming practices as a rational answer to the challenges farmers were facing, rather than as anthropological curiosities.

4.2 Three visionaries

Agronomic research on mixed cropping became fashionable in the 1970s and its striking results helped lift the peasant farmer to a somewhat higher rank on the respectability scale. But the strongest push came from three agricultural economists who pictured the peasant farmer as a rational economic person, capable of absorbing suitable innovations into his own farming system, rather than being in need of complete re-education. That happened almost at the same time in three different places, like great scientific discoveries sometimes do (and fads and fashions too). These three scientists, two British and one American worked almost unobserved for some years before coming out with their new insights and setting in motion the FSR movement. I will discuss their work at some length because they became very influential.[2]

The first one was David Norman, who worked at the Institute for Agricultural Research (IAR) at Samaru, northern Nigeria, during the late 1960s and early 1970s. Like many post-independence research stations in the former British colonies, many British scientists had stayed on at Samaru, some of them, though not Norman, still on Her Majesty's government's payroll. IAR even had a British director up to the early 1980s, a convenient arrangement for internal conflict avoidance, as several Nigerian scientists working at the institute at the time commented. Norman spent several years carrying out detailed analyses of peasant farming and showed that from an economic point of view, mixed cropping as practised in northern Nigeria was in fact the most sensible thing to do. And three Samaru agronomists, David Andrews and a little later Ted Baker and Neil Fisher, showed the same thing for biological productivity.

---

[2] The three of them were twice nominated, collectively, for the World Food Prize in 1999 and 2000.

It is interesting that a few years earlier some other agronomists at the Samaru station, together with IITA's maize breeder, had shown that, with the right variety, maize would be a superior crop for the West African savannah, provided fertiliser was applied (Kassam et al., 1975). At that time maize was just a small backyard crop, fertilised with farmyard manure. They compared sole maize with the farmers' common sorghum-millet mixture and concluded that the former was far superior. Why maize should be grown sole, however, was not obvious, except that that was how station research was done. Fifteen years later maize had become a major crop in the area, grown both sole and intercropped with sorghum or cotton and many other crops as well. Farmers had correctly concluded that the maize and fertiliser were the essential components of the package, not the sole cropping.

There were several remarks in a paper published by Norman in 1974 which presaged the way of thinking of what was going to be the 'Farming Systems' movement later on. One of them was: "Until [extension workers] can suggest changes [to farmers] that have a convincing return it is unlikely they will ever be truly effective in their work". This may sound pretty obvious, but it was far from that at a time when, in spite of much benevolent rhetoric, peasant farmers were still looked at by many, not as rational producers but as ignorant primitives in need of education. The rapid adoption of intensive maize growing by those same peasants effectively vindicated Norman's attitude.

The second of these visionaries was Michael Collinson, who worked at another famous ex-colonial research institute during the 1960s: the Western Region Research Centre at Ukiriguru in Sukumaland, Tanzania. That institute worked mainly on cotton with smaller emphasis on food-crops, maize and rice, as it still does today. A few years ago Collinson edited the *History of Farming Systems Research* in which he described his gradual conversion from a run-of-the-mill farm management economist to a student of farmers' own priorities, resources and practices, using the much more informal methods which have become the norm in FSR. He also set up a so-called Unit Farm at the Ukiriguru station, an idea borrowed from the Imperial College of Tropical Agriculture in Trinidad, the West Indies. A Unit Farm was an area on a research station set aside on the research station and managed by a local farming family. That is of course a very artificial set-up, but much less so than the scientists' experimental fields which in Africa did not even remotely resemble a real farm. And observing it must have been a great educational experience for a scientist, especially a young one without much experience. For one thing, it demonstrated in a very direct way that the usual recommendation by all crop specialists to plant their crops as early as possible did not make any sense in the real world, where farmers had to plant several crops with the very small resources of a peasant family. Collinson's emergence as one of the leaders of the FSR movement had a lot to do with his rejection of the laborious formal surveys of the farm management economists of his days, in favour of more informal ones, for convenience first and from conviction later. Those informal surveys, initially intended as 'pre-survey', before the real thing took place, eventually became *the* survey, later termed diagnostic survey. In the late 1970s, when Collinson worked at CIMMYT in East Africa, his early experiences were developed into the concept of 'recommendation domains', groups or types of farmers with similar constraints or opportunities who needed similar innovations. That is the concept for which Collinson has become most famous.

The third scientist (not in importance, the ranking is random) in our FS R pantheon was Peter Hildebrand, an American farm management economist, who worked in Colombia, El Salvador and Guatemala in the 1960s and 1970s, in collaborative projects between national and American research institutions. His research was much more experimental than that of the other two, perhaps because he had worked on real farms when growing up in a small rural town in Colorado. And may be simply because he was an American. Two of the research stations in Latin America where Hildebrand worked had both a research and an extension task, as in the American land grant college system, so: 'with an extension mandate, we were always interested in the immediate application of every experiment so we looked at it differently than many of the scientists who viewed an experiment as part of a series, each of which should be publishable', to quote from Hildebrand's memoirs. This sums up the real issue more aptly than the unfortunate phraseology about 'effective research-extension-farmer linkages' that developed in later years: if your technology is not adopted it probably is no good. Like Norman and Collinson, Hildebrand looked at what the real farmers were doing, not what some imaginary ones might do in the future, and argued that technology development and testing should start from there. For Latin American peasants, as for their African counterparts, that meant mixed cropping under often marginal conditions. Hildebrand devised a rapid method to survey peasant farming, similar to Collinson's Exploratory Survey technique, which he called 'Sondeo', but his best-known contribution to on-farm research is a clever method to interpret the results of tests carried out in farmers' fields. To explain that I must say something about the theory of field experimentation.

Suppose you want to know which of four maize varieties is the highest yielding. That of course does not only depend on the varieties themselves but also on where and how they are grown. Some varieties may do better under some conditions and others under other conditions. Let us ignore those complications for a minute and start with a simple experiment on these four varieties at the research station. We will grow them under the best possible conditions by applying plenty of fertiliser, keeping the plots free of weeds and controlling pests and diseases by spraying with pesticides, the way the breeders do it. The reason is that that is how their varieties will have been bred in the first place, and they are therefore more likely to give of their best if they are so treated. The result of such an experiment may not be very useful to farmers, but at least it is easy to do.

So how would a variety test like that be set up? A very primitive test would comprise four plots, one for each variety, and a comparison of the yields you got from each. But a sensible person, even without any knowledge of statistics, would understand that there is no way of telling from the results of this test whether the differences are not actually due only to chance – if you had planted all four plots with a single variety you might still have obtained the same results. So instead of one plot for each variety there should be several, may be five or more, to gain an idea of the variability in the field and to increase the chance of drawing the right conclusions. The plots are usually grouped in compact blocks, each block containing plots of the four varieties, as shown in Figure 4-2. That makes it more likely that the varieties in each group (block) are grown under similar conditions. Within each block the varieties or treatments must be arranged in a random order to avoid systematic errors (it is also necessary for a mathematical reason, which will not be explained here). That is why such a trial is called a randomised block trial. Now if the yield of one of the varieties, averaged over

I	V_1	II	V_2	III	V_3
	V_3		V_3		V_2
	V_4		V_1		V_4
	V_2		V_4		V_1
IV	V_4	V	V_4	VI	V_2
	V_1		V_2		V_3
	V_2		V_1		V_4
	V_3		V_3		V_1

Figure 4-2. Possible field lay-out of a randomised block trial with four varieties arranged in six blocks

all the blocks, is higher than that of the others and if it is also the best in most of the blocks you will be more confident that it is indeed better than the others. That result may still be due to chance, but the larger the number of blocks the less likely that is. There is a statistical procedure called ANOVA which is used to calculate just how unlikely your result would be if there were really no variety differences. If that chance is very low you may conclude that there are some genuine differences, which you can then examine in detail. That is the essence of the statistical argument.

But the results of this station trial would still not be very useful. The trial only tells you, for example, that one of the varieties did better in this particular field at this station in this year with this amount of fertiliser and pesticide application. You will now start to see why breeders and agronomists run such trials year after year in different places until they are satisfied that a given variety is consistently better, at least most of the time in most places. But these places are usually sub-stations which are also under the researchers' control, so the conclusions are still valid only for maize grown by researchers. And that may be quite different from the way farmers do it, especially in peasant agriculture in developing countries. But let us assume for the moment that farmers grow their crops more or less in the same way as the researchers, as they do in Europe or America, so that when the varieties are grown by real farmers you would expect more or less the same outcome as when they are grown at the research station.

After carrying out a lot of trials in many sub-stations you may start noticing that in some of the stations some varieties do better than the others and you wonder which properties of the different sites are responsible for this. If you could find specific physical parameters responsible for the differences you could tell farmers which variety to grow under their particular conditions. In the 1930s Yates and Cochran, two famous early statisticians, took the first step in that direction by devising a simple way to examine patterns in the response of different varieties. I will use a numerical example from their paper with five barley varieties (which they themselves borrowed from a group of agronomists in Minnesota). The results of their analysis for three of the varieties are shown in Figure 4-3. First they calculated the 'site average' for each site, that is the average yield of all the varieties at that particular site taken together (this site average is now usually called the 'environmental index'). They then plotted the individual yields of each variety against the site averages. The graph shows the results for three of the varieties (the site average was calculated from all five of them). The yield of the 'Trebi' variety turned out to be high in fields where average yield was high and low in fields where it was low. That is hardly surprising, but the interesting thing is that the yield of some other varieties was fairly stable across sites. Varieties may even cross over, like 'Peatland' and 'Velvet', which means that the former did better in 'poor' sites and the latter in 'good' sites. Many years later the method was revived by Finlay and Wilkinson in Australia, who called it 'stability analysis' and applied it to a large group of barley trials.

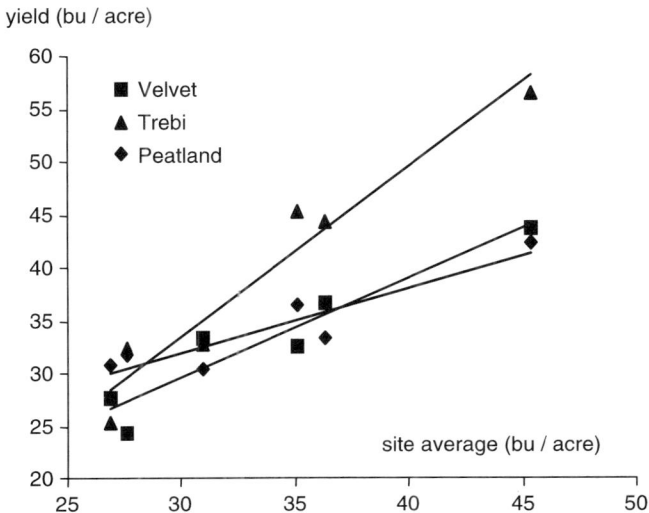

Figure 4-3. Yields of three barley varieties plotted against average yield of all varieties at the testing sites. (From Yates and Cochran, 1938.)

This is a helpful way to explore response patterns across testing sites, except that it does not tell you anything about the underlying physical factors causing the differences. For, what does the site index really measure? Knowing that a particular variety will do better if the productivity of a field is high would not be very helpful, unless you know which factors determine the productivity. In the Yates and Cochran case, management was the same everywhere, so differences between the average site yields would have to be due to physical factors like soils and climate.If the idea is to make specific recommendations to farmers, those factors would have to be analysed further. Yates and Cochran were of course well aware of that.

Hildebrand carried out such trials in farmers' fields, rather than in research stations, because he suspected that the results from the research stations could not be trusted to apply under real farmers' conditions. His genius was to realise that stability analysis was very relevant for on-farm trials. It could be applied to any kind of treatment, not just varieties (as Yates and Cochran had already remarked) and, even more importantly, he saw that differences in management were most interesting, rather than a nuisance which masked real physical differences, as agronomists might argue. Because differences in productivity between farmers were often caused less by soils and such things, than by factors which can be lumped together as 'management'. Hildebrand was also less interested in treatments which were stable across different conditions than in the way the effect of treatments would change with different farmer conditions, so that one could recommend one type of treatment to one type of farmer and a different one for another type of farmer. So he called the analysis 'modified stability analysis'. Later he regretted that and changed it to 'adaptability analysis'. It became very popular with on-farm researchers thirsting for respectable analytical methods and has been widely used.

Let us see what adaptability analysis can do for on-farm research. Figure 4-4 shows an example of an on-farm trial with two maize varieties carried out in Malawi (the data are from Hildebrand and Poey, 1985). It is a case of two varieties crossing over, the local variety doing better in poor yielding fields and the improved one in the highest yielding ones. So, according to the results you would recommend the local variety to farmers who might be expected to have low yields and the improved one to others. That is how Yates and Cochran used the technique in the 1930s, except that they only had to account for purely

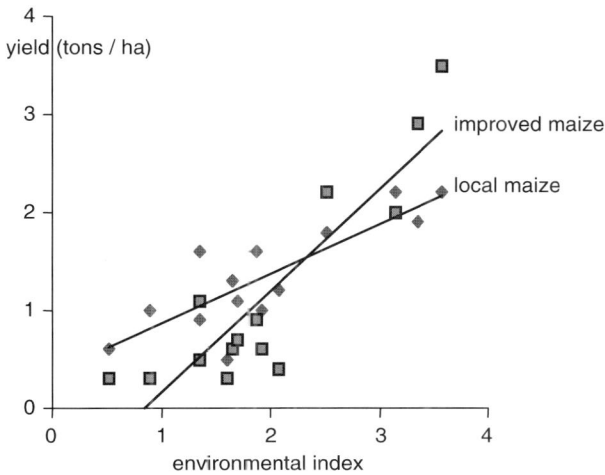

Figure 4-4. Adaptability analysis of a non-farm trial with two maize varieties. (From Hildebrand and Poey, 1985.)

physical factors. In farmer-managed on-farm trials, however, the environmental index results from the effect on a farm's productivity of an amalgam of physical and management factors, so how do you know which farmer will turn out to have low yield? That is not so easy to predict, precisely because farmers' yields are not only influenced by soil and weather, but also by the way they manage their farm, which, to make things worse, may also vary greatly from year to year.[3] Calling that 'management' is not much help really, unless you know what distinguishes good from poor management. That is the crucial question and the success in answering it will determine the effectiveness of agricultural extension anywhere, and not only in the tropics. I will come back to that in Chapter 6.

What these three pioneers did would later be called FSR, but that term only started to be used by the end of the 1970s. There were other pioneers as well, including some agronomists, but Norman, Collinson and Hildebrand were by far the most influential. FSR became the name for an entire movement, which was initially dominated by agricultural economists. For a while Norman argued that the name should be Farming Systems *Approach* to Research, which made a lot of sense, but when nobody would listen he gave up.

---

[3] In industrialised countries the management factor will be quite stable over the years, while in Africa it sometimes looks as variable as the weather.

4.3 IITA joins the FSR movement

I was little aware of what was going on in the world of great ideas about peasant farming until I joined the IITA. IITA at that time was eager to jump onto the Farming Systems bandwagon. They had convinced the Ford Foundation that it would be a good idea to promote FSR in national research institutes in West Africa and that IITA should take the lead. Or maybe it was the other way around, because at the time the Ford Foundation were already supporting FSR in several countries. In any case, I was one of the people the Institute hired to do just that. The fact that neither IITA nor I had any experience in that area did not seem to matter, and indeed it did not.

First something about IITA itself. The institute had had a Farming Systems programme since its inception in 1967, but that was an entirely different affair. Its grand design was to develop alternatives for shifting cultivation. In its original configuration shifting cultivation or slash-and-burn agriculture was a land use method practised by small farming communities who would clear a plot to grow crops for a few years and, as they moved on, let the vegetation close again behind their back so to speak. That of course hardly existed any more, but all sorts of fallow-based systems with varying fallow lengths, which probably evolved from shifting cultivation, were common in Africa at the time and so they are today. Two soil scientists, Peter Nye and Dennis Greenland had done thorough studies on shifting cultivation in the Sierra Leonean and Ghanaian rainforest areas in the 1950s and their work became the source of inspiration behind much of IITA's soils and agronomic research, particularly after Greenland had moved to the Institute as Director of Research. By the early 1980s three showpieces had come out of the search for alternatives to fallow-based systems: alley cropping, zero tillage and live mulch. They were all meant to reduce the need for fallow. In Chapter 7, I will explain how they work. Three of the scientists who developed them were to become celebrities in their profession and in the end they received prestigious honours from the American Society of Agronomy and the Soil Science Society of America. In 1982 the technologies had hardly been tested outside the research station and their originators kept themselves very busy in further refining the technologies before they would be considered ready for the real world, presumably in some fairly remote future.

Just before I joined there had been a change in leadership at the institute. The new Director General and the Director of the Farming Systems Programme had both come from management positions at the FAO of the UN. This meant a drastic departure from IITA's tradition of hiring senior staff with a strong research background. These were people of a different kind, by no means unscientific, but I think they were looking more for respectable causes to be pursued without excessive expenditure of effort, in the FAO style. They were genuinely motivated to make the institute's work more relevant for the real farmer. The new Farming Systems approach appeared to fit the bill very well. It combined the self-evident truth of its principles with the compelling logic of its approaches: analysing the real farmers' constraints and opportunities and taking those as point of departure for the development of new and better adapted technologies. And IITA already had a lot of technologies in store ('on the shelf' in the jargon), which could be used immediately, so the institute was in a unique position to marry the results of many years of hard research with a novel approach to deliver those results where they could be used most effectively. The new leadership first hired two social scientists, an agricultural economist and a sociologist and once they had secured a Ford Foundation grant they hired me on that ticket to team up with them. I am not quite sure how I matched the picture, but it was probably because I had done studies on indigenous cropping systems while in Cameroon. The three of us were expected to launch IITA as a major actor in the new FSR movement. We therefore, first held two launching workshops. The first one was an 'expert consultation', where FSR practitioners from all over the world came together and explained to us what FSR was all about. Figure 4-5 shows a photograph of all the participants, lined up in front of the conference hall.

For those interested in names, here are some of the celebrities, actual and future, as well as some ordinary people, who were present at the meeting. It reads like a who-is-who in tropical agriculture: Michael Collinson of CIMMYT, Norman, then at Kansas State University I believe, B.T. Kang and Bede Okigbo of IITA, Hubert Zandstra of IRRI, Peter Matlon of ICRISAT, Christine Okali and Jim Sumberg of ILCA, Neil Fisher from Samaru, Professor Agboola of the University of Ibadan, Braun of FAO, Louise Fresco from an FAO project in Zaïre, Clive Lightfoot, who worked

Figure 4-5. Participants at the consultative expert meeting on FSR held at IITA in 1983. (Reproduced with permission of IITA.)

with Norman in Botswana, Ron Cantrell from a Purdue University project in Burkina Fasso, Peter Walker from Zimbabwe, who was to become IITA's statistician a few years later, Andrew Ker from IDRC. I must stop there, at the risk of annoying some people for not mentioning their name. After exposure to so much brain power I think we had all qualified as FSR scientists.

Our second workshop was to launch a West African Farming Systems Network. Setting up regional networks for all kinds of purposes was becoming popular among international institutes and has remained so ever since. In the case of FSR-new style it was especially relevant for IITA, because international institutes were not really meant to work directly with small farmers, except perhaps to try out 'prototype' technology or test new research approaches. Otherwise, working with the farmers was the task of national research institutes end extension organisations. In fact, the FSR approach had been created in national research institutes, albeit by expatriate scientists working there. Even so, the international institutes thought they could become effective promoters of the new ideas because of their regional mandates, scientists of repute and lavish funding. Regional networks were emerging as the promotional vehicles. The launching meeting was attended by national scientists and research managers from West African countries as well as some people from one of IITA's sister institutes, IRRI, who had experience in organising regional research networks. You may wonder why our network was to be restricted to West Africa. One good reason would have been that Africa is simply too large to be covered by a single network. But that was not the real reason. It was that eastern and southern Africa had been pre-empted by another sister institute, the Centro Internacional de Mejoramiento de Maiz y Trigo (CIMMYT), with Michael Collinson as its local leader. Our own West African network[4] got established and a chairman and a committee (of which I was a member) were elected. The objectives were not very clear but they were expected to evolve with time. The Steering Committee was offered a tour of the Asian Cropping Systems Network by IDRC, to see how that worked. That, we thought, was an excellent idea, and we all enjoyed the trip greatly, including statistical comparisons of several randomly allocated massage parlours that we assessed along our path.

The next thing to do was design an IITA–FSR methodology, which could then be promoted through the network, thus establishing IITA's regional leadership. The term FSR did not go down well at IITA where it had meant something entirely different, so we decided to call it On-Farm Research (OFR) instead, which nobody could object to. I will use both terms indiscriminately to denote what we were doing. How does one put a methodology together? The Ford Foundation programme officer in Lagos suggested the scavenger approach: cut and paste parts of existing methodologies and put them in an IITA jacket, that is what he had done himself in Pakistan. I thought that was a bad idea and very boring too. So we organised a training workshop where we were actually going to develop our methodology – by doing, together with the participants. They came mainly from Nigeria and Ivory Coast, because that is where the Ford Foundation wanted us to pilot the FSR approach with the national institutes. We had invited the USAID-funded Farming Systems Support Programme (FSSP), based in Gainesville, Florida, to co-organise the workshop with us. They were our competitors for FSR leadership in the region and we thought that was quite a smart tactical move.

---

[4] It was called the West African Farming Systems Network (WAFSRN) and in French Réseau des Systèmes de Production en Afrique de l'Ouest (RESPAO).

Methodology was rather a grandiose term for FSR, which was or should have been based essentially on common sense. But some people think that simple ideas are taken up more easily when couched in impressive terminology. That is probably true and if it establishes respectability for its adherents there is nothing wrong with it, as long as the methodology and its terminology do not suffocate common sense. So, what were the essential ingredients for an FSR methodology? Firstly, you had to demarcate and describe 'characterise' the area where you were going to do the research, the target area, and its inhabitants, using existing information. Next you would pick some representative villages and meet farmers in their farms to find out what their problems were and what kind of research they might need to solve them. That was called the 'diagnostic' or 'exploratory phase'. Once that was done you had to find or design potential innovations which could be proposed to farmers to 'address their constraints' or 'exploit their opportunities' and which would later be tested in their fields. That was called the 'design' and 'planning stage'. And when all that was over, you were going to do the real thing, that is, carrying out trials with new technology in farmers' fields to put the ideas to a real life test, that is the 'experimentation phase'.

Our first methodology workshop would deal mainly with the diagnostic phase of FSR as well as a little bit on the choice of technology for on-farm testing. The expected output was a simple set of diagnostic procedures, something like Collinson's Exploratory Survey and Hildebrand's Sondeo, suitable for West African conditions. The procedures would include guidelines for visiting the villages and talking to farmers, extension workers and traders, looking around in the fields, finding out what were major constraints and opportunities and developing some clever ideas about what research could contribute to help improve production. The important thing was to do it in an organised way, because these things can easily get out of hand if everyone starts to chase after his own pet topics. A somewhat formalised methodology and a team leader who maintains structure can go a long way in ensuring a more or less orderly survey.

The first few days of the workshop were spent in the classroom to draft the outline of a methodology which would then be tested in the field by the participants. IITA and FSSP staff presented some introductory papers to set the ball rolling, about how to characterise the physical environment, like rainfall and soils, how to conduct interviews and what kind of technologies were already available which could later be tested by farmers. Next the workshop participants developed a checklist of things to find out and a scenario for how to do it. We, the newly qualified FSR experts, together with our FSSP colleagues, could handle the software part, because that was mainly a matter of common sense, and we asked the veterans in the IITA Farming Systems Programme (actually they were about my age) to take care of the more technical aspects, like physical characterisation and available technology. We expected, naively, that they would jump at this opportunity to put their knowledge to practical use, but I think they considered the whole thing as rather amateurish and not really worth spending their time on. This latest fad would probably blow over if wisely approached, i.e. mainly ignored. So they dug some presentations from their drawers which they kept there for this sort of occasion and did not show up again.

Otherwise, the workshop went very well and we all enjoyed it. We studied secondary data sources, designed checklists for things to find out in the field, did somewhat infantile role plays, and then went to the field in small multi-disciplinary teams for 4 days. There we talked to farmers, looked at their fields with them and in the afternoons discussed what we had seen and what kind of things could be proposed to the farmers to try out in their fields. Finally we reconvened at IITA and wrote up our findings. What came out of that in terms of farmer constraints and possible research topics was perhaps rather trivial in one sense but highly relevant in another. I have listed the findings of one of the groups in table below 4-1. It is worth analysing this list, because there are a lot of issues which recurred in different guises in the following years. It is not a list ranked by importance, so the numbers have no particular significance, except that I have regrouped the items a little according to whether they are related to the environment, pests and diseases, farm management and criteria of that sort.

So, what was trivial about this list? First look at the constraints in the first column. If you would compare them with the hundreds of such lists produced in many places over the next 20 years and up to this very day, you would notice some conspicuous similarities. Constraints 1, 4, 9 and 10 will almost invariably be there: that apparently is what farmers will tell you almost anywhere when asked what their agricultural problems are. Points 5 and 6 mention the most important crop disorders of cassava. Not a big surprise here either: cassava is the major crop in this kind of environment in Africa,

Table 4-1. Farmers' constraints identified by the first IITA OFR training workshop, and the on-farm research proposed to address them

Farmers' constraints	Proposed research
Inadequate soil fertility maintenance	Fertiliser application to maize–cassava intercrop (method, time and rate)
Unreliable rainfall in second season	Develop, screen, evaluate drought-resistant maize, cowpea varieties for late season
Uncontrolled bush fire	
Weed problems	On-farm comparison of hand weeding, herbicide and melon intercropping in maize–cassava intercrop
Cassava pests: grasshopper, green mite, mealy bug	Insecticide control of grasshoppers Screening of local varieties for tolerance or resistance to green mite and mealybug Dipping cassava cuttings before planting
Cassava diseases: mosaic, anthracnose	Mosaic-resistant varieties
Rodents Disappearance of crops from the system: rice, cowpea, groundnuts Seasonal labour and cash shortage Inadequate extension and input supply	
(no specific constraints addressed here)	On-farm evaluation of cassava varieties developed by national institutes Extend cassava cropping by applying fertiliser to last cassava cycle

wherever you go, and these pests and diseases were very widespread. IITA had a successful research programme on the biological control of mealy bug (Figure 4-6) and green mite, while its breeders had produced a very good variety which was highly tolerant to the mosaic virus. So there was hardly need for a diagnostic survey to find that out. I think the most interesting constraint is the one about the crops which had disappeared, although the empty cell to the right shows that it was not clear what to do with that information.

Figure 4-6. Cassava, heavily infested with mealy bug

Now look at the proposed research. Most of it was conventional stuff: fertiliser trials, testing drought-, pest-, disease-resistant or just any crop varieties, chemical pest control. You would think that kind of research would be done anyway, with or without FSR. It was, but only inside the research stations. The proposed research to tackle the weed problem was a little more interesting. It wanted to add a crop to the common maize–cassava mixture, namely melon, a broad-leaved creeping species, which would so to speak replace the weeds and produce something useful instead of being a nuisance. That was a technique which had already been researched for several years at the research stations by IITA and several national institutes in Nigeria, so this would be a good opportunity to take it out to the real farm. Whether you needed FSR for that is another question.

Does that mean it was all frivolity? Not at all, although that was definitely what the sceptics thought it was. We ourselves were aware of the relative insignificance of our findings but that did not bother us. Even though the results could have been written up without even going to the field at all, the redeeming feature was the shared experience among the researchers and a consensus about what to do next. If the on-farm tests which were proposed would actually be carried out, that would drag the researchers out of their stations and into farmers' fields, which was exactly what we wanted. It would not really matter what they tested in the beginning, even if it was based on the wrong ideas. By working with farmers they would soon enough find out what was really important. That was our hypothesis.

In the next couple of years I went around West and Central Africa to preach the FSR gospel. First I worked with the Nigerian and Ivory Coast teams who had participated in the workshop, to carry out their own diagnostic surveys. Later we held training workshops in other countries as well and carried out surveys with the local staff, who were usually getting financial and technical support from various donors. FSR was rapidly becoming popular, there was a lot of enthusiasm around and it was easy to fund the surveys from project budgets. But diagnostic surveys are only useful to set a process in motion, and after the survey the real work had to be done. Of course we talked about on-farm experimentation in the workshops: how on-farm trials would be different from station trials, how the farmers should be the ones to carry out the tests with the scientists as observers, how to analyse results from scrappy farmer fields, that kind of thing. In actual fact we talked about things most of us had never really done ourselves, but if we could get the national scientists to start on-farm trials we would all learn how to do it along the way. I will come back to that later in this chapter but I want to interrupt the FSR story here for a little while to talk about African farming itself.

4.4 Some amazing things about West African farming

Running around African farmers' fields as part of my proselytising job was quite an educational experience. There are things about West African farming which are really impressive. By narrating some of that may be I can make up a little for what will turn out to be a rather depressing story about the achievements of FSR.

I have already hinted at the amazingly rapid adoption of intensive maize production in the West African savannah. Maize can produce spectacular yields if conditions are right. That means enough water and sunshine, a lot of fertiliser and a good variety which does not collapse under its own weight when heavily fertilised. When fertility is low and there is no fertiliser, however, maize may well produce nothing at all, while millet and sorghum will still be able to scrape enough from the soil to give some yield. For ages maize had been sitting in people's backyards where household refuse and manure could be applied but in the 1960s and 1970s field trials and theoretical calculations at the Samaru research station in northern Nigeria showed that fertilised maize could be a very productive savannah crop indeed. Then came the World Bank-funded Agricultural Development Projects (ADPs) which employed several former British staff from the Samaru research station. They started a massive demonstration programme with the new maize varieties grown as a sole crop combined with fertiliser and they organised the fertiliser supply. The varieties came from IITA and the superior potential of this combination caused maize to spread like bush fire. But not the way it was demonstrated. Farmers picked up the varieties and the fertiliser and incorporated them in their own cropping system, which meant combining it with sorghum, cotton, groundnuts and other crops. Only the larger mechanised farmers adopted maize as a sole crop. What could be learned from this is that farmers will pick what suits them from an extension package and ignore the rest. If the opposite is also true, i.e. that those technologies which farmers do not adopt are no good, it means that research had actually produced precious little else that was of value for farmers. FSR was going to look into that presently.

The success of maize in the African Guinea savannah illustrates farmers' responsiveness to novelties with a difference. If nothing useful is on offer, however, they will just continue doing what they have always done. Which is quite an intelligent response if you look closely enough. There are three West African crops in particular which are very fascinating. They are yams, cassava and oil palm. Yams and oil palm are indigenous to West Africa, so it is not surprising that farmers have learned to do amazing things with them. Cassava was introduced in the seventeenth century from

South America but it has been taken up so completely that it might as well be indigenous. I will talk about yams here and about the other two in later chapters.

Yams belong to the genus *Dioscorea* and different species in this genus occur in all four continents. Although there is a lot of variation, they all have spreading or climbing vines and most of them produce underground tubers. Several species with edible tubers have been domesticated. Domestication is really an odd term for a plant species, but that is what plant geographers call it. It means that historically people have picked interesting specimens from the wild, grown them as a crop and over time selected and exchanged the types with the most desirable traits. Yams are propagated by planting small tubers or the top parts cut from larger ones, both called setts. In West Africa three major species are grown: white yam (*Dioscorea rotundata*), yellow yam (*D. cayenensis*) and cluster yam (*D. dumetorum*), all of them with large tubers (Figure 4-7). There is a lot of folklore and ceremony surrounding yams, but that is not what I want to talk about. I will describe an intriguing yam growing technique practised by the Yorubas in the wooded savannah of western Nigeria and the adjoining area of the Republic of Bénin.

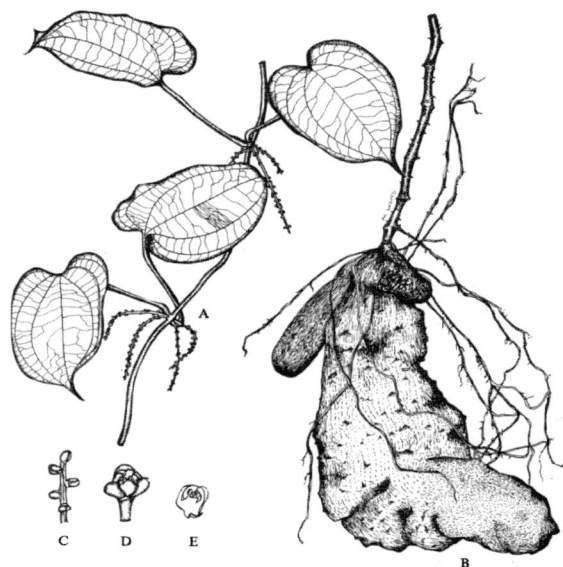

Figure 4-7. Yellow yam (*Dioscorea cayenensis*). (From Purseglove, 1972; reproduced by permission of Blackwell Science, Oxford.)

The species they use is white yam, which is the most appreciated and prestigious one and definitely a man's crops. For best results the vines must be supported by stakes or anything they can climb on.

The rains in the area usually start in earnest in April and continue until the end of October, often with a short dry spell in August. The months of November through February are almost perfectly dry. The rainy season is quite long, much longer than it takes for a crop like maize or groundnuts to mature, so two such crops could be grown one after the other in the same year, but only barely. That is what farmers sometimes do, but we are interested here in a much more efficient cropping pattern which involves yams.

It begins with clearing a field from mature shrub vegetation sometime during or towards the end of the dry season. The field will eventually become a yam plot, but not right away. In the first year after clearing the bush maize is planted followed by sorghum. Keep an eye on the rainfall-cropping pattern diagram of Figure 4-8 to get the picture. Such diagrams are convenient to show how a cropping pattern is organised in relation to the rainfall. I think Bede Okigbo, who was the Director of the IITA Farming Systems Program before my time, invented them. Clearing the future yam plot from the fallow vegetation is done very carefully. Traditional Yoruba farming is really quite conservationist. Trees, shrubs and undergrowth are cut down and burned but the perennials are not uprooted or otherwise destroyed completely. So the original vegetation can restore itself quickly after cropping has ended. The price paid for this is that everything has to be done manually, because the tree stumps hinder animal or tractor drawn ploughs. Yams do best on fertile forest soil and farmers will always look for such soil for their yams, even in areas where most of the perennial vegetation has

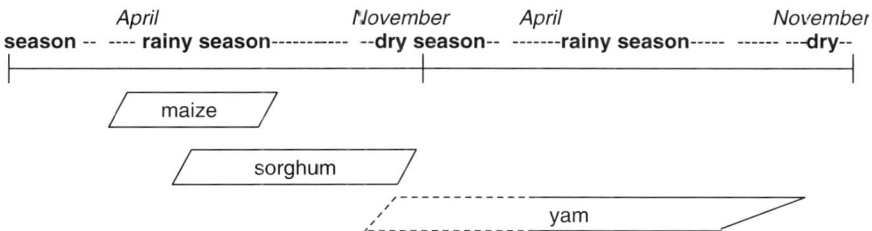

Figure 4-8. Cropping systems diagram for the Yoruba yam-based system in southern Nigeria

gone. Ask a farmer to show you a suitable plot for yams and he will take you to the richest vegetation in the area. So if you want to grow yams again later on you have to make sure the original vegetation comes back as quickly as possible.

After clearing, low ridges are made on which maize is planted first. The maize will mature by August but before that, in June, sorghum is already planted on the side of the maize ridges. After the maize harvest the ridges are refashioned so that the sorghum will now be on top of them. That operation serves at the same time as a weeding round. The sorghum will mature during the following dry season, because flowering is induced by the shortening day length after September. That is a useful trait, because sorghum panicles are very susceptible to moulds when they are hit by rain.

Now comes the most interesting part. By December when the sorghum is getting ready for harvest the ridges are refashioned again into rows of fairly large heaps and a yam sett is buried inside each of them. On top of the heap, right above the location of the sett, a cap is placed consisting of a mixture of straw, leaves and soil.

Figure 4-9. Sorghum trellises for next season's yams

This is to protect it from excessive heat and prevent drying out. After the sorghum grain has been harvested the long stalks are bent in the direction of the rows and arranged in hedges or trellises which will serve as support for the yam vines in the following rainy season (Figure 4-9). The sets sit in the heaps for several months waiting for the rains to break, when they sprout and cover the trellises with their vines. The yams stay in the field throughout the rainy season, develop a thick canopy and are harvested at the end of the rains or later. The cycle may then be repeated or followed by a different crop combination in case the soil fertility has gone down too much. This is a really ingenious system and also very well adapted to the environment too. It would be difficult to invent anything better than that, which should make us agronomists a little more humble. This account was just meant to whet your appetite for African agriculture, we must now return to the FSR story.

4.5 FSR catches on at national research institutes

The national research institutes in West Africa we worked with usually had very large areas to cater for, with large differences in ecological and socio-economic conditions. Some of them, in particular in Nigeria, were specialised commodity institutes, meaning that they worked on certain crops or groups of crops, like oil palm or cereals or root and tuber crops, and had names like the Nigerian Institute for Oil Palm Research (NIFOR) and National Root and Tuber Crops Research Institute (NRCRI). Others had a more geographical mandate, like IAR in Samaru which I mentioned earlier as the place where Norman developed his ideas. In the 1980s, with the rise of 'systems' thinking commodity institutes began to be looked on as anachronisms, except perhaps for plantation crops like oil palm or rubber where crop and system more or less coincided. Food crops in particular could not be researched meaningfully in isolation of the system they were part of and it therefore seemed logical that research institutes should have geographical mandates and study crops as part of the cropping of farming systems that farmers practised in their designated area. Some of the institutes, like IAR in northern Nigeria and the Institut des Savannes in Bouaké, Ivory Coast, already had geographical mandates and now the commodity institutes were also given wider research responsibilities for the area in which they were located, while continuing to work on the commodities they had special expertise in. An institute which had research responsibility for farming or cropping systems, rather than just individual crops, was expected to be favourably inclined towards the adoption of FSR, so this trend should also create a favourable environment for the new FSR teams which were going to be set up. It was not at all obvious, however, how the institutes would be able to do meaningful FSR across the entire range of conditions found in their areas, but that is what they thought or were told their regional mandate obliged them to do. Diagnostic surveys therefore had to be carried out in each of the different agroecological zones in the institutes' mandated areas, presumably to be followed by on-farm experiments in all those zones, but that was of later concern. The trend was now to do diagnostic surveys and that was already enough work. How they would carry on from there would be thought about later.

Whatever their shortcomings, many national researchers put a lot of enthusiasm into diagnostic surveying in those early years. I think many of them appreciated the respectability FSR conferred on their parents' farming practices which had hitherto been looked upon as something primitive which had to be rooted out as soon as possible. But the quality of the surveys was usually not very good. The scientists often failed to drop their superior know-all attitude when dealing with farmers and extension workers and a lot of the findings were rather insignificant. In later years we, the researchers, were criticised for controlling the whole diagnostic process, leaving extension workers out and not really listening to the farmers. That criticism was not entirely unjustified, although we did try to get extension workers on the teams and organised feedback meetings with the farmers to check how realistic the ideas were that we had developed. But I think the most important problem was not that the researchers dominated the diagnostic process. It was their inability to make the transition from diagnosis to a long-term on-farm testing programme with genuine farmer involvement. FSR was not meant to be just a series of farmer surveys with a high feel-good content, but should lead to a reorientation of the entire research process towards the real-life farm. Never mind if the diagnosis was not much better than what you could have written without leaving your office, as long as it was the prelude to some good on-farm experimentation, which we were confident would help research break out of the sterile environment of the research stations. But I am running ahead of the story, so let us look at how the early surveys were carried out and what was done with the results.

The idea of FSR was that the key to agricultural development was to be found in the real farm and that scientists should expose their technologies to the rigour of that farm, instead of the pseudo-reality of the Institute's experimental fields and the journal editors. This reality test in turn should lead to a better appreciation of what was really important for farmers and eventually to more relevant research, both on-station and on-farm. And since practically all agricultural research was applied research, FSR should be every-body's business, not just that of a few FSR adepts. FSR had logic on its side but its advocates, the social scientists and agronomists were not the most accomplished scientists in the stations and they were often looked down upon by hard core researchers like plant breeders and soil scientists. The fact that they exhibited the arro-gance typical of true believers also did not help in bringing the others around. But many donors started to advocate FSR so there was new money coming its way and the institutes started to think about setting up FSR programmes and assigning staff to capture the new cash flows.

On-farm research[5] is by its nature multi-disciplinary, so there had to be teams including at the very least crop specialists, social scientists and soil scientists working together. Research institutes did not work that way. Even those with a geographical mandate were organised along commodity lines and each commodity programme would have its own crop breeders, entomologists and agronomists. There might also be a separate soils programme with soil scientists of different flavours doing their own thing, and an economics programme if the institute employed economists at all, or the economists were assigned to different commodity programmes. Where then would FSR, which was multi-disciplinary and multi-commodity fit and how could you get people from different programmes working together across all their established disciplinary and organisational boundaries? I know all this must seem pretty boring, but the way FSR eventually got organised had a major bearing on its achievements or lack of them, so unfortunately I cannot avoid the subject. I will first use Nigeria as an example, other countries will turn up as the story unfolds.

In Nigeria we worked with three large institutes in the southern part of the country. It was not because of our powers of persuasion

---

[5] I use the terms Farming Systems Research and On-Farm Research interchangeably, there was not really that much difference, in spite of much debate about terminology.

that all three of them were ready to give FSR a try. Even before IITA held its FSR launching workshops in 1983 the director of agricultural research in Nigeria, D.E. Iyamabo, had held a national FSR launching workshop with a large number of scientists from all over the country. They presented interesting and not so interesting papers, some of them claiming that they had in fact been doing this FSR thing all the time, as scientists will in order to be left alone. At the end of the workshop it had to be decided where to go from there and since that was not quite clear, my proposal to actually go and do something in the field, like some diagnostic surveys to start with, was welcomed. Some national institute staff volunteered or were hand picked from different programmes by the institutes' management to form adhoc OFR teams. These were the ones who participated in the training workshop later in the year where we developed the IITA diagnostic methodology.

You must have noticed my indiscriminate use of the word 'workshop'. In fact, it is only at this point that I became aware of it myself and I counted 14 occurrences in this chapter so far. Originally, in analogy with the technical interpretation of the word, it meant a place where people got together to do real work, albeit intellectual, but the word has become loaded with so many meanings that in the end it denotes not much more than 'meeting'. Today the word stands for almost any gathering, especially the kind where development workers get together to babble about their latest intellectual *tours de force*. That is a pity, since it must have been a nice metaphor when used for the first time. Linguistic corruption is not the monopoly of FSR but it has definitely made a significant contribution. But in those early days there was still an element of work in the workshops.

There was certainly no problem to get the few economists in the research institutes on board the OFR teams. They ranked low on the prestige scale of what were essentially technical Institutes and saw this new international movement as an opportunity for emancipation or perhaps even dominance. The agronomists were also not hard to convince. They must have been tired of years of inconsequential trials on their stations with planting dates, planting density and fertiliser rates which nobody took particular notice of, least of all the farmers. The breeders were a different story. In all four Nigerian OFR teams together there was only one breeder and he was approaching the end of his career. Breeders actually had something to show for their work. New crop varieties could make a real difference. Even if the most successful ones had actually been bred by IITA, national breeders could be proud of their association with the process. They were quite happy with the way breeding was done and in no hurry to join the FSR movement. So the OFR teams finished up as loose associations of economists and agronomists together with an occasional entomologist or weed scientist, usually four or five altogether for any one team.

The IAR, Samaru, in northern Nigeria, where Norman had worked was a special case. They had an active on-farm programme dating back to Norman's days and were very proud of it. When I came to visit them for the first time they laughed at me and wondered whether it would not be better if they came to IITA and taught us some FSR. But in the end they did participate in all kinds of activities and played a major role in the National Farming Systems Network which was set up later on with Ford Foundation funding, under the leadership of George Abalu and James Olukosi.

The next hurdle on the road to active OFR programmes was money. The Ford Foundation grant on which I had been hired allowed us to hold meetings and training courses and to travel around, but not to fund on-farm research, at least not in Nigeria. That was a blessing in disguise. In many countries OFR teams were created with lavish donor funding, usually in splendid isolation from the rest of the research establishment. That has not helped their durability. In Nigeria we had to look for local money and the logical sources were the ADP, funded by World Bank loans. I have to say a few words about those ADPs, before continuing with the FSR story.

The ADPs had been set up in the 1970s in several areas to promote so-called modern technology for peasant farmers. They had created massive parallel organisations run by expatriates with local staff seconded from government. They meddled in everything, from land clearing with heavy equipment through agricultural extension, supply of seed, fertiliser, insecticides, produce marketing to on-farm research. Expensive irrelevancies funded with other people's money has been the World Bank's trademark up to the present day, at least in agriculture, and the ADPs were among them. There was one ADP success story, however, the rapid expansion of maize growing in northern Nigeria, but that could have been done at a fraction of the cost by simply putting the seed and fertiliser supply in order. By the mid-1980s the ADP concept really had become an anachronism and so had many of their often numerous expatriate staff. The money had become scarce because it had to be shared among the large number of ADPs which had been set up over the years, most of the expatriates were gone and the ADPs were now expected to integrate with the local institutions. And perhaps most important of all, they lacked new technologies which might have the same spectacular impact as maize in northern Nigeria, so they were looking to the national research stations and to IITA to provide them.

That was the situation when we at IITA and the scientists at the Nigerian national institutes embarked on OFR. The new IITA leadership thought it would be a good idea if its OFR team would work directly with the ADPs, because they were still comparatively well organised and could probably ensure rapid dissemination of IITA's technologies. But after making a few exploratory visits I felt that would not be a good idea. Providing research services to the ADPs was the task of the national institutes, not of IITA. Our role would be to assist them, rather than working directly with the ADPs. That was also the concept of the Ford Foundation grant I was employed on. So, together with some of the national scientists, I went to talk to the ADPs' coordinating body, the Federal Agricultural Coordination Unit (FACU), a sort of parallel Ministry of Agriculture, and asked them whether they were willing to sponsor OFR to be conducted by the national research institutes in the ADP areas, with technical support from IITA. They were, but the money would have to come from the ADPs themselves. FACU designed a prototype contract which was then worked out between each of the ADPs and the research institute closest to them.

During the next year or so, I took part in the diagnostic surveys carried out by the OFR teams of all the major Nigerian institutes under ADP contracts. The ADPs provided transport and off-station allowances and some of their field officers participated as well. Survey reports were prepared and they were published nicely by FACU. It was all very promising but I was looking forward to the real thing, which was carrying out on-farm tests. That of course was also what the ADPs wanted, because they needed technology, not just survey reports. So, once the surveys were done the teams designed on-farm test proposals and submitted them to FACU, who would then advise the autonomous ADPs whether or not to fund them and how much that should cost. Since I considered myself as the operation's technical adviser I thought FACU should seek my opinion on the quality of the proposals. I must admit that I had as little real experience in on-farm research as any of the teams and calling myself an adviser did not necessarily mean I was listened to. The FACU man who supervised the operations, himself an agronomist plucked from the University of Ibadan, probably felt he did not really need me any longer, because he thought himself as qualified as anybody to pass judgement – which he was not. Since I had been instrumental in setting the whole operation in motion, at least I was asked for comments on the first round of on-farm test proposals submitted by the three southern institutes. It was all rather conventional stuff, mainly things they had been doing for years on the stations, with land preparation, planting densities, varieties, fertiliser and herbicides. But that was OK. They should first empty their shelves and see what it was really worth when applied on the real farm and then later on may be there would be some more creative thinking. And a better variety is always welcome, if it is still really better when grown by farmers. The important thing now was to carry out the tests properly And that is where the trouble began.

When you start working with a farmer he must believe that what you have on offer might be of some interest to him. Even if that is not immediately obvious, he may still agree to give it a try, because you never know, there may be some incidental benefits, like free fertiliser for instance. Once a trial has been planted he must recognise it as something that he might actually do himself. Trials conducted in the research station are not like that. They are laid out in a rigorous geometric block pattern, the plant rows are straight and exactly dimensioned, the plots are weeded on time and all on the same day, if there is an insect attack it is controlled by insecticides, all plots are harvested on the same day (unless it is a planting date trial, of course) All these things most farmers do not usually do. That is how the statistical textbooks tell you to do it. If you conduct trials on-farm in that way they are bound to fail one way or the other, because farmers will see them as the scientists' affair, not theirs. And that is what happened.

Apart from the statistical textbooks there were other influences that prevented on-farm researchers from doing the obvious thing. Suppose you want to start with something simple like a new maize variety. You may ask the farmers to plant the new variety in their own way along with their own variety and see what happens. If you are not an expert you would think that makes a lot of sense. But things were not as simple as that. First, the breeder of the new variety would object that the *superior potential* of his variety would not be *expressed* under such inferior conditions, and second, the agronomists would argue that the farmer had to be shown what modern technology was capable of, rather than just letting him do his own antiquated thing and perpetuating his backwardness. So he should follow the *recommended practices* and plant the maize in rows at the *correct* density with the *right* amount of fertiliser. All this is baloney of course, if the thing you want to know is whether a new variety when grown by a farmer is better than the one he already has. If the researcher insisted on all these things the farmer would politely express consent, while trying to steer the tests to a worthless corner of his farm, as far as possible from the fields where he had to produce his family's food.

Farmers are often quite clever in outwitting scientists who they fear may put their crop in jeopardy. I like to narrate a beautiful example of that. One of the national teams in southern Nigeria wanted to carry out a yam trial planted on ridges instead of the usual heaps. The ridges would later be 'tied', by connecting them across the furrows at short intervals, thereby forming little basins to prevent rainwater from running off. This should improve water availability to the crop which often suffered from drought. There were also some other treatments which I forgot. I think only four or five farmers had signed up. If there is one thing farmers do not want to mess around with it is their yams. Yam setts were very expensive, so they first asked the researchers to buy the setts for them, which they did. Then they pointed out a corner of a field where the trial was to be planted. None of them had planted anything in the trial field yet when the researchers arrived with the setts, but after the yam trials had been planted *all of them* planted the rest of the field with maize and cassava, not yams. So the trials turned out to be located in fields which were actually not intended for and, as it turned out, unsuitable for yams. The farmers had effectively steered away the researchers from their real yam fields, got free setts and conducted their own trial, which consisted of testing whether the researchers could grow yams in a field where they could not, and at the scientists' own cost.

In a lot of the early on-farm work farmers were treated as free labour in the scientists' trials, the purpose of which they failed to see. Not willing labour of course, because soon they would start complaining that all this land preparation, planting along planting ropes and timely weeding was a lot of work for which they actually did not have the time, so could the scientists please provide the money so they could hire some extra hands. The scientists, fearing that their trials would be overgrown by weeds, would give in and in the end they were conducting very bad station trials in farmers' fields, which further damaged their scientific prestige when their fellow scientists or the director came to visit the fields. But that was in the early and mid-1980s, when all of us were still learning from experience. We will have occasion in a later chapter to see where the learning process led and whether in the end common sense prevailed. For now, I will continue my narrative about the early expansion of the FSR movement in Nigeria and in Ivory Coast, the second target country of the Ford Foundation grant.

I believe my earlier role in organising training and exploratory surveys had been genuinely appreciated by the Nigerian teams but once on-farm testing started I think they did not really want me around any more. Of course I was welcome to visit the sites and they would listen politely to my comments. I would also be invited to read papers, even keynote ones, at their workshops, but that was it. Meanwhile I saw that the on-farm trials did not even remotely resemble what they had been meant to be, but there was little if anything I could do about that. It was logical that the national scientists, who were often quite senior, did not want an IITA scientist on their backs who had as little experience as they had. Our relationships remained quite cordial, but I decided to stay at a distance, hoping that with time and through their contacts with the growing international FSR community, their programmes would gradually overcome their teething troubles.

About the Ivory Coast I can be brief here: the time was not yet ripe there for Anglo-Saxon style FSR. I think the reason why the Ford Foundation wanted us to work in the Ivory Coast was not entirely free of cultural imperialism. The French were still very much in control of agricultural research in the country, as well as of many other things, and the Ford Foundation felt it was time for some good Yankee pragmatism there. What we should have done is convince the Savannah Research Institute in Bouaké, the largest foodcrop research institute located in the centre of the country, to put a group of people together and start some FSR work, as had happened in Nigeria. The Minister himself, however, decided other-wise and hand-picked five researchers from three different widely dispersed institutes who formed a sort of nomadic FSR team, with-out any institutional basis. I think that was part of his strategy to break the French hegemony in agricultural research and develop-ment. If I have learned one thing it is that FSR cannot flourish in isolation. The team obtained its own Ford Foundation grant and we organised a training workshop with a contingent of Nigerians in attendance who were already more advanced and would bring in their experiences. A few months after the workshop the team car-ried out a few diagnostic surveys in which I also participated. I will tell the story of those surveys in Chapter 8. But they never managed to get serious on-farm work going. The apotheosis of their failure was an experiment which they reported at a workshop at IITA, in 1989. They had recruited some young unemployed secondary school graduates, hired a piece of land, put a barbed wire fence around it and made them practice *modern agriculture* there. That finally put an end to our relationship. I think the team eventually fell apart because of internal frictions.

In retrospect, we overestimated our powers of persuasion in trans-mitting FSR concepts to national institutes. I think the only way to do that effectively is by being part of an institute and transforming it from the inside, the way Hildebrand did in Latin America and Mark Versteeg in Bénin, and even in those cases I wonder how long their influence persisted after they left. Nevertheless, I have seen national scientists all of a sudden discovering what FSR really meant. After that there was very little need for training or coaching, it simply became second nature to them. So it can happen, but turn-ing around an entire institute is a different business.

After a few years of talking and scarcely being listened to we decided that it was time for a change. First of all we were going to write a book about OFR, perhaps that would help. Books provide visibility and prestige to their authors, which help when you try to convince others. And secondly we were going to set up our own OFR sites in some villages in southern Nigeria and do some proper on-farm experimentation there, because we started feeling a little ridiculous talking about OFR without doing it ourselves. The book came out in 1986 and our OFR sites were started around the same time, in collaboration with the University of Ibadan. I will come back to that also in Chapter 8.

4.6 IITA's tenuous relationship with FSR

During the 1980s, when the FSR movement went through its most vigorous development, IITA was probably the international institute where the FSR concepts had the least impact. The word 'ownership' had not been invented yet, but we failed from the start to establish ownership of our new approach at IITA.

For one thing, we were probably too conceited. While having little to show to match the achievements of the established technical programmes, we thought that the logic of the FSR ideas and the strength of our rhetoric would compensate for the lack of content. The institute was very successful on various fronts. It developed good looking, though in the end inconsequential, alternatives for fallow-based agriculture, bred excellent maize, cassava, soybean and cowpea varieties, and its biological programme for the control of a devastating cassava pest, the mealy bug, was very successful. Nobody had therefore much time for or interest in this FSR thing which they felt consisted mainly of hot air and in any case should be the business of national institutes. On the first score we were yet to prove them wrong and on the second they were right: the institute's task was to develop prototype technologies and methods, not going around farmers' fields to test or demonstrate them, that was the task of the national scientists and the extension services. We maintained, however, that international institutes would only be successful if the national ones were capable of doing their part of the job and we had to help them in improving their skills, particularly in FSR. Contrary to some of its sister institutes, however, IITA never really accepted that educational role for itself and FSR remained at the fringe of the institute's activities.

There was one exception, though: the 'outreach' programmes in Cameroun, Ghana, Zaire and Rwanda, which were carried out by local and IITA contract scientists and sponsored by USAID and the World Bank. The Americans were the first donors who got sold on the FSR ideas and promoted them vigorously through a large number of research projects in Africa, Latin America and Asia, carried out by mixed teams of national and international scientists. Some of them were contracted with American universities, but some of the larger ones in Africa with IITA. IITA in turn hired researchers on the international market on a contract basis and provided backstopping from Ibadan. The outreach projects were not

really taken very seriously by the landed IITA nobility in Ibadan
and outreach staff were seen by some of them as largely second
rate. If the donors wanted to spend money on FSR that was fine
with them, as long as it did not interfere with the real science, so the
outreach programmes provided a convenient and harmless outlet,
an opportunity for an occasional visit and otherwise a minimum of
nuisance.

Through the years, IITA remained an essentially technical institute, with several major achievements to its credit, some of them for
the benefit of the African peasant. Especially the results coming out
of the crop improvement and the biological control programmes
had a direct and considerable impact. Really good crop varieties
actually need very little effort to be adopted and the natural enemies of crop pests imported from elsewhere also find their way
around without much assistance. So, for these programmes the need
for on-farm research was not felt very strongly. Distribution of
planting material through national institutes and extension services
for the varieties and releasing the natural enemies at a sufficient
scale would do the job.

The technologies generated by the Farming Systems Programme,
however, were of quite a different nature. As we will see in the later
chapters, if adopted, they would have affected the farmers' entire
way of farming. So you would think those technologies should be
tested together with farmers and right from the start to avoid wasting time on what might turn out to be unacceptable to them in the
end. Having an OFR team inside the institute which could help in
testing all those nice technologies in farmers' fields would seem to
be an asset. But that is not how it was looked at by the Farming
Systems Programme's soil scientists. Their main motivation seemed
to be to satisfy their scientific curiosity and enhance their reputation in the eyes of their peers. The flow of publications coming out
of the programme was considerable but adoption of the technology
by the farmers was practically nil. Following one's interests as a scientist and disregarding possible applications is all very well for
basic science, but agricultural research is mainly applied research
and without applicability and adoption there is little or no justification for it. The research was funded from public money with the
intention to benefit peasant farmers, but since these farmers had no
voice to complain or make demands, research had to develop its
own ethics and discipline by exposing its results to the conditions of

the real farm as early as possible. That, we told the other scientists, was the purpose of OFR. But we had not achieved or proved anything yet, we mainly talked. As early as 1983 an unfortunate amount of jargon, methodological fetishism[6] and casuistry was already developing around an essentially sound and simple concept, which was enough to put the more scientific scientists off. And the fact that FSR was dominated by social scientists or others who talked like them, did not help either. So the research at the IITA station went on and on, until the last of its creators retired and little more was heard about the technologies which were once celebrated as triumphs in the struggle for uplifting the African farmer. But that was much later. In the mid-1980s the FSP scientists were at the pinnacle of their careers and absorbed by their science, while we were yet to prove the power of OFR by setting up our own on-farm research sites and hopefully provoking the station researchers to come out and put out their wares in front of the farmers. That story, however, will have to wait until Chapter 8. I will first spend sometime on farming itself and on the technologies which farmers used and those that scientists proposed.

---

[6] The term is due to Sigmund Koch in a 1981 paper in the *American Psychologist*, quoted in an academic lecture by the Dutch psychologist Jaap van Heerden.

Chapter 5. Forests, Fallows and Fields

I have spent a lot of time discussing the rise of FSR and there is much more to come. I think that is justified, because after all FSR has been one of the defining features of tropical agricultural science in the second half of the twentieth century. But FSR is software, it deals with methodologies, approaches and attitudes. The hardware is agricultural production: farming practices used by farmers and of course improved technologies invented by scientists. To maintain good balance I will devote the next three chapters to that, before continuing with the FSR story.

In this chapter we are going to look at the technicalities of farming in the African forest area. That may sound boring to some, but it is actually quite fascinating to examine what a casual observer (and some people who should know better), may see as a messy and even hopeless affair, in need of a complete overhaul. I have already given examples of highly ingenious farmer practices which would be difficult to improve upon as long as the social and economic conditions do not change drastically, and more will occur in what follows.

During the last century the agricultural sciences have, perhaps unavoidably, been broken up into numerous specialities. That has gradually distanced from the specialists' view the fact that farming is more than the simple addition of numerous disciplines. It is about assembling those pieces into an integrated operation, which is what farmers do with more or less success. Even though agronomy has also suffered from fragmentation it is still the broadest of the agricultural sciences and should be capable of putting together a picture of the whole, which is what I shall try next. I will talk mainly about the tropical zones with wet climates, which are called the humid and sub-humid zones, because those are the areas I am most familiar with. I will first explain what these zones are, but if you are not particularly interested in the climatic details you can skip Box 5-1.

5.1 Fallow-based cropping

Most cropping in lowland Africa is based on fallow as a means of restoring the fertility which is taken out by the crops. The land is used for a few years and it is then abandoned for some time until

Box 5-1. Humid and sub-humid climates

The most commonly used classification of African climates distinguishes five
zones: pre-humid, humid, sub-humid, semi-arid and arid. These are not very
precise terms and in particular the sub-humid zone embraces climates with very
different characteristics. The map on this page shows the five zones as they
occur in West Africa.
Around the equator the climate is humid, with mean annual rainfall
around 2,000 mm and no pronounced dry season or a short one of one or a
few months.

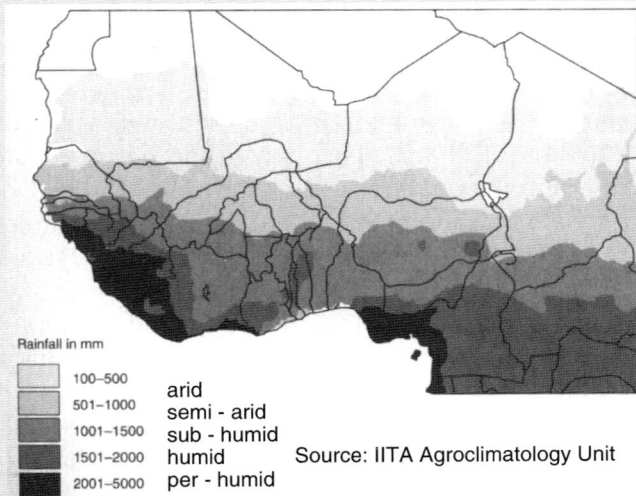

Rainfall in mm

100–500	arid
501–1000	semi - arid
1001–1500	sub - humid
1501–2000	humid
2001–5000	per - humid

Source: IITA Agroclimatology Unit

As you move away from the equator the climate becomes monsoonal, which
means that there is a clear alternation of rainy and dry seasons. Some monsoonal climates in West Africa are very wet (pre-humid or super-humid), even
wetter than those around the equator, with more than 2,000 mm of annual
rainfall, as you can see on the map. There is one place in western Cameroon, at
the foot of the Mount Cameroon volcano, with as much as 10 m of annual
rainfall, which together with the Chocó province in Colombia is the wettest
place on earth. The climax vegetation in the humid tropics is evergreen forest.
If the forest is not there, something must have caused it to disappear, or there
may be peculiar conditions like swampy or very thin soils which hamper
tree growth.
With distance from the equator the climate becomes dryer, going from humid
through sub-humid to semi-arid and arid (although the boundaries across
Africa zigzag a lot), the rainy season becomes shorter and the total amount of
rainfall decreases.
Humid climates with 1,500–2,000 mm annual rainfall and a dry season of 3 or
4 months are found in a fairly narrow belt running parallel to the coast from

west to east, with a gap in Central and Eastern Ghana, Togo, Benin and western Nigeria, where the coastal climate is sub-humid. The sub-humid zone, with
1,000–1,500 mm rainfall, is a transitional band between the humid zone on one
side and the semi-arid zone, with 500–1,000 mm and a savannah vegetation on
the other side. The sub-humid zone is a hotchpotch which covers a vast area
with very different vegetations and cropping systems. It is usually subdivided
further on the basis of vegetation types: the semi-deciduous forest, the forest-
savannah transition and the southern Guinea savannah zones.[1] All these sub-
zones fall within a fairly narrow rainfall range and their natural vegetation is
determined by the rainfall 'modality' rather than the annual total amount.
I need to explain what that means and why it is so.
At lower latitudes the sub-humid zone has two rainy seasons, interrupted
by a short period with a little less rain, and a dry season of 4–5 months. This
rainfall pattern is called bimodal, which means that there are two rainfall
peaks. As you move further north, the two peaks merge and the pattern
becomes unimodal. The transition from bimodal and unimodal rainfall also
marks the transition from forest to savannah. Under a bimodal rainfall regime
the climax vegetation is semi-deciduous forest, while a unimodal sub-humid
rainfall regime results in a savannah vegetation, called Southern Guinea
Savannah. In order to see why that is so, look at the following two graphs,
which show the rainfall patterns of Ibadan in Nigeria and Sokodé in Togo.
Ibadan has an average annual rainfall of 1,250 mm, while Sokodé, which
is located in the little dark wedge in the centre of Togo on the rainfall map.

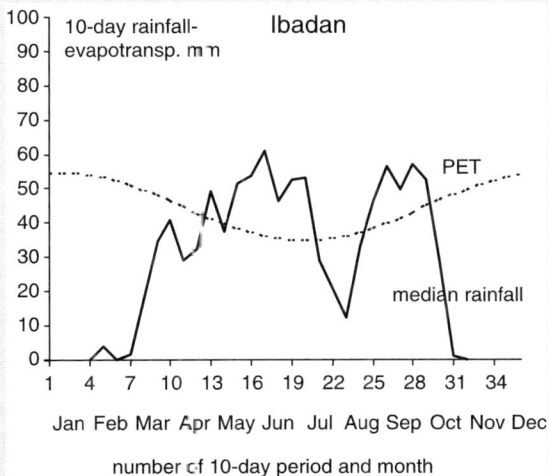

(continued)

<hr>

[1] This vegetation-based classification is quite useful and extends into the semi-arid
zone, where we meet the northern Guinea savannah (800–1,100 mm) and the
Sudan savannah (500–800 mm).

Box 5-1. (continued)

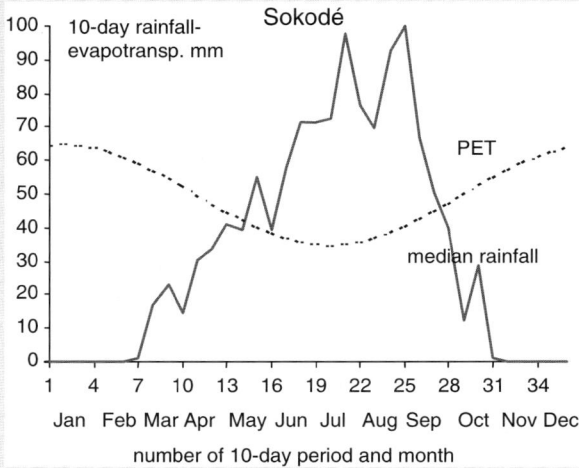

has 1,350 mm.[2] Yet Ibadan is at the edge of the semi-deciduous forest zone while Sokodé, with 100 mm more rain is definitely savannah. Why is that? Look at the two rainfall graphs.

The time axis is subdivided into periods of 10 days and the curves are for 'median rainfall' and average potential evapotranspiration (PET) for each 10-day period. 'Median' means that half the years (out of a long series of, say, 20 years or more) had more rain than the median value and half had less. PET is the amount of water which a full vegetation would lose to the air if there were no water shortage in the soil. At the end of the rainy season the rain falls short of what the vegetation can use, but there is still moisture in the soil, stored from previous surplus rainfall. Once that is used up the soil becomes dry. In Ibadan that happens after the middle of November and lasts until the middle of April, i.e. there are about 5 months when the vegetation suffers from water stress. The dip in rainfall in late July and early August has little effect on deep-rooted trees and shrubs, because enough moisture has been stored from the previous period to make up for the shortage. In Sokodé, however, the period of stress extends from late October until late April, i.e. about 6 months. The very large amount of rain in August and September does not help, because the soil cannot store that much for later use. So, in spite of higher total rainfall the stress period in Sokodé is 1 month longer than in Ibadan. That is the reason why the climax in Ibadan is semi-deciduous forest while it is wooded savannah in Sokodé.

---

[2] The dark-shaded wedge is supposed to have more than 1,500 mm of rain, but in Sokodé it is actually about 1,350 mm.

its fertility is restored sufficiently for another cropping cycle. The number of years the land is cropped and the length of the fallow depend on several things, most importantly the soil's native fertility and the number of people the land must feed. The larger the number of people the shorter is the fallow, but the extent to which the fallow can safely be shortened depends on how much time the soil needs to restore itself. And that is related to its inherent fertility. The better the soil the more quickly the fallow vegetation will develop and the earlier the land will be suitable for cropping again. That is a very crude description which, though true in a general sense, ignores the fact that in some areas with quite poor soils population density is surprisingly high. So farmers seem to have found a way out of nature's straightjacket. I will come to that. For now let us assume that the simple scheme of low fertility-long fallow-few people holds true.

It is reasonable to assume that today's fallow-based systems are all derived from shifting cultivation, so it will be interesting to look at that age-old system in some detail, because it will help us understand today's cropping practices. But rather than starting with shifting cultivation, which is probably the oldest type of land use on all continents, I will move back in time from the present to the past. First I will describe a typical fallow-based system used today in south-western Nigeria and then look at shifting cultivation as the mother of all.

5.1.1 Food crop growing in south-western Nigeria

South-western Nigeria is part of the former cocoa belt which stretches across the sub-humid forest area of West Africa, from Cameroon to the Gambia. Land use and crop production practices of the local people, the Yorubas, are strikingly similar to those around Yaoundé in Cameroon, which I have described in Chapter 3. Permanent cocoa orchards are found on the heaviest soils, while food crop fields are scattered around in a patchwork, alternating with secondary bush plots which are recovering from earlier occupation. I will first say a few words about cocoa, the crop which lent the area its name, at least among agriculturists, before describing food crop growing.

Cocoa is a native of South America. It is not clear when it was first introduced to Africa, perhaps as early as the seventeenth century, but it only started spreading rapidly as a smallholder crop in West Africa in the late nineteenth century. That is a fascinating story in itself, but I will resist the temptation. The heaviest soils are used for cocoa, interplanted with colanuts, a mild stimulant with a similar effect as coffee. Cocoa and cola belong to the same botanical family, the *Sterculiaceae*, but cocoa originated in South America while cola is indigenous to West Africa, indeed a happy international marriage. The two are grown in permanent mixed orchards dominated by cocoa, with no other crops underneath, as long as the cocoa canopy remains dense. There are always some large trees as well, retained from the previous forest or nurtured into maturity by the farmers, like bitter cola (*Garcinia kola*) and the hardwood species Obeche (*Triplochyton scleroxylon*), the latter also belonging to the *Sterculiaceae* family. The most magnificent tree of all is the Iroko (*Milicia excelsa*), a hardwood species, belonging to the mulberry family (*Moraceae*). And finally there are oil palms dotted about all over. In pre-(mineral) oil boom times cocoa was Nigeria's most important export commodity, but today it is in a sorry state because the oil boom of the 1970s resulted in neglect of extension, input supply, replanting schedules and marketing system. Most of that used to be handled by the government through government-controlled cooperatives and marketing boards, which gradually became mired in corruption and mismanagement. The organisation collapsed during the 1970s and most of the cocoa plantations are old now, with very poor pest and disease control. Nowadays, Nigerian cocoa yields are among the lowest in the world.

Food crop production is completely separate from the cocoa orchards, except that when the cocoa is in the final stage of decline food crops are slipped into the open places where cocoa trees have died. The main food crops are cassava and maize and the way they are grown is very interesting. The fallow vegetation is slashed between December and March, stacked in small heaps in the field and burned, to make room for the crops which will be planted in April with the first rains. While slashing and burning the vegetation the farmers take care not to damage saplings of useful species like Obeche, Iroko and oil palm. The oil palm is indigenous and occurs as a volunteer species everywhere in the humid and sub-humid parts of West and Central Africa, as long as the vegetation is rich enough to inhibit bush fires which would kill the seedlings and young trees. The natural sparse stands of oil palm are actually an important crop in their own right and there is an entire village industry to process the fruits into red palm oil, which makes delicious meals, once you have acquired a taste for it. The annual crops are grown under a light oil palm canopy in what you may call an indigenous agro-forestry system, which is the word scientists have coined for systems which combine annual and perennial crops (Figure 5.1).

After clearing a plot for food crops, one of two things may happen. Some farmers plant maize immediately with the first rains without any tillage, sometimes as early as late March. That is risky because the early rains may not persist and several weeks of drought may follow before the rains become steady. If that happens the young maize seedlings will wither and the crop must be replanted, but if the rains continue the farmer will prepare heaps to the side of the young plants and plant cassava cuttings on top. So, this early maize planting looks like a calculated risk. If it succeeds it brings a price bonus because the fresh cobs will be available before the bulk of the crop comes to market. If it fails the loss is light because hardly any labour was invested in planting the crop in the untilled soil. I am sure that is the rationale, but it is unlikely you will get this explanation when interviewing farmers. They may tell you that there was no time for heaping, or that this is the way their fathers did it. Like many things

Figure 5-1. Food crop fields after burning the eupatorium fallow

in African farming, and in other traditional societies as well for that matter, the foundation of evidently rational practices is often shrouded in tradition, not something in need of explanation.

The other, more common practice, is to heap the field before or after the rains break and plant maize at the side of the heaps once the rains appear to have reliably set in. A week or two later cassava stem cuttings are buried in the centre of the heaps. Fields which have been cleared from secondary forest are quite clean and there is not much need for weeding. Once the crop is planted it will take care of itself because it outgrows and shades the weeds. After a shorter fallow, however, there is much more weed growth and weeding is necessary, otherwise the crops will be overgrown instead of the other way around.

Today the main fallow species is *Chromolaena odorata* (Figure 5-2), a fast growing semi-perennial which according to legend was introduced from Asia by soldiers who had fought in their colonial masters' wars and accidentally carried the seeds home in their boots. It has

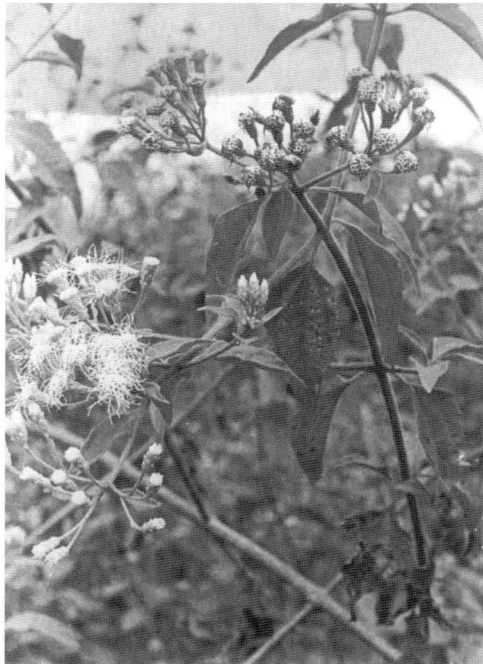

Figure 5-2. Chromolaena odorata

spread all over West and Central Africa and is considered a nasty weed by some and a beneficial species by others, the latter's argument being that it does quite a good job in restoring soil fertility. You will remember the *Chromolaena* controversy which I touched upon briefly in Chapter 3 when talking about food crop growing in Cameroun. *Chromolaena* seedlings make it necessary to weed the maize–cassava crop at least once, which is one of the reasons why some farmers dislike it, even though it is much easier to remove than many other annual and perennial weed species.

Maize is the first food that becomes available in the new season and when the cobs start maturing part of them are picked to be eaten as boiled green maize or sold. The rest is left to mature. After the maize harvest the field is lightly hoed and the heaps around the cassava are refashioned. The cassava, which has been sitting below the maize canopy will now take over rapidly in what de Wit called a flying start and exploit the short second rains, from late August to the end of October, more efficiently than a crop planted in August would. The plants persist across the following dry season, and into the next rainy season. Cassava roots are dug up when needed for food or money. That may start as early as December of the year it was planted, but most varieties can stay in the soil for up to 2 years. By that time the fallow vegetation is well on its way back in[3] and lasts on average for 4 years.

As long as the fallow is long enough to restore fertility and other conditions remain unchanged, it is hard to see how this system can be improved upon for environmental friendliness. In the real world, however, things do change, sometimes for the better but usually not (according to the pessimists). The advent of cocoa was a major change, but one which was easily absorbed by the system, without much effect on food crop production, because there was still enough land around and the cocoa occupied soils which were difficult to till because of their heavy texture. The next newcomer was *Chromolaena*. I do not think that species could have taken over the fallow as rapidly as it did unless there was already some unbalance due to mounting population pressure and intensification of land use. It would stand

---

[3] The same cropping cycle may also start at the beginning of the second rainy season instead of the first, but the second rains are quite erratic and maize is more likely to fail. Instead of maize, cowpeas are sometimes planted with cassava. Otherwise, everything happens in the same way.

no chance in real forest regrowth. Under the circumstances, the species was a qualified blessing. It grows rapidly, forms a thick canopy and is easy to remove when you need the land again. In the 1970s that was not the accepted opinion among the experts and some even wanted to introduce natural enemies to root it out. That was a bad idea which fortunately was not put into practice. *Chromolaena* does hamper the regeneration of the original forest vegetation, though. At IITA there were several hundred hectares of regenerating forest which used to be village land but have remained untouched since the relocation of the population when the institute was established in 1967. Twenty-five years onwards there were still large patches of *Chromolaena* which effectively smothered other species. The question is, however, what would have happened if *Chromolaena* had not been present in the farmers' fields. That is hard to tell, but I suspect that, with the rather short fallows of 4 years which are now common, the decline of soil fertility would have been more rapid and in the end the food crop fields would have been invaded by other, much more nasty species, in particular speargrass, *Imperata cylindrica*. And that may still happen in the end, even with *Chromolaena* around. A little further north, in the forest–savannah transition zone close to the large city of Ibadan, you could actually see it happening.

The area north of Ibadan is at the edge of the forest zone. The forest vegetation is fragile and once it is removed, it takes a long time to re-establish itself. There is much pressure on the land, due to the nearness of the large city of Ibadan, and fallows are therefore relatively short. The combined action of these factors eventually results in an open fallow vegetation which is dominated by *Chromolaena*, grasses and other annual species. This makes it sensitive to bush fires, which in turn destroy fire-sensitive seedlings of trees and *Chromolaena* and favour colonisation by speargrass. That is bad for the regeneration of soil fertility, increases the weed problem in the crops and, finally, kills the oil palm seedlings. I have converted the process into the cause-and-effect diagram of Figure 5-3, the kind of graphics which FSR researchers like very much and which sometimes do clarify things (at least if it is kept simple, which is not always the case).

In the 1980s and 1990s you could see all stages of the transition side by side in the area: some remaining secondary forest, decrepit cocoa orchards, patches of *Chromolaena* fallow with oil palm, some mixed with speargrass and fields heavily infested with speargrass

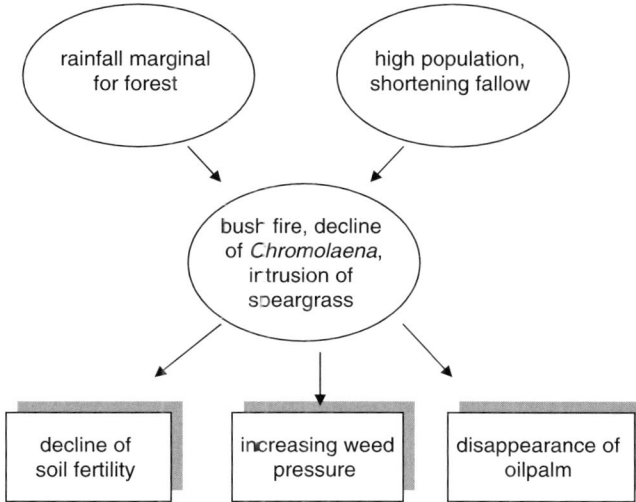

Figure 5-3. Cause-and-effect diagram for three problems in a forest–savannah transition area

where the oil palm had disappeared. Speargrass is probably the most dreaded cosmopolitan weed species which has infested large tracts of previously fertile land in the wet tropics, from Brazil to Sumatra and from Zaire to the Gambia.

A lot of the early agronomic work by IITA, in its search for alternatives to shifting cultivation, revolved around the maize–cassava cropping system of south-western Nigeria. That system is actually not shifting cultivation, it is fallow-based or 'recurrent' cropping, but it must have evolved from shifting cultivation. That has happened everywhere in Africa where population density had become too high to allow a really long fallow. Letting the land revert to fallow is quite convenient because nature will do the cleaning up after the farmer has left and it is essential because it replaces the nutrients which the crops have taken out of the topsoil, by drawing up a fresh supply from the subsoil. But once the fallow becomes too short for perennial species to establish themselves, the system breaks down, unless farmers or scientists perhaps, find ways to stop or slow down the decline.

When numerically literate agronomists look at today's maize–cassava system, they will probably come to the conclusion that it cannot last because more is taken out of the soil than can be put back in by nature in a short time. The question is how long it will last. You might think that such an obvious question had been answered conclusively by now, but that is not the case. The prophets of doom have repeatedly predicted the collapse of African peasant farming but the peasants continue to scratch a living out of their poor soil, so the prophets' arithmetic must be flawed. I think it is worth examining how far we can get in prophesying the future of fallow-based systems in the humid and sub-humid tropics, which are presumed to be fragile systems and therefore candidates for decline and eventual collapse. I will start with shifting cultivation. It is now the received wisdom among agriculturists and geographers that shifting cultivation was the precursor of today's fallow-based systems, like the one in south-western Nigeria, and that shifting cultivation was inherently stable. If that was indeed so, then we must try to understand the nature and causes of that stability in order to foretell the future of today's systems derived from it. The kind of data which are needed for that analysis were collected and analysed systematically for the first time in the 1950s by Peter Nye and Dennis Greenland in their epochal work on shifting cultivation. We shall shortly look at that work in detail, but before doing that I shall go back a little further in history and present some interesting work on shifting cultivation done in Indonesia in the 1930s. It yielded a pleasingly comprehensive analysis which showed how the system worked, why it was stable and what problems were to be expected when it evolved towards higher land use intensity. It does not really matter that it is from a different continent. Shifting cultivation was surprisingly similar all over the world, suggesting that the amount of choice for low intensity land use was small or nil, resulting in very similar systems everywhere.

5.1.2 Shifting cultivation, the mother of all systems

(a) Diagnosing the 'shifting cultivation problem' in the Dutch East Indies

In the second half of the nineteenth century the colonial government in the Dutch East Indies had introduced new liberal land policies which allowed European entrepreneurs to acquire or rent land and led to the rapid establishment of a highly successful plantation sector. In order to protect peasants against exploitation by the aggressive export crop industry laws were enacted which regulated access to land and the conditions under which European entrepreneurs could acquire land or rent it from peasant farmers. Furthermore, around the turn of the nineteenth century, the government initiated research and extension activities to better understand indigenous agriculture and stimulate its development. Studies were carried out on the way peasants managed their farms, how traditional land rights were administered, what was the nutritional status of the rural population and things of that type. Government set up a research station to study 'indigenous crops' and an extension service to bring the research findings to the farmers and show them how to improve their production. In that context an interest arose in shifting cultivation, which was still widely practised in the thinly populated so-called outer islands ('outer' meaning relative to Java, of course) and formed the direct interface between indigenous farming and the ever expanding plantation industry.

In the East Indies there were mainly two indigenous crop production systems: an intensive one based on irrigated or rain-fed paddy rice on terraced land in the densely populated islands of Java, Bali and Lombok, and the extensive fallow-based system of the outer islands of Sumatra, Borneo and Celebes. The latter, called *ladang*, was the same as what was elsewhere known as shifting cultivation. The main ladang crop was upland rice,[4] which could be grown twice before the land had to be returned to fallow, because the decline in soil fertility and the increased weed problems made it unattractive to continue. Ladang on cleared forest plots and irrigated paddy in terraced land may be seen as the start and end point of the evolution from very extensive to highly intensive land use.[5] Ladang in its original form was still widely practised in the 1930s. During a conference of extension workers which was held in 1930 a commission was formed by the colonial government's department of agriculture to study what was seen as the 'ladang problem'. Extension supervisors were worried that the system would not be sustainable with the rapidly increasing population. Also, part of the land was being taken out of the ladang cycle by conversion into plantations with perennial crops like rubber, coffee and pepper,

---

[4] Upland rice is grown as a dryland, rainfed crop, like wheat or maize, while lowland rice is grown in flooded paddy fields.

[5] How that evolution actually takes place is another matter, but a simple linear progression from extensive to intensive as some authors claim is not necessarily the way things happen.

both by European companies and by the smallholders themselves. A questionnaire was sent around to the districts' extension supervisors to gather quantitative information and a paper was written by J.A. van Beukering on how ladang could be transformed into a more intensive cropping system (this paper was published only after the war, in 1947). Not much seems to have been done during the remaining pre-war years and after the war Indonesia's struggle for independence intervened. But the paper contains some interesting observations which remain relevant today.

Van Beukering first estimated how many people could be supported by a stable ladang system. In order to do that he needed to know three things: how much rice is needed to feed a person for a year, what is the average rice yield from a ladang field, and how long are the permissible cropping and necessary fallow periods to keep the ladang system going. The survey had shown that the average rice yield from ladang varied from 500 to 1,200 kg/ha across the archipelago. In the first year much higher yields were sometimes obtained, but for his global analysis van Beukering adopted 1,000 kg/ha/crop, for an average of 1.5 crops/cycle, because about half the farmers would grow a rice crop in two successive years. Next, he assumed that an average person consumed 150 kg of rice per year, a figure which was based on rural diet studies.[6] Finally, he needed a figure for the minimum number of fallow years the soil needed to regain its previous level of fertility and make the ladang cycle stable. He put that at 12 years, without giving much justification. The questionnaire survey must have been the basis for the estimate, it was probably what the field officers saw around them. With an average of 1.5 years of cropping and 12 years of fallow/cycle, 9 ha of suitable land would be needed for each hectare of actually cropped land to practise ladang without danger of fertility decline. Furthermore, part of the land would be unsuitable for cropping because of steep slopes, paths, water bodies and settlements. Let us assume conservatively that 70% is suitable. Then the gross land need would be 13 ha per hectare of cropped land and 1 km² of land could support about 50 persons/km². In case you lost track I have summarised the calculations in Box 5-2.

---

[6] That looks like a poor diet, but then the ladang farmer would always interplant the rice with a variety of minor crops, including pulses, he would hunt for wild animals and catch fish in the local streams and perhaps his wife would grow vegetables around the house.

Box 5-2. Carrying capacity under ladang culture, according to van Beukering

---

1. With 1.5 years of cropping and 12 years of fallow/cycle, a total of $(1.5+12)/1.5 = 9$ ha of *net* agricultural land is needed for each ha of actually cropped land.
2. If 70% of the land area is suitable for agriculture the *gross* land need for ladang is $9/0.70 \approx 13$ ha/ha of cropped land.
3. An average person consumes 150 kg of paddy per year, for which $150/1000 = 0.15$ ha of crop land is needed, that is $0.15 \times 13 \approx 2$ ha of gross land area.
4. Hence, 1 km² of land can support $100/2 = 50$ persons.

Some authors mention much lower numbers, like 10–15 persons/km², but I think those are actual population densities observed in some shifting cultivation areas rather than potential densities.

Now suppose the population density increases beyond the threshold of 50 persons/km², what will happen? The study's statistics showed that the 50 persons mark had already been passed in some areas and van Beukering estimated that with current trends the system would breakdown almost everywhere in the following 30–50 years. Signs of imminent collapse were already visible in some places: uncontrolled fire, disappearance of fire sensitive trees, and, indeed, intrusion of speargrass. In some areas the original vegetation had already been completely replaced by large expanses of that grass. So what to do? The study came up with an elaborate long-term programme to steer the system away from its disastrous path, towards a new productive equilibrium. In the vision of the author that would be mixed farming, combining field crops, fodder production and farm animals, with their manure used for fertilising the land. That was a long shot, too long in fact as the author was well aware, since he quoted E.B. Worthington in the motto of the chapter which outlined his ideas for the system's future, as a kind of self-admonition:

It is now coming to be realised that drastic methods rarely achieve their object, and that improvements are more likely to be attained by gradual development from existing methods.

Another surprisingly modern point of view from the colonial East Indies, and a warning that was not heeded by the development experts of the 1960s and 1970s who wanted to make the transition from traditional to modern farming in one stride. But stating the need for gradual changes is one thing, charting a realistic, adjustable course from the present towards a desirable future is quite another. Van Beukering made a brave attempt but when you read it you knew it was a hopeless task. First he wanted to map the ladang areas into individual watersheds. Watersheds (see Box 5-3) are useful units for the planning and management of land use and van Beukering wanted a land use plan to be designed for each of them. Farmers would be allowed to continue their ladang practices for a while, but in designated locations instead of the rather unstructured way it was traditionally done. Or may be there was a structure but not of the kind the agriculturists were aiming for. Meanwhile, measures were to be taken for the conversion of the watersheds into stable land units. Forest reserves would be planted in vulnerable places such that the ladang lands were protected against erosion and the crop land should be rotated more systematically around the ladang area. Van Beukering thought that would be as far as one should go in the beginning in regulating the system. Assent and collaboration should be sought from the people and their leaders 'without which the measures cannot be expected to be carried out successfully'. But once agreement was reached, he felt it should be enforced by the indigenous authorities, in the typical dualistic fashion of the Dutch colonial administration.

Regulating the existing system was just the beginning, the final aim was a permanent production system to replace ladang. That future system would be mixed farming, integrating crop and animal production. The model was not worked out in great detail, but it was clear that one major role of the animals, next to producing milk and meat, would be to produce manure to fertilise the crop fields with. They would feed on crop residues and fodder crops grown in the farm and graze in the wider area around the settlements. In that way, fertility from the periphery would be transferred to the farm land. Without this source of nutrients permanent crop growing would be out of the question on the chemically poor soils which

---

Box 5-3. What is a watershed?

A watershed is all the land from which water drains to a particular point or waterway, for example one side of a hill which drains into a stream at the foot of the hill. What happens in one location in a watershed will be felt in another, for instance if the trees at the upper slope are cut there will be increased run-off washing away the soil from fields further down the slope.

were most common in the ladang areas.[7] Transfer of nutrients from the grazing lands to the crop fields was an interesting concept, which had been the basis of mixed farming with cattle and sheep in Europe, in some areas up to the end of the nineteenth century.

Although the road ahead would be long and arduous, and even the final destiny was uncertain, the agriculturists were told to always keep the target system in front of them, as a beacon for any intervention. Van Beukering's paper put up some milestones along the road to this Holy Grail. Once the ladang areas had been demarcated a start should be made with protecting the land. Fire lanes were to be cleared to prevent uncontrolled bush fires, sloping land would be protected against erosion by terracing and bunding. Later on the fallow would be shortened and fast growing trees would be planted to replace natural forest regeneration, while creeping nitrogen-fixing legumes were to be grown as part of the cropping cycle in the farmland. The extension service was to promote the adoption of all these measures by demonstration and advice. Experimental farms would be set up where the mixed farm model was worked out in more detail by trial and error. Studies would be carried out about the proportion of the land which had to be under fodder crops or pasture for the animals, what kind of grass and fodder species and silage methods were suitable and other such details. Many years later all these ideas would turn up again in numerous projects for soil and water conservation and ecological farming in Asia, Africa and Latin America.

The colonial extension supervisors were sceptical. When asked during the 1931 survey whether the creation of watershed-based management units, the point of departure for the whole operation, was a realistic idea most of them answered it was not, because it would be impossible to control. Van Beukering waved their objections away and said it was possible if one did not attempt to implement it to the letter. Whatever the case, no further steps were taken to carry out the ideas either during what remained of the colonial era or later. And in the end, after independence, something completely different happened. Rather than becoming mixed farmers, many shifting cultivators launched into profitable smallholder

---

[7] The fertile paddy soils of Java and Bali are mostly of volcanic origin, but the soils in the 'outer departments' as they were called, where ladang was practised, are mostly geologically old leached soils.

rubber production, while continuing their ladang fields for subsistence. The vast expanses of land still available in the outer districts also attracted land-starved farmers from Java, who brought with them their experience with semi-intensive *tegalan*[8] cropping and transplanted that into the ladang areas. As far as I am aware, the mixed- farming concept was not adopted at any appreciable scale and today's semi-intensive system consists of a combination of smallholder plantation crops and tegalan, proba- bly with the use of chemical fertiliser. Clifford Geertz (1968), an American anthropologist who worked in Indonesia during the 1950s, described how this came about in a very readable study, if you can disregard the rather pedantic style. The study, called 'Agricultural Involution' is recommended reading if you want to see how Indonesia's two contrasting systems, ever more intensive paddy rice growing (that is what 'involution' refers to) and ladang, evolved during the colonial era and immediately after.

(b) Quantifying shifting cultivation

One thing was conspicuously missing in van Beukering's work: neither the ladang system nor the model for the future farm which he thought should replace it were quantified. Of course there was a guesstimate for the maximum number of people that a ladang system could support, but that was essentially based on what the people had reported from the field as the actual fallow length where ladang was still a stable system. Nor was there an indication about the yields which could be expected on the future mixed farm, the farm size that would be needed to support a family, whether the soil could actually support continuous cropping, whether fertiliser would have to be applied and similar concerns. All these details he thought were going to be found out in due time in the experimental farms, by trial and error. Today's agronomists would be worried about the nutrient budget of van Beukering's future farm and I would be surprised if it did not worry those of the 1930s. The nutrient-budget concept is important for the understanding of the productivity and sustainability of agricultural production. It will

---

[8] Tegalan is a semi-permanent cropping system with short fallows, practised by Java's paddy farmers on their dry landsa, which are not suitable for rice paddies.

come back repeatedly in this and Chapter 6 and for those who are not of the profession I will first explain what it means.

A crop will take up nutrients from the soil, some of them in fairly large amounts, in particular nitrogen, phosphorus and potassium, others in small or very small quantities. Part of the nutrients leave the field with the harvested produce, part of them remain there with the crop residues which are left in the field, while some are washed down by the rains, out of reach of the crop roots. The nutrients are replenished from the soil's own stock and some may be added by the farmer in the form of fertiliser or manure. The whole process of nutrients circulating through vegetation, crops, humans, animals and back and shunted out with the produce or by leaching, can be analysed in the same way as the cash flow of a business. You can draw diagrams showing all the movements of nutrients and then prepare a balance sheet to show how much is lost or taken out, how much the soil can supply and how much needs to be replaced from outside sources if you want to remain a farmer. That is the nutrient budget of the system. Nutrient flow diagrams became quite a fashion in the 1990s. The one shown in Figure 5-4 is a good looking specimen, from a book published by the International Institute for Environment and Development (IED) some years ago. Of course the devil is in the details: even today it is very difficult to get reliable figures for all the 'In' and 'Out' terms of the equation, except for a few well-studied and relatively simple cases, but with a big thumb, suitably wrapped in a reliable looking computer program, you get a

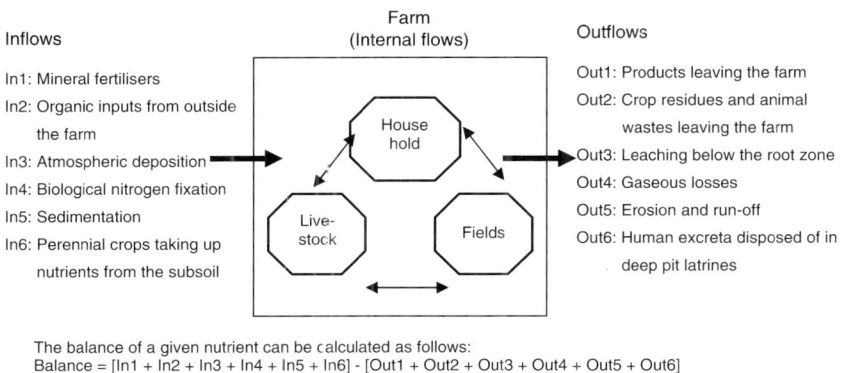

The balance of a given nutrient can be calculated as follows:
Balance = [In1 + In2 + In3 + In4 + In5 + In6] - [Out1 + Out2 + Out3 + Out4 + Out5 + Out6]

Figure 5-4. Analysis of nutrient flows on farm. (From Hilhorst and Muchena, 2000; reproduced by permission of the Int. Inst. for Environment and Development, London.)

long way. I will come back to all that later on. Let us now go back to our review of shifting cultivation, the mother of all farming.

Even at the time van Beukering carried out his study in the Dutch East Indies it would have been possible to work out at least rough nutrient budgets for shifting cultivation and for the hypothetical continuous cropping alternatives, but van Beukering, being an agricultural economist, was not equipped to do this and I do not know if anybody else did. Quite probably not I think. Nye and Greenland were the first to gather the necessary data and carry out the calculations for shifting cultivation. That was in the 1950s, when they worked at the University College of Ghana in West Africa. They estimated how much of key nutrients is stored in the forest vegetation and the soil and what happens to it when the vegetation is cut down and burned and the soil is used for a few years of cropping. A lot of measurements had already been carried out all over Africa since the 1930s, which they used in their comprehensive essay, along with measurements and observations of their own. The book treats shifting cultivation in both forest and savannah ecologies, but I will limit myself to the forest.

When setting out to write this book I had forgotten most of Nye and Greenland's work. I only retained some general notions such as that the highly weathered and chemically poor soils of the humid and sub-humid tropics need at least 10–12 years of fallow to be restored to the original level of fertility, thus allowing stable shifting cultivation. That incidentally agreed with van Beukering's conjecture. So I read it again and it turns out to be an impressive piece of work which touches on practically everything that is important to understanding the system. Although soil fertility and the nutrient cycle were the main topics of the book's 150 pages, its scope was much wider. It showed that shifting cultivation is the outcome of rational choices and concludes, as did van Beukering, that when land is abundant, shifting cultivation is the sensible thing to do and gives the highest returns with the least effort. Today that is conventional wisdom, but I think Nye and Greenland's work did a lot to make it that. The whole treatment in fact reminds one of van Beukering's works, enriched with a thorough quantitative review of the dynamics of soil fertility, which was going to form the basis, 15 years later, for IITA's search for alternatives to shifting cultivation. They did not quote van Beukering's work, though, probably because it was written in Dutch.

Most refreshing is Nye and Greenland's clear statement of the questions they want to answer and how they will do that, while their data analysis uses a minimum of scientific jargon, which makes it understandable for those, in the authors' words, 'with a smattering of soil science who are interested in the subject'. And whenever scientific terminology is used, it is explained carefully and clearly. For a rigorous scientific text that is quite exceptional. I will try to reproduce the gist of what they say about soil fertility under shifting cultivation and then make some calculations of my own in Appendix 2.

A mature forest is in equilibrium: there is no net biomass production or loss, because as much is dropped to the ground as litter as is produced in new growth. The processes in the soil under a mature forest are also practically in equilibrium. Litter is attacked by the organisms living in the soil which set free the nutrients and convert organic material into new humus. At the same time humus is broken down by other organisms, until the point is reached where as much is added each year from fresh litter as is broken down by natural turnover. The nutrients which are set free from decomposing litter and humus are taken up again by the vegetation. There are some losses by deep leaching out of reach of the roots, but they are small and partly compensated by nutrients which come in with rain and dust. The equilibrium is a dynamic one. There is a continuous flow of nutrients which maintains a characteristic distribution of nutrients over the different parts of the system, the vegetation, the litter, the upper part of the soil where the humus has accumulated, and the soil lower down. Nye and Greenland's simple chart shown in Figure 5-5 illustrates all that. Small disturbances, like a fallen tree or seasonal differences in biomass growth make the system oscillate but around a stable equilibrium.

The interesting questions are: how much of various nutrients do each of the stores hold, in what forms and, most importantly for an agriculturist, what will happen when the system is jerked out of equilibrium by burning the vegetation and growing a crop? I have summarised the results of Nye and Greenland's very careful data analysis in Appendix 2. That is not so easy for a rich text like theirs, and even the summary takes a lot of space. It is a rewarding story, though, which I hope readers will take the trouble to study in depth.

After much juggling of numbers I concluded (Appendix 2) that the primary limiting nutrient for a mixed crop of maize, cassava and

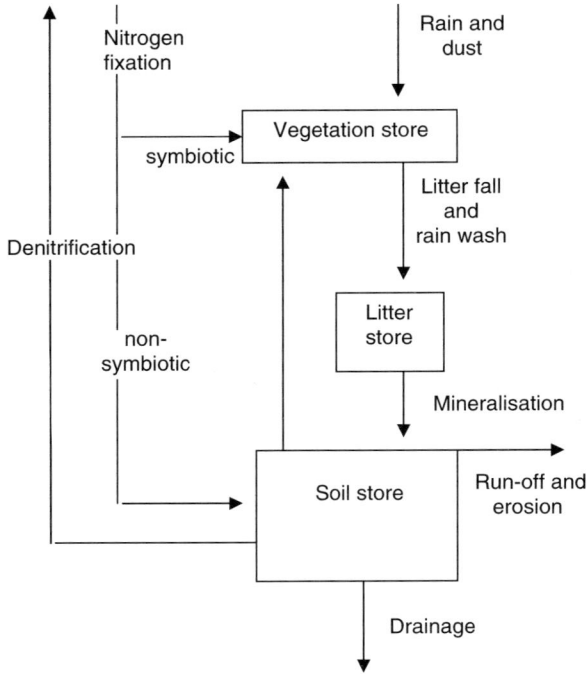

Figure 5-5. Nye and Greenland's nutrient flow chart for shifting cultivation

plantains after 10–12 years of fallow will be N. The only source of N is decomposed humus, because N in the vegetation is lost by burning. Also, an important part of the N is released at the wrong time, in the so-called N flush at the beginning of the rainy season when the crops are too young to make use of it all. Thus a part of the N is leached out of the topsoil, out of reach for the crops. The decomposing humus also releases some phosphorus, but most of the P available to the crops comes from the fallow vegetation and is set free by burning. Although some of it may subsequently be converted into less available forms, it is unlikely to be limiting during the first cropping cycle after forest fallow.

A large part of the K which was present in the fallow vegetation is also deposited on the soil surface with the ashes and becomes available to crops. Part of it will be washed down further, dragged along by leached nitrate, and equally out of reach of the crop roots. Cassava and plantains are big potassium consumers and

Table 5-1. Predicted potential yields in a long fallow shifting cultivation system

	Maize	Cassava	Plantains
Yields (kg/ha)	1,750–2,500	14,300–20,500	4,300–6,100

although their yield is also constrained primarily by N, K is predicted to be only just adequate, at least in the chemically poor soils of the humid tropics.

In Appendix 2 a calculation is made of the yields of maize, cassava and plantains which can be expected after a 10- to 12-year-long forest fallow, using a simple spreadsheet model. The predicted yields, in rounded figures, are shown in Table 5-1.

These are indeed the kind of yields farmers may get after a long forest fallow. The reality will usually be somewhat less when other factors – such as heavy shade from standing trees and from the surrounding forest if plots are small[9]– play a major role.

What would happen if a second crop of maize, cassava and plantains were grown instead of letting the field return to fallow? Farmers are unlikely to do that, but many researchers in their quest for continuous cropping have carried out trials with several crop cycles following long fallow. The results are of interest to us because we should now start to see shortfalls of other nutrients (apart from nitrogen). Let us see what the nutrient budget predicts and then compare that with real, measured data.

First the humus and N. There will be changes in the soil humus content because some of it is decomposed while the first year's crop residues are partly converted into new humus. Building fresh humus from maize residues consumes more N than the residue contains, so some of the N released by humus decomposition is incorporated into newlyforming humus. As a result there will be less N for the crops and the maize yield is predicted to be some 25% lower than the first year's and that of the cassava and plantain almost 50%, because of shortage of N.

[9] Also, the potential yields were calculated for planted area, which may be considerably smaller than the total surface area if there are large trees or tree stumps lying around or standing in the field.

What about P and K? There is less of it available now because it has been partly depleted by the first crop, but that is not expressed in the yield because N remains the most limiting element, but now only just. If the maize yield were boosted to 3 t/ha by applying N fertiliser, P would become the limiting element and a response to P fertiliser can be expected. The same is true for the cassava and the plantain: by increasing their yield by N fertiliser, shortages of both P and K will arise. In a third crop shortage of P for maize and P and K for cassava and plantains will even occur without the application of N.

I think most careful observers of African agriculture will agree that the predicted yields of Table 5-1 for a long forest fallow system are quite reasonable, and so are the expected nutrient deficiencies in case the occupation period were extended. The inquisitive reader, however, will not be satisfied by my assertion and will want to see hard evidence. So let us look at some real, observed data.

(c) Some real data

There are many more measured data available for cereals than for root and tuber crops, so I will pay most attention to the former. Nye and Greenland quoted yields for 'cereals' ranging from 900 to 2,300 kg/ha, which practically covers the range predicted for maize by my model calculations. Some authoritative studies, however, carried out since the 1950s, have reported maize yields which were well outside the predicted range, in either direction. What was responsible for that?

On the low side are a large number of FAO 'Freedom from Hunger' trials which were carried out during the 1960s. The average reported maize yield without fertiliser in the forest zone was 1.22 t/ha, substantially less than the prediction from my analysis. At the other extreme there were trials carried out at the IITA research station in fields freshly cleared from forest which gave (unfertilised) maize yields ranging from 2,000 to as much as 4,000 kg/ha. That is more than my fertility analysis would allow. In fact, there is generally very little consistency in yields reported by different authors for shifting cultivation. The data quoted in Pedro Sanchez' well-known textbook on tropical soils in particular give a confusing picture. On the one hand it quotes cereal yields 'in shifting cultivation areas' ranging from 1,000 to as much as 5,800 kg/ha, without comment. On the other hand, it states: 'Crop yields under shifting cultivation are extremely low. They range from 0.5 to 1.5 t/ha of cereal grains', apparently forgetting what had just been said in the same chapter. So, what explains the large differences among measured data and what do they imply for our model predictions?

Let us look more closely at the FAO data first. I suspect that the low overall maize yield had something to do with the way those trials were conducted. As I remarked in Chapter 4, farmers are very good at shunting scientists' experiments to harmless places if they are less than confident about their outcome. I have seen many so-called on-farm tests in fields where the farmer himself would never have grown the test crop, or indeed any crop. I would be surprised if many of the trial fields had really been cleared recently from long fallows. Farmers were probably conducting a test of their own, to see whether the scientist could grow a good crop where they themselves could not; I have given an interesting example of that phenomenon in Chapter 4. In the 1960s most workers had too little experience working with small farmers to be aware of this kind of distortion. That could explain the low average yields. In any case, in our own, fully farmer-managed tests in south-western Nigeria, carried out between 1985 and 1988, the yearly averages for the local maize variety without fertiliser ranged from 1.34 to 1.90 t/ha, which is quite consistent with the model predictions.[10] The yields of individual farmers ranged from 0.3 to 3 t/ha, the low yields being from fields which were simply neglected by the farmers and became overgrown by weeds for some reason or other, while 3 t would be a little outside the 'permissible' range.

How did it happen that maize yields were sometimes as high as 3 or 4 t/ha, as in our on-farm trials and at the IITA station? If my calculations are correct 2,500 would be the maximum, because there is simply not enough N available for more maize. A clue comes from another IITA trial which compared ploughing and no-tillage. When the field was left untilled (and the weeds were controlled by herbicides) the yield was about 2,000 kg but after ploughing it was 3,500 kg/ha. The effect of ploughing is to turn and loosen the soil and work in the litter left on the surface (the vegetation had been removed, not

---

[10] At the time of the trials the fallow was usually much shorter than 10–12 years, but the average soil humus content, in one village 1.51% and in the other 2.13%, suggest the soil conditions were not yet far removed from those under the shifting cultivation.

burned), all of which would stimulate microbial activity and accelerate humus decomposition. Nye and Greenland had already drawn attention to published data which showed much higher humus decomposition rates than the 3% they assumed and ploughing the field would probably result in such higher rates and therefore more N being released during the first season. What farmers themselves will normally do is closer to zero tillage, so we should expect farmers' yields to be closer to those in the non-tilled treatment. The occasional higher yields of up to 3,000 kg in farmers' fields may have been due to more thorough tillage such as the occasional farmer does, or by an exceptionally fertile field. Another contributory cause is the maize variety, which in the IITA trials was an improved one with a better grain/straw ratio and therefore a higher grain yield from the same total biomass production. When a humus decomposition rate of 4% plus a higher grain/straw ratio of the improved maize variety are assumed the model will calculate grain yields which are in the same order as those observed in the IITA trials. The predicted range of maize yields then works out at 2,600–3,800 kg/ ha, quite close to the field results. So I think we can confidently stick to the figure of about 1,750–2,500 kg/ha as the yield good farmers may get from a local variety without fertiliser and light tillage and something like 2,000–2,900 for an improved variety which puts more in the grain and less in the stover. Bad farmers may of course get any yield below that, depending on how bad they are. I will come back to the issue of good and bad farmers in Chapter 6.

So, after a long fallow farmers can get reasonable maize yields without applying fertiliser, but they are not getting anywhere near the crop's potential – if conditions are ideal and solar radiation is the only constraining factor potential yield is more likely to be in the order of 7–8 t/ha with the right variety, even in the humid tropics. Even taking account of reductions caused by shading by surrounding trees (which can be considerable when fields are small) and the use of less responsive varieties, the potential yield of a well-fertilised crop with little or no water stress will still be as high as 5 t/ha as the best yields of fertilised maize in our on-farm trials showed. So, let us look more closely at the experimental record for the effect of fertiliser. The nutrient budget says that the yield shortfall of farmers' unfertilised maize is due to N shortage. You would therefore expect a significant yield increase when applying N to the crop, but research carried out in the 1960s showed quite erratic responses in maize and other cereals. For instance, the FAO 'Freedom from Hunger' trials carried out in the 1960s showed only a very small average fertiliser effect. In more recent trials, however, much larger effects have been recorded. At the IITA station, for example, fertilised maize grown after forest fallow yielded up to and sometimes even more than 6 t/ha. And in our on-farm trials in south-western Nigeria and those by Mark Versteeg and co-workers in Bénin in the 1980s the yields of 75% of the farmers in most years ranged from 2,000 to 4,000 kg/ha for maize at a modest fertiliser rate.[11] So what explains the low N-responses in the earlier work? Well, I think it must have been due either to experimental sloppiness, or to experimental conditions being very different from those we have in mind here, or both. For example, the experiments may not have been on freshly cleared land but rather on land which had been used for some years, so P, K or micronutrient deficiencies may have developed. As I said earlier (twice), farmers are unlikely to take you to their best land when you ask them where you can put your trial. Whatever the cause, when 25 kg of fertiliser-N only increased average maize yields by 180 kg as was the case with 343 FAO trials in the 1960s, that is an efficiency of 7 kg of grain/kg of N, something must have been wrong. The more recent evidence, both from research stations and from farmers' fields, shows that quite considerable N responses can be expected on land which has been freshly cleared after a long fallow.

So much for maize yield when N is the most limiting nutrient, but what about the experimental evidence for P and K? Reviews by Nye and Greenland, Ahn and Sanchez showed that responses to P and K observed after a long forest fallow have been quite erratic. If there was a response at all it occurred most often in the second or third cropping year. That is what our nutrient analysis predicts. It is rather frustrating to try and make sense of those trials, though. They rarely seemed to address the kind of questions we want to ask. That is the trouble with a lot of agronomic trials: they were often set up without a precise hypothesis, sometimes with no obvious hypothesis at all. In Pedro Sanchez words: 'Much

---

[11] Their average N-response was around 20 kg of grain for each kg of applied N, with a maximum of 35 kg, quite respectable responses as you will see in Chapters 7 and 9.

of the research has been conducted without attention to soil characteristics or soil test data. It is disappointing to see reports on
hundreds of lime trials on soils with pH levels between 6 and 7.'
What about more recent work? Research at IITA showed that
P effects on maize yield can be expected quite soon after clearing
forest fallow, even in the second year, provided N is applied as
well. Early K effects on maize were less common, but they did
occur with cassava.

These admittedly rather scant data are consistent with the model
calculations, which is quite an exciting conclusion in itself, at least
for me it is. The simple spreadsheet model of Appendix 2 turns
out to be a useful auxiliary research tool and it is surprising how
little use has been made of this kind of tool (we will have occasion in Chapter 7 to look at another demonstration of the power
of spreadsheet modelling). Of course, spreadsheets have only been
around in the general research community since the rise of desktop
computing in the early 1980s, but the opportunities they have
offered for more than 20 years since then have hardly been exploited
by rank-and-file agronomists. Perhaps that was because modelling
in general had barely gained respectability in the profession.

Most of what I have said so far about crop yields under shifting
cultivation has been for maize. What about cassava? According to
our calculations the expected yields in the first cropping cycle
would range from about 14–20 t/ha. What do the real data say? Nye
and Greenland's essay mentioned a yield range of 13.5–45 t, while
Sanchez quoting an unpublished inventory (by a student presumably) gives a range of 8–59 t. These data are pretty erratic too and
no indication was given by the authors of the conditions where particularly high yields were obtained. In our own on-farm trials yields
of unfertilised local cassava ranged mostly from 5 to 21 t, with an
average of 13 t/ha. If we exclude the yields at the low end due to
farmer neglect (and cassava is a crop easy to neglect) the range is
again close to the calculated yields. So let us say that the model
predictions for cassava yield were also quite realistic.

(d) Carrying capacity and sustainability

If we accept, not too audaciously I think, that the expected yields
in Table 5-1 are realistic for a single cropping cycle alternating with
10–12 years fallow, what can we say about the carrying capacity of
the land, for humans that is? And for how long can the system be

sustained? Let us start with the former. Earlier on, I mentioned a figure of 50 souls/ km^2 for unadulterated "ladang" in the East Indies according to van Beukering's analysis. That was based on pure subsistence farming, but that hardly exists any more. Even in areas where land use can still be classified as shifting cultivation, the people cannot avoid exposure to the general economy and they will want some surplus produce to sell in order to buy consumer goods, pay school fees, etc (perhaps even taxes). I have therefore assumed that a family will need the same amount for sale and barter as they consume directly as food.[12] Furthermore, I have lumped together the yields of maize, cassava and plantains by defining a 'maize equivalent', which is of course unity for maize and 0.2 for cassava and plantains, whose produce contains a lot of water and peels.[13] Let us see where we get then, using the productivity figures obtained so far. The arithmetic in Box 5-4 gives a figure of 55–78 persons/km^2, surprisingly high, considering that

---

Box 5-4. Carrying capacity under shifting cultivation

1. With a 2-year crop cycle and 12 years of fallow, a total of 14 ha of agricultural land is needed to start a new crop cycle on 1 ha each year.
2. If 70% of the land area is suitable for agriculture the *gross* land need is 14/0.70 = 20 ha/ha planted annually.
3. A farming family plants a food crop field every year, so they will always dispose of one field in the first stage of the cycle and one in the second. Together they produce about 1,750–2,500 kg maize, 14,300–20,500 kg cassava and 4,300–6,100 kg plantain/ha, that is a total of 5,470–7,820 'maize equivalents' (never mind the air of precision of these and the following calculations).
4. Actual yields will be lower than potential, because of scattered trees and tree stumps, shading and less than perfect management. I will discount 20% for those factors.
5. If an average person consumes 200 kg of 'maize-equivalent' per year and the same amount is bartered or sold, then they must plant 400/4,376 – 400/6,256, i.e. 0.0639–0.0914 ha of crop land annually, that is 1.28–1.83 ha of gross land area is needed.
6. Hence, 1 km^2 of land can support 100/1.83–100/1.28 = 55–78 persons.

---

[12] I am aware of a logical snag here. The extra amount produced will also feed people, but let us simply assume that they live somewhere else, so the calculated carrying capacity is for rural people living directly off the land.
[13] For cassava 0.2 is also close to the price ratio, for plantains, which is a high value crop its is not.

half the produce is assumed to be sold. In the rural area around Yaoundé in Cameroon, which I described in Chapter 3 the population density was 83 persons/km^2 in the mid-1970s, just over the threshold for 12 year fallow. The fallow length of most food crop fields was much less than 12 years, though. The fields were quite intensively cropped, in most cases with fallows of little more than 4 years. At the same time there was a considerable area permanently under cocoa while some land was left under fallow for much longer. When doing our survey we asked the people why they left the more forested land uncropped for so long while they complained at the same time that the food crop fields were losing fertility. The reason they gave was that clearing the heavy vegetation needed a young man's strength, but many of the younger generation had turned their backs on agriculture and drifted to the towns. I doubt whether this was the whole explanation, though: this kind of dichotomy of field usage is actually common where people have settled down in permanent villages. I will come back to that later.

Let us now see whether we can say something about the likely durability of a system with an average of 2 cropping years and 12 years of fallow. The calculations in Appendix 2 showed that large amounts of nutrients, in particular K, leave the field with the produce, and if they do not somehow return to the land there may be an end even to a low intensity system like this. That is the crux of the matter: how much of the nutrients are irreversibly extracted from the soil? Real shifting cultivators were themselves part of their environment and the nutrients passing through them would eventually return to the land again, so the system was practically closed if you considered the total territory where shifting cultivation was practised. In principle it could go on forever. Nutrients were continuously cycled through the fallow vegetation, the humus, the animals and the humans with no or very small net losses. That is also the reason why, contrary to more intensive types of land use, the nature of the soil had little effect on the system's stability. The soil was merely the matrix in which the nutrient cycles were embedded. Shifting cultivation in its original form is practically extinct now and most of today's fallow-based farming, even those with a relatively long fallow, is much more 'leaky' – there is a net loss of nutrients, because even traditional communities are to some extent integrated in the wider economy and some of the produce is sold outside the area to earn a cash income. Furthermore, the people do not move their homesteads around now but tend to stay in permanent settlements, so there is a transfer of fertility from the outlying fields to the homesteads, because that is where the residues and the human wastes are concentrated. Finally, fields near the settlements are farmed more intensively and their fallows will be shorter than further away so that they will be less effective in pumping up lost nutrients from the sub-soil. All these factors increase the system's leakiness and the soil's nutrient stock must eventually get depleted, even if it takes a long time. The calculations of Appendix 2 showed that a system with 2 years of cropping followed by 12 years fallow where the nutrients which leave the field are not restored may last as long as 180–340 years. When the population density exceeds the threshold for stable shifting cultivation the fallow is shortened and depletion will of course be more rapid.

(e) How hard is the evidence for a minimum fallow length of 10–12 years?

Did Nye and Greenland actually provide solid proof that 10–12 years of fallow is the minimum for a stable shifting cultivation system? Well, more or less: their evidence was partly direct and partly circumstantial. They did show convincingly that 10–12 years of fallow would normally restore secondary forest and maintain the soil humus and nutrient contents of the topsoil to a high level. But they could not say at how short a fallow length the fallow vegetation would start to deteriorate into a grass savannah. Nor could they have done so in a general sense, because that depends very much on the nature of the soil and the way the farmers farm. Even after their thorough quantitative analyses of the nutrient cycle the evidence remained largely as it was before: circumstantial. In their final synthesis they sum it up as follows:

In the superhumid forest regions . . . 1–2 years of cropping are *normally* followed by 10–20 years of fallow [while] cropping periods of 2–4 years followed by 6–12 years of fallow are *common and apparently stable* in both forest and savannah regions [of Africa].

And, a little earlier:

where the periods of crop and fallow are described as normal *it is usually implied* [by the authors, that is] that cropping at that intensity could be prolonged for many cycles without noticeable erosion of the soil, a change in the type of fallow, or a marked decline in yields.

And that is it: a stable system is a system that is stable. So things had not progressed all that much since van Beukering, except for three things. First, Nye and Greenland had collected from many sources factual evidence that the fallow vegetation would indeed degenerate and be invaded by annuals and speargrass if the fallow became too short, thereby cutting off its pumping function. And they put some rough figures to that: less than 10–12 years in the pre-humid tropics and less than 6–8 years in the sub-humid regions. Second, they found that in inherently fertile soil the fallow would degenerate more slowly than in poor soils. And finally, their work contributed the figures which allowed quantitative prediction, however crude, of the nutrient status of the soil and attainable yields under shifting cultivation, using today's more sophisticated calculation tools. That is what I have tried to do here and what I am going to do presently for more intensive fallow-based systems.

5.1.3 The closed cycle undone

(a) Which options do farmer have?

What can shifting cultivators do when the number of people the land must feed exceeds the maximum for shifting cultivation? In the short run there are only two options. They can use the land a little longer before letting it go back to fallow again, or they can clear a new piece of land, before its fallow vegetation is mature and should have been left alone for a few more years. Let us start with the first option: it has been studied intensively, because scientists like permanent cropping.

After one crop cycle the soil's humus content would have declined by only 5–6 %, so N-release will be similar as in the first cycle. At first sight you would therefore expect that N will not be an immediate problem. The calculations in Appendix 2, however, predict rather acute N-deficiency in the second cycle, resulting in quite a steep maize yield decline. Research at the IITA station has shown that to happen over and over again, even when fertiliser is applied to the maize. N-deficiency is caused by the conversion of the crop residues from the first cycle into humus, which immobilises a lot of N.[14] If N-fertiliser were applied to boost maize yield to 3 t, P has

---

[14] The reason is that the crop residues' N-content is low, so the micro-organisms which break them down must take up N from the soil to build fresh humus.

to be applied as well, otherwise the 3 t cannot be attained because there will not be enough P. Cassava and plantain yields were also predicted to be depressed by N-shortage. If the farmer would try to increase their yield by applying N-fertiliser, he would trigger both P and K shortage, because there has been no fresh addition of these nutrients, unless the soil is rich in native fertility and it can dig into its reserves. Otherwise the farmer has to apply N-P-K fertiliser.

Another major problem is weeds. After a long fallow the land is practically clean, but it will inevitably be invaded by weeds during the first cropping cycle and weed control becomes much more difficult during the second. And plantains may have to be dropped because they are very sensitive to poor growing conditions and may succumb to pest attacks, so it would probably be better not to plant them at all. Finally, using the land for a second round is harmful for the next fallow. The soil has to be tilled again and more intensive weeding is needed, which together will destroy some of the remaining root stocks and young seedlings of the fallow species and prevent their rapid re-installation. So growing another crop cycle looks like a bad idea. Farmers are unlikely to choose that road to intensification, they will rather shorten the fallow. What will be the consequences of that choice?

As the fallow gets shorter the soil's humus content will go down slowly, but it takes a long time to reach a new equilibrium. If the change from 12 to 4 years took place abruptly, it would take an amazing hundred years before the humus content reaches its new equilibrium, according to Nye and Greenland's analysis. That is to say, if the farmer continues his usual environment-friendly, though perhaps not very profitable manual cropping practices, which cause a minimum of soil degradation. The slow decline in the soil's humus content will only have a small effect on N release at first,[15] but we expect P and K shortage to start showing up soon, because less will be accumulated by the fallow vegetation and released upon burning. Whether P and K deficiency will occur and when it happens depends on how much P and K can be accumulated in the vegetation and the topsoil in just a few fallow years. Contrary to shifting cultivation, that also depends on the soil's native nutrient stock.

---

[15] Although supply is reduced initially by N-immobilisation due to decomposition of the crop residues.

Furthermore, in a short fallow the vegetation will be less effective in pumping up nutrients from the subsoil and, finally, since intensification and increased marketing of the produce usually go hand in hand, more of the nutrients leave the system completely. In other words, the original system whereby nutrients were continuously recycled and little of them left the system is now converted into one which shunts nutrients out of the soil, to the subsoil and to consumers outside the area. In today's parlance that is soil mining.

Shortening the fallow will of course also affect the botanical composition of the vegetation. Large trees take a long time to establish, so the vegetation will become dominated by thicket and if the soil is fragile and farmers do not take care to protect perennials, eventually grasses will take over: guinea grass, *Andropogon*, and in the end speargrass. Today *Chromolaena* will be the first to invade the fallow, but if things get really bad speargrass will still make its appearance. If that happens the decline process will take much less than the hundred years predicted by Nye and Greenland's humus model.

So, if we want to say something about the productivity and sustainability of recurrent cropping, that is cropping with short fallows, we must estimate how serious the associated nutrient losses are and how long it takes before productivity starts spiralling down. It has been fashionable during the last four or five decades to conjure up-doom scenarios whereby Africa's soil resources are being rapidly depleted and the whole continent is in danger of becoming a barren wilderness. Even shifting cultivation was suspect. The FAO stated in 1957 that 'at each turn of the [shifting cultivation] cycle the soil becomes more depleted of nutrients and its productivity is less, and more short-lived' (quoted, disapprovingly of course, by Nye and Greenland). That opinion, it was not more than that, was of course not vindicated by the facts and even in more intensified systems the rate at which fertility was lost seemed to be slower than expected. H. Vine, one of the most insightful soil scientists in pre-independence West Africa already had doubts about the accepted opinions of his day, when he read a paper at a conference in 1954, entitled: 'Is the lack of fertility of tropical African soils exaggerated?' Today, farmers still seem to be able to extract crops from land which according to the doom scenarios should long since have been destroyed by overexploitation. So what are the facts about fertility decline? That is what we will look at in the next few pages. For those who are not interested in the nuts and bolts of nutrient cycles I will only sketch the broad outline here while the arithmetic can be found in Appendix 2.

(b) Shorter fallows, lower yields

For N things are fairly straightforward again. Practically all the N on which the crops feed comes from decomposed humus. When the land is being cropped there is a net loss of humus but after it has returned to fallow and given enough time the process is gradually reversed. Free-living and symbiotic bacteria which live in association with the vegetation start to fix atmospheric N which is taken up by the vegetation and returned to the soil by litter fall. Part of litter and decaying roots are turned into new soil humus and if there is enough biomass, more humus is formed than is broken down. It was that same process by which the large store of N in humus-rich soils was built up during the many years or even centuries the land was under undisturbed forest. Under a production system of 2 years of cropping and 4 years of fallow the soil's humus content will gradually decrease until a new equilibrium is reached, which will be about 50% of the maximum accumulated under a mature forest. At that level the humus would still release between 70 and 100 kg/ha/year, which the simple spreadsheet-based nutrient budget of Appendix 2 predicts to be enough for (unfertilised) maize yield of between 1.1 and 1.6 t/ha. In the following year and a half there will be enough N for 12–17 t of cassava, but in many cases there will not be enough K for that kind of yield as we will see in a moment. Plantains will no longer be a serious option at all in a short fallow system. They tend to do poorly when soil fertility is low and when the crop languishes it becomes sensitive to pests.[16] Farmers may still plant them at favourable locations but otherwise the cropping system is likely to be just maize (sometimes associated with a variety of minor crops) followed by cassava.

For P and K the predictions are much more uncertain than for N. A major source of uncertainty is the botanical composition of a 4-year fallow vegetation: it makes a lot of difference whether the fallow consists of a well-developed shrub vegetation with

---

[16] Especially the larvae of the plantain weevil which tunnel into the corm (the underground swollen stem) and nematodes.

small trees, *Chromolaena*-dominated thicket or man-made grass
savannah, in the worst case consisting of speargrass. And that in
turn depends very much on the soil's native fertility. Only if the
fertility is high enough there will be enough nutrients left at the
end of each cropping cycle to allow quick restoration of a lush
fallow vegetation. And the richer the vegetation the better it will
be able to pump up nutrients from the subsoil and restore the
topsoil's nutrient content. It is a self-strengthening process. Let
us assume for now that we have such a stable system. Our model
then predicts that maize yield is limited by N-shortage alone and
that cassava yield will be limited by K-shortage also. The yield
range was therefore predicted to be 9.5–15.5 instead of 12–17 t
which would be feasible if only N were limiting. The model calculations say further that P should be just adequate to match the
available N and K. In soils where P-fixation is strong P deficiency
may also develop, however. Later on we will see what will happen
if we bring in nutrients in the form of fertiliser or manure or
whatever to increase crop yields.

In chemically poor and acid soils there may be so little P left
after the cropping period that some fallow species have difficulty
getting established and are pushed out by others, especially hardy
grasses. If it happens the fate of the system is quickly sealed. It will
spiral down to a man-made savannah. That has happened in many
areas, for example in south-western Benin in West Africa and in
the two Congos in Central Africa where you find serious speargrass infestation.

(c) Carrying capacity and sustainability

We started from the assumption that farmers' response to increased
population density would be to shorten the fallow and grow maize
and cassava as the major crops, one crop per cycle. Does that help
to increase the land's carrying capacity and if so, how many more
people could the land carry when the fallow is reduced to, say, 4
years? That is an interesting question which we will tackle now.

Yields may be low under a short fallow regime, but as long as the
fallow vegetation does not degenerate the system is pretty well
buffered, by the humus stock which guarantees some maize and by
cassava's ability to extract nutrients even when their concentration is
low. Obviously, yields will be lower than under long fallows, but
cropping frequency is more than twice as high. Is the net gain

enough to drastically increase the land's carrying capacity and how long can such a system last? It is straightforward to estimate the carrying capacity for recurrent cropping with 2 years of cropping and 4 years of fallow with the figures we have found so far. Box 5-5 shows that the land can now support about 70–110 people/km^2, compared with 55–78 for a system with 12 year fallows. That is only about one third more, and at the cost of a lot more hard work for weeding, because short fallows will result in more weeds and the less vigorous crops will also be less capable of suppressing them. So, instead of one or two light weedings as in shifting cultivation, probably three will be needed and more awkward ones, because there is likely to be a shift to weeds which are more difficult to control. These figures, 70–110 persons/km^2, are for land that has reached the new humus-equilibrium corresponding with 2 years cropping and 4 years fallow. As long as it has not got there yet the situation will be a little more favourable, because crop yields will be somewhat higher.

What about the sustainability of recurrent cropping systems? We have seen that shifting cultivation will last for centuries, even if the nutrients in the produce leave the land for good. Recurrent systems are less stable and their durability will be strongly affected by the quality of the soil, contrary to highly buffered shifting cultivation. How long can a system based on 4-years fallow last? We have seen that the soil's reserves of nutrients become important when the

---

Box 5-5. Carrying capacity under recurrent cropping

1. With a 2-year crop cycle and 4 years of fallow, a total of 6 ha of agricultural land is needed to start a new crop cycle on 1 ha each year.
2. If 70% of the land area is suitable for agriculture the *gross* land need is 6/0.70 = 8.6 ha/ha.
3. One crop cycle produces 1,100–1,600 kg maize and 9,500–15,500 kg cassava, that is a total of about 2,990–4,770 'maize equivalents' (once again, ignore the air of precision).
4. Average yields will be lower than potential, because of differences in weed pressure, shading and less than perfect management. I will discount 20% again for those factors.
5. If an average person consumes 200 kg of 'maize-equivalent' per year and the same amount is bartered or sold, then 400/2,392–400/3,816, i.e. 0.167–0.105 ha must be planted annually, for which 1.436–0.903 ha of gross land area is needed.
6. Hence, 1 km^2 of land can support 100/1.436–100/0.903 = 70–111 persons.

fallow is shortened. Good soils have fairly large amounts of K adsorbed to the clay and humus, which are not called upon as long as fallows are long. Furthermore, extra P and K may be released by decomposing minerals. With more intensive cropping these surplus nutrients are gradually extracted by the crops. How long that can go on is difficult to tell. It depends on the size of the soil's native stock of nutrients and on the ability of the fallow vegetation to recover nutrients from the subsoil and to unlock P which has become immobilised and restore them in accessible form to the topsoil. As long as the vegetation remains dominated by deep-rooted perennials, nutrient cycling can take place effectively, even though there will be some leakage to deeper soil layers. As the nutrient content goes down the vegetation will gradually shift to oligotrophic species, that is species which can grow under low nutrient concentrations, like many grasses including speargrass. The system will then spiral down to a grass climax and a much lower level of productivity. It is difficult to say how long recurrent cropping with 4 years of fallow could last, because we do not know when the system will enter the downward spiral. I have therefore taken the easy way out and simply assumed, unrealistically of course, that the system remains intact until all the available nutrients are gone. With those caveats Appendix 2 calculated a lifespan of 90–130 years for a system with 2 years of cropping and 4 years of fallow. That is still a reasonably long time, but as I said, the system may start degenerating much earlier if the soil's native nutrient stocks are small. It has already happened in many areas in Africa.

By now all these figures for yield, carrying capacity and durability at different fallow lengths and fertility conditions must have become a confused tangle in most people's heads, so I have put together a summary in Table 5-2.

(d) Increasing yield by applying external nutrients

If a farmer is no longer satisfied with the meagre crops he gets from recurrent cropping, what can he do to boost them? For maize the first limiting element is N. Fertiliser N may be applied to correct that, but if a lot of N is applied, it will have to be combined with P and in many cases with K as well. If maize yield is to be increased to 3,000 kg or more, for instance, either or both of them will become limiting. And if you want to raise cassava yield, K-fertiliser will almost always be required. The large differences in native soil

Table 5-2. Summary of calculated features for shifting cultivation, recurrent cropping and transition soil conditions, no nutrients added

Number of years		Potential yield ranges, kg/ha				Carrying capacity	Durability
Crops	Fallow	Maize	Cassava	Plantain	Maize equivalent	Persons/km²	Years
Shifting cultivation; equilibrium soil conditions, with nutrient export							
2	12	1,750–2,500	14,300–20,500	4,300–6,100	5,470–7,820	55–78	180–340
Recurrent cropping; equilibrium soil conditions, with nutrient export							
2	4	1,110–1,630	9,400–15,700	–	2,990–4,770	70–111	90–130

fertility and previous land use history make more precise predictions impossible and the responses to P and K observed in fertiliser
experiments are quite erratic. Another option is to let nature help
and grow a nitrogen-fixing legume ahead of the crops. Legumes not
only fix N, they are also better than maize at extracting immobilised
P and by leaving part or all of their biomass in the field its N and P
will be set free for use by the crops. That is why so much research
has been done from the early part of the twentieth century onwards
to find leguminous species which farmers could grow to beef up the
N and the available P in their soil. That is called green manuring,
about which Nye and Greenland said that no farmer had been seen
to actually adopt it. Green manuring has been a constant component of the alternative cropping systems scientists have recommended ever since, with as little success as in Nye and Greenland's
days. Why that is so is an interesting question, which I think has not
been answered satisfactorily. If the reasons were understood, either
green manuring would now be widely practised, or the researchers
would have stopped bothering. Neither is the case.

(e) What if the system gets unbalanced?

Just a few words about the likely sequence of events when the system becomes unbalanced and starts spiralling down, either because
the bush fallow is too short to maintain the fallow vegetation's
integrity or because the soil has become depleted, or both. The first
thing that will happen is incomplete regeneration resulting in thinner ground cover. Once that has set in it will accelerate itself. Bush
fires make their appearance and destroy young fire-sensitive plants
and seedlings. A new climax will now develop dominated by grasses.
In the humid zone the final stage will be speargrass. The soil's
humus content will decline and nutrients are no longer drawn from
the subsoil so the amounts of nutrients being cycled through the
system are much smaller, aggravating N, P and K deficiencies.
Furthermore, N deficiency will be made worse by decomposition of
the grasses at the beginning of each crop cycle which will temporarily immobilise N. Without fertiliser growing maize will no
longer be an option and farmers will instead grow cassava or
another root or tuber in association with a legume like groundnuts.
Cassava, which is capable of extracting nutrients where other crops
are not, will be the last crop to produce something. It is likely that
any gain in planted area realised by further shortening the fallow

will be completely undone by the decline in productivity of the land, so I think that the maximum rural population density that a fallow-based system can accommodate will not be much more than the 60–120 persons/ km^2 calculated earlier and will decline as the system slips out of equilibrium. Dramatic decline has indeed occurred in some areas but not as much as one might have feared. Farmers have often found ways to prevent or mitigate degeneration and safeguarding some form of nutrient cycling and maintenance of the humus, before decline became irreversible. We will look at some of these creative solutions later on.

5.2 Where are fallow-based systems heading?

It is now almost a platitude to say that the natural fertility suggested by the luxurious appearance of a tropical forest is mainly an illusion. The wealth of the forest, which has been accumulated during centuries is cycled around through the vegetation and the soil and the farmer can siphon off some of that wealth, provided he does not destroy the overall equilibrium. If he does, by taking out too much too quickly or by large-scale removal of the vegetation, the age-old heritage will rapidly be squandered leaving just a sterile skeleton without the ability to restore itself, except perhaps over another several centuries. People living in unity with the forest reaped its products or the nutrients in its soil and left the heritage intact by moving elsewhere when there were signs of decline. When communities are forced to feed more people due to whatever societal developments, they must find ways to produce more while preserving the heritage for as long as possible. That is the challenge of intensified land use, which is harder to meet in Africa than in other continents.

  The most common approach to the intensification of land use has been to reduce the duration of the fallow, but if that is not accompanied by additional measures, the fallow vegetation will eventually shift in the direction of savannah. Nye and Greenland described an example from Ghana, at the edge of the semi-deciduous forest zone, where the vegetation had degenerated to 'man-made' grass savannah which would eventually become dominated by speargrass. The same thing was happening in the 1980s in a similar environment north of Ibadan in Nigeria as I described in the beginning of this chapter.

The spread of *Chromolaena* may postpone the 'grassification' of the fallow for a while but eventually it will also be pushed out. Along the road from forest to grass fallow things will happen to the crops. The first crop that will be dropped is the plantains. As the humus content declines and therewith the N supply, maize will also go, unless the farmer can afford to apply fertiliser. And eventually, nothing but cassava may remain.

There are farming societies in Africa whose cultural practices are much more destructive than others and who will therefore reach the point of no return earlier. Perhaps they migrated from savannah areas and brought farming habits with them, which were alright there but which are disastrous in the forest. Otherwise I cannot understand why they would try to rid their fields completely of trees as they do in some areas in Bendel State in Nigeria and in Congo–Brazzaville, sometimes immediately behind the timber loggers, and turn the land into a savannah as soon as possible. Compare that with the Yorubas and many other people who jealously spare the tree stumps and seedlings to carry them over to the next fallow.

5.2.1 Living with the decline of fertility

An interesting cropping technique is practised by farmers in some man-made savannah areas to get around the severe early-season deficiencies of N and P which normally occurs after the grassy vegetation has been cleared. It occurs in many areas in West and Central Africa in different variants, but my examples refer to the western Cameroon where it is practised in severely degraded land. Towards the end of the dry season the land is cleared of the speargrass which is then covered with soil buried in ridges or narrow beds and burned slowly. When the rains break groundnuts are planted on the beds with a low density of maize added (Figure 5-6). The system is called *écobuage* in French. I do not know whether there is an English term for it, there must be because the technique is used extensively in the English-speaking north-western part of Cameroon. The system seems to help to release otherwise immobilised P and keep the speargrass away at least for a while. The groundnuts fix their own N while the maize is just a bonus if it works and does no harm if it does not. The system is controversial among agriculturists because it is thought to burn the little organic

Figure 5-6. Groundnuts with low density maize in *écobuage* field (left) and spear-grass fallow (right). Western Camer⊂on

matter there is, destroy the soil's structure and make it more prone to erosion. In other areas all over the continent similar practices are found, sometimes without controlled underground burning. In that case the speargrass is burned on top of the soil, ridges or beds are made and groundnuts (or Bambara nuts or both) are grown, with or without a low density of maize. The idea is essentially the same. I have no idea what the long-term prospects of the system are, but I think it is quite a creative solution.

5.2.2 Delaying the decline

Are there other, perhaps more effective ways to avoid, or rather slow down the decline? Nye and Greenland had some other interesting examples of farmers escaping from nature's straightjacket and attaining a much higher human population density than you would expect, even under severe fertility constraints. Let us see how they were able to defeat the rather bleak prognosis that emerges from our analysis so far.

In some high rainfall areas quite high human populations are supported by the land, for example in the Igbo area of eastern Nigeria with 140 and in Kabale, Uganda, with 200/km^2. That is more than the 75–150/km^2 which I estimated to be the absolute maximum for fallow-based crop production without some other measures to maintain the land's fertility. How come? Kabale has fertile, deep red loam soils, which could be part of the explanation, but the soils in eastern Nigeria are not particularly fertile, in fact rather poor. The authors tried to understand what exactly these people did to feed so many mouths.

A few things are clear. First, in both cases farmers combined intensively cultivated compound- or nearby fields with more extensively cultivated outer fields with fairly long fallows. Second, perennial plant species played an important role, in the outer fields as fallow species and in the near fields as crops. In Igbo land oil palm and some other trees were cropped, in Kabale bananas. Farm animals would be grazed in the outer fields, and they brought back some fertility with their manure which was used to fertilise the near fields, together with household refuse. In both areas there was therefore a net fertility movement from the outer to the nearby fields, through both crop residues and animal manure. The perennials in both field types would recapture some of the leached nutrients, the fallow species in the outer fields and the perennial crops in the inner fields. In the Igbo area of Nigeria there was one hidden source of nutrients in, though. In the 1950s one fourth of the region's food was imported, partly financed from remittances from relatives working elsewhere. The proportion is likely to be even higher today.

Vine described another intriguing case from south-eastern Nigeria (south of Igbo land) with more than 2,000 mm of annual rainfall and poor soil, where farmers had managed to practise a system with 1.5 years of cropping and 2.5 years of fallow for a long time. In the 1990s it was still there. What they did was keep the grasses under control, do very little burning and, most importantly, carefully preserve and even replant the very deep rooted indigenous shrub species *Acioa barteri*, which was trimmed during the cropping season and coppiced easily during the fallow period. So the shrubs were permanently present, as a kind of pulsating fallow vegetation, restrained during the cropping season and taking off when cropping ended and acting as a nutrient suction pump to bring back up what had been washed down. IITA would pick up that species later on in its work on alternatives to shifting cultivation, which will be one of the topics of Chapter 7.

Another example is the banana-based agriculture found in the Great Lakes area of Central Africa. This is a mid-altitude zone, where temperatures are more benign than in the lowland tropics, but I still want to mention it because this is an exceptional example of how people have found ways to push agriculture to the limit of the possible, without the use of external inputs. In some parts population densities are very high, especially in Rwanda and Burundi, up to 500 persons/km^2, whose existence is made possible by the combination of fertile volcanic soils and ingenious cropping systems. I will describe one system from the Bukoba district in Tanzania, however, which is adjacent to Rwanda. The soils' natural fertility is very low, but people have nevertheless managed to cram 75 people into a km^2 of land. People live in scattered family settlements with three concentric shells around each settlement, each with its own distinct land use. The homesteads which are located within a permanent banana grove form the core. It is surrounded by a ring of annual crop fields with short fallows and another large outer ring of very acid, nutrient-poor grassland. The grassland is not really a shell attached to a particular settlement but rather a communal wasteland (which would have been called 'commons' in Europe), which is shared with neighbouring families. There is a steady flow of nutrients from the outer circles to the banana groves. The people live inside the groves, they keep their animals there which are fed with produce from the outer circles and bury everything that is organic (including their own dead) among the bananas. Again, the productivity of the intensively cultivated fields is maintained by the transfer of fertility from the periphery to the centre and by its very careful management and prevention of losses.

All the successful solutions involve transfer of fertility from one place to another and the presence of a tree component in the system, which appears to be essential to conserve and effectively exploit the limited fertility there is. But whatever ingenious methods farmers have invented to deviate the course of nature, the soil's inherent fertility will put a ceiling to the number of people the land can support. And eventually the system's productivity must go down, no matter how slowly

5.2.3 Boserup and Malthus

Agronomists and soil scientists were not the only people to look at fallow-based systems and try to develop a vision about their future. We have already seen the work of van Beukering the agricultural economist in the Dutch East Indies, which did not attract the attention it deserved, except perhaps in the Netherlands where it has also now been forgotten. A highly influential study was published in 1965 by a Danish agricultural economist, Ester Boserup, about the causal relationship between population pressure and intensification of agriculture. The part of her book which describes fallow-based systems reads like a commentary of Nye and Greenland's work, although there it is not mentioned at all, nor is any of her work quoted by them. Greenland's 1975 paper in *Science*, entitled 'Bringing the green revolution to the shifting cultivator' does not mention Boserup's work either. The world of the social and the technical scientists apparently were entirely separate.

Boserup's programme was to prove that population pressure drives agricultural change, rather than agricultural change determining population density. That was her central tenet and she set out to refute the opposing view which apparently was held by most social scientists, and which she called Malthusian. Boserup's opponents argued that population growth was constrained by an 'inherently inelastic' food supply, which means that food supply does not respond flexibly to population increase because it is limited by environmental and technical factors. Only when innovations are introduced which allow agriculture to breakthrough the production ceiling can population grow again. So, population is the dependent variable and is determined by the limits to agricultural production. Boserup, however, argued that it is the other way around: population pressure forces people to intensify and thereby extract more from the land, so population is the independent factor which drives agricultural change. Her proof that the others were wrong is quite clever. Like van Beukering and Nye and Greenland she states that it is only human nature to try and get the highest output with the least effort and under low population density shifting cultivation turns out to be the best choice. According to the Malthusian view, or what she held it to be, population density will only increase beyond the limits imposed by shifting cultivation once a new more intensive production system *has been* adopted, otherwise it will remain stagnant. Boserup argues that it cannot be like that, because why would farmers adopt a more intensive production system, unless they are forced to by population increase? They would certainly not because a more intensive system based on manual labour would invariably demand more labour per unit of output (she brings together a lot of supporting data for this thesis), so it would be contrary to human nature to adopt such a system. *Ergo*, population pressure is the driving force of agricultural intensification, *quod erat demonstrandum*.

I have considered this argument for quite some time and even now I cannot quite decide whether it is trivial or not. To be honest I think it is. Malthus pictured an apocalyptic future where the populations of entire countries or continents would eventually reach the limits imposed by the carrying capacity of the land, which in Malthus' days was not impressive. Whatever people try, they cannot break through that barrier. Only a dramatic change which moves the carrying capacity beyond its previous ceiling would allow renewed population growth. That kind of change would be beyond the innovative capacity of ordinary farmers and therefore further population increase does become dependent on a technical breakthrough. However, as long as that point has not been reached people can still get more food from the land by intensifying land use if there are more mouths to feed. Take an area where the population approaches the carrying capacity of a stable shifting cultivation system. The intensification which will be needed to feed more people does not require a dramatic breakthrough, the farmers will simply have to shorten the fallow and put in some more work to do the trick. And they will, even though it means more work for the same output, because their survival instinct is stronger than their abhorrence of hard work. So Boserup is right here, but I fail to see why Malthus has to be brought in to be argued against. So let us blame it on (some of) the social scientists' obsession with hypothesis building, paradigms and analytical frameworks and their habit of engaging in endless quarrels about them.

If Boserup's rather laboured quarrel with Malthus were all there is to the book it would not deserve all this attention, but there is much more. The book has been quoted extensively, especially by social scientists, to support the view that traditional land use systems will inevitably move towards higher levels of intensification, but that is not what Boserup is saying. First, she makes a very careful analysis of intensification patterns and their negative effect on labour productivity, in the absence of technological innovations. Then she simply says that successful farming societies must have gone through some sort of intensification process to deal with increasing population. But the simplistic idea of intensi- fication as a linear process for which Boserup is often and unjustly brought to the stage is almost certainly incorrect. There is no doubt that many agricultural societies have devised intricate and often very elegant systems which allowed them to attain quite high- population densities, by recycling and harvesting nutrients in an effective way and reducing losses as much as possible. But there must have been countless instances where the process failed and the societies succumbed or had to move elsewhere in search of pristine soils. And whether today's farming communities who are struggling to scrape a living from their poor soil will repeat the intensification process is highly questionable. Boserup makes it clear that modern developments have broken the chain of events which in traditional communities could have led to intensification of agricultural production:

[with increasing rural population and declining labour productivity] people in rural areas, instead of voluntarily accepting the harder toil of a more intensive agriculture, will seek to obtain more remunerative and less arduous work in non-agricultural occupations. In such periods, large-scale migrations to urban areas are likely to take place. . . . The ensuing rise of food prices *may* provide the needed incentive for an intensification of agriculture and be followed by a rise of rural money wages which helps to keep migration within bounds.

But Boserup thought that was unlikely to happen, because national governments preferred to 'increase food imports [as] a means to avoid the political and social trouble in the urban areas which would be likely to follow rising prices of food in terms of urban wages'. Today's society offers the illusion of better oppor- tunities outside agriculture to the young people who do not want to spend their days toiling the soil for a meagre income or just to feed themselves. As a result, many agricultural areas have become stagnant and mired in poverty, with neither the incentive nor the able-bodied young men to search for new ways to maintain or increase the productivity of the land. Perhaps this is what Malthusianism looks like in Africa's least developed countries at the beginning of the twenty-first century: stagnant agriculture and a population which, instead of starving in the villages does so in the large urban centres or survives on international aid. Meanwhile the youths who should be the main hope for change are loitering around in towns and cities and are either thinking about ways to swell the ranks of immigrants in Europe or end up in the criminal circuit. That is really a gloomy picture, too gloomy perhaps. People are becoming conscious of the danger that several countries in Africa may be slipping into massive chaos, which could be the beginning of a serious search for a way out. Well, may be. I will come to that in the final chapters, but for now let us go back to the early 1960s when things did not yet look so desperate.

During the early post-colonial days many agriculturists were aware of the need for drastic changes if agriculture were to become an occupation that young people would choose voluntarily for a living. Boserup characterises the agronomists' solution as follows: 'to modernise and increase food production by means of industrial input, mechanised equipment as well as chemical fertilisers'. But she was not optimistic:

[S]tepping up agricultural output by the introduction of modern industrial inputs cannot be realised unless a rise in agricultural prices relative to those of industrial goods is allowed to take place. The modest increases in output per man-hour which can be obtained by the use of industrial products or scientific methods may not be sufficient to pay for the scarce resources which they absorb.

And, she argues, that applies *a fortiori* to societies which have not yet attained a certain degree of industrial development, because:

it seems somewhat unrealistic . . . to assume that a revolution of agricultural techniques by means of modern industrial and scientific methods will take place in the near future in countries which have not yet reached the stage of urban industrialisation. It is not very likely, in other words, that we shall see a reversion of the traditional sequence, in which the urban sector tends to adopt modern methods a relatively long time before the agricultural sector undergoes a corresponding transformation.

That means that modernisation of the urban sectors must come first, before there can be hope for modernisation of agriculture. Note that Boserup does not say that industrial development must necessarily precede agricultural development, although I think that was the prevailing opinion among development economists in the early 1960s, including my agricultural economics teacher in Wageningen, Professor Joosten. What were agronomists to do in the meantime?

5.2.4 The optimism of the agronomists

An easy way to refute the accepted opinion is to point to the highly successful industrial crop sectors, in particular the plantations, long before there was any sign of urban industrial development. That is not quite fair, however: the plantation companies were small states within the state. They would import both the industrial equipment and the expertise from the industrialised world in a set-up which was typical for the 'dualistic' colonial and early post-colonial economies, as the Dutch economist J.H. Boeke (1930) called the Dutch East Indian version. The only role the locals played in the traditional pre-independence plantations was that of labourers and foreman, not supervisors or managers.

But even in respect of peasant agriculture some influential people must have held opinions in the 1960s which were entirely different from the mainstream economists' view. For around the time Boserup wrote her essay the idea emerged of setting up a series of international agricultural research centres to generate technologies for smallholder agriculture in developing countries. That idea was based on the premise that there was hope for lifting agriculture out of stagnation by technical innovation and that agricultural development could be the driving force for development in other sectors, not necessarily the other way around. The creation of the international centres was not the only example of that optimism. They were preceded by many development projects which were started in the early post-colonial years of the 1960s on the same premise. They advocated the use of western type technology like mechanisation and chemical inputs and setting up farmer organisations and marketing boards in the hope that these would give productivity a major boost. And they thought that by adopting all those innovations there would be no need for fallow at all.

A typical product of this line of thought were the so-called Farm Settlements in western Nigeria which started in 1960, the year the country became independent, and similar initiatives in the francophone countries. In Smyth and Montgomery's book (1962) on soils and land use in south-western Nigeria the ideas are explained eloquently:

Prospective settlers are trained in modern farming in an Institute before being placed on a Farm Settlement where they are assisted financially under a supervised credit scheme until their crops come into bearing and they are self-supporting. The . . . advantages, in addition to the benefits of good housing and welfare of all kinds, are the pooling of transport, heavy equipment and tools, the common services of processing, storage and marketing and equipment maintenance and the bulk purchase of fertilisers, insecticides and fungicides. In addition, expert tuition and advice from Ministry of Agriculture officials are at hand.

This is followed by the following sentence which testifies to the supremely optimistic outlook of the ex-colonial and the first batch of Nigerian officers:

The other less obvious advantage which is of the utmost importance is the setting of a new standard of living in the rural areas, a general improvement in the lot of the farmer and the new regard, by the population at large, of farming as an enviable occupation, to be sought after by young educated school leavers.

These people, most of whom I daresay had been trained as agronomists and soil scientists, believed that a permanent and profitable form of agriculture with a combination of food crops and cash crops like cocoa was possible in the more humid areas, provided the farm model and the technology were right and a number of reliable services were available. The concept was clearly technology-driven agricultural development. That concept would soon be vindicated by the successes of the green revolution in Asia, which was based on a new generation of varieties of wheat and rice, the so-called High Yielding Varieties, or HYV, bred by the International Research Centres in Mexico and the Philippines, combined with large amounts of fertiliser and pesticides. But things were different in Africa and an African green revolution did not occur, at least not on the same scale as the Asian one. I will try to explain why, though I am not entirely satisfied with what I am going to say.

    In the first place there was no single dominant crop in the wet parts of Africa. In Asia the newly bred short straw varieties of paddy rice showed a spectacular response to fertiliser in the uniform and controlled environment of the irrigated paddy field. Maize could have played a similar role in Africa, and it did to some extent in the drier areas, but much less reliable growing conditions, among other things, stood in the way of a real maize-driven green revolution there. And in the wetter parts the physical conditions were also less favourable for such a thing to happen. The small plots are usually scattered among plots at different stages of fallow regeneration and valuable trees are left standing in the crop fields causing a lot of shade and reducing the effectiveness of fertiliser. Furthermore, a large application of fertiliser to a maize–cassava intercrop favours the maize and hence suppresses the cassava. The best way to harvest a lot of maize would be to grow it as a sole crop in the first rainy season in large contiguous fields without trees, and grow something else, like a grain legume in the second season. That was actually the model promoted by the early post-independence agriculturists. It turned out to be very risky and could result in the total loss of the fragile topsoil because the torrential early rains would wash it away. This would not happen as long as farmers stuck to their habits of mixing annual and perennial crops planted in small fields scattered around the bush. But the new ideas attracted new people who wanted to launch into so-called modern farming in the forest area, probably with money earned in some other business. They would uproot the trees and shrubs, plough up a large piece of land with a tractor and plant maize. But not for long. It soon became clear that simple models borrowed from temperate agriculture would not work and scientists started thinking about entirely new cropping systems, more suitable for the use of modern inputs, which was needed if a green revolution were to take place in Africa. One of the main reasons why international institutes like IITA were set up towards the end of the 1960s was to develop new technologies and even entirely new farming concepts, suitable for African conditions, in support of this new breed of emergent farmers. But developing new farming concepts is one thing, getting them adopted is quite another, as we will see in Chapter 7.

Another factor which was of great importance in Asia's green revolutions and which was all but absent in Africa was effective agricultural services. In Indonesia, one of the countries where the green revolution took hold, the introduction of high yielding rice varieties was accompanied by a massive and sustained promotion campaign at all levels and by a regimented organisation which ensured that the seed, fertiliser and pesticides were delivered in sufficient amounts and in a timely manner. In Africa these services have seen a steady

decline to the point where they are practically non-existent or
entirely irrelevant in many countries today. In the 1960s, however,
the situation was still relatively favourable and there was hope that
an invigorated extension service could help promote the emergence
of a new generation of farmers who could bring about an African
green revolution. But the absence of truly successful technology
combined with lack of institutional competence eventually made the
services collapse under their own weight.

5.2.5 Nye and Greenland's vision

I conclude this chapter with the options Nye and Greenland saw for
intensified agricultural production in the forest area. Their ideas were
going to have a strong influence on IITA's Farming System's Program
in later years. First they asked themselves whether there were lessons
to be learned from the way European agriculture had evolved from
shifting cultivation to today's highly intensive production systems.
Since time immemorial farm animals had played a crucial role in the
European farm: oxen and horses as draught animals and cattle and
small animals as sources of animal products as well as producers of
manure. In the Middle Ages production relied on the continuous
concentration of fertility from large areas of common or wasteland
to the much smaller areas of arable land, where two cereal crops were
grown followed by one year of fallow. Animals were grazed on the
commons and stabled at night and during winter. On the wasteland
farmers also gathered grass for winter feeding and sods and litter as
bedding for the animals. Everything revolved around the production
and conservation of as much manure as possible, which was transferred to the farmland. Having an adequate supply of manure was
crucial, especially in areas with poor sandy soils. The principle of collecting fertility from a wide area and concentrating it in the much
smaller farmland area has been surprisingly universal in traditional
agriculture across the continents and it still underlies today's peasant
agriculture in Africa's more densely populated areas. And unfertilised
crop yields of African peasants are also very similar to those of
Europe's up to the mid-nineteenth century.

The first major breakthrough in Europe was the development of
the so-called Flemish method in the seventeenth century. Its main
innovation was the production of fodder crops, in particular turnips
and red clover, which were grown in the farmland as second crops

after cereals. This allowed an increase in the number and improved feeding of the farmers' cattle, which produced more manure permitting elimination of the fallow year and cropping the farmland more or less permanently. In spite of this innovation plant nutrients remained the major limiting factor and the transfer of fertility from the wasteland to the farm continued as before, while farmers tried to supplement this from whatever other sources of nutrients they could lay their hands on. In the poor sandy soil areas in the Netherlands, for instance, farmers who were close enough to the town would go and empty the pit latrines in the small hours of the morning before their suppliers woke up. The next breakthrough came from artificial fertilisers, after the middle of the nineteenth century, which reduced but did not end the importance to crop production of the manure produced by domestic animals. Even though the pit latrine continued as an additional source of nutrients for a long time, the combination of fertiliser, improved fodder and manure production and the improvement of the soil's organic matter raised the system out of its subsistence level. This laid open the road towards today's highly efficient industrialised farm.

Nye and Greenland then asked themselves whether a similar chain of events could take place in the humid tropics. The first thing that came to mind as an important step towards more efficient production was animal traction. In the savannah its adoption took place quite smoothly but in the forest zone it would require the removal of the perennial fallow species' roots and stumps, thereby destroying them completely. By doing that a great part of the topsoil's nutrients would be washed down irretrievably by the surplus rain water and the entire topsoil might even be removed if the fields were large and the soil was left exposed too much, for instance by growing two short cycle crops annually. Perhaps fertilisers could solve the leakage problem to some extent and by applying them cautiously and at the right time, a large part of the fertiliser-supplied nutrients would end up in the crop. Permanent fields could also be protected against erosion, by planting the crops along the contours, planting grass bunds to break run-off and similar manoeuvres. But the soil's organic matter would slowly disappear and in the long run the soil would lose its ability to tie up nutrients and water. The best option in Nye and Greenland's view was to do none of those things and leave the forest area to produce what it does best: perennial crops like cocoa and rubber, and import foodstuffs from areas which are more suitable for producing them. That of course would make a lot of sense, but the kind of ideal world where that would be possible was nowhere in sight, nor is it today, in Africa or anywhere else. So a next best solution had to be found.

Nye and Greenland then came up with a model which was similar to van Beukering's for the East Indies: a mixed farm with intensively farmed plough land, where both crops and fodder were produced, fertilised with animal manure, combined with grassland to graze the animals. Contrary to van Beukering's model the grazing land would not be permanent: it would rotate with the crop land around the farmer's property, so that the grazing land at the same time served as fallow phase for the crop land and ensured that the humus stayed at an acceptable level. In other words ley farming. Nye and Greenland had doubts, however, whether a ley system without trees would work, apart from the severe problem of animal diseases to be expected in the forest zone. The deep-rooted trees in the traditional fallow-based system are much better nutrient pumps than the shallow-rooted annual species in grassland, and humus production from tree debris is also much better. Perennial cash crops like cocoa and coffee would remain part of the system but not in association with the food crops. Intensified food crop production in the forest zone itself would somehow have to include a tree component. They mentioned the corridor system which was tried in the Belgian Congo as a technically possible model. That was a communal land use system consisting of 100 m wide strips in all stages of cropping and fallow. Each farming family would have a holding at right angles to the corridors, so that they would have a plot in each of the strips. It was essentially a regulated fallow-based system, a typically clever colonial idea which was swept away after independence along with other bad as well as good ideas from those times. Another more realistic example was the permanent very deep-rooted Acioa trees of eastern Nigeria (*Acioa barteri*, formerly classified in the *Rosaceae*, now in the *Chrysobalanaceae* family), which were jealously protected by the farmers and coppiced during the cropping phase.

And finally there was the novel idea by C.F. Charter in 1955, who suggested to plant freely coppicing species in rows and prune them back when a crop was grown in otherwise permanent food crop fields. That was going to be one of the major technologies researched for two decades by IITA, as you will see in Chapter 7.

Chapter 6. Farmer Skills, an Elusive Property

After completing a long tour of the technicalities of farming in Africa I must now pay attention to the farmers themselves. They were there in the previous chapters all right, but in a rather impersonal way. In this chapter I will redress the balance a little by looking at individual farmers as operators of their farms and how skilful they are in that.

With the arrival of FSR in the 1970s and 1980s the peasant farmer supposedly moved to centre stage and henceforth his needs were going to determine the scientists' research agenda. We had occasion in Chapter 4 and will have occasion again in Chapter 8, to talk a lot about FSR methodology which was developed to help scientists help farmers help themselves, but let us stop here for a while to see who those farmers are, what their present skills are and whether they can be improved. You will have to wade through some technical detail again, but I think it is worth it. If we do not understand what farmer skills are about I do not see how we can talk about agricultural development and the kind of research which is needed to bring it about.

What makes a good farmer is an important question but difficult to answer. It is important because, if you know what they do differently from those who are less successful, you can try to convince the latter to change their unprofitable habits to those of the former. In African peasant agriculture, being a good farmer is a rather elusive quality. When you have worked with the same group of farmers for some time you may come to the conclusion that some of them are really good and others are not, only to find out that suddenly one of those you had considered as good has slipped several rungs down the quality ladder. That raises two interesting questions. The first is the one I started with: what makes a good farmer, and the second, how consistent is that quality. I will use four examples, three from Africa and one from the Netherlands, to see if they can be answered.

In south-western Nigeria and probably everywhere else in West Africa crop yields are very variable and, interestingly, the differences among farmers in a particular year are much larger than the

% of farmers

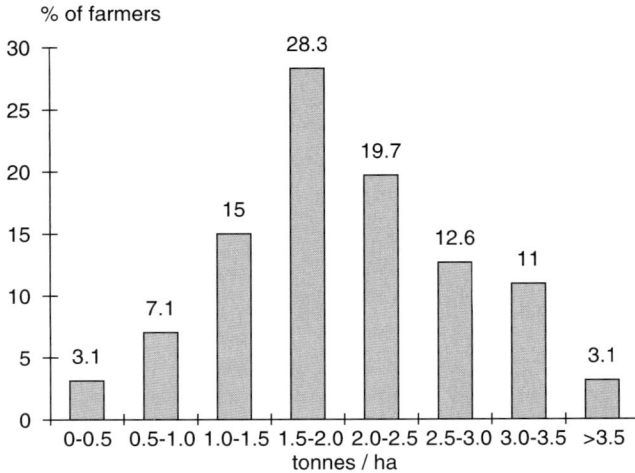

Figure 6-1. Distribution of maize yields in on-farm trials in south-western Nigeria

differences between years and locations. Figure 6-1 shows the distribution of maize yields in our fully farmer-managed on-farm trials, carried out from 1986 to 1988. The largest group of farmers (28.3%) obtained yields of 1.5–2 t/ha. The average yield of the 10% lowest producers (the two left-most bars in the diagram) was about 700 kg/ha, while that of the top 10% was 3.3 t/ha, almost fivefold.[1] Annual average yields between 1986 and 1988 on the other hand ranged from 1.18 to 1.85 kg/ha over two locations with quite different ecologies. The trials were hardly distinguishable from the surrounding farmers' fields. When we made an excursion to one of the villages, IITA's Director General was annoyed that he could not tell the trials from ordinary farmer plots, so I am sure that the yield ranges in the trials were quite similar to the farmers' own. We measured many things and did a lot of statistical analyses to find out what caused the large differences among farmers, with quite disappointing results. Chemical soil parameters for instance explained little or nothing, perhaps with the exception of phosphorus. In

---

[1] The yields were from over 100 farmers, averaged over two common treatments, one with fertiliser and the other without.

most analyses I have seen from other areas it was very much the same, which raises serious doubts about the expensive habit of agronomists of taking soil samples in all their on-farm research plots. There are probably many tons of samples from on-farm trials lying around in many institutes waiting for their turn to be analysed, which will never come. That is just as well since if they were to be analysed even more money would be wasted. Another factor in our tests which explained next to nothing was the number of years the field had been under fallow. At first sight that is strange, after all that has been said in the previous chapter about the build-up of fertility by the fallow and you would think that a field would produce better if the preceding fallow had been longer. That is generally true, but I will explain later why you will observe that kind of effect only rarely in farmers' fields.

Two factors which together explained about 50% of the differences in maize yield in our 1986 trials were the fields' shadiness and the number of plants surviving to harvest. Shadiness is related to the presence of oil palms and other trees and these will obviously have an effect on the crops growing beneath them, but the other factor, the number of plants at harvest, needs further scrutiny. The number of maize plants at harvest is of course not the same thing as the number planted. There were, in fact, only small differences in the density at which the farmers planted their maize; nor was there any change in yield when the density was increased in the trials (it was one of the treatments, unless fertiliser was applied at the same time). That in itself is quite remarkable, at least for agronomists. Many of them have an obsession with planting density and they like to declare the farmers' density 'sub-optimal', because it is usually lower than the density which gives the best results in their station trials. But surely, it would be surprising if the farmers had not sorted out something as simple as the spacing at which to plant their crops, as indeed they usually have. That is to say, until something else changes, for example when they start using fertiliser. Then their traditional density may no longer be the best and increasing it may result in higher yield. In statistical jargon, there is an interaction between fertiliser and planting density, which is also the reason why the optimum density found by researchers in trials carried out at a research station is higher than the density at which farmers usually plant their crop.

Let us now turn to plant stand at harvest, which did correlate strongly with yield in the on-farm trials. Since there was not much difference in planting density among farmers something must have happened between planting and harvest which caused some fields to lose a large part of their original stand, such as competition by weeds, damage by rodents or termites or other disorders. When young plants are lost they can still be compensated for to some extent by more vigorous growth of the remaining ones, which now have more space. But the later a plant is lost, the greater the damage because the remaining plants gradually lose their ability to exploit the extra space. And since stand at harvest is the cumulative result of losses during the season it is not surprising that it correlates strongly with yield. L.C. Zachariasse, in a study of farmers' yield variability in the Netherlands, called that a 'result variable'. It does not explain yield differences in the same way as independent variables (such as the number of weeding or the amount of fertiliser that was applied) but results from a variety of things a farmer does, almost in the same way as the yield itself. It is in fact a direct component of yield, so instead of asking how to get a good yield you might as well ask how to avoid stand loss, which will be highly, if not perfectly, correlated. In 1987 and 1988 we therefore looked more carefully at what happened during the season, how well the fields were weeded, how much damage was caused by rodents, termites and stemborers, how much shade was there, etc. The first three had some effect on stand losses and yield, so farmers would be best advised to plant immediately after land preparation to avoid early weed competition,[2] avoid termite-infested fields, keep the field well weeded and slash the borders to keep rodents out, etc. None of these factors themselves correlated very strongly with yield when taken individually, but their cumulative effect in the form of stand losses was quite strong (Figure 6-2). So, the farmer who did all these and probably a few other things correctly would maintain the best stand and eventually get the highest yield. In other words, the best way to get a high yield is being a good farmer; some farmers simply do most things right while others do not. That is about all you can say.

---

[2] That is not the same thing as planting early in the season. The early rains are notoriously unreliable in West Africa and farmers sensibly stagger their planting dates to spread the risk.

Figure 6-2. Maize–cassava in south-western Nigeria, with large gaps between maize plants due to stand losses

In a similar, more elaborate study in northern Cameroon in the early 1990s, Bart de Steenhuijsen-Piters analysed the variability of farmers' sorghum yields, in an area with two ethnic groups, both growing the crop in their own ways. For both groups, the yield differences between years and between farmers were surprisingly similar to those for maize in south-western Nigeria. The only explaining factors that stood out in the Cameroun case were planting date and what the author called 'plant density', which actually was plant stand at harvest (the publication, de Steenhuijsen-Piters' Ph.D. thesis, does not say so explicitly, but the author told me). And that represents the farmers' skill in maintaining a good stand through to harvest. So, again the farmers who got a good yield were those who did everything or most things right, they were good farmers in short, interesting but not very useful from an extension point of view. Extension services need to offer explicit and well-tested recommendations, such as which variety to grow in what season or how much fertiliser to apply, or they should be able to counsel the farmers about the way they manage their farms and help them make the right decisions. If they cannot do either, extension will quickly degenerate into meaningless ritual as it often does in Africa, until both the farmer and the extensionist get bored and give up, while continuing to go through the ritualistic motions to satisfy the latter's job description.

Why are things like soil fertility and duration of the fallow rarely found to be significant factors when one tries to explain yield differences among farmers? Let us look at fallow length. As we saw in the previous chapter, fallow length has a strong effect on soil fertility, by replenishing the nutrients in the topsoil, building up organic matter and improving soil structure. If you were to carry out a trial at the research station and compared the yield of, say, maize grown after fallows of different lengths, everything else being equal, you would *always* find that yield is higher when the fallow has been longer. So, you would expect the same effect to show up, when comparing the yields in farmers' fields where the length of the fallow has been different. But things are not as simple as that, because the 'everything being equal' condition does not hold. Incompetent managers may still get poor yields after long fallows, while some skilled ones may obtain better yields with short fallows. Even so, if the number of farmers in the study is large enough you may still be able to sort out these two effects, by using suitable statistical techniques. But there is another complication which will further obscure the effect of fallow length. Farmers will choose a field for their crops which they expect to produce a reasonable yield, and a field which is naturally fertile may produce the same or even a higher yield after a short fallow than does a less fertile one after a longer fallow. So the effect of the fallow length would not show up at all, or may even be reversed. That does not invalidate the theory that longer fallows will lead to higher fertility, the effect is simply obscured by the differences in farmers' technical skill and by their judgement about the fertility status when choosing a field. Something similar happens with other field properties, like chemical soil fertility.

What have been the experiences in the Netherlands in respect of variability among farmers? Growing conditions are much more uniform and physical differences are smoothed out to a large extent by the use of fertiliser and pesticides. L.C. Zachariasse did a variability study in the late 1960s in a young polder area with a group of farmers, uniform in terms of size of their farms, soil type and crops. In spite of this uniformity he found considerable differences in farm income which were explained to an important extent by physical yield differences, as the (rounded) figures in Table 6-1 show, although not as large as in Africa. Surely, given the farmers' uniformity, it should be much easier to find the causes of the yield differences. It was not.

Table 6-1. Average yields and yield ranges in Zachariasse's study

	Yield in kg/ha	
	Average	Difference between max and min
Winter wheat	5,350	1,800
Sugar beets	60,000	18,000
Ware potatoes[1]	52,500	19,000
Seed potatoes	32,000	19,000

[1] Ware potatoes are for eating.

Zachariasse found that yield correlated with the early growth of the crops and with their condition at different times during the season, but the latter are typically result variables, as Zachariasse called them, they integrate the effect of various things farmers do, and perhaps what they did in earlier years, as yield itself does. What he was really after, as we were in Africa, were specific independent variables which had the strongest effect on yield. Some of those turned up in the analysis,[3] like soil structure in autumn for winter wheat, quality of the seedbed for spring-planted crops, planting density within the row, proper fertilisation, but all with only a weak effect on yield when considered alone. So the best you can say again is that the better producers were those who did many small things right, especially early in the season.

An interesting observation in Zachariasse's work was that 'usually the same farmers realised high or low yields in successive years'. De Steenhuijsen-Piters is silent about this, but our tests in southwestern Nigeria also showed some degree of consistency in the year-to-year performance of individual farmers, but much less than in the Dutch polder area. That does not mean that there were no differences in skill among the Nigerian farmers. There certainly were, but I think the reasons why the farmers' yields were much less consistent over time than in the Netherlands were of an entirely different nature. For many farmers, and I think especially for the more

[3] The author used factor analysis; see Appendix 3 for an example.

skilful ones, farming had little prestige and they were continually looking for alternative gainful activities. That could be trading or off-farm employment or something like that. So a farmer who you thought was exceptionally accomplished could all of a sudden start neglecting his crops and take up something else. In the Dutch polder area on the other hand farming is always a profession and most often a vocation and all attention and effort go to farming year after year. It is the poor performers who are more likely to drop out than the skilful ones, the opposite of what happens in the West African forest area.

Not surprisingly, the fascination of researchers with yield variation in Africa has continued in recent years. Agricultural extension in Africa has been notoriously ineffective and if you could put your finger on what makes a good farmer, you could tell the extension officers and perhaps render them more effective. But recent studies have hardly met with more success than the earlier ones. A study by Robert Carsky and his colleagues in northern Nigeria came to essentially the same conclusions as de Steenhuijsen-Piters in Cameroon. They found that maize yield was most strongly associated with stand density at flowering and somewhat with striga incidence and date of first weeding, while chemical soil parameters did not correlate with yield at all. So we seem to have been going around in circles.

At the end of his study Zachariasse argued that 'a further dissection of the result variables in the underlying independent variables' should be carried out in order to sharpen future extension messages. But I doubt if that would help much. Even though you may find a few factors with a statistically significant effect on yield, the precise identification of the causes of farmer variability is likely to remain an elusive goal, because of the relatively small contribution of individual factors to yield and the fact that they may be different from year to year. It is the farmers' skill to make the right mix of choices and decisions which result in consistent over-performance by some and consistent under-performance by others.

So, should we give up trying to understand what makes some farmers better producers than others? Surely, this kind of studies done so far in Africa do not seem to get us beyond the rather obvious conclusion that the better producers are the better farmers. Are there better and more powerful methods which would help? Perhaps. The three African studies I mentioned used multiple regression techniques which clarified little, if anything, about the characteristics of the successful farmers. Zachariasse used factor analysis with somewhat more success, at the price of collecting much more detailed information of the kind that is difficult to get in Africa. The sample size would probably have to be considerably higher as well, because of the much larger variability. And more attention should be paid to farmers' management practices rather than to the physical field parameters (chemical soil fertility, soil type, length of preceding fallow, etc.) which researchers find easiest to measure. Future studies should perhaps make more use of multivariate techniques like factor analysis, although I am not really convinced that that would greatly increase our understanding. But if you are lucky and smart you may find a few robust factors which are important in most years and which can help extension workers give some useful advice to farmers. For interested readers, Appendix 3 gives a worked example of factor analysis for farmer maize yields in south-western Nigeria. It shows that in one particular year weeds, rodents and termite incidence explained about 78% of the yield differences, while the rest had to be attributed to undefined farmer skills and pure chance. But no two years are the same and the factors which explain yield variation can vary over years. It is the genius of a good farmer to know what is important in a particular year and make the right choices, and the bad luck of the researcher that he can only reconstruct the decision process afterwards. The factors which explained a respectable 78% of the yield differences in that particular year may not be important in the next but the successful farmer will know which ones are. That is what farmer skill is about, knowing what is important during a particular season, not retrospectively as scientists do with their analyses. But even though farmer skills cannot be precisely defined, certain things may show up year after year, like the field conditions early in the season in Zachariasse's case, and the importance of weed control in ours. If done well, variability studies may identify such major factors and could still help to formulate more precise extension messages. But they require several years of study, with a considerable number of farmers and a lot of data collection – typically good material for Ph.D. theses. And you may still do little more than confirm what you already knew, such as the importance of keeping weeds under control.

Perhaps it is more important to tap into the skills of successful farmers, to the benefit of the others, rather than trying to precisely define what these elusive skills are. Is it possible to transfer them to others and gradually scale up all or most farmers' productivity? My son, who studied at an agricultural college in the same polder area where Zachariasse did his research, told me an anecdote from one of his teachers, which is quite relevant here. There were three farmers with the same type of farm, all growing the same crops. One of them always had considerably lower yields than the other two. They met regularly in a farmer study club and the good producers promised to give the less successful one a phone call every time they were going to carry out an operation. At practically every instance the two of them called at about the same time and the other one did exactly as they did. And lo and behold, that year he managed to do as well as the others. But when they stopped calling the next year he did as badly as ever. I do not know how genuine this anecdote is, but it does show the dilemma. Good farmers know what to do at the right time and they can help others by telling them, but that does not mean that by doing so they can effectively transfer their skills. Every year is different and the best way to do things in one year may not be the best in the next, so the conversion of farmers into good producers will take more than a single year. The farmer study clubs which have sprung up in the Netherlands during the last few decades are perhaps a medium where this time-consuming learning process can take place gradually, and the Farmer Field Schools, which have become popular in the tropics may play the same role in developing countries. I will come back to that in Chapter 11.

Chapter 7. Mainly Technology

7.1 What is technology?

The word technology does not mean what it seems to. Cosmology, entomology, physiology, anthropology are the study or science of something, the cosmos, insects, life processes, etc. Not so with technology, it denotes the techniques themselves. When agronomists speak of technology it can mean practically anything that farmers use or do to produce a crop: a maize variety, a threshing machine, fertiliser, a recommended planting density, counting insects to decide whether a crop needs spraying or growing crops in mixture. Technology is a collective name for all those things. But it can also be used with an article: *a* technology. A new maize variety is *a* technology. For someone with an exaggerated sense of linguistic propriety that is perhaps distasteful, but it provides the kind of shorthand which students of human behaviour find convenient.

There are two kinds of technology: farmers' technology and researchers' technology. The farmers' type is usually looked upon by scientists as something that needs to be fixed, whence they call it 'farmer practices' and their own technology they call 'improved' or 'modern'. I have already given several examples of non-modern yet ingenious and interesting farmer practices, like shifting cultivation, Yoruba yam-growing techniques, maize–cassava cropping in Nigeria and groundnut-based cropping in Cameroon. In Appendix 4, I have brought together several more examples of mostly sensible and always fascinating farmer practices, but this chapter is going to deal mainly with the scientists' type.

If you want to change something, presumably for the better, you must first understand very well what you want to change and then decide which parts are in need of fixing and what is better left unchanged. If you accept the modest yields they produce, I think there is actually quite little to improve upon in traditional African cropping systems, as long as they have not gone out of balance because of some external factors. Population increase is the most important one if you accept that as an external factor. Stable traditional cropping systems, with their balanced combinations of different kinds of crops and fallow, are usually hard to beat, until they start degenerating. Several generations of scientists have been concerned with what they saw as the problem of fallow-based cropping and tried to find ways to convert it into intensive production systems, with continuous cropping of the land. Most of what IITA's Farming Systems Programme has been doing since the Institute's creation in 1968 is in that category. We are going to have a closer look at that now.

7.2 Agronomists' technology: Alley cropping, zero tillage, live mulch and more

Many of the questions which were studied by IITA's Farming Systems Programme in the 1970s and 1980s had already been put forward in Nye and Greenland's work. Perhaps that was because Greenland was the programme's first director, but I think it was actually the other way around: he was probably appointed because the institute's founding fathers recognised the validity of the questions he raised. They were all essential to be answered if a solution was to be found for the unresolved 'shifting cultivation problem', as van Beukering had put it and which was felt to be an evermore pressing need for developing countries in the 1960s, especially in Africa. That is to say, shifting cultivation itself was not the problem, but rather its breakdown under population pressure and the expected decline in productivity of the fallow-based systems which came in its place.

From the start the Farming Systems Programme was dominated by soil scientists and rightly so, because the main physical problem with fallow-based crop production is loss of soil fertility. In the beginning a lot of exploratory studies were conducted about the soils in south-western Nigeria, where the IITA main station was located, their nutrient stock, how fragile they were and that sort of thing. I am not a soil scientist and I will not venture to judge how much really new knowledge the work produced, but I know that one problem has been its poor accessibility for the general agricultural public, that is, people like myself. Nobody ever took the trouble to bring it all together, you have to dig it up from journal papers, IITA annual reports and proceedings of conferences. In the more leisurely old-colonial and immediate post-colonial days in Nigeria things had been done quite differently. Scientists would spend many years and even decades to classify and map the soils in different regions, study soil fertility and land use and develop land development plans. They had a different attitude in respect of the purpose of agricultural science, perhaps because they were attached to the colonial ministries of agriculture and had to contribute to agricultural development. However that may be, they wrote comprehensive treatises with a high utility content, even Nye and Greenland did that for shifting cultivation although they were academics, at the University College of Ghana. That attitude had largely been lost since the scientists' main audience had shifted to their journal-reading international peers and their future career prospects have become highly internationalised. I learned most of what I know about the soils of south-western Nigeria from the book *Soils and Land Use in Central-Western Nigeria* by A.J. Smyth and R.F. Montgomery, published in 1962, the synthesis of 20 years of soils research by the colonial agricultural service. In the 1980s it could still be bought at the Government Printer in Ibadan for next to nothing, and perhaps still can be today.

In south-western Nigeria, which is part of the West African cocoa belt, soils are actually quite good by West African lowland standards and the rainfall is moderate so leaching is not so rapid. The two are in fact related. So the conditions were not really representative for the African forest zone where shifting cultivation and its derivatives were found. But it was a convenient place to work and the proximity of the University of Ibadan was another reason why the institute was established there. Later on part of the soil studies were moved to the south-eastern corner of the country where the soils are really poor and acid. As the main station was in Ibadan, most of the experiments were done there. And many of these were very interesting and smart experiments.

Before setting up an experiment you have to define the problem which you want to address and the hypothesis you want to test. In this case the problem was, roughly, that population growth had led to intensification of land use, the soil would not be able to sustain the resulting cropping systems with short fallows for very long and, since the population was not going to shrink, something had to change in the system to bring it back into balance. The first part of this problem statement is actually an assumption: that the system in its current form cannot be sustained. I have examined that question in Chapter 5 and although the assumption was probably correct in the long run, it was supported less by real observation than by rhetoric, supplemented with research results like those of Nye and Greenland. I also think that what really motivated the scientists was the vision of a completely transformed agricultural production sector. What they tried to do was design a radically new farming concept based on continuous land use, not a gradual transformation of current systems. The initial hypothesis underlying all the experiments was that it should be possible to farm the land permanently without loss of productivity if you did it right.

The Institute developed three major technologies, each of which dealt with one or more of the problems you could expect when trying to introduce continuous cropping. I will describe each of them because they were all quite innovative, they dominated African agronomy for two decades and in the end they got nowhere. These technologies were alley cropping, zero tillage and live mulch.

7.2.1 Alley cropping

As long as there are trees on the land there is continuous cycling of nutrients from the top of the soil to the lowest soil layers. Their roots spread through the soil where they form a kind of three-dimensional tangle which catches the nutrients on their way down before they pass out of reach. If you cut down the trees you open up a serious leak in the system. A shifting cultivator will only clear a small patch and leave the stumps which will slowly resume their growth right from the start. He will crop the plot for a short while and then move on to the next patch. When burning the vegetation the nutrients are deposited on top of the soil and as they are washed down, the mobile cations K, Ca, Mg are captured and bound by the clay particles and the humus. As long as they remain there they can be extracted and taken up by the shallow-rooted crops, but with time they will gradually be leached to lower soil levels, dragged along by nitrate ions, out of reach of the crop roots. They would be lost if it were not for the trees. Their active roots died when the trunk was cut down, but as new foliage is formed new feeder roots will also be formed by the old ones which remain in the soil. So these roots will be there in time to capture most of the leached nutrients before they have been washed down too far. The story of phosphorus is different. It is much less mobile but it is gradually converted into less soluble forms from which the crops can no longer easily extract it. The fallow trees, however, will remobilise it again and put it into their tissue from where it will be released when the vegetation is burned.

The agronomists of the 1960s, however, were after continuous cropping, so the trees had to be put out of the way. If they would grow in neatly spaced rows perhaps you could leave them there and drive the oxen or the tractor between them, but then they would still have to be cut down because they produced too much shade for most crops and their roots just below the surface sucked up much of the nutrients so there was little left for the crops. So even though the trees did such a good job catching all those nutrients, there was really no way they could be left standing and still grow crops every year. Or was there? B.T. Kang, the IITA soil fertility scientist, thought so. Shortly after the creation of IITA, 500 ha of secondary forest had been cleared away[1] to make room for the research farm where a large part of the field research by all the institute's departments was done. B.T. Kang, BT as he was popularly called, planted rows of trees or hedges of different species in one of those experimental fields. The hedges were planted 4 m apart,[2] and crops were grown in the alleys between them, in the way Charter had already suggested in the 1950s. That is why he called his system 'alley cropping'. As long as the trees were small they did not interfere with the crop, but after a year it was necessary to start pruning them at the beginning of the rainy season and again once or twice during the growing season, to prevent them from affecting the crop plants. And the soil in the alleys had to be tilled rather well to force the roots down and leave the upper soil layer for the crop. Once the crop was harvested the trees were let go and they would form a lot of branches and leaves and roots and start pumping up the nutrients. At the start of the next cropping season they would be pruned again and the prunings were cut up and worked into the soil so that they could decompose and release their nutrients for the crop to take up. The schematic picture of Figure 7-1, from the IITA 1997 Annual Report, shows how all this is supposed to work, although, curiously,

---

[1] Well, not really forest, it was the kind of landscape I described in Chapter 5, with food crops, cocoa and secondary bush at different stages of recovery.

[2] Initially he used 2 m spacing but soon found out that that was no good.

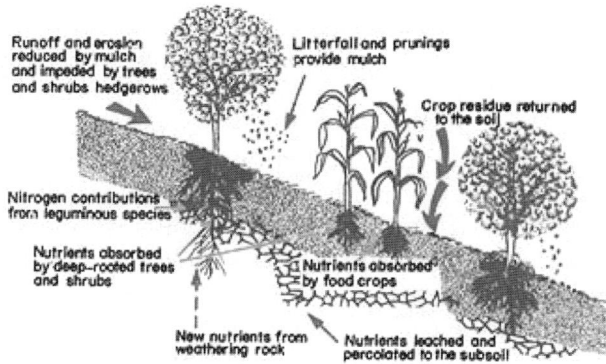

Figure 7-1. Schematic representation of alley cropping (but see the text, because this is not really alley cropping). (From IITA Annual Report 1997; reproduced by permission of IITA, Ibadan.)

the picture does not actually show alley cropping. If it did, the hedges would be trimmed while there is maize and there would be more than two rows of maize in each alley. I guess the picture was borrowed from a document about agroforestry, whereby trees and crops grow simultaneously in the field.

Alley cropping was a really good idea. It divided the tree cycle, which in shifting cultivation takes up to 20 years, into annual-truncated cycles, with a crop grown quickly between them each time tree growth had been set back by pruning. Once such a promising idea is born its originators, like over-concerned parents, tend to nurse and pamper it and postpone its confrontation with the tough real world for as long as they can. In fact, alley cropping was nurtured for so long in the protected environment of the research station that when it ventured outside it succumbed to the harshness of African farming. I will come to that later. Let us first indulge a little in the interesting scientific questions alley cropping raised and do some arithmetic.

(a) Alley cropping with maize and cowpeas

Suppose you want to grow a crop of maize in the alleys every year during the first rainy season, and cowpeas during the second. That does not make much sense to a farmer in the sub-humid zone who wants to grow maize and cassava, not cowpeas, for reasons I have explained earlier, but it does make sense for a scientist who wants to study the biological and soil processes in the simplest possible way. Maize, being very responsive to good and very intolerant to bad growing conditions, is the agronomists' favourite test crop for studying the effects of their technologies. And the reason why cowpeas were used is that they mature early, at least the type bred by IITA, and leave enough time for the hedgerows to recover. In Appendix 5, I have developed a fairly simple spreadsheet program for alley cropping and used it to simulate a maize–cowpea rotation grown in the alleys on a chemically poor soil but with a high humus content because of a long history of shifting cultivation. It calculates how much maize and cowpea yield you can expect over the years and whether the yield would stabilise at an acceptable level, without bringing in nutrients from outside. The conclusions are quite interesting. I will briefly describe them as well as the experimental evidence for their validity. If you are interested in the technical details you are urged to read Appendix 5 as well.

Maize yield was predicted by the model to peak around year 5 after which it would decline very slowly because available P was getting depleted, in both the topsoil and the subsoil. That would also reduce cowpea yield and biomass production by the hedges. The system would maintain maize yields in excess of 2 t/ha for 20 years, because the hedges fix N and extract P from sources which are inaccessible to crops and put it in the topsoil in available form. After 20 years maize yield was predicted to fall below 2 t/ha, cowpea yield below 250 kg and pruning yield was only one third of its potential. The system would probably have declined too far and the land should have been left to fallow for a while after 10–12 years of cropping to replenish the available nutrient stocks.

All that made sense, but making sense was not enough. Hard data were needed to support, (or to refute) the predictions. They were provided by a 12-year trial carried out at the IITA station, also with maize followed by cowpeas, where enough data on soil changes and yields had been collected to test the model. Predicted and measured maize yields and soil parameters for the trial's 'control treatment' are shown in Table 7-1.[3] The researchers had started off

---

[3] The data on the yields of cowpeas and hedgerow prunings in the paper were less detailed. What was there was reasonably close to the simulated results.

Table 7-1. Measured and predicted maize yields from 12 years of alley cropping with maize and cowpeas; IITA 1982–1993 (Kang et al. 1999)

Measured and predicted crop yield, t/ha, by year

Year	Maize										Cowpeas
	1	2	3	4	5	6	7	10	11	12	11
Measured	1.20	2.70	3.32	3.40	2.72	2.63	3.20	3.20	2.42	2.34	0.45
Predicted	1.24	2.54	3.06	2.76	2.77	2.65	2.58	2.50	2.49	2.48	0.50

Measured and predicted soil parameters

	OC%	Available P, ppm	Exch. K meq/100 g
Initial, measured	1.24	6.18	0.35
After 10 years, measured	0.97	5.70	0.49
After 10 years, predicted	1.33	4.85	0.54

hopefully, applying no fertiliser at all to the control treatment, but when the first year's maize yield turned out to be only 1,200 kg/ha apparently they got nervous and decided to henceforth apply fertiliser; N, P and K in the next 2 years and P and K afterwards. The model also predicted a fairly low maize yield in the first year, but not as low as the observed yield.

The low maize yield in the first year was probably caused by immobilisation of N, an interesting phenomenon which happens when grassy vegetation or residues of the previous year's cereal crop are incorporated in the soil. I will come back to that phenomenon later on. For the following 9–10 years the calculations showed that alley cropping with a moderate amount of fertiliser should be able to maintain maize yield at well above 3 t (and cowpeas above 450 kg). The subsoil, however, was predicted to be getting depleted of available P and although the hedges could still extract P from less accessible sources, their biomass yield was predicted to decline slowly, dragging down the crop yields as well, in spite of the fertiliser.

And what did the real data say? Maize yield started to decline a little earlier than the model predicted but otherwise simulated maize (and cowpea) yields were quite close to the real yields. The average pruning yield given in the paper was also similar to predicted yield. The model predicted a slow and steady decline in yield of prunings, which may also have occurred in the trial, but the paper did not show pruning yields over time, nor did it mention a decline. Nutrient content of the subsoil was not measured so we have no reality check for the predictions about P-depletion. In any case, the nutrients must have come from somewhere so I think it is fair to assume that they came from the subsoil, as the inventors of alley cropping always assumed. Still, it is surprising that nobody took the trouble to measure it.

Generally speaking I think the model predicted the outcome of the long-term trial quite well, although the word 'predict' is a little pretentious in this case, because it involved some tinkering as I have explained in Appendix 5: I had to adjust one or two model parameters before the model behaved the way the real thing did. Nevertheless, I was quite proud of its power.

(b) Alley cropping with maize and cassava

Maize followed by cowpeas may have been a convenient combination for the researchers, but it was not common among farmers in the forest zone. In the early 1990s, the then Director of IITA's Farming

Systems Programme, Dunstan Spencer, therefore decided that alley cropping ought to be tested with maize and cassava, and without any fertiliser, in a long-term trial at the IITA station. Perhaps we should have known from the start that it was not going to work. The maize–cassava combination, as I have already argued several times, is a highly effective but gluttonous crop combination which leaves little room for anything else. The hedgerows were bound to suffer from severe competition by the cassava and that crop's high consumption might rapidly deplete the K-stock, unless there was a lot of it in the soil initially. Let us see what the data showed and what the model predicted. The key figures are shown in Table 7-2.

Predicted maize and cassava yield in the first year were 'normal', because the vegetation had been burned and the nutrients in the ashes added to the topsoil and there were no hedges yet. Unfortunately, the paper did not give the first year's crop yields, so it was not possible to see whether there was a yield dip in the second year as the model predicted. In the third year yield was relatively high and much higher than predicted, but in the fourth there was a pronounced dip. The entire sequence of measured maize yields was rather strange and trying to explain some of that away becomes a little strained, but I will still explain why a dip would be expected and why it did occur in the model (which is of course saying the same thing). Maize yield was predicted to be depressed in the second and third years, because some of the N released by decomposing humus was immobilised by the conversion of maize and cassava residues into fresh humus. That phenomenon is common in savannah areas where the fallow vegetation is mainly grassy, but the model predicts that it should also occur in alley cropping in the forest area as long as the contribution of N by hedgerow prunings is small. Later it is compensated by the N-yield of the hedges.

In the field trial the maize yield went down gradually from around 2 to 1 t in year 7 and then collapsed to 300 kg. The model on the other hand, after the initial dip, predicted a gradual decline from 1.7 to 1.3 t between year 5 and 12. The yields of cassava and of the hedgerow prunings, however, were 'predicted' quite well, although cassava yield in the trial seems to have collapsed also in the last 3 years while the modelled yields remained at around 5.5 t. So something must have gone wrong with the crops in the field after 7–8 years which was not taken into account by the model. I suspect that the yield decline must have been caused by a factor other than

Table 7-2. Measured and predicted maize, cassava and pruning yields and soil parameters from 12 years of alley cropping with maize and cassava; IITA 1989–2000 (Tian et al., 2003, 2005)

		1	2	3	4	5	6	7	8	9	10	11	12
Maize yield	Measured		1.90	2.47	0.71	1.49	1.28	1.02	0.29	0.32	0.47		0.50
	Predicted	1.49	1.12	0.90	1.17	1.42	1.30	1.14	1.05	0.94	0.94	0.91	0.89
Cassava yield	Measured			5.20	4.70	4.60	3.20	7.00	5.20	5.20	2.40		2.60
	Predicted	4.89	3.68	4.07	4.99	5.38	5.34	5.22	5.15	5.10	5.06	5.03	5.00
Pruning	Measured		0.81	3.80	4.80	3.70	3.80	3.00	2.80	1.10	2.50		
Biomass	Predicted	0.34	1.27	4.62	4.54	3.58	3.04	2.70	2.48	2.33	2.23	2.15	2.10
OC(%)	Measured	(1.25)			1.30								
	Predicted				1.17								
Available P	Measured	(12)			1.49								
(ppm)	Predicted				1.08								
Exch. K	Measured	(0.32)			0.12								
(meq/100g)	Predicted				0.18								

Soil figures in year 1 (between brackets) were measured before the start of the trial nutrient supply. However that may be, the model predicted severe depletion of available P in the subsoil, which caused the simulated pruning yields to go down slowly. The decrease closely matched the actual decrease in the field. This must eventually lead to a decline of crop yields below the economic threshold.

Subsoil nutrient contents were again not measured but the authors concluded from indirect evidence that subsoil P was indeed being depleted. Topsoil exchangeable K-content remained adequate throughout the trial (even after 20 years according to the model), but subsoil K-content was getting dangerously low as well, affecting the growth of the hedgerows. The paper is silent about this. If correct, continuous alley cropping was leading to serious impoverishment of the subsoil which, after 12–15 years of cropping, would require a long fallow for replenishment of the nutrients from bound sources. That is, of course, unless fertiliser were applied, and, in the case of P, placed in the neighbourhood of the hedgerow roots, in the subsoil.

These presumed solutions can be studied in long-term field trials, of course, but such trials get quite expensive if you want to look at many different things at the same time, as researchers usually do. And, as also happens most of the time, after a while certain phenomena turn up which you had not even thought of and which are more important than the ones you did consider when designing the trial but can no longer be included. And if by that time the money is finished or the leadership has turned its attention somewhere else you may not even be able to set up a new trial to answer the new questions. So the conclusion, which you were probably already anticipating, is that research in alley cropping can only be done effectively with the help of some kind of modelling.

That is not to say that intelligent individuals cannot come up with the right kind of questions and even correct answers, without modelling. Just one interesting example of that in relation to alley cropping, or rather to trees as soil improvers. Gavin Gillman, who succeeded Tony Juo as IITA's soil chemist and who was himself an authority on the amendments for acid soils, argued that conditions in the lower strata of acid soils in the humid tropics were so poor that trees needed help to set them off on a flying start. So he advocated a technique which he had pioneered in Australia, using dust from ground basaltic rock as a soil amendment. If applied in sufficient quantity it had been shown in Queensland and elsewhere to increase the soil's adsorption complex and reduce P-fixation which is especially severe in the forest zone's old highly weathered soils, rich in iron and aluminium. Applying a massive amount of basaltic dust would be prohibitive for the African smallholder, but putting it just at the bottom of the plant hole, close to the roots of the tree seedlings, would reduce the cost. And once the trees were established and had formed a vigorous root system they might just overcome the harsh conditions outside the man-made soil niche. If it worked that would be just the thing we needed for those soils to help trees or hedges to get established, unlock the soil's native P and improve the efficiency of applied P. Without such amendments alley cropping might not even make sense because the hedges' roots would just remain in the topsoil and compete with the crops, rather than performing their most important role of bringing up nutrients from lower down. I do not know what became of this (Gillman left the institute after a few years), but it was one of the first fresh ideas since the invention of alley cropping itself. It is the sort of thing which advances science, not modelling, although modelling can help to explore the long-term implications of the new ideas. Hard core physical scientists often consider modelling as second rate, which it is in a sense, and they are not easy to convince that they should pay attention to it. Gillman for one fell asleep during a talk on modelling I gave at IITA in the early 1990s, although he said afterwards that the beginning and end of the talk had been quite good. I think the importance and power of modelling have been both underrated and overstated and I will spend an entire chapter on it (Chapter 9), in order to explore its potential and limitations, as well as to indulge in what I think is an exciting subject.

7.2.2 Zero tillage

Zero tillage is the second major technology pioneered at IITA to which I want to devote some time. As the name says, it means that the soil is not tilled at all, which may sound surprising because the first thing most people think of at the word farming is a plough with a horse or a tractor in front of it. It was not exactly a new idea, but in Africa it had never been tested, or at least never by agronomists. African farmers themselves do practise zero or minimum tillage if they can get away with it, for example, after clearing a fresh forest plot, because there are practically no weeds at that stage and the soil has a beautiful porous structure and does not need any tillage. Or they plant first and do the tillage and weeding later, as I described a little earlier for south-western Nigeria. But the agronomists' zero tillage usually means a production system where there is *never* any tillage, as a matter of principle. That is what was tested at IITA.

Some preliminary trials were carried out on small plots during the institute's earlier years, followed by a large-scale experiment, for which forest land was especially cleared in three different ways. One was to disturb the soil as little as possible by shaving off the small and medium-size trees at soil level with a bulldozer-mounted 'shear blade', cutting up the rubble and leaving it on top of the soils. Another method was the way it was done in plantations by knocking down the trees with heavy equipment and shoving them together with large rakes and leaving them in so-called wind rows. And finally there was the farmers' method: cutting down the trees, gathering the branches, burning them and scattering the ashes in the field. Then half of each plot was conventionally tilled by plough and harrow and the other half was not tilled at all. In the no-till plots the seeds were injected in the soil by a clever contraption, called a rolling injection planter, which had been designed especially for zero tillage. It was a wheel with equally spaced beaks (Figure 7-2) which opened a little hole, dropped the seed in and closed the hole again. The machine was pulled or pushed through the field or several units were mounted on the toolbar of a tractor. After harvest the crop's stover (that is the residue) was left lying on the soil surface and the planting machine could plant through the litter.

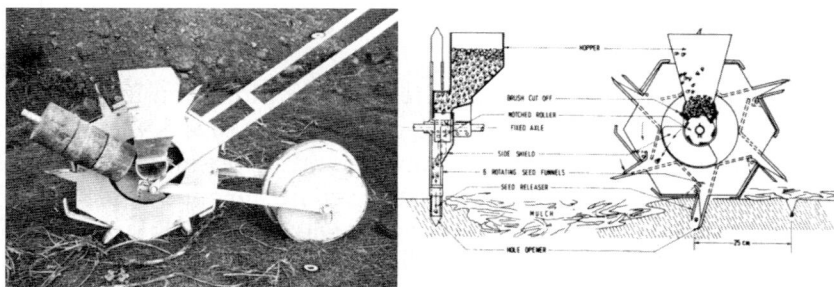

Figure 7-2. The IITA rolling injection planter. (From IITA Annual Report 1977; reproduced by permission of IITA, Ibadan.)

One of the reasons farmers till the soil is to get rid of weeds, or at least bury them so that the crop has a chance to make a head start. In zero tillage you obviously have to use another way to suppress the weeds, and that was by the use of herbicides.

But I have not told you yet why IITA thought that zero tillage would be a good idea for farming in Africa, although you may have started guessing what hypotheses were being tested in the experiment from the treatments I just described. I think the argument ran as follows. If you clear a forest patch and remove all the trees the soil becomes exposed to the deleterious effects of the flash storms which are common in the tropics. These destroy the soil structure and carry the topsoil away down any existing slope. Ploughing the land makes it worse and if it has also been cleared in a rough manner with tree pushers and then ploughed, the soil will really be messed up beyond repair. But if the vegetation is cleared cautiously and the soil is not tilled it will form a sponge-like structure with worms moving around and boring holes where the water can infiltrate. And the crop residues scattered in the field will prevent the little rivulets which will form during a heavy rainstorm to merge into a big water avalanche. So in summary, two major factors were studied in the long-term experiment. The first was the effect of land clearing, comparing the methods used for establishing a new rubber or oil palm plantation or a large commercial farm, the farmer's manual method and a method conceived by IITA's soil physicist Rattan Lal, who was the experiment's lead scientist: clearing with minimal soil disturbance by shear blade and conservation of the vegetation residue. The second factor was the tillage method, zero tillage or conventional tillage (conventional in the western sense) superimposed on the clearing methods. So each large land clearing plot was divided into two sub-plots, each with one of the tillage methods.

Of course this was a well-designed trial which had to be run for many years, because all kinds of measurements had to be carried out about how much soil was lost each year by erosion from each treatment and what happened to crop yield over the years. Lal's plots acquired some international fame, because it had become very rare to have well-designed and well-run long-term trials of considerable size in Africa. And the results were very interesting too. They showed how much soil could be lost from erosion by failing to protect the soil from direct impact of the rain, and that after some

years of growing grain crops like maize and cowpeas soil compaction would become a problem and the yield would go down
whatever the clearing or tillage method. But the question which to
my knowledge was never asked was who was expected actually to
use the results of these trials. IITA's work was supposed to benefit
small-scale producers and the institute's explicit aim was to find
alternative cropping systems which would allow the farmers to
practise permanent or semi-permanent cropping. Clearing forests
with tree pushers or shear blades was not exactly what farmers
would consider doing, so that part of the tests was apparently not
meant for them. Zero tillage then perhaps? Well, if they would only
grow grain crops, maybe, and if they could use herbicides. But the
crop combinations practised by farmers almost invariably included
cassava which formed an excellent combination with maize or
groundnuts but which was not really suitable for zero tillage,
because cassava has to be planted on heaps or ridges to get well-
formed roots. And since this combination provided excellent soil
cover for up to 2 years there was not much problem with erosion
anyway. Nobody seemed to be very much concerned about practical
application, except in a hypothetical sense of some ideal farmer
who might emerge at some future time, for whom all this would
turn out to be miraculously relevant. Or perhaps the target group
which was being served tacitly, were large-scale farmers in Brazil or
southern Africa for whom the work could be useful but for whose
benefit the institute had not been set up. And of course there was
the agronomists' own peer group who were doing similar things in
other enclaves elsewhere in the world and who were competing for
space in the research journals.

7.2.3 Live mulch

Live mulch was another really elegant concept. It meant that you
planted a kind of carpet of a non-aggressive creeping plant species,
preferably one which fixes N, and plant the crop through it. If that
would work it would be really spectacular. The live mulch keeps the
weeds away, it protects the soil when there is no crop and it fixes N
into the bargain. A species that does all that and at the same time
does not smother your crop and also survives when that crop is
growing vigorously would be a gift from heaven. It must have been
great fun to work on this technology. Its originator, Okeizi Akobundu,

collected many species of creeping legumes and tested them to see which ones would combine all the necessary features. But perhaps that was really an illusion. A species which is vigorous enough to suppress weeds would also tend to suppress the crop, while one that leaves the crop alone would probably not be vigorous enough to suppress weeds or it would be shaded out by the crop. But the agronomist's box of tricks is well stuffed, so you can spray some herbicide where the crop will be planted and when the legumes climb on the maize you slash them off or you may even try some growth regulators to keep them down. The work went on for many years and then started declining. In 2001, several years after Akobundu had left the institute, an IITA publication about herbaceous legumes for West Africa made no mention at all of live mulch.

To my surprise, however, when I was doing a consultancy job in Mali in 2005 I came across a technology promoted by CIRAD, the French international research organisation, which they called *semis direct sous couverture végétale*, abbreviated in the good French tradition as SCV. It embraced both live and dead mulch (the latter also a hot topic at IITA in the late 1970s and early 1980s) and had been undergoing testing in many countries since 1998. In southern Mali it was being tried with cotton. A very good idea indeed, and I was more than a little ashamed I had not noticed earlier that live mulch was being carried forward by the French. It was presented as a joint product of French intellect and Brazilian farmer skills, without any mention of IITA as the pioneer of the same ideas, more than 25 years earlier. But perhaps this was just another demonstration of IITA's inability to disseminate its products, or rather its reluctance to go out and mount a vigorous testing and promotion programme in the real world.

7.2.4 *In situ mulch and the control of speargrass*

In situ mulch was another important though less spectacular technology. It involved growing a leguminous cover crop, killing it with a weed killer or letting it die naturally, in situ, during the dry season and then planting a crop of maize through the litter in the next rainy season. It resembled green manuring which already had a long history in tropical agronomy, but this time the green manure crop was not ploughed into the soil but rather left on top of it. It served the dual purpose of contributing some N and perhaps a little organic matter and protecting the soil from erosion. The most successful leguminous species was *Mucuna pruriens* or velvet bean which was very vigorous but died off naturally during the dry season. There were some problems with the plant reseeding itself but I will leave that aside. The idea was good, but nothing much was done with it once it had been shown to work at the research station. Then the weed scientist, Okeizie Akobundu whom we have met in connection with live mulch got interested in mucuna because it might be able to suppress speargrass, an aggressive grassy weed of the humid and sub-humid zones (Figure 7-3). Speargrass was becoming a nuisance at IITA's doorstep, in farmers' fields just north of the station. Akobundu had been running some preliminary trials in farmers' fields (there was no speargrass at the IITA station itself) which caught the attention of Mark Versteeg. Versteeg worked with the national research organisation in Bénin to which he had been

Figure 7-3. Young suckers, underground stems (stolons) and flower of speargrass (*Imperata cylindrica*). (From www.invasive.org; reproduced by permission of L.M. Marsh, Florida Department of Agriculture and Consumer Services.)

seconded by IITA to help them develop an on-farm research programme. They had set up a team in the province of Mono in the south-west of the country, where farmers had serious problems with speargrass. Some fields were so badly infested that it was almost impossible to grow a crop at all. And if they did grow maize, the main crop in the area, the yield was pathetically low. So Versteeg and his Béninois colleagues argued that they needed a plant which was aggressive enough to suppress speargrass without itself becoming a nuisance. Mucuna looked like an excellent choice and they set out to test it with the farmers in their fields. Not by forcing it upon them, but by a circumspect approach which left it entirely up to the farmers to decide whether they wanted to try mucuna or something else. The way they did it was also interesting, but I will not go into that here.[4] Mucuna worked quite well in getting the speargrass down to a manageable level and in the end it became the single most successful technology introduced in Mono by the research team and was vigorously promoted for a while by the local extension organisation.

There is something magical about legumes; they have always attracted a wide range of people, from hard core scientists to believers in natural healing in the broadest sense, as a cure for many ills. I would like to tell you an edifying anecdote about that. In the early 1970s, just after I had joined the tropical crops department at Wageningen University, a small scandal developed in the Dutch press about an agronomist, G.F. van der Meulen, whose brilliant ideas were being ignored, presumably because he was not a member of the club, or so he thought. He had experimented for decades with cover crops to smother speargrass and improve soil fertility, first in the great plantations of the Dutch East Indies, starting as early as the 1920s, and later in Brazil, where he had been associated with a development project run by an NGO. But he had never published his ideas or results outside the gray literature of plantation records and advisory reports. The key to the control of speargrass, van der Meulen argued sensibly, was to suppress the grass for long enough so that its underground stolons would shrivel.[5] Those stolons are stuffed with reserves for the plant's survival. If you

---

[4] A handsome little booklet was written about the work in Mono province by Valentin Koudokpon (1992) and published by the Royal Tropical Institute in Amsterdam, from where it can probably still be obtained.
[5] Stolons are underground shoots from which the plants can regenerate.

plough up a field infested with speargrass, what you actually do is redistribute the weed by cutting its stolons into pieces and scattering them in the soil so that they can take off again with the next rains. Van der Meulen's method was similar to the one promoted later on by IITA but much more elaborate. First you had to plant a mixture of two legumes into the speargrass swath, an aggressive one like *Mimosa invisa*, or perhaps *Mucuna pruriens*, together with a species that can survive the dry season, for which he thought *Centrosema pubescens* was particularly suitable. The aggressive species will keep the speargrass suppressed during the rainy season, but when it declines or dies off in the dry season the grass will quickly recover because of its stolons, unless another species takes over. That is where the other legume comes in. It takes over from the *Mimosa* or *Mucuna* and survives the dry season, keeping speargrass down. Next, the field, which has been enriched with N fixed and P set free by the legumes, is turned over to *Eupatorium*, a species which we have already met at several occasions. If the conditions are right it forms a thick cover and rejuvenates the soil through its extensive root system and deals speargrass the final blow. I do not know how exactly van der Meulen managed to get the *Eupatorium* established through the remnants of the preceding legumes but never mind, there must have been a way. I think the whole thing was a rather clever succession of species playing precise roles at particular times and culminating in a climax vegetation of *Eupatorium*, which was not recognised as a potential soil improver in West Africa until much later. All this sounds plausible enough, so what was the uproar about? That is very interesting and had little or nothing to do with agronomy or the merits of the method, even though the dispute was triggered by claims about some properties of one of the plant species, *Centrosema*. The matter was that van der Meulen asserted that *Centrosema* could absorb moisture from the air and transport it downwards into the soil, keeping the soil under a *Centrosema* cover, moist during the dry season. That of course had met with a lot of scepticism among the men of learning. Van der Meulen had approached several aid agencies including the Dutch government with proposals to promote his method on a large scale but nobody seems to have been interested. So he and a University professor, I believe a civil engineer, who had joined his case went to the press and declared that the officialdom, incited by 'Wageningen', were boycotting his ideas. Wageningen, they said, just ridiculed van der Meulen's claims about *Centrosema* as those of a nutcase not worth listening to, the real reason being that he did not belong to the Wageningen mafia and therefore had to be marginalised. They were blocking the wide-scale application of a brilliant method which could revolutionalise tropical agriculture, or something to that effect. Their complaint was not entirely unfounded, there is always an innate reluctance in the establishment to listen to outsiders. But Professor Ferwerda of the tropical crops department at least did give a balanced seminar on the merits of the van der Meulen method and invited the man himself to do likewise, which he did. But his style was pretty awful and I do not think he made any converts. Even so, I think the department would have given support if a proposal for testing the approach had reached it, which in the end did not happen. The whole thing never rose above the level of squabbling and soon the interest died off without a trace. Nothing was ever published about the method and if you search for any reference today through the powerful Google machinery you find nothing. There are probably few people who even remember there ever was something called the 'van der Meulen method', in spite of its undoubted technical merits.

7.2.5 And more . . .

The most prestigious agronomic research done at IITA was that on alley cropping, zero tillage and live mulch. But there was also work of a more conventional type. Research agronomists often carry out applied research in support of the crop breeding programmes and some of the agronomy at IITA was in that category. It was concerned with finding the best planting density for new varieties, how much fertiliser they should be given, whether they tolerate late planting and things of that sort. The reason is that new varieties may respond differently to such management practices than existing ones do. If they were really ambitious the agronomists might study a crop's physiology to find the traits that make a variety successful and then tell the breeders what they should be looking for. The Root and Tuber Crop Improvement Programme in particular always had an agronomist or crop physiologist on its staff who worked on questions like what makes a cassava variety successful in mixed cropping with maize, or whether there were differences in photosynthetic rates between varieties. Although the issues may

have been pertinent, I think their complexity exceeded by far the competence of the researchers involved, and not only at IITA. With some
exceptions, which I will come to in the chapter on modelling, crop
physiology in support of breeding has been notoriously unsuccessful,
so I will say no more about it for now.

Most of the applied agronomy work at IITA related to breeding
was eminently forgettable. As I have argued repeatedly, research
stations in Africa were so different from a real farm that it made no
sense at all to do detailed studies on crop management under
conditions which only remotely resembled those of the farmer.
I think anyone, especially if he has not been trained in the agricultural sciences, would come to the conclusion that that kind of
routine agronomic work in Africa should be carried out in farmers'
fields, not in the research station. But logic is not the only and
certainly not the most powerful factor when it comes to making
such decisions. The other breeding programmes apparently did not
want to waste their resources on something as trivial as crop management agronomy and the little agronomic work they thought was
needed was usually done by the breeders themselves. They
employed other disciplines which were thought to be more relevant
for their crops. For instance an entomologist in the Grain Legume
Improvement Programme would screen new breeding material for
sensitivity to insect damage or test insecticide formulations or
spraying equipment, because cowpea, its major target crop, is
among the most sensitive in the world to insect damage, in the same
league as cotton. Much of the work on chemical pest control, however, was only slightly more relevant than routine agronomy: chemical insect control was only rarely carried out by farmers outside the
subsidised trials they hosted for the scientists, since they did not
have the means or the skills for such methods.

IITA's agronomic research which was not associated directly with
the Crop Improvement Programmes was done by the Farming
Systems Programme, later renamed more appropriately as Resource
and Crop Management Programme. We have already seen the work
on alley cropping, zero tillage and live mulch, all of them essentially
agronomy. Then there was back-up research carried out in support of
those technologies, most of it engineering. The engineers were also
based in the Resource and Crop Management Programme and
worked closely with the agronomists. They came up with all kinds of
clever devices which would increase the efficiency of the technologies,

for example a tractor-mounted bush cutter to prune the hedgerows in alley cropping. And the famous rolling injection planter which could plant crop seeds through the rubble which remained on top of the soil in zero tillage and in the in situ mulch system (Figure 7-2). From the perspective of the Institute's professed clientele, the small African farmer, their usefulness was conditioned by that of the technologies they were meant to support. And that, unfortunately, was not much.

A lot of other conventional agronomy work was done as well, like trials on fertiliser rates, biological, chemical and mechanical weed control, mixed cropping with different densities of the component crops, climatology, crop physiology and that kind of thing. A mildly interesting case was the use of melon as an intercrop in maize. The idea, which I think originated with the Nigerian Root Crop Research Institute, was that the intercrop would replace weeds while producing something useful at the same time. Quite sensible really, but not much different from other kinds of intercrops and the one which farmers in eastern Nigeria practised traditionally.

And then there was agroclimatology, the science of weather and climate and how they affect agricultural production. The climatologist's task included routine weather measurements but he would also be called into other scientists' trials to measure light penetration and air movements in the crop canopy and perhaps moisture patterns in the soil, to explain what happened to the crop, all of it of doubtful relevance. The Programme Director scathingly called the climatologist's job the Institute's biggest sinecure, but that reflected as badly on him as on the incumbent, I think. In the early 1990s, he was succeeded by a new type of climatologist, who launched the Institute into crop modelling, with results of equally doubtful relevance, as we will see in Chapter 9.

7.2.6 *Return of the legumes*

In the early 1990s, two of us in the RCMP, Georg Weber and I (but mainly Weber) wanted to make a fresh start trying out legumes as dual-purpose crops in farmers' fields. That work was somewhat more useful than most of the conventional agronomy had been, an opinion which I hope was not entirely biased by my own involvement in it. We thought that it was time to look again at legumes as 'auxiliary' crops, grown for the purpose of N fixation, soil protection, weed control, or all three of them. You will remember that a lot of work had been carried out with legumes in the colonial era and again 10 years later. It looks as if the interest in legumes follows a cyclical pattern, with each new generation of agriculturists and development workers getting excited by their as yet unfulfilled promises. Each generation also has its own explanation why it had not worked in the past but why it would this time round. And indeed, there was some justification for a repeat because soil degradation was more advanced now, at least so it was thought, and perhaps farmers would be more willing to invest in technology for soil improvement, rather than letting nature do the job, as the work in Bénin Republic with *mucuna* had shown. And we had learned that growing a legume only for the sake of its beneficial effect on the next crop was hard to sell, so perhaps if it also produced some edible or otherwise useful grain for humans or animals, then that would help and hence the word 'dual purpose'. This time most of the work had to be carried out at the real farm, we did not want to waste more time at the station with an essentially simple technology which farmers could perfectly well handle if they wanted to. In that way the scientists would get first-hand information about adoption or the reasons why it still did not take place. Finally, there was the computer. Georg Weber developed a computer program to help the user choose legumes which best fitted his needs, a so-called expert system. The program was named LEXSYS and Weber and I went to work with some national research teams to test it. As it happened, both of us left the institute before LEXSYS was fully operational and it was taken over by another agronomist, Robert Carsky (1998, 2001), who took it to the IITA substation in Cotonou in Bénin Republic when he was transferred there. Some more development was done there but I think the program never reached the stage where it became what we hoped it would: a practical back-up tool which could help on-farm researchers and extension agents to choose the best possible legume in dialogue with farmers. Instead of setting up a vigorous participatory on-farm testing program around LEXSYS, the agronomists who took over from us reverted back to researcher-controlled trials, carried out in farmers' fields this time, to rediscover things which had been found out countless times before or were in no great need of being found out anyway.[6]

---

[6] For example, like the N contribution of soybeans to following maize crop, the effect of P applied to a legume cover on the following maize, the effect of soybean on *Striga* infestation in maize.

7.2.7 *Were the new technologies actually adopted?*

Up to the mid-1980s the new technologies invented at IITA remained very much confined to the institute's experimental farms. When it came to testing them in farmers' fields the scientists were surprisingly conservative. There was always some more work to be done before the technology was ready for application. So, come back in a few years' time and then surely we can go on-farm. Perhaps this is even too much credit. In fact, I wonder whether the interest of most station scientists, with the exception of the breeders, went much beyond the technology itself as a technical and biological challenge, rather than something that had to be transferred to real farmers' conditions as soon as possible. Uncountable publications, presentations at international conferences and M.Sc. and Ph.D. theses have been written about minimum tillage, alley cropping and live mulch, answering evermore detailed questions which originated mainly in the scientists' own minds, assisted by imaginary conversations with some phantom African farmers. It may be legitimate for basic research to set its own targets without concern about application, but this was supposed to be research with immediate relevance for farmers. The technologies eventually did leave the IITA gate, but mainly to be repeated in other research stations run by national scientists in various countries.

During the latter part of the 1980s, some of us took alley cropping to the real farm, only to find out that the precise management which was needed was simply not applied. When a farmer failed to prune the hedgerows in time, which he was almost certain to at one time or another, his maize yield would tumble. Root competition would also be more serious than it was at the station where the roots of the hedgerows were forced down by intensive tillage between the rows. And if a farmer would leave a leucena alley field uncropped for some time the hedgerows would seed abundantly and the seedlings became very hard to uproot if left to grow for a year or so. Very soon the farmer would try to get rid of the alleys altogether, giving all kinds of funny explanations why he had tried to set fire to the hedges when a researcher stopped by. Minimum tillage and live mulch were never tested on farm at all, except perhaps by some small development projects which were interested in the ideas and tried them out in their own corner.

Over the years rather ludicrous claims were made in the IITA Annual Reports (and in those of other institutes as well) about

imminent breakthroughs, without any foundation in reality. Such
claims were challenged by no one, in spite of the annual ritual of
the Board of Trustees meetings and the quinquennial programme
reviews by high calibre international scientists. Perhaps they were
all fooled by the technical and biological merits these technologies
undoubtedly had, and by the scientists' assurances, repeated year
after year, that a breakthrough was around the corner. Minimum
tillage and alley cropping were practised only at the highly mechanised IITA farm, and with considerable success for that matter. But
they were entirely out of reach for the professed target groups of
small African producers and made no contribution whatsoever at
the real farm. They were cycled practically entirely around national
and international research stations and when one of them, alley
cropping, was taken out to the real farm it failed. By the mid-1990s,
when all the scientists associated with alley cropping, zero tillage
and live mulch had left the institute, reference to these technologies
dropped almost completely from the institute's publications.

Like most of the agronomists, the engineers at IITA also worked
for an imaginary future farmer who would run his farm as a business. I have already mentioned the equipment they designed to
facilitate zero tillage and alley cropping, but that was not the only
thing they did. Even though modern commercial farmers were
nowhere in sight in West Africa, they went on merrily cranking out
a stream of gadgets which those farmers were going to need. The
apotheosis of this work was the motorised so-called farmmobile
which could do about everything from chiselling the soil, planting
the crop, hauling it out of the field and taking it to the market. A lot
of the equipment just circulated in the virtual world of research
institutes and foreign-funded development projects without ever
entering the real world of the West African farmer. Until most of
this engineering work halted in the early 1990s.

Speargrass control by Mucuna fared a lot better. It was taken up
by a dedicated and effective on-farm research team in Bénin
Republic who offered it to farmers in an area where speargrass
was a very serious problem, threatening the farmers' entire crop
production system. Once they had shown that it could work
mucuna was taken up by the government's extension service and the
Non-Government Organisation Sasakawa Global 2000. The *mucuna*
story shows the power of good, genuinely participatory on-farm
research. In the late 1990s, an impact study was carried out which

estimated that 14,000 farmers had used or were using *mucuna*. That is impressive and it would be interesting to look again now, almost 10 years hence, to find out whether it is still there today. I am a little worried, though. There were several unresolved problems, like the sensitivity of dry *mucuna* to bush fire, the fact that the grain was barely edible for humans and the lack of an assured seed supply. Perhaps the surest sign that all is not well with a technology is when farmers continue to complain that they 'lost the seed' or the planting material or some other essential component, and cannot get fresh supply. Except in war situations I have never heard farmers complain that they lost the seed of crops which were important to them. But when scientists or extension workers desperately want to believe that their technology is appreciated by the farmer they will just swallow that kind of nonsense and come back every year with a fresh supply. In Bénin complaints continued to be heard about *mucuna* seed, even after the farmers had had enough time to learn to produce their own. Hence, again, I would be curious to find out what happened in the case of *mucuna* after the extension stopped supplying seed.

7.2.8 How to enhance the chances of adoption: an example

The story of legumes as 'auxiliary' crops is a good example to demonstrate where things can go wrong when it comes to translating a good idea into something that farmers can and will adopt, and how perhaps it can be well done. But it is always easier to know what was wrong than to do what is right. Nobody can fail to be impressed by the promise of legumes as cover crops and soil improvers and generations of agronomists, development workers and eco-freaks have tried to convince peasants in Africa of the same, so far with very little success. And every time the interest waned again without any explanation as to why it did not work. Agronomists have preferred to run their interesting trials with legumes on their stations, adding infinitesimally to the already vast body of knowledge, rather than doing the obvious: finding out why farmers will not adopt them. That is because most agronomists like working with plants and cropping techniques better than working with farmers. That is all right, as long as there are others who work with the farmers and tell the researchers what is wrong with their

technology and how it can be improved. But the extension organisations in Africa, who should do that, have been notoriously ineffective and the record of the non-governmental organisations which
have sprung up like mushrooms everywhere has not been much better.
In fact that was the reason why the Farming Systems movement
emerged. We were going to work directly with the farmers, pulling
along the station scientists and the extension agents and knocking
some sense into their heads. Not very successfully, though. I will come
to that in the next chapter, so let us restrict ourselves to the legume
story. A good step forward in the promotion of legumes was the
LEXSYS computer package. It forced the scientists and the extension
agents to think about the kind of legume that would be most suitable
for the farmers' existing production system and whether it would be
profitable for them to put one there. But LEXSYS could not do the
most important thing: starting a dialogue with the farmers to see what
they thought about it and whether they wanted to give it a try.
LEXSYS only helped to make the scientists better prepared for
the dialogue. And it also satisfied the scientists' justifiable taste for
sophistication while still putting application by the farmer first.

But in the end we did not get very far with the package. Perhaps
we did not put enough effort into it or we were simply naive, expecting that the tool would find its way to the user anyhow. And
legumes alone were perhaps too little to build a programme around.
So Weber and I tried to convince IITA that we should set up a support facility to help national research teams do better on-farm
research. The work on legumes, aided by LEXSYS, would fit well
into that, since legumes could be an interesting option in many
ecologies. We even coined a name for the support facility: Support
Group for Adaptive Research Cooperation (SPARC). But IITA was
not really interested in working with national teams in this way.
Management opted for an entirely different orientation following
the trend of the times, biotechnology and other sophisticated
research for which national institutes were thought to lack the
capacity. The LEXSYS package was eventually taken over by the
Forestry and Agroforestry group at the University of Bangor in
Wales from whose ftp site it could be downloaded, free of charge.[7]

---

[7] The ftp site for downloading (June 2006): ftp://ftp.bangor.ac.uk/pub/
departments/af/LEXSYS/

7.3 Plant breeders' technology: Crop varieties

Plant breeding was a different matter entirely. New crop varieties
have undeniably had the greatest impact on agricultural production
of all technologies contributed by scientists, both in developed and
underdeveloped countries and whether for peasant or commercial
agriculture. Being a good breeder and seeing one's varieties spreading on the sole basis of their merit must be very gratifying. Neil
Fisher, an accomplished agronomist whom I have introduced
earlier (in Chapter 4) and whom we will meet again, once confessed
his regrets not to have chosen to be a breeder and actually produce
something tangible and useful.

Plant breeding mimics the way farmers historically have picked
interesting specimens from their crops and continued to grow those
that suited them best. And crossing different specimens to increase
the variation to choose from, as breeders do, is not much different
from the way nature itself creates interesting variation from which
you can select. Breeding was just a clever way to speed up a process
that is as old as farming itself. Maybe that is why it has been so successful. But there is a limit to what you can achieve by just making
crosses between individuals from the same species, so the breeders
have invented new tricks. If you want to add something really new
you may have to look outside the species to find it. An example is
resistance to mosaic virus in sugarcane and cassava. The key to
success in both species was finding resistance in related species and
then getting around the natural barriers against inter-species crosses
to combine the resistance of one with the economic quality of the
other. The barriers were not absolute in this case: the agricultural
species could be crossed fairly easily with their cousins and such
crosses even occurred in nature. Remember from Chapter 2 the
Kassoer cane of Indonesia which was a natural cross between noble
cane, *Saccharum officinale* and the grass *Saccharum spontaneum*. In
cassava something similar was the case (Hahn et al., 1980). The economic species *Manihot esculenta* crosses naturally with tree cassava,
Manihot glaziovii which is resistant to cassava mosaic and the breeders continued to backcross the hybrid with *M. esculenta*, until they
got plants which combined tuber quality with mosaic resistance. The
actual breeding process was a little more complicated than that, but
this is essentially what happened. The story is very similar to that of
sugar cane, for one thing because both are vegetatively propagated.

Two types of breeding work have contributed most: breeding for resistance to important pests and diseases and breeding for high yield. For the first category I have already mentioned resistance to mosaic viruses in sugarcane and cassava. IITA also bred maize varieties resistant to the maize streak virus and plantain varieties resistant to black sigatoka disease. Both programmes were honoured with the King Baudouin prize for international agricultural research. The attractive thing about pest and disease resistant varieties is that they are usually robust: if they are good at all they will be good whatever the setting they are used in. And farmers can multiply their own seed or planting material without loss of the resistance traits. Unless the resistance breaks down, which is another matter. That is why some breeders' varieties, especially of cassava and maize have made such inroads in peasant agriculture in Africa.

7.3.1 Breeding for high yield

In the category of high yield potential we have hybrid maize and the HYV of rice and wheat which triggered the green revolution in Asia. The interesting thing is that their creation was based in each case on a single clever concept: heterosis for hybrid maize and short stiff straw for the others. It is worth looking a little closer at that.

First heterosis, also known as hybrid vigour. It had been known for a long time that when two lines of a normally cross-pollinating species are repeatedly selfed and the inbred lines, once stable,[8] are crossed, their progeny may show unexpected vigour in addition to a high degree of uniformity. This phenomenon has been used with considerable success in maize where the modern varieties used in industrialised and some developing countries are all hybrids. Their drawbacks for peasant agriculture are the need for high rates of fertiliser to bring their superior potential into expression and the need to purchase new seed every year, because in the second generation much of the vigour and all of the uniformity are lost. I will say no more about hybrid varieties here, but come back to them in later chapters.

---

[8] In each inbreeding generation the undesirable progeny is eliminated.

The story of the high yielding so-called semi-dwarf varieties of wheat and rice is even more interesting and their impact has been much greater for smallholder farming in the tropics (I will only occasionally yield to the bad habit of calling the varieties HYV, but that is how they are commonly referred to). The key to their success was their short stiff straw and short erect leaf blades. That was important for three reasons. First, by having their leaves mainly pointing upwards the sunlight could penetrate deeply in the canopy and canopy photosynthesis was high. Second, since a smaller proportion of the biomass went into the straw, more grain was produced for the same amount of biomass. Third, and probably most important, the shorter and stiffer the straw, the more fertiliser the plants could take up without becoming too lush and collapsing under the weight of their panicles. Norman Borlaug, the later Nobel laureate, started work on semi-dwarf wheat in Mexico in the 1950s and by the 1960s his varieties had revolutionised wheat production, especially in India, a country which had been believed by many to be on the road to starvation on an apocalyptic scale. IRRI in the Philippines followed suit in the 1960s and came up with a (continuing) series of very successful short-straw rice varieties. Figure 7-4 shows their first variety of global fame, IR8, with its two parents. The IRRI varieties did the same for irrigated rice as their

Figure 7-4. The most famous IRRI paddy variety IR8, together with its two parents. (Reproduced by permission of the International Rice Research Institute.)

counterparts had done for wheat: revolutionising cereal production in formerly half-starving countries like India and Indonesia.

7.3.2 What comes next?

Here are just a few words about modern developments in plant breeding. First, there is of course the promise of genetic engineering, which is producing varieties with traits that had been out of reach for breeders until quite recently, because they simply did not occur within the botanical species to which the crops belonged or within related species with which they could be crossed. Genetic engineering has enabled the breeders, or rather the genetic engineers, to isolate genes from completely unrelated organisms and transfer them into the crops. In that way crop varieties have been bred with resistance to important pests or to weed killers like the omnipresent glyphosate (better known by the trade name Roundup), so that the stuff can be sprayed at any time without affecting the crop. There is even talk of breeding rice varieties with the capacity to engage in symbiosis with N-fixing bacteria, as legumes do, to enhance the crop's yield when farmers cannot afford to apply high rates of fertiliser. I am not sure how realistic the last example is, but the rise of genetic engineering is quite certain to have a huge impact, in temperate as well as tropical countries, provided consumer resistance (much of it metaphysical) to genetically modified organisms, or GMOs as they are now commonly known, can be overcome.

The second development I would like to say a few words about is the attempt to further boost the already very high yield potential of rice. The quantum jump in yield potential which led directly to the green revolution of the 1960s and 1970s had come from a stroke of genius on the part of some breeders who picked exactly the right combination of traits to break through the yield barrier. In the 1980s rice breeders were desperately looking for ways to repeat that feat. They felt that a new breakthrough was needed because populations in Asia kept growing at a high rate while agricultural land area remained stable or even went down, as in China where it was converted into urban settlements. So yields had to go up substantially if the teeming millions were to be adequately fed in the future. This time the breeders did not have a single or a small number of traits in mind for the new varieties, as they did in the earlier days, but rather a whole complex of traits which together was called an 'ideotype', a term invented in the 1960s by an Australian breeder, C.M. Donald. In fact, the combination of traits of the green revolution varieties also formed a quite simple ideotype, and a very successful one. The new ideotype of rice the breeders were trying to design since 1989 was called, not very imaginatively perhaps, the 'New Plant Type' (NPT). Not much progress was made in the early years, but when I visited IRRI in 1998 there was a lot of renewed excitement about the NPT, because the breeders thought they were now on the right track by using some hitherto unexploited genetic resources from Indonesia. The new generation of varieties with the ideal make-up for superior yield potential had already been given a name: VHYV, for Very High Yielding Varieties, the successors to the HYV. In 2004 breeders at IRRI and in China reported that their best lines outyielded the best HYV by up to 30% in their trials.[9] If those claims hold good that would qualify as another major breakthrough, perhaps the last one as far as yield is concerned, because yield was now approaching the physical ceiling imposed by solar radiation, as we will see in Chapter 9. Unless genetic engineering succeeds in radically redesigning the crop's photosynthetic apparatus, converting it into one similar to that of maize[10] and putting its yield potential in the same class was one of the longer-term objectives of IRRI's rice improvement programme. I cannot judge how realistic that was, only time will tell.

Meanwhile the other international institutes had not been idle. The Africa Rice Centre WARDA, now in Bénin Republic has made a breakthrough of its own by crossing the Asian rice species *Oryza sativa* with the African species *Oryza glaberrima*, thereby opening up entirely new vistas because of the enormously increased range of properties now accessible for incorporation into a new generation of rice varieties, especially for African upland conditions (remember that upland rice is a rain-fed dry land crop). And of course these varieties have a name: *Nerica*, for new rice for Africa. You may wonder why it took so long for the breeders to start crossing these two species, after all crossing anything crossable comes naturally to breeders. The problem (challenge for Americans) was that

---

[9] Virk et al., 2004.
[10] Rice has a so-called C_3 photosynthetic pathway while maize has a C_4 pathway.

the species were very hard to cross, and if it succeeded the young embryos would abort. Furthermore, in the few successful crosses the offspring would be mostly sterile. Those were formidable problems which stood in the way of hybridisation until in the 1980s new techniques were invented to get around the incompatibility. One was embryo rescue, whereby a young embryo was dissected from the ovary and transferred to a test tube. There was much more to the creation of Nericas, like the use of molecular markers and anther culture, but embryo rescue meant the first major breakthrough. If you are interested in the details, consult a modern text on breeding or biotechnology. The new techniques used in the creation of the Nericas were biotechnology, but not genetic engineering, because the changes in the genome resulted from 'natural' genetic exchange processes, aided by manipulative techniques. Nerica varieties have spread rapidly throughout Africa and they seem to be making a real difference for the productivity of upland rice, a typical poor man's crop.

So, improved (non-hybrid) varieties have done a lot of good in Africa and the beautiful thing about them is that once they are in the hands of the farmers they can be maintained without much cost, although farmers would be well-advised to refresh their seed stock every few years. And the end of the breeding saga has certainly not been reached yet. But there is a limit even to what breeders can do. However good a new variety may be, the full expression of its potential will always be constrained by other factors, natural, societal or managerial, which breeders have no influence on.

7.4 Pest and disease control

I am not going to say much about plant diseases, where plant breeders have usually managed to stay ahead of the disease organisms by turning out new varieties more rapidly than the organisms could change their genetic make-up. But I must say a few words about some of the technologies developed by scientists to control insect pests, obviously a permanent concern of farmers and scientists alike.

Since the Second World War the chemists have always been able to find new powerful chemicals to kill the insects which had become resistant to the previous ones, until concern about the environmental costs of chemical pest control resulted in ever stricter rules imposed on its use. Scientists have therefore increasingly opted for non-chemical control methods, especially those working in and for tropical agriculture, where the cost of chemical control also weighed heavily on the crop budget. In most of Africa non-chemical methods were even the only feasible ones because the cost of chemical control was prohibitive, except in crops like cotton. I will briefly describe two of the most successful non-chemical methods here.

7.4.1 Biological control

Biological control lets nature do most or all of the work to keep harmful insect at a low level. The most common form of biological control is the use of one organism, the predator, to control another, the pest organism, by pushing down the latter's population to a harmless or at least a less harmful level. In nature that happens all by itself because there is usually a fine balance between plants, the organisms which attack them and their predators. But crops are not really part of nature, they are highly artificial elements which cannot survive on their own – they would quickly disappear if the farmers were not around to protect them. That applies a fortiori to crop species which originated elsewhere, cassava for example, which was imported into Africa from South America, originally without the pest–predator complex associated with it. When a cassava pest from the crop's area of origin manages to invade its 'new' area of colonisation there are no natural enemies to keep it under control the way it happens back home, and disaster will result. That is what happened with the cassava mealy bug (and the green mite) which came into Africa in the 1970s. IITA scientists therefore went to South America to scout for natural enemies and, to cut a long story short, eventually found a highly successful one, a small wasp, and introduced it into Africa where it managed to cause a significant reduction in mealy bug populations. And another good thing is that agriculturists did not have to be concerned about adoption because the predator spread merrily without needing participatory methods, just a sufficient density of release sites. There have been other successes of a lesser magnitude with biological pest control in Africa, which I will leave aside. Watch out for more achievements in this area in the future.

7.4.2 Integrated pest management

IPM was another post-war breakthrough, pioneered in California cotton in the 1960s in response to excessive and ever-increasing pesticide use. In some central American countries at the height of the chemical control era in the 1970s, up to weekly sprays with highly poisonous chemicals were needed to keep an increasing number of pests under control. You may remember my story about cotton and paddy in Indonesia in the late 1960s which described the same trends. IPM was invented in the 1960s, if that is the word, to drastically reduce pesticide use and re-establish at least a partial balance between the pests and their natural enemies. That balance had been thoroughly disturbed by killing predators along with the pests, resulting in outbreaks of attacks by insects and mites which had previously been effectively controlled by their enemies.

IPM was conceptually simple, as is often the case with brilliant ideas, which does not mean that it is easy to implement. Four essential ingredients were the establishment of economic thresholds for the major pests, below which no control measures should be taken, parsimonious use of selective insecticides, that is insecticides with a minimum of collateral damage to beneficial insects, the use of pest resistant or tolerant varieties, and finally stimulating the natural enemies as much as possible. Here are just a few words about each of them.

Economic thresholds. Much field research was needed to set the economic thresholds and then translate them into observable quantities, such as *if there are fewer than x active caterpillars per plant on average, then there is no need for spraying.* The method had to be simple enough to be handled by farmers or perhaps by professional scouts, armed with some training and simple scoring boards to record their observations. Insect scouting has been vigorously promoted in cotton and paddy and it has probably been the single most effective method to reduce pesticide use, but I doubt that it has thus far had much success in Africa.

Selective insecticides. Ideally, one should use insecticides which only kill the target pest organism, but that is an illusion. In the last 30 years, however, there has been much progress in the use of more selective insecticides. Some of them were so-called botanicals of long standing, such as *derris, neem* and *pyrethrum*, others were of recent development, such as less toxic synthetic chemicals. Most interesting was the use of suspensions of the bacteria *Bacillus thuringiensis* which produces a substance which is toxic to a range of insects. In the industrialised countries the trend towards less toxic substances has been much strengthened by bans on a lot of earlier very toxic and persistent insecticides.

Resistant varieties. The search for resistant varieties has been strongly boosted by modern developments in genetic engineering. The earliest and most spectacular achievement was the introduction into the cotton and maize genomes of the Bt gene from *Bacillus thuringiensis* which codes for the bacteria's toxin. There is no doubt that insects will in due time also develop resistance to such toxins (as they already have for Bt), but genetic engineering will speed up the breeders in their race with the pests. This is definitely going to be a hot area in the future, in spite of public opposition in some circles, especially in Europe.

Stimulating beneficial insects. The most sophisticated way to stimulate beneficial insects is increasing their numbers through regular releases, which can only be done by well-organised research and extension organisations. There are less spectacular methods which are within reach of farmers in less sophisticated societies. An interesting example is sticking perches in paddy fields from where birds can catch the moths of the stem borers, which has had some success in Bangladesh.

There are many other possibilities for IPM. Some are sophisticated, such as genetic engineering of the pests themselves, others are simple, like growing trap crops and killing the insects there, growing daisies to reduce nematode populations, or even sending out the kids to collect egg masses from the leaves. Obviously, in view of its variety, there is no way IPM can be couched in general prescriptive terms, beyond a broad definition of principles, illustrated by examples, as I have done here. Specific IPM packages have to be worked out for each crop and for each area, which obviously demands great skills on the part of the farmers and the extension personnel. No wonder the method has been most successfully applied where these skills were well developed, in particular in Asia's intensive paddy-production areas. And it is wise to associate the farmers directly with the development of local IPM methods, a method pioneered by FAO through its Farmer Field Schools.

7.5 Has agronomic research been useful for the African farmer?

I think we can safely say that plant breeding and biological pest control have been most effective in tropical farming, including in Africa, with IPM as a third, except in Africa. But what about agronomy? I am afraid agronomy, *sensu stricto*, has produced preciously little that has been of much use to the African farmer. Of course, a wealth of information was gathered about biological, chemical and physical processes, all buried in international journals and at best taught in University courses in the USA and Europe, even in Africa, but the return to small-scale farming from decades of expensive international research in agronomy has been next to zero. For scientists there were three possible responses to this dismal situation.

The first was to simply ignore the fact that the technology was not being applied in the real world, or not even asking oneself any questions in that respect. In pure science sovereign indifference about application has even been more or less the standard attitude and it must be said that it has been vindicated over and over again by the fact that most breakthroughs in basic science have ultimately found application in most unexpected ways, especially in physics. But agronomy is typically an applied science whose explicit goal is not to explore the scientific frontier but to come up with something useful for farmers. That is what the international institutes' sponsors were paying for, so this option would not be acceptable for its scientists.

The second option was to admit that research had to come up with something useful, but to define an imaginary beneficiary who would emerge later and who was going to be very happy to find out that all these technologies already waiting on the shelf. I think that was the attitude of several of my IITA colleagues. One of them, when I asked him whether the zero adoption record of the Institute's agronomic technologies did not worry him, answered that the African farmer was not ready for what he was doing and that he was quite happy to contribute to the body of knowledge which would turn out to be very useful later on, either in Africa or somewhere else. Although the sponsors, if they had known, would not have been particularly pleased, the expectation was not entirely unjustified. Take zero tillage. Twenty-five years have elapsed since it was first proposed, without any impact in African smallholder agriculture, but some large farmers and companies in southern Africa are now coming around to see its merits and have started using, if not zero tillage , at least minimum tillage, using so-called ripper tines to open up narrow planting furrows and leaving the rest of the land untilled. And alley cropping, or alley farming as it is now commonly called, though it has turned out to be a failure with African peasants, is still attracting a lot of attention in other continents, including south-east Asia (Philippines) and even Australia.

The reasons why scientists have felt justified in continuing to develop technologies which did not work in Africa are many. I will come back to them in some detail in following chapters. At this point I will just sketch a brief historical perspective. As I have argued repeatedly before, the prevailing vision of pre- and post-independence agriculturists was that African peasant farming had to evolve rapidly towards semi-commercial family farms, as it had done in Europe. Those farms would need all kinds of support services which were going to be provided by strong government, farmer-led cooperatives and an efficient private sector. Practically all early development projects worked from those premises and made efforts to create conditions which were conducive to a rapid transformation of agricultural production. And it was the role of research to develop the technologies which were going to boost the farms' productivity. But while the ideal receded with every step taken towards it the agronomists, rather than admitting their mistake, continued to work for a farmer who was increasingly becoming a figment of the imagination.

There was another way for the agronomists to improve the chances that their work would be useful to the real rather than to some imaginary farmer. That is what the visionaries of the 1970s had in mind when they started advocating FSR – working for, or rather with, today's farmers with all their limitations and those of the socio-economic and institutional environment they face. If more of us had indeed chosen that road, our work might have had more visible impact where it was meant to: in the real African farmer's fields.

Chapter 8. Follies and Sanity of Farming Systems Research

I will now pick up the story of FSR at the point where I left it in Chapter 4.

FSR was the first movement in the post-colonial era which argued that research should support the development of existing peasant agriculture, rather than trying to replace it as soon as possible by farming models from the west. It has had an enormous influence, not only on agricultural research but on agricultural development generally and, through its offshoots, on practically every other branch of development as well. So it is worth further examining how it developed and eventually became corrupted, in spite of its essential sanity.

In the 1970s and 1980s a number of enthusiastic people rallied around the FSR ideas and put their considerable forces to work to promote them. In quite a short time FSR became something of a mass movement and FSR teams were set up in many countries, mostly with technical and financial support from foreign donors. Soon, however, there were indications that the ideas were not always understood by the neophytes at a more than trivial level. Many of them, though able to utter all the fashionable phrases and formulas, did not make the rather profound mental change which was necessary to grasp the essence of FSR and apply it effectively. In spite of its undeniable logic FSR, couched in formal methodology by the best of minds and generously supported with a lot of taxpayers' money, remained thin when it came to concrete achievements. The movement's leadership responded by tirelessly generating new methodology which they thought would be a substitute for true understanding. But mindless application of increasingly ritualistic methodology has, more than anything, polluted FSR's healthy principles and eventually brought it to the verge of insignificance.

That is a sad conclusion. Why did such a good idea come to so little in the end? Even brilliant ideas may of course get corrupted beyond salvation if they are appropriated by large numbers of people without the necessary minimum understanding. That is not an inevitable process, though. The FSR concepts are as valid as ever and, ironically, they have been applied with more success in developed than in developing countries. So it is worth examining what happened to FSR in Africa and what exactly went wrong. Perhaps it can open the eyes of a new generation which is now deluding itself with the same quasi-scientific rituals FSR has succumbed to, instead of staying with both feet on the ground. The International Farming Systems Association (IFSA) cheerfully held its eighteenth global meeting in November 2005, so there must still have been quite a number of people around who had not given up. That could be because people do not want to admit that the plug should be pulled, but it is more likely that the idea simply refuses to die because of its vitality in spite of abuse by its practitioners.

8.1 The pathology of diagnosis

Science is reductionist, and breaking down complex phenomena into manageable parts has been very effective in exploring the laws of nature and to a lesser extent those of the human condition. That is why there are different scientific disciplines, sub-disciplines and sub-sub-disciplines and today few people can claim to completely master even their own discipline, let alone have more than cursory knowledge of adjoining ones – or of the whole for that matter. The agricultural sciences are no exception and when a scientist looks at a farming system he[1] will most clearly see that part which he is most familiar with. Nor is that all bad; some disciplines have been very successful in applying the reductionist approach to the solution of difficult problems, or the creation of new technologies such as more productive crop varieties. But the reductionist approach seems to have been less successful in Africa than in other continents. The early FSR scientists thought that this was because in Africa farming was a much more complex affair than in the industrialised countries and that by looking at just one element in isolation, like soil fertility or the farmer's crop varieties, you would lose sight of the whole and fail to see how changing one element would affect the rest of the system. And if you did not take all those interactions into account you were unlikely to find something that would be really useful for the farmers. FSR therefore wanted to look at farming as

---

[1] Remember that 'he' stands for 'he/she', unless stated otherwise.

an integrated system, the way farmers themselves experienced it, rather than treating it as a collection of different disciplines. In other words, FSR was 'holistic' and interdisciplinary. But stating the need for a holistic approach is one thing, applying it in a fruitful way is another. There were no established methods to do that, so they had to be invented first. I have already introduced some FSR methods in Chapter 4, but I will review the field a little more systematically here.

The methodology developed by the early FSR workers was based on a conceptual framework like the one shown in Figure 8-1. FSR scientists are very fond of flowcharts. We have that in common with

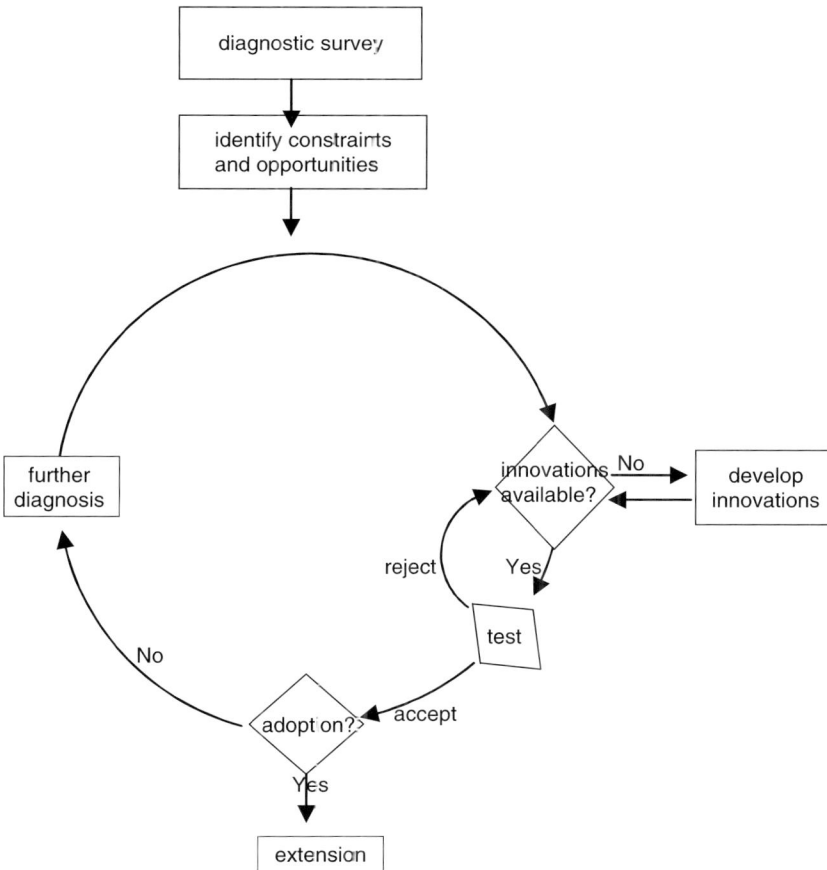

Figure 8-1. Flowchart of farming systems research

other system thinkers like computer modellers. The chart shows that
diagnosis plays an important role: first to learn enough about the
farming system to be able to start the real work – choose innovations,
carry out on-farm tests and interpret their results. And later on, after
on-farm testing has started, more diagnosis is needed to find out
more about the farmers and their farming practices, correct earlier
misconceptions and try to understand why things do not work the
way they were expected to. Nobody can deny that all that makes a lot
of sense. It is not really different from the way applied research[2]
works in other areas, whether reductionist or holistic. So there was
nothing wrong with the FSR concepts, but when they were converted
into activities the trouble started. One of the first symptoms that
things were not going well was the pathological development of
'diagnosis', so let us look a little more closely at that.

8.1.1 What are diagnostic surveys for?

Diagnosis is meant to find out why things are the way they are,
whether there is room for improvement, and if so, what needs to be
done to bring that about. That sounds simple enough, but of course
it is not. When two people, say a soil scientist and a sociologist, talk
to the same farmer they will come back with different conclusions,
each according to his discipline. That is inevitable when they visit
the farm on their own, so FSR wanted people from different disciplines to go to the farm together and come back with a shared
assessment of the farm or farming system. That would help to
design an intervention programme which addressed the farmers'
real concerns and took into account the complex interactions which
were at work in the farm. We had to look at the farm or the farming system 'as a whole', not just at its constituent parts, like the soil,
the weeds, the animals or the farmers themselves. So we formed
multidisciplinary teams who would meet farmers together and come
back with a consensus about what the *leverage points* were (the term
is due to that prolific inventor of FSR terminology, Michael
Collinson) to move the farming system in the right direction. That

---

[2] I am using the term 'applied research' here to denote all research which aims at
the application of innovations, which often involves some kind of adaptation to
the users' conditions.

did not really help. The soil scientist still wanted to do something about soil fertility, the economist about marketing and credit, the agronomist about weed control, while the sociologist wanted to conduct studies about *intra-household decision making*. But never mind, you cannot change human nature by one diagnostic survey. The important thing was that they would start doing some on-farm trials as soon as possible, because that would force them to work with real farmers and to stop fooling themselves. They were bound to find out soon enough what was really important and once they reached that point they might start doing something really useful, or so we thought. But we underestimated mankind's inclination to avoid hard work and keep busy with trivialities of a high feel-good content instead, such as working on the further improvement of diagnostic methodology.

In the 1980s diagnostic methodology was known under different names, because different schools of thought felt they had to put on their own label, even though they were all doing practically the same thing. So there were *diagnostic surveys* (DS), *exploratory surveys* (ES), *sondeos*, *rapid rural appraisals* and probably a few more. In the end *rapid rural appraisals* carried the day and became widely known as RRA, probably because RRA sounds better than DS or ES and sondeo is too outlandish. I have come to realise the power of eloquent terminology and melodious acronyms and RRA indeed conveys the idea better than the others. All the original methodologies contained guidelines to ensure that the surveys were conducted in an orderly way and would provide the information which was needed to formulate an on-farm research programme. Our IITA methodology, for example recommended starting with a study of existing information, and then conducting a brief recon-naissance tour of the area, also known as a windscreen survey. For the actual survey we proposed to use a checklist of things to be looked at and to be discussed with farmers and a data sheet to record information about each field that was visited, including its history. That should help to understand why certain cropping patterns were practised in certain soils and by certain farmers and how farmers decided to do what in their fields.

Why did it not work the way we wanted? There could be two reasons for that. The first one is that the methodology was simply no good and did not bring out what you want to know, even when used as intended. In that case, improving the methodology should have led to better results. The other one is that the methodology was all right but its practitioners missed the point of the whole thing, even though they mastered the vocabulary of FSR-speak. In that case, their understanding would never improve significantly, whatever methodology they used, unless there was a spark of enlightenment, after which the methodology really does not matter much any more. Let us diagnose the diagnostic process a little further and see what actually happened.

8.1.2 A diagnostic survey in the Ivory Coast

I will use for an early example a survey conducted in 1984 by a national team in the Ivory Coast, in which I participated. You may remember from Chapter 4 that the Ivorian team had been trained in diagnostic methodology during a workshop which we had held in the Ivory Coast mainly for their benefit earlier in the same year. They were going to apply their newly acquired skills in the former cocoa belt, an area to the south-east of Bouaké.

The methodology said that before starting the survey the team should collect existing, so-called secondary information, thus benefiting from the things other people had already found out before them. The Ivorian team skimped that stage, which was regrettable. Wherever you go in Africa, you are unlikely to be the first to study the local farming system and its evolution may be as important as the way it looks today. You will remember the example in Chapter 3 about the food crop system in Cameroon, which had already been studied by the anthropologist G. Tessmann in the beginning of the nineteenth century. Writing up what other people have seen also helps to focus the survey. But digging up old publications from dusty files may be asking too much and a lot of it could no longer be found anyway, except perhaps in European libraries. The Ivory Coast team did not bother. They simply took the prototype checklist which had been put together at the IITA workshop in 1983, changed it a little and then went straight to the survey area.

Our team leader was fond of sitting under a tree in the village and chatting with farmers. He also liked to teach them a few lessons about what they should do to improve their lives. That included working hard, listening to the extension agent, whom they probably had never seen, forming cooperatives and getting credit from the bank. That of course was very important but not the intention of the survey. He also liked to look at all the varieties of eggplant the farmers were growing and asked them to display them on the table we were sitting behind. Since he was the leader the rest of the team dared not challenge him so we wasted a lot of time in these sessions, even after we had reviewed the process and concluded we should be spending more time in the field, instead of just sitting under a tree.

The survey area was part of what had been the cocoa belt, before the cocoa growers moved to the famous Tai forest in the south-west of the country and started cutting down the trees there to plant cocoa. Cocoa was in the final stage of decline in the survey area and so was the local economy. The only crop that looked promising was swamp rice which was grown in the waterlogged inland valleys by Mossi immigrants from Burkina Faso. Some of these people would be curious enough to venture into our village meetings, but always remained standing at the fringe of the farmer groups. The survey report contained the usual trivia, with a lot of descriptions of varieties, plant spacing, pests and diseases and farmers' complaints about lack of markets, cooperatives, labour, inputs and of course lack of money. An interesting observation was that farmers were searching for alternatives to the cocoa which had moved westwards, taking with it the most enterprising members of the community. Those who had stayed on dabbled a little in vegetables, but there was no support or marketing structure, as there had formerly been for cocoa, and they did not get very far with that. The survey report contained one innovative methodological element, which I think was contributed by the team leader – village maps, showing the location of different field types relative to the settlements. Quite a few years later village mapping appeared in the literature and became one of the tools in what would later be called Participatory Rural Appraisals (PRA). So we missed an opportunity to make a splash in diagnostic methodology. Professor C.T. de Wit used to say that an apple should be polished to bring out its lustre, but I must admit to our failure to see that we even had an apple in the first place.

You would expect that our team developed some interesting ideas about practical research that could be carried out with the farmers. That is really what these surveys were meant for, but there was very little of that in the report. It talked only about improved varieties of rice, maize and yam which should be tested and that the farmers needed fertiliser and pesticides. Surely you do not need a diagnostic survey to come to that kind of conclusion, so I went back to my field notes to see whether that was really all there was to say. In fact a few mildly interesting things had been proposed which did not make it into the report. Box 8-1 describes one of them.

Another observation from the survey was that farmers needed alternative cash crops to substitute for the lost cocoa. The report even coined the term 'negative cocoa syndrome' for the effects of the decline of cocoa on the local economy, but stopped short of giving ideas on what to do about it.

The whole thing may have been a little meagre, but at least the survey did put the finger on some real problems in the area. And it should have done some good for team building and for a shared appreciation of the farmers' constraints and opportunities. Perhaps so, but I think the major shared experience was that of indulgence in activities of a very different nature after working hours, with little relation to the tasks at hand. And team building would be difficult anyway for scientists who normally worked in different institutes, hundreds of kilometres apart and who had little more in common than their age and the minister's instruction to consider themselves a team.

Box 8-1. Chromolaena, vegetables and nematodes

The main fallow species in the survey area was *Chromolaena,* as was to be expected in this environment. The nematologist (the team leader himself) observed that okra plants grown in some freshly cleared fields were severely infested by root knot nematodes. That is a tiny worm-like animal which lives inside the roots and causes their malformation and poor growth. Nematodes build up in the soil if you grow sensitive crops a few years in succession. Any gardener knows that you should therefore not grow potatoes after tomatoes or vice versa because of cyst nematodes which attack both crops. They also know that growing marigolds may keep infestation with certain types of nematodes down. *Chromolaena* belongs to the same botanical family as marigold (*Compositae* or *Asteraceae*) and several species in that family have a nematode suppressing effect. If anything, you would therefore expect *Chromolaena* to *reduce* nematodes, not increase them. One of the ideas from the survey which did not make it into the report was to do a literature scan to find out what was already known, and then perhaps carry out a simple farmer-managed test with okra and some other vegetables to see whether *Chromolaena* did indeed raise nematode infestation. The idea was not spectacular, but at least it was based on real, even though non-holistic field observation, and if it turned out to be correct you would have a simple recommendation which cost nothing: not to grow okra immediately after *Chromolaena* fallow.

Judging from what they decided to do later on I think the team never really grasped the essence of FSR, or were out of sympathy with it. As you will remember from Chapter 4, after the survey they decided to carry out a life-size test of modern agriculture behind a barbed wire fence with high school dropouts. The diagnostic survey apparently had been little more than a ritual, perhaps meant to satisfy the donor, and what followed was a politically desirable on-farm testing programme, to convert loitering youngsters into modern producers. Perhaps that is what the team leader had in mind all along, which may explain why he did not bother to write up research ideas which he probably found irrelevant from a political point of view. The proposed barbed wire trial was the last I heard from the team. I think after that the Ford Foundation also gave up and did not renew the funding.

8.1.3 Further methodological embellishments

The experiences of other FSR groups with diagnostic surveys were quite similar. Most of the time the findings were rather trivial and did not help much in generating better research ideas. A *felt need* therefore emerged among influential FSR scientists to further improve the diagnostic methodology. If there is one thing that has done most harm to the FSR movement I think it is the obsession with diagnosis. Compulsive methodology embellishment resulted in a stream of evermore elaborate diagnostic tools which were brought into practice in more surveys, with the same trivial results. As we will see, the embellishment of diagnostic methodology continued right into the twenty-first century, while the record of successful technology adoption remained as pathetic as ever.

I for one do not believe that diagnostic methodology matters all that much. Many people walk around with blinkers on anyway and they will always come back with the same trivial results, whatever methodology they use. I have worked with people who were very good at finding out things about farmers without using any apparent methodology. One such was Neil Fisher of Samaru, Nigeria (mentioned in Chapters 4 and 5). Apart from being a solid station researcher he had a profound grasp of indigenous farming, without needing quasi-scientific prescriptions on how to find out. But he said farewell to Africa in the mid-1980s, never to return. Anyway, for those with less acute minds it might help if they have some good guidelines to work with. And if the methodology were really good it could even help dummies to develop the right attitude and come up with something useful, or so it was thought.

(a) Holistic perspectives

The first improvement proposed by the pundits was strengthening the *conceptual framework* of FSR. We had to drop our disciplinary bias and start looking at the farm as a *system*. It had to be studied from a *holistic perspective*, which means that everything is related to everything else. Obviously this is best illustrated by a *diagram*. The specimen I have put together here (in Figure 8-2), although already quite impressive, I think, can be further embellished by adding double arrows showing two-way interactions and dotted lines representing information flows. In fact, every box can be linked up with almost every other one. If team members would understand

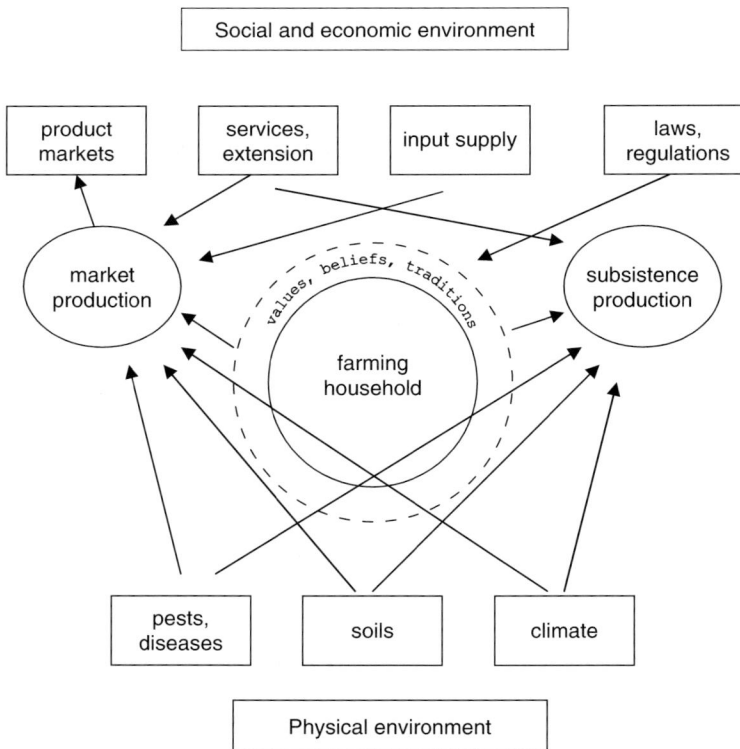

Figure 8-2. The farm and its environment

the holistic concept they could be expected to transcend the narrow confines of their disciplinary biases.

Participatory approaches

The next thing was to redress the *top-down* nature of the surveys. The survey teams were in the habit of firing questions at farmers, jotting down their answers and then going home to design solutions, which most of the time they had in mind all along anyway. A new generation of thinkers, most of them from academia this time, came to the conclusion that we were not really listening to farmers, and launched the *participatory* subculture which in due time was to eclipse (or swallow) the FSR movement. The participatory approach meant that you should not just see farmers as objects of study but rather work with them to analyse their constraints and find solutions. That sounds like a good attitude. Several books were published about that with titles like 'Farmer back to Farmer', 'Farmers First' and, the most creative of all, 'Beyond Farmers First'. They contained a lot of examples showing how important it was to treat farmers as real people. But the mere recognition of that important truth was not enough, it had to be enshrined in the methodology. So diagnosis was enriched with *participatory tools*. Many clever tools were developed, some of which are briefly described in Box 8-2. Using all those tools was actually a lot of fun

---

Box 8-2. Some participatory tools

Venn diagram: interlocking or overlapping circles showing how different groups in the village were related to one another and to the world at large.
Social Village Map and *Physical Village Map* showed where everything in the village was located; they had to be drawn in the sand by one of the villagers with assistance from others and then transferred to big sheets of papers.
Time Line of Events showed what had happened in the village earlier on and why.
Transect Walk was a line drawn across the Physical Village Map along which the survey team would walk to look at the crops and other things they met on the way
Focus Group Discussions were held with groups in the village about topics like maize growing or gender issues.
Matrix ranking: the team would make a list of problems and ask individuals to rank them in order of importance. By adding up the individual scores you would arrive at a final overall ranking.

and it gave the practitioners the illusion that they were acquiring real skills and getting significant information, contrary to the messy surveys we used to do in the early years.

Then came the analysis of the findings. The team would get together in a room with a blackboard or a pad of large sheets of paper attached to a tripod, called flip charts. If they were very participatory some farmers would be there as well. The team would write the most important problems on the flip charts and brain-storm about the *underlying causes*. That required another tool, which was called *problem tree analysis*. You have already seen a problem tree in Chapter 5, but that was a very simple one, and not very multidisciplinary. If you look at a problem with a multidisciplinary team you will get something quite complicated. I just invented one here for the sake of illustration, say for the problem of fertility decline in a moist savannah area in West Africa. It is shown in Figure 8-3. This problem tree is still quite simple and can be under-stood at a glance. You can also make a big tree with many branches for several problems at the same time, with a lot of arrows crossing over, showing common causes for different problems. That can result in a really beautiful piece of holistic artwork, which, however, does not mean that the useful information load increases in proportion to the charts' complexity. It was quite the opposite, in fact.

Once these tools were enshrined in the methodology they were to help us to avoid the kind of sloppy job we had been used to do before, such as not going to the field at all and just sitting under-neath a tree talking and missing a lot of essential information. It did not really help much, though. The reports which came out of these surveys, now called PRA, were much embellished with diagrams, maps and charts, but the conclusions were mostly as obvious as ever.

Meanwhile a new movement had been born. I only started to realise that the participatory approach had become the new aca-demic *paradigm* at a workshop held at the International Service for National Agricultural Research (ISNAR) in The Hague in 1993. The meeting was facilitated by a group of young scientists from the Institute for Development Studies of the University of Sussex. They gave introductory lectures about participatory research which was presented as an entirely new concept invented by Robert Chambers and his followers in Sussex. I could not believe my ears. At the end of the workshop there was a panel in which I was invited

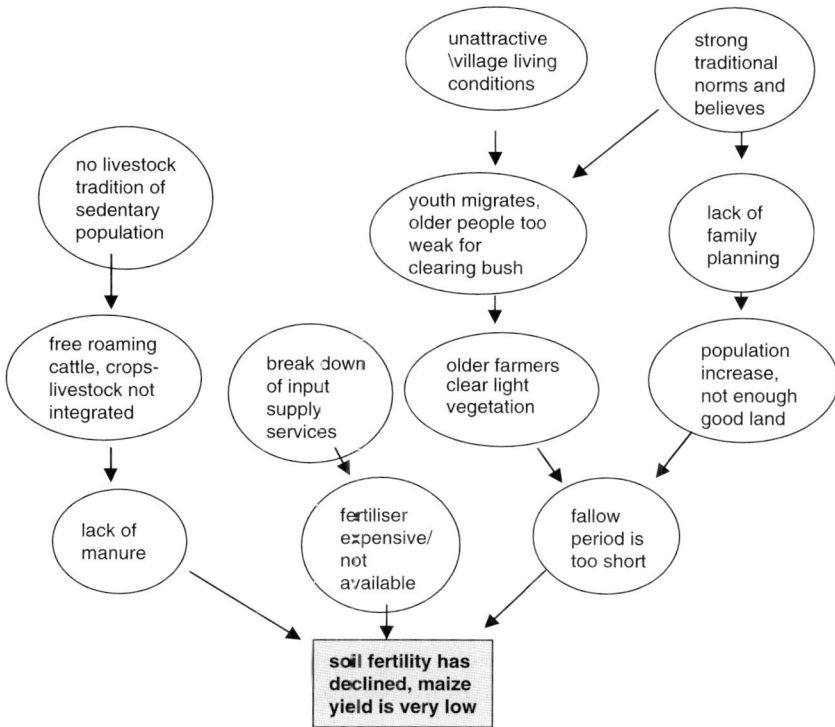

Figure 8-3. An (imaginary) problem tree for the decline in soil fertility

to take part. I ventured the suggestion that this was a useful improvement of the FSR approach but that perhaps it should not be presented as something entirely new. Many people in the research institutes had already been educated in FSR and would it not be better to convert them to a more participatory attitude rather than starting another fad and confusing everyone? That of course did not go down well with the young Turks, while the international scientists who were present had decided that it was not in their interest to swim against the tide and were silent or expressed their admiration for this revolutionary concept. So I think I looked like a pitiable dinosaur who had survived from a long-gone age.

And indeed, I must have been half asleep most of the time while the participatory paradigm swept through the ranks of development thinkers and field workers. By the early 1990s PRAs were no

longer confined to agricultural research and had become an essential
part of the general development *tool kit*. It became tremendously
popular in rural development circles and particularly among
donors of development funds. Very soon, all development programmes, whether dealing with education, health, agriculture or
anything at all had to swear allegiance to the participatory dogmas
and stuff their documents with participation speak. Of course I do
not deny the value of participation, far from it. It is a potentially
very useful and powerful notion, which, if properly understood and
applied, will make development workers aware of the real ambitions and needs of their avowed clientele. Regrettably, as many
other good ideas, it suffered under the onslaught of mediocrity.
Rather than changing people's attitude, 'participation' and its
jargon became ritualistic terms, conjured up in development documents as powerless incantations. And *participatory methodology*
became another pseudoscientific discipline, taught all over the
world by scientists of uncertain description. Just attend a training
course on development planning and implementation, or something of that nature in Ethiopia, Mali or the USA, or search the
Internet with 'participatory development' as key words and you will
see what I mean.

(b) Sustainable livelihood analysis

The participatory approach had been pioneered by agricultural
scientists and development workers, but when its methods began to
be used in other sectors as well, it was felt that the approach was not
sufficiently encompassing to account for the complexity of rural
societies. The British Department for International Development
(DfID) therefore introduced an entirely new analytical approach in
the late 1990s: the *sustainable livelihoods framework*. The framework was based on the insights of Robert Chambers and his school,
whom we have already met as leaders of the participatory movement. The argument was that whereas agricultural production may
be important for rural households, there are many more things
which affect their lives. They have all kinds of resources, physical,
human and environmental as well as various coping mechanisms
(their terminology) which allow them to confront the many challenges they face. An analytical framework could therefore not be
limited to agriculture alone, as the one of Figure 8-2, there were

complex networks of interrelationships which effectively related everything to everything else, and they were not just agriculture-related. And understanding all those networks, resources and coping mechanisms was necessary before effective measures could be devised to foster development. I will spare you the details, but this was the truly holistic apotheosis of diagnosis. It made me for one wonder in desperation whether its practitioners would ever be able to make a soft landing in the real world where farmers were waiting for research to come up with something useful.

8.2 On-farm experiments

For FSR the excessive appetite for diagnosis and participatory appraisals was little short of disastrous. Tools and terminology were in a continuous flux and they were let loose on the same poor farmer, while distracting the practitioners from what they should really have been doing all along: trying out innovations and letting the farmers judge their usefulness. I do not mean that thorough analysis of farmers' production systems is necessarily useless. It can help an institute or even the government to formulate better research and development policies. It may also produce excellent teaching material for academic and other kinds of training. But the surveys done by most FSR practitioners were not meant for that. Their intention was, or should have been, to help researchers and farmers choose promising technologies for on-farm testing. The quicker that was done the better, as I have argued repeatedly. The real lessons would only be learned once the researchers started exposing their ideas to the reality of the farm.

Fortunately, however complicated the FSR scientists made their diagnosis, sooner or later the time would come when they or somebody else decided that they had studied enough and should start doing something practical, like testing some new ideas or materials in farmers' fields. They might still postpone that for a while by participatorily designing interventions and then doing *ex ante analyses* of those interventions and asking farmers what they thought of them and so on. And there were of course methodologies for that also, but I will skip those and assume that now the researchers were ready for the real thing: on-farm testing of innovations.

8.2.1 Simple principles . . .

When you talked to agronomists about on-farm trials many of
them showed particular professional reflexes which, if not corrected, led directly to failure. They had been conditioned by many
years of college training on the principles of proper statistical
design and more years of applying them at the research station.
These were extremely hard to unlearn, as I have experienced
through many years of training courses and lectures on on-farm
experimentation. You had to tell established agronomists that they
should drop some of the most treasured principles of field experimentation, when the institute's scientific director and statistician
thought the institute's scientific prestige was at stake and that these
scrappy on-farm tests could never lead to publishable results. As an
example, let us look in a little more detail at one of the simplest
types of on-farm tests, variety comparisons, and how easily their
on-farm variants could be spoiled by the unconsidered application
of statistical principles.

Suppose we have chosen the varieties we are going to test on-farm,[3]
how do we proceed from there? The first question the agronomist
will raise is how to choose the farmers who will carry out the tests.
I will first look at that question in the naive way of one who only
disposes of common sense. Since we have no idea whether the varieties will in fact be any good when farmers grow them you may
think it would not really matter much which farmers do the tests in
the beginning. The most important thing is that they are real farmers who will grow the varieties in their own way. So word could be
passed around that anybody would be welcome to join the tests. All
farmers are interested in new varieties, but the most curious ones
will of course come forward first. If the varieties work for them you
can try later whether they also work for others. And if they do not,
then forget about them. But, surely, that is highly unscientific and
ignores well-established statistical principles of random choice? It
does not, really, but it is hard to convince people who consider themselves experts in field experimentation as most research agronomists
do. Because, what does the theory say about choosing a location for

---

[3] There are of course also participatory methods to do that, but I will not go into
that.

the tests? First you have to define the conditions for which the conclusions must be valid. The agronomist will think of soils and climatic conditions. Soils can be quite variable, both because of their physical and chemical nature and because of the way they have been used before. The textbook tells you that the trial plots are best stratified for soil conditions, which means that there should be an equal number of test farmers for each soil type. Second, there is the climate. Even if there are no systematic differences in climate across the area the agronomist may still insist that there can be important differences in rainfall events, because the intensity of tropical storms can be highly variable, even within a short distance. You cannot stratify for that, so the farmers in each soil class must be scattered around as much as possible to capture the rainfall variability. Some diehards may even want to have different, equally sized groups of farmers, who plant the test at different prescribed times to capture any effect of planting time on the performance of the varieties. Next the social scientists want to have their say and will argue that there are farmers in different wealth or farm size classes who must be equally represented in the sample.

When everyone is happy with the different classes of soils and farmers you now have to find farmers in each class who are willing to take part in the tests. So the research team will be very busy before the season starts talking to candidates and making appointments for when they will come to plant. Some of them may actually not be genuinely interested, just polite as most farmers are, but there is no way to find out until the tests are actually started. The chosen trial sites will be scattered across the target area and, when the planting time arrives, the FSR team will spend a lot of time and fuel running around meeting the farmers on the appointed days. The first time the team arrives a farmer may be absent because he forgot the appointment, next time he has decided that he actually wants to plant a little later and when you come later he had changed his mind again and has already planted the field so you have to find another farmer who is representative for his class. Perhaps you think I am exaggerating, but I assure you that I am not.

Suppose that, with a lot of headache, the tests have been planted more or less satisfactorily. The reluctant farmers whose arm had to be twisted will think they are doing you a favour, as in fact they are. There is a limit to their cooperativeness, though, and when they are overstretched you can be sure that first thing to be neglected will

be your trial. Maybe they will ask you to pay them to hire additional labour. If you decide to do that then other farmers will find
out very soon and will make the same requests. In that way even the
simplest on-farm tests become very expensive and more unrepre-
sentative than if you had just let any farmer run them the way they
wanted. By all this arm-twisting a distortion has been introduced
into the test, which will affect its representativeness much more
than soil or farm size classes: the way the test fields are managed
will have nothing to do with the way farmers would normally do it.
So, what is the conclusion? The conclusion is that the 'naive'
approach, mentioned at the beginning, was in fact the only sensible
one. Just let it be known that any farmer who wants to do the test
should inform the team when he will be ready for planting. The
team will come and deliver the seed and perhaps help the farmer to
subdivide his field into plots and assign each of the varieties to a
randomly chosen plot. A simple drawing of the field shows where
each variety is planted. The farmer will plant the seed of all the
varieties, including his own, exactly the way he would if there were
no trial. That is all. It is up to the farmer how he will manage his
field, which will not be different from the way he would normally
do it. And that is exactly what you want. Of course there are many
practical complications, but if the researcher is convinced this is the
way to go they can be solved. The most important thing is to keep
things as simple as possible and let farmers run the show. That is the
only way you will get meaningful results.

   But how can you draw conclusions from that kind of trial, with all
those differences among farmers, not just in soil and rainfall, but also
in planting date, spacing, weeding, pests, and much more? First, you
have to substitute careful observation for control over the trial conditions. Keeping track of when farmers plant, at what density and how
much fertiliser they apply is much simpler than trying to control those
things and end up with a test that has no relationship with the farmers' own way of doing things anyway. And then use statistics to sort
out the differences among farmers and how the performance of the
varieties or other technologies varied with those conditions. But that
is a different kind of statistics from the one that can so easily fail with
on-farm trials. Luckily IITA's statistician, Peter Walker, was sympathetic, and helped us with the analyses, besides safeguarding our credibility. When reading a draft of this book he confessed having been a
reluctant convert from the old school himself.

A simple yet powerful kind of analysis for farmer-managed trials is Hildebrand's 'adaptability analysis' which I introduced in Chapter 4. I will use it to analyse an on-farm test comparing two maize varieties, a local and an improved one, each of them grown with and without fertiliser.[4] The test was carried out in south-west Nigeria by 21 farmers (there were actually 25 but in some of the plots no yield data could be collected). The only thing we had bought them at the time of planting was the seed of the improved variety, the fertiliser for the fertilised trial plots and a tin measure. Once the maize cobs from each plot had been weighed (in the field of course, and in the presence of the farmers who kept them afterwards because they were theirs) the average yield of all four plots in each farmer field was calculated first. The average yield is called the 'environmental index' for that field. That is reasonable because average yield will be higher when the 'environment' is better. Environment here means everything that affects yield, including the farmers' management. Next, the yields of particular variety–fertiliser combinations, obtained in each field, were plotted against the environmental index of that field. The graph of Figure 8-4 shows the plot for two combinations: the local variety with and without

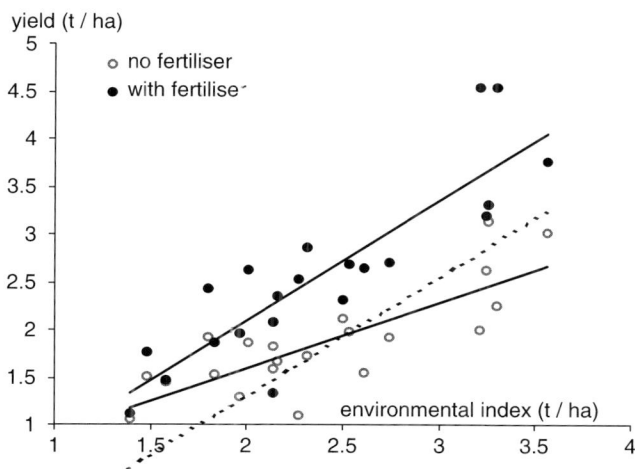

Figure 8-4. Adaptability analysis of maize yield in an on-farm trial with and without fertiliser

---

[4] In all cases the maize was intercropped with cassava, as was usual in the area.

fertiliser. You can also of course make a similar graph for the
improved variety with and without fertiliser, or you can plot the
yield of each variety, averaged over the two fertiliser levels, or each
fertiliser level averaged over varieties. But let us look only at the first
case, the yield of the local variety with and without fertiliser, shown
in Figure 8-4. What can we conclude from that figure?

The drawn lines are called the best-fitting linear relationships
between the yields for a particular treatment combination and the
environmental index of the fields where the test plots were located.
The top line is for the plots with fertiliser, the bottom line for those
without. In all but one of the fields the yield with fertiliser was
higher than that without, but the really interesting thing was that
the slopes of the two lines were very different, so they moved
further apart as the environmental index, i.e. the field's average
yield, was higher. Hence, the better a field, the higher the benefit
from fertiliser. The expected yield increment due to fertiliser in a
field with a particular environmental index is the vertical distance
between the two drawn lines at that point. If you stopped there you
would think that fertiliser was a good thing for all farmers, irrespective of their yield, but that is not the case because fertiliser does
not come free. At the time when these trials were carried out the
cost of the fertiliser per hectare at the applied rate was equivalent
to about 800 kg of maize. The yield line for the plots with fertiliser
was therefore shifted downward by that amount to get the net yield,
the dotted line in the graph.[5] It crosses the line for the non-fertilised
plots when the environmental index is about 2.5 t/ha. That means
that the 50% of farmers whose fields had an index of less than 2.5
would have lost money by applying fertiliser to their local maize.
No wonder then that farmers in this area were unlikely to apply
fertiliser to their local maize; it was too risky. Perhaps the situation
was more favourable when using the improved variety? It was
indeed. Only 10% of the farmers would have lost money from
applying fertiliser to the improved variety, because it responded
better to fertiliser, even at a low overall yield level of the field. I have
simplified the situation slightly for the sake of clarity, for instance
by ignoring the price of the improved maize seed[6] and the better

---

[5] The idea was contributed by Mark Versteeg (Versteeg and Huijsman, 1991).
[6] Farmers could replant their own seed for several seasons as it was an open
pollinated, and not a hybrid variety.

quality of the seed (from the research station), which may have somewhat exaggerated its yield advantage. Anyway, I think this is a convincing example of a simple analysis of a series of tests which were managed entirely by the farmers.

The conclusion from the tests was that for certain farmers it was a good idea to apply fertiliser to their maize, while for others it was not, but how do you know which is which before the maize is planted? If you could be sure that a good farmer would also be good next year and you knew all the farmers individually, you would have some basis to advise them whether or not to apply fertiliser. But a farmer who was in the high yield category in one year may end up at the bottom of the league in the next. That is typical of peasant farming in Africa as I have shown in Chapter 6: being a 'good' farmer is not a stable property, the way it is in industrialised countries. I have done some ranking of the same farmers in different years which showed that the consistency in performance was rather weak. I am not saying there was no consistency at all, but some personal calamity may befall a farmer which causes him to neglect his farm and pushes down his ranking. Or he may decide that this year he will spend more time on some other, more profitable business and pay less attention to his field crops. So, that is not going to work.

But perhaps by collecting a lot of information about the things farmers do we can find the most important factors which caused some farmers to have good yield and others to have poor yield. If that were the case you could tell the farmers how to do things right so that their yield would come out higher. That turned out to be an illusion as well, as we have seen in Chapter 6, and the only things the poor extension worker can tell the farmer with confidence is that the improved variety will practically always yield more that the farmer's own and if he does a good job and gets a reasonable yield, fertiliser will pay for itself, otherwise it will not.

8.2.2 . . . and/how to make simple things complicated

We have seen that one sure way to ruin on-farm trials is to impose the statistical principles of on-station trials, as discussed above. By doing that the tests are likely to degenerate into poor replicas of station trials, the outcome of which will be looked upon with rightful scepticism by farmers and station researchers alike. The other way, with similar effect, is to impose all kinds of restrictions on how

farmers may apply the innovations proposed to them. I will again
use the simple example of an improved variety to clarify this. It may
be hard to believe for the uninitiated that such a simple thing as a
variety trial can be the source of so much confusion, even after
30 years of on-farm experimentation, but I assure you it is, even today.
And for more complex technological innovations it applies *a fortiori*.

During one of IITA's Board of Trustees meetings, somewhere in
the late 1980s, I was asked what I thought about the potential for
hybrid maize in the forest zone. I replied that in our on-farm trials
the recommended hybrid variety did not do any better than the
local variety and significantly worse than IITA's open-pollinated
TZSR.[7] Then the hybrid breeder stood up and said that he had seen
those so-called on-farm trials and that he doubted my ability to
conduct proper experiments. Fortunately my reputation with the
Board was better than that and this did not significantly damage
my credibility, but it was exemplary for the conflict between some
technology generators, especially crop breeders, and some on-farm
experimenters, which continues to this very day.

The problem was as follows. As I have said earlier, breeders
tended to breed their varieties under exceptionally favourable conditions. The reason was that they wanted to maximise the probability that good traits are expressed as clearly as possible. First, if
breeding lines were grown under less favourable conditions, like
poor fertility, a high yield potential might not show up and a potentially good line could be missed. Second, there were often considerable differences in soil fertility over short distances, even at the
research station, which would obscure differences between breeding
lines and reduce the chances of identifying genetically good material. The only way to escape from this dilemma was to eliminate the
variability as much as possible by applying a high rate of fertiliser
to smoothe out local differences. These arguments for breeding
under favourable conditions were plausible, but once a variety had
been selected they obviously no longer applied and the variety now
had to be tested under real conditions, hoping that its potential
stood up to the test. But that is not how breeders saw it, or some of
them at least. We had endless discussions in workshops and training

---

[7] TZ stands for Tropical Zea (*Zea* is the botanical name for the genus maize
belongs to) and SR means Streak Resistant. Streak is a viral disease.

courses about this, where logical arguments failed to bring the breeders, national and international alike, around to what I thought was an inevitable conclusion: that new varieties which were meant for farmers should be tested under their conditions and with their management. The breeders, although accepting the former (as long as 'conditions' only meant physical conditions in the farmers' fields), would reject the latter and wanted to impose management practices like those they applied at the station: prescribed densities, recommended fertiliser, timely weeding, etc., to ensure that the varieties' superiority could be expressed. I am not implying that the breeders were lacking in intelligence, but their approach was motivated by quite conventional ideas about the purpose of applied research in general and on-farm variety tests in particular: show the farmers how much they would gain, not just by growing the new varieties, but by using all the other recommended inputs and practices as well. That is the kind of struggle FSR workers had to engage in continuously with scientists from other disciplines, but their chances of prevailing would have been better if they had been entirely clear themselves about what they expected from their on-farm experiments, which many were not.

FSR's intellectual leadership of course quickly noted the problem of incomprehension among their flock and started looking for a solution. There were those among them who believed that ritual was an acceptable substitute for real understanding. In FSR ritual was called methodology and it reached exceptional heights, not just in diagnosis. It was also introduced into on-farm experimentation. I will illustrate that by the way the most crucial aspect of on-farm trials, the management of trials in farmers' fields, was thus ritualised.

The only way on-farm trials can be successful is when farmers consider them as their own, with the researcher only providing things the farmer does not have, such as the seed of an improved variety. However simple that principle may sound, even today it is rarely respected. David Norman, one of the fathers of FSR and a stern advocate of methodological correctness, observed in the mid-1980s that FSR practitioners, though well trained in diagnostic methodology, tended to lapse into their old habits when making the transition to on-farm experimentation. He therefore decided that it was necessary once more to take the practitioners by the hand and he introduced a *classification* of on-farm trials, which he hoped would steer them away from their pernicious habits. He distinguished two functions in on-farm experimentation: trial *management*, that is who is the boss, and trial *implementation*, or who does the work. Since there are two parties involved, the researcher and the farmer, four combinations result, in theory at least, as Table 8-1 shows.

At one extreme we have researcher-managed–researcher-implemented trials, RMRI, which is a sort of station trial conducted in farmers' fields, and at the other extreme FMFI where the farmer does everything. Then there are RMFI trials, where the researcher is the boss and the farmer does the work. The fourth combination, FMRI, is not there for obvious reasons, although it would be interesting to try that for educational purposes, the scientist's education that is. RMFI trials are the ones many station-trained agronomists like best and which have been the source of most problems, as I think I have explained at length.

A lot has been written about these different types of trials, when to choose which one, what would be the farmer's and the scientist's contributions in each, and much else, enough to fill many hours of classroom lectures. The reason I find the classification dangerous is that it introduces the kind of methodological flummery into on-farm experimentation, which has been so damaging for diagnostics. Instead of presenting the principles of on-farm experimentation as mainly common sense, it suggests that once you have learned the magical definition of each class of trial you understand its essence. It has just added another layer of jargon and further obscured the essentially simple truth: that farmers should test innovations in their own way, with an absolute minimum of interference by scientists. And, what is worse, it sanctioned excessive control by the researchers over the on-farm trials, by giving it official recognition as part of the approved OFR methodology.

Table 8-1. Types of on-farm trials according to D.N. Norman

	Trial implemented by	
Trial managed by	Researcher	Farmer
Researcher	RMRI	RMFI
Farmer		FMFI

If it was so difficult to get these relatively simple ideas across even to the more intelligent scientists that makes you wonder how the rank and file could ever have been converted. Probably the most effective way was to go and work within a national institute and convert its scientists little by little, from the inside. Peter Hildebrand did that in Latin America, David Norman in Botswana and Mark Versteeg in Bénin Republic. I do not know how much lasting impact Hildebrand and Norman had in that respect, but in Bénin I witnessed the conversion of one of the local agronomists, the late Valentin Koudokpon, who suddenly understood the whole thing and henceforth did not need any more training or methodological drill, it had become part of his mental texture. But you can only work directly in so many institutes in a lifetime and in each of them you have to generate a critical mass of scientists who have gone through the same conversion, otherwise they will be crushed under the weight of contrary opinions once they are on their own again. So, some other mechanisms were needed to reach the commonwealth of scientists. It is natural to think that writing books on FSR is the answer. Several have been published, including two by myself and my colleagues, but the question is whether they really helped, and I doubt it. In the final reckoning I think many, perhaps most agriculturists in developing countries continued to believe deep down that the early post-colonial model for development was the right one: transforming agriculture as quickly as possible into western-type high-input commercial farming. If that is so, the painstaking process of improving peasant agriculture step by step, advocated by FSR, probably had but little appeal to them, even though they may have pretended otherwise. And whether peasant agriculture, particularly in Africa, will really be capable of transforming itself, slowly but steadily, is a question that kept haunting most of us from time to time. For if the answer is no, we have all wasted an enormous amount of time and should instead have continued to vigorously pursue the earlier technocratic model of the Farm Settlement schemes (which I introduced in Chapter 5) and their counterparts in the francophone countries, the Unités Expérimentales. If we had, perhaps there would now be a vibrant generation of middle class farmers taking care of food production in Africa. But I very much doubt that also. The issue cannot be ignored, though, and I will return to it in the final chapter of this book.

8.3 Francophone approaches

I have paid but little attention so far to what was happening in the francophone world and whether something similar to FSR was going on there. I will look at that now before continuing my mostly anglophone chronicle. In fact, a lot of things were going on in the francophone world. Most of us had little or no notion of that and whenever the two sides chanced to meet in the early 1980s, in workshops and similar events, it was not only language that stood in the way of mutual understanding. For both, the other side's ideas, if understood at all, sounded like voices from a different world. The French and their cultural allies (*la Francophonie*) profoundly disliked lack of structure. The informal and rather scrappy surveys which were used in the anglophone world were not their kind of thing, nor were most of them convinced that peasant agriculture could or should be improved stepwise by small incremental innovations. And I think the anglophone side felt that the French attempts to promote the emergence of middle class farmers were really a remnant of the colonial past, which would never work. The French counterpart of FSR was called *Recherche sur les Systèmes de Production* (RSP) or simply *Recherche-Système* which indeed means literally FSR, but in the early years similar terms concealed quite different contents.

An interesting paper comparing the anglophone and francophone approaches was published by Louise Fresco in 1984. She gave a thorough factual description of the differences and similarities which she felt 'must be interpreted in the light of the colonial and post-independence history', but then decided she would not do that and rather stick to the facts. Perhaps that was a good idea, because it avoided unnecessary irritation which could have resulted from her first objective, and would not have helped her wish to bring the two sides closer together. I will draw extensively on her paper, but I would also like to do what Fresco decided not to: look at the different mind set of the two sides and the different roles the French and the British played in their former colonies and how that was reflected in their contributions to FSR. The Americans eventually played a dominant role but their contributions had similar intellectual roots as those of the British.

Before independence agricultural research in francophone Africa was carried out in strongly centralised institutes, which were organised along commodity lines and which had their administrative and intellectual headquarters in Montpellier and in Paris. There were institutes specialising in oil crops, fibre crops, fruits, livestock and several more. Then there was an institute for general tropical agriculture (IRAT),[8] which worked mostly on smallholder food crops and cropping systems, and an institute for more basic research, ORSTOM.[9] IRAT was the first to launch into systems research. That was in the 1960s and the work they did in Senegal became their showcase, in particular the Unités Expérimentales. They were a kind of real life field laboratory where researchers, extension staff and farmers worked together to collect information and test new ideas. The main philosophy was that farming had to undergo a profound transformation, the final outcome of which would be viable middle-class farms producing for the market with a degree of subsistence production (producing food for your own consumption). The scientists developed blueprints for future farming systems, and an *itinéraire* or pathway to get there. Piecemeal adoption of technologies like new varieties or fertiliser was not thought to contribute much to the desired transformation of the production systems, but they were still tested because that might whet the farmers' appetite for the real thing. I think van Beukering in the Dutch East Indies would have liked all this.

In order to work out the blueprints a lot of information was needed. The idea was to classify the farmers in different categories, draw up detailed profiles for each of them and assess their capacity for change. Obviously, that involved a lot of socio-economic surveys, interviews, database building and analysis. Station research was also needed to test prototype systems under controlled conditions for each farmer category before venturing into the real farm.

In the 1960s the ideas in the Anglophone world about where agricultural development was to go were actually quite similar, but things changed quickly, perhaps because of the much looser links between Britain and its former colonies and the failure of early attempts like the Farm Settlements and the Tanganyika groundnut scheme (which I will say something about in Chapter 10). The conviction emerged that development of peasant farming was to take

---

[8] Institut des Recherches Agronomiques Tropicales et des Cultures Vivrières.
[9] Organisation de Recherche Scientifique et Technique Outre-Mer.

place in a stepwise fashion rather than in a few big leaps. Detailed socio-economic surveys and cropping systems research at the research station therefore went out of fashion and the informal methods of 'Anglophone FSR' gained ground rapidly as the favoured toolset.

So, in the early post-independence decades rather profound differences evolved in the basic assumptions, the philosophy and the methods of Francophone and Anglophone FSR, which made communication quite difficult. In the 1970s the FSR pioneers in the Anglophone world had rejected the idea of introducing entirely new farming models to peasant farmers as an impossible goal, even though it lingered on in the World Bank-funded ADPs (and in IITA's Farming Systems Program) for much longer. But the French clung to the idea, for as long as they maintained control over the African countries where they could experiment with it. In the end they also gave up and gradually the two approaches began to converge, as the French distanced themselves from their former colonies and more or less tacitly abandoned the idea of a radical transformation of the African farm.

8.3.1 Francophone contributions to FSR methodology

That is not to say that nothing worthwhile came out of all that work, quite the contrary. Francophone FSR made three important methodological contributions which eventually found their way into the Anglophone world as well: *typologie des exploitations, conseil de gestion* and *gestion de terroir*. I will briefly explain each of them.

(a) Typologie des exploitations

The English translation of *typologie des exploitations* would be something like *farm household typology*, although that may sound a little un-English. There was no real equivalent in Anglophone diagnosis. Collinson's 'recommendation domains' came closest, but the domains varied according to the technology you were looking at, while 'typology' was more like a basket term for the overall characterisation of a farm.

A farm may be characterised at different levels of detail. At the simplest level we could for example distinguish small, less small and somewhat bigger farms (I am not saying 'small, medium and large', because really large farms are exceptional in West Africa).[10] The bigger ones have enough resources to buy fertiliser and draught animals, while the small ones can certainly not and the less small ones only sometimes. In the francophone system, however, typology usually was much more elaborate. A lot of detailed socio-economic surveys were carried out, about how exactly the different types of farmers farmed, how many animals they had, what was the seasonal labour profile of their farm, how they treated their wives and much more. All that information was thought to be needed, because for each type of farmer there was expected to be a different pathway (called *itinéraire*) to the holy grail of efficient market-oriented production. And in order to chart them out you had to know a lot of things about their condition today, because that was the point of departure of their *itinéraire*. And the more you find out the more new questions you will have. That is perhaps interesting but all that detailed information does not really get you much farther than a thoughtful person could get by going around for a week or two talking to farmers. Still, farmer typology can be useful if not pushed too far. It brings some structure into the thought process about different categories of farmers and what kind of improvements could be useful for each of them.

One of the protagonists of farm household typology was Paul Kleene, a Dutchman in long-time French service. Actually, I think Kleene developed the concept, together with Benoît-Cattin while they worked in the Unités Expérimentales in Senegal during the 1970s. I had known Kleene while we were both students in Wageningen, but our paths never crossed again until we found ourselves participating in a training workshop in Bénin Republic in the mid-1980s, both of us as resource persons. The workshop used the IITA methodology, which Kleene found too simplistic and he tried to steer us towards farmer typology. Now it is very difficult to change the process of a workshop in mid-course and besides I, as the principal resource person, did not agree. So there were some mild clashes and in the end we continued in our loosely organised IITA way. At that time Kleene was the leader of an FSR team in

---

[10] There are some very large farms owned by (former or current) politicians and by large companies like breweries and oil companies, but they have rarely been successful.

Mali, where he had the opportunity to apply his ideas to the full. Many years later, when looking back at the Mali years, he concluded that the period of painstaking analyses of farmers' conditions had taken far too long. But Kleene was also a shrewd observer with an excellent rapport with farmers and the observations he made early on in the Mali project stood the test of time. The added value of the detailed data which were gathered later on was not at all obvious, except that they confirmed what Kleene knew all along. Backing up one's assumptions with hard data is of course justified and even necessary, if only to convince others, as long as data collection does not become an aim in itself. And that is what happened in the Mali project afterwards: increasingly complex characterisation and unmanageable loads of data, without Kleene's ability to separate the millet from the chaff.

(b) Conseil de gestion

Conseil de gestion, which means farm management counselling (the full name is *conseil de gestion aux exploitations agricoles*), was one of the best things invented by Francophone FSR. It was really an extension rather than a research tool, whereby the extension worker would sit together with individual farmers or groups of farmers to analyse their farms and see how they might be improved. It involved more than just sitting and talking together of course. During the first session a crude analysis was made of the farm and after that the farmer had to do some bookkeeping for a while, perhaps for a full season. With that information a more precise analysis was made and a better farm plan was worked out which should be more profitable or otherwise more attractive to the farmer. That could involve some new investment, for instance for an additional pair of oxen, or perhaps replacing one crop by another. Management counselling was not an easy thing to do. The extension worker had to be well trained in the method, nor was that enough: he also had to have a genuine understanding of the concepts, which was not necessarily the same thing. Today, the *conseil de gestion* continues to be vigorously promoted, but mostly in development or research projects, where a lot of coaching can be given to the people who deliver the method. I do not know to what extent it has become part of mainstream extension, nor whether the whole concept has ever been evaluated.

The successful use of the *conseil de gestion*, at least in research and development projects, stimulated the development of similar tools for more specific purposes. An instance is the *conseil de gestion de la fertilité* (soil fertility management counselling). It works in a similar way as farm management counselling but focuses on the farm's soil fertility instead of its monetary budget. The extension worker and the farmer try to map the flows of nutrients in the farm and explore options to increase their efficiency. These exciting ideas have also attracted attention outside the Francophone world[11] but again, I do not know to what extent they have matured into real practical development and extension tools.

(c) Gestion de terroir

An extension of the *conseil de gestion* concept beyond the individual farm was the *conseil de gestion de terroir,* which is difficult to translate. It means something like terrain management (although that may sound too martial) and deals with the management by a community of the area which they consider as their own. It includes agricultural and 'waste' or forest land, water sources, quarries, anything that is of economic or social importance for the community. In English the term watershed management is often used, but that is not quite the same. A community's territory may cover several watersheds, or two neighbouring communities may share the same large watershed. I think the idea should be getting clear now. *Gestion de terroir*, or even the more limited concept of watershed management, did not get the same attention in the Anglophone world initially, because of the prevailing opinion that development should start from individual farmers or groups of farmers and that they would become interested in issues at a higher level as they became more prosperous.[12] And there was of course the difference between the French and Anglo-Saxon cultures again. The French like hierarchical systems, which is well illustrated by their even broader notion of *système agraire*. That is the overarching system

---

[11] For example, the NUTMON (for nutrient monitoring) approach which was vigorously promoted from Wageningen University for a while until it was folded into the broader MONQI (monitoring for quality improvement) package.

[12] A lot of watershed management projects have, however, been carried out in East Africa, especially since the 1990s, many of them with assistance from Scandinavian countries.

encompassing all the agricultural and related land use in a region. It is more than just the sum total of individual farms and farming communities sharing the same production practices, much as *gestion de terroir* is more than just the management of a collection of watersheds. Today the broader *terroir* concept is also getting more attention from non-francophone agriculturists. One reason is that it is indeed important: for farming to be successful in the medium and long term it needs to be embedded in a stable environment where the soil is not being washed or blown away because protective vegetation cover has been removed, for instance by irresponsible tree cutting or because large herds of cattle walk across twice a year. Setting and enforcing rules against the destruction of natural vegetation and creating special corridors for passing cattle herds are all part of *gestion de terroir*. On the other hand I think that development workers have often retreated to a higher level of intervention to get away from the frustration of dealing with farmers who did not respond to the innovations proposed to them at the farm level. In case this is getting too abstract, I will give a concrete example, from Mali.

Mali-Sud, the country's cotton belt, is exposed to serious erosion hazards. That is at least in part due to the successful expansion of cotton growing, driven by the wide-scale adoption by farmers of animal traction for soil tillage and other operations. I will not go into the details, I am sure you will see the link. The FSR team based in the area, the one which was set up by Paul Kleene in the late 1970s, picked up the erosion problem at once. They argued that the root causes of the problem should be tackled first, that is to say the unchecked flow of water, originating from the plateaux and escarpments in the landscape. Control structures had to be built there first, to break the water flow and gently guide water that did not infiltrate down through natural or modified drainage channels. Within the fields the farmers had to construct stone bunds, later changed to strips of grasses and shrubs, along the contours. Figure 8-5 gives the general idea. That required the consent and cooperation of everyone who farmed along the slope, and if it was to be done really effectively, several watersheds had to be tackled together and eventually an entire community or even several communities had to be mobilised.

At first the research team set up village labour gangs to haul the stones and dig the trenches (and sometimes helped them with a

landscape units: plateau   escarp- co luvial   cultivated land   valley   valley
                                     ment   area   slope   bottom

stream

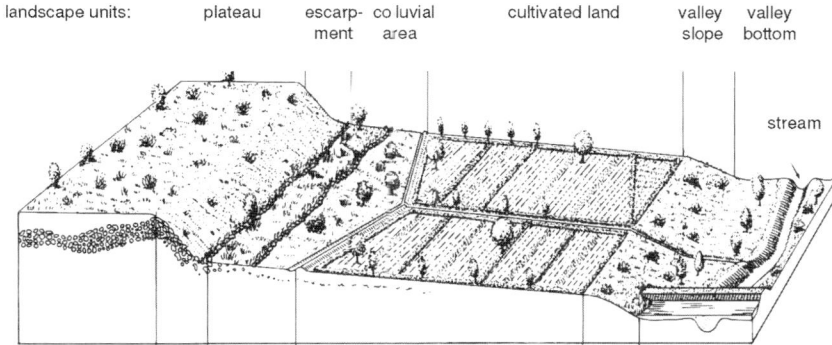

Figure 8-5. Erosion control measures in a watershed in Mali Sud. (From Hijkoop et al., 1991; reproduced by permission of KIT, Amsterdam.)

lorry). At the same time they worked with individual farmers to install anti-erosion measures within their fields, like horizontal bunds and ditches or grass strips or low shrubs planted along the contours to interrupt rainwater run-off. It was very difficult to convince farmers to do that, at least as long as the erosion problem was moderate as in Mali-Sud. And convincing them is really the only way to make things stick. It is practically impossible today to impose erosion control measures or any land use measures for that matter, because imposition is linked in people's minds with colonialism, especially development workers' minds, I think. And if that were not so, it would still be impossible because any rule which requires close supervision and sanctions in case of non-compliance will quickly be corrupted. I do not remember ever hearing the suggestion to deal with erosion in that way, except in Ethiopia perhaps, and in Rwanda and Cameroun before independence. So farmers must be convinced or lured into protecting their land, for example by bringing in shrubs and grasses which are also otherwise useful, such as for human or animal consumption. Development workers then speak of multi-purpose species. It is quite a challenge, and very interesting too, and it even appears to have worked in some areas, like the Machakos hills in Kenya and some hilly areas in the Philippines. I have never been there myself but the stories sound convincing. And there were also cases where farmers themselves used age-old methods to protect their land from erosion or other kinds of exhaustion. I have described some of them in Appendix 5.

But in Mali-Sud the problems were perhaps not yet serious enough, or the FSR team did not have enough patience, or both. In any case, after a few years of rather inconsequential tests the team concluded that farmers were not yet convinced of the need to take measures against erosion in their fields. They withdrew to the watershed and *terroir* levels where the effects of erosion were becoming visible to the naked eye, in the form of erosion gulleys.

Was that the right decision? I do not think so. Even if gullies are the most spectacular signs of erosion, farmers will be aware of the less spectacular signs of erosion within their fields. Especially in the beginning of the rainy season when the fields are most exposed the damage done by rain water running down the long slopes and carrying along fine soil particles and crop seedlings can hardly be missed. Much of the erosion problem starts right there, inside the farmers' fields. I think the team simply gave up too soon, perhaps because the researchers who were responsible for the work were agricultural engineers who like to move earth and stones, and social engineers who like to mobilise communities. But more likely because one quickly gets tired of talking to what appear to be deaf men's ears. And rather than trying to find out why the message was not getting across it was tempting to move up one or two levels where other mechanisms could be used to mobilise the people. The question is how sustainable the results at that level turned out to be. In fact, by the early 2000s enough time had elapsed to answer that question. It could have been observed how much of the earlier erosion works remained and how many new structures the people had built on their own initiative. A survey to that effect was carried out in 1996 which showed positive results on both scores, although the term 'people's own initiative' did not quite reflect the reality since the CMDT, the all-powerful development organisation, had adopted erosion control as one of its thrusts.[13] The situation changed dramatically when the CMDT stopped its erosion control programme along with a lot of other things which were not directly related to cotton and I heard rumours that things have not been well with erosion control since. I think a good and especially an honest study

---

[13] The *Compagnie Malienne pour le Développement des Textiles* (CMDT) was responsible for all agricultural development in Mali-Sud, until they withdrew into their core business, cotton, in the early twenty-first century.

of the history and present status of erosion control in Mali-Sud would bring out important lessons for everyone concerned about the great problems of land degradation in Africa.

(d) Methodological rapprochement?

As I said earlier, the continued association of the French with research and development in their former colonies allowed them to set long-term objectives, such as the transformation of peasant farming into modern market-oriented production, and patiently work towards their attainment. With time, however, French influence waned and the long-term commitment was gradually lost and with it went grandiose visions about development pathways and research in support of them. Many Francophones must also have come to realise that the conclusions from their endless studies were not much better than what could have been achieved by much simpler and less time-consuming methods. Those who meanwhile had learned English had also learned about PRA and thought it was not such a bad idea after all. They called it *MARP*, for *Méthode Accélérée de Recherche Participative* (I do not need to translate that) and started using it hesitatingly. The anglophones in turn learned about *typologie*, which was somewhat similar to but not quite the same as the 'recommendation domains' which Collinson had invented in the 1970s and which meant that different technologies were suitable for different types of farmers or soils. The 'typology' concept was probably easier to grasp, but I am not sure whether it made any more headway than the recommendation domain idea, which had never really caught on with FSR practitioners, whether Anglophone or Francophone. I think the *Conseil de gestion* and *Conseil de gestion de la fertilité* on the other hand were really outstanding tools which could enrich any extension method. I am doubtful, however, whether they had made serious inroads in extension yet by the early 2000s, except in some pockets of activity in francophone countries where they were piloted with technical and financial donor support. Or, to put it bluntly, by expatriates. The NUTMON group mentioned earlier was an example of fertility counselling in other countries and the Farmer Field School approach, pioneered by FAO (which I will say more about in Chapter 11) also contained elements resembling farmer counselling. I think these are hopeful developments which may result in innovative yet down to earth extension tools even though there is a long way to go before they can be effectively used by the average extension agent.

8.4 What about technology adoption?

The loud methodological noise on both sides of the linguistic divide can make us forget what the ultimate goal of Farming Systems and On-Farm Research was: adoption by farmers of new technologies which will improve their productivity, their well-being, or both. 'Adoptability' has always been an implicit and often an explicit concern in diagnosis and on-farm experimentation, but that is not the same as adoption. Research programmes have rarely looked seriously and systematically at adoption of the technology they were testing in farmers' fields, which, in my opinion, should have been part and parcel of on-farm research itself. The question is therefore, how did the on-farm researchers know whether their technologies were actually being adopted, and if not, why not? Well, they rarely did, either because they did not know how to find out or because they thought it was not really their job. That is surprising. In business, neglecting the market and the sales records will quickly lead to bankruptcy, but agricultural research does not have such unforgiving clients. So the researchers must force themselves to ask the difficult questions, or be forced in some other way by their clients or by other agencies, any of which rarely happened. Let us look at the record.

In the early years when FSR was on the rise we did not pay too much attention to adoption. We were too busy building capacity, studying farmers and their production systems and getting experience in carrying out on-farm tests in a realistic way. I think that in those years the sponsors of FSR, the international donors, were already very pleased that researchers were finally venturing into the real world and felt that questions about adoption were premature. Even at that time, however, many FSR workers were well aware that it was not enough for a technology to be 'good' for it to be adopted. Most innovations needed some kind of institutional or commercial support: a reliable supplier of fertiliser and pesticides, for example, honest and solid traders for the export of so-called organic products, artificial insemination services, credit facilities to purchase equipment, training in the use of the technology, the list can be extended almost indefinitely. FSR workers therefore sought collaboration with government extension services (the notorious *research-extension linkages*) and development projects. Others decided to work only on innovations which did not need support which was not available. Open-pollinated crop varieties, for example, can be adopted without the help of extension or regular seed supply, if they are good, even though extension assistance does help to speed up the process. So, a lot of attention was paid to the conditions which had to be satisfied for adoption to take place, but very little to the question of whether it was actually happening.

One reason I have often heard why so little attention was paid to adoption in those years was that there was no accepted methodology for it. As you will know by now, I have been very critical about the usefulness of a lot of FSR methodology, but it had one major advantage. A particular activity would acquire a kind of official status by having a methodology attached to it, which told the practitioners that it was part of the game and that their job was not complete until they had done it. Regrettably that has not been the case for adoption, which is surprising in view of adoption being the ultimate goal of FSR. Some of us did realise early on that monitoring of adoption should be part and parcel of on-farm research, but we probably did not make enough noise to convince others. In IITA's on-farm research villages in south-western Nigeria, however, we did monitor adoption, although in an admittedly rather rudimentary way. Box 8-3 explains how it was done for a maize and a cassava variety and fertilizer.[14]

The findings turned out to be quite revealing, even exciting, for three reasons. First we had to face the truth about the maize variety and fertiliser which were dropped by the farmers as soon as we had turned our back while the cassava variety was actively adopted. Second, by tracing the spread of the cassava variety by farmer exchange we saw that it was most rapid within a 2 km radius and rapidly petered out beyond. And that of course had interesting

---

[14] You may start thinking we only carried out variety trials, but that is not so. There were also tests with cowpeas, soybeans, yams, planting density, alley cropping, oil palm and other things. The maize–cassava tests were used as examples because they were the easiest to explain.

Box 8-3. Monitoring technology adoption; an example from south-western Nigeria

---

Two years after farmers had tested a new maize and cassava variety and fertiliser, we revisited them to see what had happened to the technology. They knew that we supplied seed, planting material and fertiliser only once, so if they fancied the varieties they had to keep their own seed, as they would normally do, and if they liked the fertiliser they had to buy it through the usual channels. It turned out that only a minority kept seed of the improved maize; apparently a yield advantage of 15–20% was hardly noticed or did not seem worth the trouble. None of the farmers went out of their way to buy fertiliser. As we saw earlier this was not surprising because the risk was too high, but you would expect that at least some of those who got a considerable yield increase would buy fertiliser at least once, which was not the case. Economists explain this by saying that the transaction cost for obtaining the fertiliser were too high. That of course is a circular argument: just increase the transaction costs until your model 'predicts' what you see, in our case non-adoption. But yes, the transaction costs were high. The farmers complained that the people at the (government) fertiliser depot, at 20 km from the village, were not helpful, they told them to come back tomorrow, asked for bribes, would not supply small quantities, the usual things. We suggested that they could organise themselves, hire a pick-up and ask the extension agent to accompany them. They thought that was a brilliant idea but did not bring it into practice and asked us to deliver it to them instead, which we would not do. The question remained why *nobody* took the trouble to buy fertiliser. Compare that with the northern savannah where farmers would queue up for hours to get their supply. But in their case maize would yield practically nothing without fertiliser, whereas in the forest the crop could still yield up to 2,500 kg/ha, plus a considerable amount of cassava. The cassava story was interesting too. We went around to ask the farmers to whom they had given or sold stem cuttings and then we went to the recipients and asked them the same question. In 2 years the maximum spread of the variety was about 2 km from the points of introduction. The spread would of course continue afterwards but for rapid self-extension one would have to 'seed' the variety at intervals of no more than 4 km.

---

implications for extension. The third lesson was that economic analysis had no explanation to offer why even the successful farmers would not go out of their way to buy fertiliser after seeing its effect in the tests. Except in a tautological or circular sense, by explaining non-adoption from the fact that no adoption was taking place (Box 8-3 has the details). This obviously brought out some interesting observations which could not have been made without looking specifically at adoption.

Most on-farm research programmes I have known carried out an economic analysis of the test results, followed by some kind of participatory assessment with the farmers. If both were positive they would declare the technology 'adoptable' and then 'transfer it to the extension service', which meant that they stopped bothering and considered the rest as the concern of the extension service. The consequences of not paying serious attention to adoption have been quite grave. I will give just one example, from an FSR programme in East Africa with which I have been (loosely) associated for years (as an evaluator as well as an adviser), where a lot of interesting technologies were tested by Farmer Research Groups (which were themselves an innovation introduced by the programme). Now, we all know that farmers are usually friendly people who do not want to offend their visitors, so the evaluation of innovations like milk churners, wheelbarrows and donkey carts invariably came out assessed as 'favourable' or 'very favourable'. During an evaluation mission I raised the question of whether there was any real, active adoption by the farmers, independent of the presence of the researchers. The answer was that the question was premature and would be answered once the farmers and the manufacturers of the equipment had gathered enough experience. After some years a new team leader took over who reorganised the programme and discontinued the work with the milk churner and the other technologies. That would have been an excellent opportunity to find out what happened to the technologies but the researchers paid no further attention to their adoption. I suspect there was really no need either because it must have been obvious, from casual observations, that the farmers dropped them as soon as the researchers did. Besides, a new generation of researchers is not necessarily keen to complete the work of their predecessors. In any case, nothing was heard about those once highly regarded technologies again. That has been the fate of many if not most technologies, which after having been declared adoptable by the on-farm researchers, disappeared without a trace, simply because their presumed adoptability was based on self-delusion on the part of the researchers.

There were of course counterexamples, of researchers who did continue to follow the fate of their technologies once on-farm testing had been completed. In 1991 Robert Tripp of CIMMYT brought together nine case studies of successful on-farm research, most (though not all) of them involving some *post-mortem* adoption studies. They were all joint projects between national and international institutes. The cases of proven adoption I found most convincing were climbing bean varieties in Rwanda, dry-seeded rice in the Philippines and diffused light storage of seed potatoes in Peru. All three of them were well-known technologies which were successfully tested on-farm and adapted by farmers and researchers to make them more suitable. I do not know what has been the sequel of the climbing bean story, after the initial successful expansion reported in the study, but dry-seeding of rice and diffused-light storage of potatoes have been genuine success stories within and far beyond the countries of initial testing. These were hopeful examples, although Tripp in one of his papers cautioned that 'the relatively few documented cases where OFR has led to substantial adoption give cause for concern'. Tripp's papers, apart from analysing the cases, also took stock of Farming Systems and On-Farm Research at that time, and he arrived at conclusions which were similar to those expressed in this book.

I am not aware of recent studies on the impact of on-farm research which have the same scope and thoroughness as Tripp's collection. Most of the papers on the history of FSR edited by Michael Collinson and published in 1999 were a little too self-congratulatory to be really convincing. But there have been local cases of well-studied adoption which have not been publicised widely enough, like the success story of *mucuna* in Bénin Republic which I mentioned in Chapter 7. I am sure more can be found. It is a relief when one sees occasional examples of good honest adoption studies, which, however, do not balance the abundance of ill-founded or even plainly false claims made about the success of on-farm research.

8.5 Client orientation

The story about FSR methodology has taken us to the beginning of the present century, but we must now move back a decade and a half and examine FSR's relationship over the years with the farmers and with those who claimed they could speak on their behalf: extension agents, leaders of professional organisations and cooperatives, NGOs, etc., in short their direct and indirect clients. Through such relationships the client should ensure that research works for their benefit and comes up with the expected results.

By the late 1980s, while many of us were busy refining diagnostic and experimental methodology and getting more participatory, some thoughtful people started wondering whom all this increasingly complicated FSR was actually meant for. You would think that was obvious, because FSR had been invented to make research more relevant to farmers. But, paradoxically, the obsession with methodology was obscuring more and more the original objectives. Instead of being a sensible attitude to applied research, FSR was degenerating into a discipline of its own, with a lot of phoney methodology which could be taught in training courses and universities. I know this is not entirely fair and there were still many good minds around who did not take part in this game, but the signs of decline were real. So let us ask some basic questions about what we were doing and why it was urgent.

8.5.1 What is client orientation?

ISNAR, one of the international institutes, which for some obscure reason had its headquarters in the Hague,[15] took the lead in this quest in the late 1980s. They hired a young consultant, as yet unaffected by the methodological virus, Deborrah Merril-Sands, who concluded that we had indeed forgotten whom we were supposed to be working for. And, of course, she coined a new name for what she thought FSR should be: *On-Farm Client-Oriented Research* (OFCOR). She had a point; that is what FSR should be: research carried out on-farm and oriented towards the clients, who were, in most cases, the small farmers. The new term caught on, after the first part had been dropped, so it became COR, Client-Oriented Research. The implication of dropping the 'On-Farm' part from the name was that client orientation should be everybody's business, not just that of the on-farm or farming systems teams, who were doing it in their own corner, in isolation of the rest of the research establishment. This was a great leap forward. The next question was, how do you make sure that all of applied research indeed becomes more client-oriented? That is where the profession showed its inexhaustible capacity for corrupting sensible ideas by substituting ritual for common sense.

---

[15] It was later transferred to Addis Ababa in Ethiopia.

For ordinary people, like businessmen or industrialists, being
client-oriented is a condition for survival, because the clients will
only buy what they like and leave the rest on the shelves. If that happens too often the company will go broke. For agricultural
researchers there were no such powerful and pitiless customers.
Research was funded from public means or, in the case of developing countries, its substitute, foreign donors. In both cases there was
only the semblance of accountability. The farmer–clients had no
voice nor was it their money that paid for the research anyway. The
question whether research was useful or not was essentially
answered by the scientists themselves, or by other scientists who
were on the editorial boards of the scientific journals or on the evaluation teams sent by the donors. So, what could be done to make
FSR more accountable to its real clients? Obviously, that was a very
important question, and one which became the major concern by
the closing years of the last century. It had obviously not been
enough that FSR by definition should be client-oriented, because
many practitioners had strayed from the straight path, so there was
need for a powerful mechanism which would steer them back. But
agricultural technology is quite different from manufactured products, which clients can choose to buy or not to buy, so a different
kind of mechanism had to be found to discipline the researchers
into producing what their clients wanted. Participatory methodology did not seem strong enough for that, because technology adoption remained far below expectation. Some donor-funded projects
had therefore been experimenting with new ways to give farmers a
more direct say in the research process. Being staffed by development workers who often bore a strong resemblance to bureaucrats,
committees would be set up where researchers, extension workers,
local bureaucrats and farmers would sit together to discuss the
results obtained so far, whether they were satisfactory and what new
research was needed. In that way they hoped the research agenda
would become more relevant.

8.5.2 Enter the World Bank

Before there was even a shred of evidence that this new client
orientation was making any difference the idea was picked up by
the World Bank (officially known as International Bank for
Reconstruction and Development, and often simply but reverently

referred to as the Bank). That great institution had the means and the power to push through anything it fancied, without effective mechanisms to put on the brakes before things went too badly wrong. How had they got involved in agricultural research in the first place? That did not seem to be their area of competence. In order to explain that I have to give a brief historical account of the World Bank's involvement in agriculture, part of which I have already related in earlier chapters.

The Bank of course had always been interested in agricultural development, because in third world countries it was and remains the most important economic sector. In the early years they funded ADPs in many countries and later, when the times of stand-alone projects were over, they invested massively in national agricultural bureaucracies, especially the extension services. The funds were used to re-equip their offices and train their staff in the Bank's favourite extension methodology: the Training and Visit system, popularly known as T&V. I will come to that in Chapter 10, but I must say a few words about it here, because research was expected to play an important role in it (I now realise that this is a two-level hierarchy: World Bank within FSR and T&V within World Bank; I hope you can keep track of the different levels). T&V was designed to reach as many farmers as possible by working with contact farmers, each of whom would transmit the information they received to a number of other farmers. T&V involved different kinds of training, one of which was the Monthly Technology Review Meeting, where research scientists would come to educate extension personnel about the latest technologies. The latter would then in turn teach the farmers in fortnightly training sessions. It soon turned out that research had precious little to offer and the Bank officials therefore decided that the research system also had to be overhauled if it were to come up with the kind of technology that was needed. That was the only way to ensure that all the money which had already been spent on the extension services would not be lost. So the Bank started to invest in agricultural research as well.

Once the World Bank had decided to put its weight behind some cause they would scout around for ideas, convert them into bankable projects (their terminology) and talk the governments of poor countries into asking for a loan to carry them out. That process was called loan negotiation. The projects they designed in support of agricultural research were quite conventional affairs in the beginning.

They would simply make considerable resources available, backed up by technical assistance, to enable the institutes to reformulate their research programmes and refurbish their facilities, hoping that they would then start to crank out innovations. The ISNAR was usually contracted to provide technical support for programme reformulation. That of course had been their stock-in-trade all along, but this time they introduced a new methodology, called ZOPP. That is a German acronym[16] which means Objective-oriented Programme Planning. It had been developed by the German development organisation GTZ in the 1980s. ZOPP involved a lot of meetings and workshops spread out over a whole year in which all the scientists and research managers as well as other so-called stakeholders participated. The process was facilitated by ISNAR scientists shuttling up and down between the Hague and the country in question. During those workshops all scientists would compete to get their pet projects included in the new programme and the final outcome was usually quite conventional shopping lists of research topics with only marginally higher relevance than what was being done before. These new programmes, however, had the distinction of being the result of a participatory process and bore the stamp of approval of a respected though rather hollow international research institute. On this basis the World Bank loans were approved, the money started flowing and the work could start.

After the refurbished research programmes had been running for some time, however, the Bank became aware that in spite of the solid planning process and the considerable amount of money and new vehicles and equipment now available, nothing much seemed to have changed. So, something must have been overlooked. Perhaps because of their association with the World Bank, ISNAR had come to the same conclusion and argued that the lack of client orientation was the reason why there was so little progress in the Bank's projects and why researchers were just continuing with their bad old habits. Research should become more client-oriented, and not only FSR but all research, so ISNAR took the lead in developing the inevitable methodology for client orientation and the Bank hired some new people on their Washington staff who understood

---

[16] Ziel Orientierte Program Planung.

the new ideas and who were going to make them operational. The main challenge was to make sure that farmers would actually come forward and tell the researchers what kind of technology they really needed, and, as importantly, that the researchers would actually listen to them. That, of course, had been the basic idea of FSR, but its record so far had shown that it was more difficult than we had thought, in spite of all the participatory methodology and farmer research groups and all. But how would another basket concept like *client orientation* change that? How would these ideas reach not only FSR scientists but even those who used to work in the quiet seclusion of the research station and who felt perhaps that their occasional farm visits largely satisfied their appetite for client orientation?

The natural thing for civil servants or development bureaucrats to do when they are faced with a challenge like this is to set up one or more committees. And since client-oriented research was gradually being taken over by development bureaucrats, committees became the essence of the new approach promoted in the next generation of World Bank-funded research projects. There were going to be 'user committees', at least at two levels, one regional, associated with the regional research institute or institutes, and one national. All the stakeholders had to be represented on the committees of course and the research clients would be in the majority. These committees are not to be confused with the already existing ones, like the Institute's steering committee and the scientific committee, of which the former also included various stakeholders. Then there were going to be Agricultural Research Funds, a national one and one in each region, each naturally with its own steering committee. Those funds were meant to introduce competitiveness into the system, as well as further strengthen its client orientation. The institutes could bid for research funds by submitting research proposals which would then be assessed by the Fund's scientific committee (not to be confused with the Fund's Steering Committee). The proposal had to be developed together with the client and the latter would certify by his signature that this had indeed happened. There were differences in the model according to the country, but the principles were the same everywhere. I could of course have included an organigramme here to show all those committees and their roles and mutual relationships, but I doubt that would make things any less obscure or depressing.

8.5.3 And did it help?

Well, no, of course it did not. You can set up any number of committees or other talk shops, but unless the farmers or their representatives wield real power over the researchers' agenda, there is unlikely to be significant change. The tacit assumption on which all those committees were based was that all would be well if farmers sat on a forum where they could express their views and request the scientists to find solutions for their constraints and problems. But in fact the problem was not so much with the farmers as with the scientists who paid lip service to participation and client orientation while just continuing to do what they had always done: set their own goals and carry out the research the way they saw fit, preferably at the station. There were of course individuals who took client orientation seriously and made genuine efforts to adapt their research to the interests of their clients, but I think the majority played along in the feel-good committees, mainly in order to obtain approval to pursue their pet projects for another year. And the farmers who were elected or appointed on the committees were not necessarily motivated most by the interest of the people they were supposed to represent. So, the whole thing developed into a farce, because decision power about research content was not put in the hands of the clients, but remained in those of the researchers.

The so-called competitive research funds which were invented and mostly sponsored by the World Bank in several countries were another farce. The idea was that researchers would submit proposals for funding which would then be examined by an independent committee for scientific rigour and relevance for the intended target group. The funds' key innovation was that the researchers had to design the proposal in consultation with the intended target group and the proposal had to bear their signature for approval. Now that may sound like a sensible arrangement, but only to people who are unfamiliar with the many ways these things usually get distorted in developing countries, in Africa in particular. Very soon the whole thing would degenerate into the usual ritualistic phantom processes and result in collections of research projects with even less underlying structure and only marginally more relevant than those of the old days. Many of the bilateral donors apparently felt intuitively that the funds did not serve any useful purpose and only few of them responded to the Bank's requests to contribute.

The only way to really make researchers listen would have been by giving real power to farmers and letting them decide what kind of issues (applied) research should address. And that would only have been possible if they or their representatives had control over the research money. The implication would have been that most applied research was no longer financed directly through untied government or donor donations but that the clients or those who represented them purchased (or were enabled to purchase) research services and contract whatever institution would serve them best. That was a bridge too far for the decision makers. I will return to it at the end of this chapter.

8.6 Development expertology[17]

This historical account of FSR and its offspring must have bewildered many readers, in spite of my attempts to give it some structure. That cannot be helped, it is in the nature of the subject, but I will now briefly sum up the story, before looking beyond the past to the future.

We have seen how a simple and sound concept started to revitalise applied research in the 1970s and 1980s and motivated its pioneers to leave their research stations and work in the real world with the real farmer. In a matter of years of vigorous growth FSR developed into a global movement attracting many adherents and a lot of donor money. Simplicity of an idea is no guarantee that it will be understood by people who have all sorts of reasons not to and FSR neophytes were often seen to continue their previous ways, even though they mastered the new FSR speak. The FSR leadership therefore kept trying to steer them to the right path by providing ever more methodology for diagnosis and on-farm experimentation, to the satisfaction of those who were more attracted by methodology than guided by common sense.

The methodological toolbox underwent a major enrichment in the late 1980s and early 1990s with the addition of participatory approaches, followed by the *sustainable livelihoods* analytical

---

[17] The word 'deskundoloog' was coined in Holland in the linguistically creative 1960s, meaning somebody who claimed to be an expert ('deskundige') in an area of uncertain description. Today expertology is more alive than ever and the word itself seems to be getting a new lease of life.

framework around the turn of the century. These two resulted in significant embellishment of the research reports with charts, diagrams and matrices, but they hardly affected the quality of the work itself. In particular technology testing and adoption remained the orphans of FSR, in spite of the numerous committees and platforms which were created to strengthen client orientation by providing forums where stakeholders could make their voices heard, though not necessarily listened to.

The participatory/livelihoods development movement had its roots in agricultural research but the now much extended scope of its methods made them applicable to a wide range of human endeavours. So the movement set itself free from the bonds of agricultural research and its concepts and methods were raised to the status of a comprehensive development paradigm. Obviously the name FSR was no longer appropriate, but no new name emerged in its place. I will refer to it by the term *livelihoods development*, which must be understood to imply the use of participatory approaches. Much like the original FSR concepts, the *livelihoods development* ideas made a lot of sense, but by cutting itself loose from agricultural research and development the movement lost its direct links with practical farming. And by wallowing in ever more complex analyses it became an academic discipline at best and sterile voyeurism at worst. Several institutions offered training courses to teach participatory and livelihoods methods and launched themselves into the consultancy market as experts in such methods. Meanwhile, the experiential content of what they taught had decreased in inverse proportion with its complexity. Whereas the early FSR thinkers had at least stood with both feet in the field, the new generation were mainly academics, widely travelled consultants, ISNAR staff, and World Bank bureaucrats, many of them with only a cursory exposure to reality, through frequent but superficial visits, often in the service of the aid agencies.

The movement was eminently successful in generating idiosyncratic language, however, which turned out to be very useful for an entire generation of development consultants. I will come to the consultants in Chapter 10, but I want to say just a few words here about their language. An amusing table was put together some years ago, I do not know by whom,[18] which allows you to generate

---

[18] Someone gave me a photocopy of the table with the author's comments but without his/her name or the reference.

the kind of vocabulary development thinkers and consultants had become infamous for. I have added a few entries of my own which reflect the methodological developments since the original table was published.

Centrally	motivated	grass roots	involvement
Rationally	positive	sectoral	incentive
Systematically	structured	institutional	participation
Formally	controlled	urban	attack
Totally	integrated	organisational	progress
Strategically	balanced	rural	package
Dynamically	functional	growth	dialogue
Democratically	programmed	development	initiative
Situationally	mobilised	cooperative	scheme
Moderately	limited	ongoing	approach
Intensively	phased	technical	project
Comprehensively	delegated	leadership	action
Radically	maximised	agrarian	collaboration
Optimally	consistent	planning	objective
Genuinely	*participatory*	*analytical*	*framework*
	client-oriented	*livelihood*	*analysis*
	sustainable	*holistic*	
		stakeholder	

Pick any one word from each column going from left to right and in three out of four cases you will have a recognisable title for a development programme. But, in the author's words: '[I]f two or three people were each to write a paragraph explaining one of these phrases to the masses, on behalf of the government of Ruritania, their different interpretations should bear further witness to the malleability of such language.' I admire the author's mild way of putting it.

This table reminded me of a long car trip I made to eastern Nigeria in 1991 or 1992 when out of boredom I started jotting down the names of evangelical churches displayed on roadside signboards. My list has 41 entries but I am sure the real number runs into the hundreds. I converted the list into a four-column table similar to the one for development programmes, with this result:

(Jesus) Christ('s)	Prophetic	Gospel	Church (Int.)
Jehovah's	Crusaders	Spiritual	Ministry(ies)
God's	Faith	World	Mission(s) (Int., Inc.)
True	Chosen	Apostolic	Crusade
Holy	Lamb	Bible	Movement
Greater	Charismatic	Sabbath	Congregation
Divine	Vessel	Christian	Assemblies

Sacred	African	Adventist	Society
United	Apostolic	Cherubim & Seraphim	Order
First	Redeemers	Fishers of Men	Witnesses
Eternal	Life	Headstone	Disciples
Deeper	Tabernacle	Believers'	Brotherhood
Zion	Evangelism	Cross and Star	
Master's			

For example: The Sacred Lamb Spiritual Ministries, or The Divine Vessel Sabbath Congregation. The church compositions are definitely more poetic than those from the development vocabulary, and though both are linguistically consistent they are equally devoid of meaning. Or, in the style of the churches, God's gift of thought come to nought, unless I am missing something.

What started off as a sensible and highly practical approach to applied agricultural research had now become a powerless game of words around an increasingly holistic concept of development, with ever decreasing reality content which truly qualified it to bear the name of expertology.

8.7 Does FSR have a future?

Had the pathology of FSR by the end of the last century progressed to the point where a cure was no longer possible? So many things had been piled upon it that it could barely be recognised as the descendent of that simple but cogent idea which attracted so many good people in the 1970s and 1980s. Perhaps things were not as bleak as I thought, though, and FSR, having survived for so long, might just purge itself of the elements which were poisoning it and obtain a new lease of life. Let us look at the themes of the 2005 symposium of the IFSA and the papers presented there and see whether there was hope that the illness was not terminal after all.

8.7.1 Themes and topics of the 2005 IFSA Symposium

The title of the symposium, which was co-organised with the FAO and the International Fund for Agricultural Development, was 'Farming systems and poverty: making a difference'. That does not sound promising, but let us not jump to conclusions just on the basis of a title. The themes of the symposium should reveal the current concerns and the direction in which FSR was moving. They are shown, somewhat shortened, in Box 8-4. The first thing that will

Box 8-4. Themes (*shortened*) of the 2005 IFSA Symposium 'Farming systems and poverty: making a difference'. (From IFSA GLO website: www.ifsaglo2005.org)

Some major thrusts for the Symposium: mainstreaming of good practices, partnership strengthening and mobilizing resources for sustainable agricultural and rural development (SARD). Other key contributions would include: improved Civil Society Organisation (CSO) interaction with Science and Technology community on SARD; enhanced local-knowledge swapping, knowledge pooling and learning; local peer-to-peer communication for innovation and learning at national and local grass roots levels; further collaboration between CSO, Inter-Governmental Organisations (IGO), private sector and Governments. The themes add value to ongoing dialogue on the effectiveness of development approaches and provide platforms and options for launching field-level actions.

Theme 1 – Food, Agriculture and Rural Development Policies in a Globalising World
Decentralisation and privatisation processes have contributed to the collapse of services to farmers. Farming systems practitioners can report impacts on smallholders, exciting R&D work in the field, debate good practice and the 'State of the Art'.

Theme 2 – Trade and Market Linkages
A richer understanding is needed of the complexity of market chains for small producers and urban consumers; the growth of local, organic and fair trade food systems; the markets for seeds and other farm inputs. The synthesis of field experience on impacts on smallholder farming families, the rural poor, would contribute to the identification of appropriate targeted policy responses.

Theme 3 – Knowing and Learning Processes
Understanding how rural people learn and share experiences has replaced the paternalistic client orientation of rural development during the 1970s–1980s. Strengthening peer-to-peer learning for better decision making in poor farm households, local multi-stakeholder collaborative learning at the field level in SARD, NRM, marketing, demand-driven advisory and research services.

Theme 4 – Development Strategies, Pathways and Synergies
Based on the characterisation of farming systems, differentiated development pathways can be identified around which multi-stakeholder alliances and public–private partnerships (e.g. CSO, intergovernmental organizations, governments, private sector, etc) could coalesce. Such alliances are required for managing ecosystems and landscapes, intensification strategies, environmental services, conservation of the natural resources and biodiversity, livelihoods diversification, conservation agriculture, eco-agriculture, urban agriculture, and linking mitigation, risk and response for vulnerable systems and populations, including HIV-AIDS.

strike you is their breadth and the small role played by traditional issues. Apparently it was no longer agriculturists, and certainly not agronomists, who dominated the association. That is also reflected in the association's name from which 'research' had been dropped long ago. Further, it was obvious that the movement had definitely left the farmers' fields and moved up to the level of institutions, policies, markets, in short, the non-physical environment in which farming was embedded. That is the cosy world where international donors and consultants and the large international institutions which co-organised the symposium felt most at ease.

It is true that the defective institutional and socio-economic environment, including poor access to international markets, can be blamed to a large extent for the stagnation of agricultural development, especially in Africa. But that was not really the issue here. The question was whether the FSR movement should also join the already large number of people who were busy looking into institutional and marketing problems, or at least pretending to. If FSR had indeed been successful in realising its original goals and was now looking for new challenges that would perhaps have been all right. But that was far from being the case, as we have seen. I think its workers had quietly retreated from the harsh conditions of the real farm and sought shelter in areas where irrelevance was less visible. The new catchwords were 'livelihood systems', 'empowerment', 'food security', 'public–private partnerships', 'farmer–market linkages', 'stakeholder platforms', and more of such niceties. All of them topics which traditionally were the concern of agricultural economists and policy makers and which were now being appropriated by those who failed to make an impact at the farm level. I am not just talking theory here, I know from direct observation that the 'traditional' FSR programmes in the field, as far as they still existed, had moved in the same direction.

The trends can also be analysed in another way, by grouping the topics of the Symposium into the traditional FSR categories. The chart of Figure 8-6 shows the classical elements of the FSR process: diagnosis, choice and on-farm testing of innovations, adoption and extension or dissemination (the chart looks a little different from the one at the beginning of this chapter but it is essentially the same thing displayed differently to make the points I wish to make here). They do not capture all the topics treated in the papers, so I added as a fifth category: 'social, economic, institutional, policy

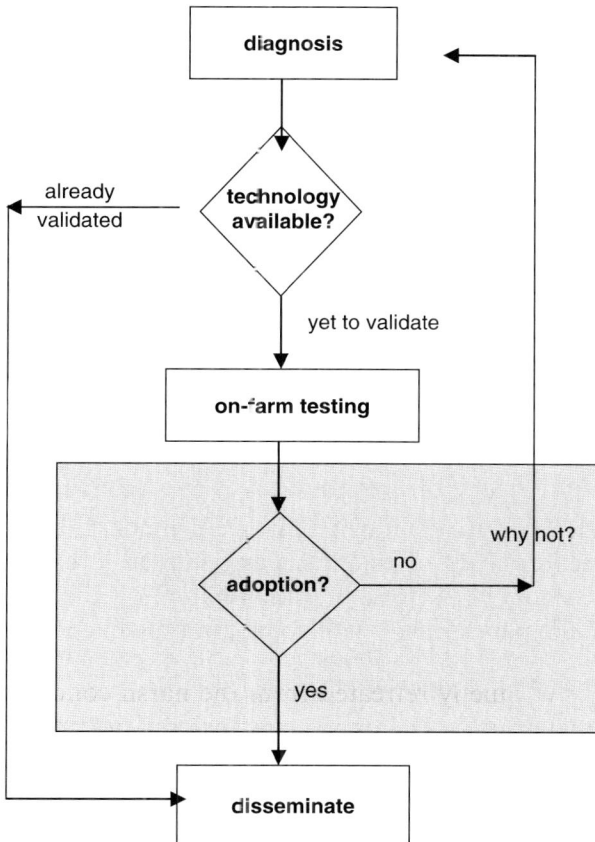

Figure 8-6. Another flowchart of farming systems research

environment'. Table 8-2 shows the numbers and percentages of papers in each of the five categories, while the full list of titles are in Appendix 6, if you are interested in the finer details. What can be concluded from the subject matter of the papers in each category?

Diagnosis continues

Diagnosis was continuing unabated, although the topics had changed a little with the trends of the times. New topics were: sustainability, food security, livelihood analysis, biodiversity, and there was a definite shift to broader subjects at the expense of the traditional farm-level issues. Otherwise it was business as usual.

Table 8-2. Categories of papers submitted for the 2005 IFSA Symposium

	Category	Papers Number	%
1	Diagnosis, policy analysis	48	27
2	Choice, testing, evaluation of innovations, technology	27	15
3	Extension, knowledge, learning methods	43	24
4	Adoption, impact studies	8	4
5	Socio-economic, institutional, market, policy environment; 'platforms'	55	30
	Total	181	

The 27% devoted to diagnosis may seem modest, but that is an illusion, because a lot of diagnosis was hidden in category 3 and some in category 4 as well. I think that altogether as much as 45–50% of the papers dealt with some kind of diagnosis.

Choice and testing of innovations

The percentage of papers dealing with the choice and testing of technology was quite low. There had also been a shift in emphasis here, with few papers on conventional technologies like crop varieties, fertiliser, weed control and topics like that and many more on water and soil fertility management and soil conservation. Surprisingly, a number of papers even dealt with such grandiose subjects as the introduction of new farming systems, which were almost anathema to the early FSR workers.

Extension, knowledge and learning

We have seen earlier in this chapter and in previous ones that the relationship between FSR and extension has been chequered, with research blaming extension for not disseminating their technologies and extension complaining that what was called 'adoptable' was only so in the eyes of the researchers themselves. And indeed, on-farm researchers had often declared their technologies proven before genuine adoption was observed, even on a small scale, with disappointing results once the sheltered conditions of the tests no longer applied. However, instead of becoming cautious and making adoption studies an essential part of their research protocols, FSR scientists had instead gravitated into the realm of extension itself and started meddling in extension affairs, as the topics in Appendix 6 under category 3 clearly demonstrate. This was perhaps understandable, given the rigidity and sometimes lack of competence of the government extension services, especially in Africa. And it was tempting, because many promising ideas had emerged in the closing decades of the last century about the way extension could be improved, such as the Farmer Field Schools pioneered by FAO around integrated pest management (also known as experiential learning), farmer counselling and farmer-to-farmer learning. You may think that I am being rigid by insisting on a strict separation between research and extension, but that is not what I mean. Both research and extension of course have to reach into each other's realms and work closely together, that has always been the position of FSR, even if it was often not brought into practice. But that does not mean that the distinction should be completely blurred. Research and extension do have their own roles, the former to identify and test innovative technologies, the latter to disseminate them and create methods and structures to enhance dissemination. If research stops producing technologies farmers can and will adopt, no amount of schooling, counselling or learning can make up for that. So, research which meddles in extension instead of doing its own job first does not solve anything.

Technology adoption: still groping in the dark

What about technology adoption, surely the only proof that FSR had lived up to its promise? Were farmers seen to be actively adopting the innovations which they had tested in their own fields? Not really. As we have seen, FSR had kept itself busy with its endless diagnoses, participatory or otherwise, as well as with on-farm testing and participatory evaluation of a wide range of innovations, from lowly varieties to quite complex technologies like alley cropping, but, surprisingly little was known about the actual adoption of those innovations. From the 2005 Symposium papers it is clear that things had not changed for the better, quite the contrary. The number of papers about adoption and impact submitted to the symposium was very small. Their paucity could have been due to the chosen themes which perhaps did not attract the more practically minded on-farm researchers, but I doubt it. I think there was simply very little to report and I know something about that, because as a consultant I have annoyed a lot of people with questions about adoption in the last 12 years or so.

The socio-economic and institutional environment

The heavy load of topics in this category showed that FSR workers had come to realise that the success of much of their work depended to a large extent on the limitations imposed by the socio-economic and institutional environment. The question was, what kind of role, if any, could researchers play to overcome or mitigate those limitations? A lot of thinking had gone into that and the results showed through in the papers. Apparently many of the symposium participants, mostly FSR workers I presume, had come to the conclusion that they themselves had to take action to mend even those defects which should have been the responsibility of others. That was a typical response of well-intentioned but impatient and often naive foreigners by whom the world of development programmes was dominated. The new catchwords were therefore *farmer–market linkages, public–private* and other *partnerships, platforms, networks* and *interaction*. Those terms had almost eclipsed the *client orientation* of only a short while ago and platforms were pushing aside the committees which had been set up under that label. At the same time the original concern with technology had, to a large extent, been eclipsed by institutional concerns, and technology-oriented agriculturists had definitely been pushed aside by development bureaucrats.

8.7.2 *Looking ahead*

By looking at this collection of papers and talking to some of today's FSR workers one gains the impression that FSR's offspring had moved very far away from the original ideas and evolved to an important extent into a general development movement, more concerned with diagnosis, systems and institutions than with technology. I think that many researchers (not all) continued to ignore the painful truth that the goals FSR had set itself remained as remote as ever and that they preferred to fool themselves with the latest methodological fashions: sustainable livelihood analyses, vulnerability analyses, food security analyses, multi-actor/stakeholder platforms and building partnerships and linkages all over the place.

But there were also signs of vitality, such as the fact that the IFSA was still there in 2005 and attracting a lot of enthusiastic people. So, let us see what can be done to reach back into FSR's healthy core and restore to the movement its original sense of purpose and its drive to attain concrete results, for the benefit of the smallholder farmer. I will not dwell further upon topics which have already been exhaustively (or perhaps excessively) covered, such as diagnosis and on-farm experimentation, but rather insist on three points which I think are paramount to the future success of applied research:

- The use of FSR concepts and methods should become everybody's business in applied research.
- Active adoption of new technology by agricultural producers should be a primary concern of the researchers.
- The relationship between applied research and its clients should be urgently overhauled.

(a) FSR as everybody's business

Setting up permanent FSR teams was not a good idea. At best, they helped researchers to find their bearings and establish credibility, at worst they isolated them from the rest of the institute and made them harmless in their cosy corner. The fathers of FSR, although they did promote the creation of FSR teams, did not intend them to be permanent as I have discussed earlier. But once the teams were there they turned out to be very hard to dismantle again, nor was that surprising, given the international attention and the lavish funding of the early years. Even by the end of the century many of them still survived, even though the glamour and most of the funding were gone. Fossilisation of the FSR teams may have been one of the reasons why fresh ideas, like the need for more participatory approaches and more client orientation, did not really catch on, either with the teams themselves or with the rest of the research establishment. All these are good reasons to dispense with the FSR team idea entirely and make the FSR ideas and their offshoots the preferred operational mode of applied research at large.

(b) The primacy of technology adoption

To the uninitiated the primacy of technology adoption may seem so obvious that it does not really need all the words I have already devoted to it and am still going to devote. After all, adoptable innovations are the justification for applied research in general and FSR in particular and if the innovations are not seen to be adopted, research has failed. It is as simple as that, I think. Still, many FSR workers have found and continue to find reasons why that criterion does not apply to them: the time has not yet come, because more diagnosis is needed or the technology needs further testing and adaptation, or other people, especially the extension service, have to get their act together first and the farmers need training before they can properly apply the technology, etc. FSR workers have got away with that kind of argument for more than two decades, primarily because their clients were not able to clearly articulate their need for innovations, force the researchers to respond adequately and throw them out if they did not deliver. And quite often the question about adoption was not even raised, because by the time it became relevant the project had finished, or the researchers had rushed on to yet other technologies. One way to make sure that the issue is at least raised is by making the analysis of adoption (not 'adoptability') a compulsory part of FSR/OFR methodology. If done honestly and skilfully, adoption monitoring is a double-edged sword: it shows which technologies are really adopted and, as importantly, which are not and why not. If they were to obtain that kind of information FSR workers might finally have some real impact and be taken seriously by their peers, the people at the research stations who create the technologies.

(c) Platforms and partnerships or clients and service providers?

The third issue which is tightly linked with the previous one is the nature of the relationship between research and its intended beneficiaries. Traditionally that relationship had been very loose and informal, in the industrialised world as much as in underdeveloped countries. Research was mainly financed by government or its proxies, the donors, which trusted the researchers to have the farmers' interests at heart and work for their benefit. FSR was invented because the model did not work well, especially in the tropics where the researchers appeared to pay more attention to the editors of the international research journals than to the peasant farmers for whom they were expected to work. The history of FSR can be interpreted as one long struggle to bring the clients' views and interests to bear upon the content of (applied) research. We have reviewed that history in great detail and observed how the struggle mainly produced an ever swelling stream of methodologies and jargon as well as various smokescreens to hide the painful truth that quite little progress was being made. In the early 1990s the FSR movement admitted that there was no genuine 'client orientation' and, under the guidance of the World Bank and the ISNAR, created all kinds of committees to give a voice to research's clients, as a substitute for real client power. That did not work either, of course. It was naïve to expect researchers to step down from their lofty position at the top of the intellectual pyramid and accept to play the much humbler role of service provider, responding to demand from its clients and carrying out the research they needed. And yet, that is what FSR had pretended to do all along.

By the early twenty-first century it had become clear that drastic steps were needed if the public research institutes were to be salvaged and restored to their intended role. Creating mixed committees, setting up competitive research funds and preaching client orientation did not lead anywhere, the main reason being, in my opinion, that the clients had no real power over the research funds. There had been some lukewarm attempts by donors in the 1990s to link the development projects they sponsored with the research institutes which they also sponsored and to provide the former with funds to purchase research services from the latter, especially in Tanzania by the Dutch and the Irish. Bart de Steenhuijsen-Piters of the Royal Tropical Institute (Amsterdam), who worked at the Selian Institute, also in Tanzania, made brave attempts to position the institute as a research contractor and concluded research contracts with several organisations. I do not know what eventually came of these attempts, but I doubt they had much impact. For one thing the projects which pioneered them did not last long enough and second, the attempts were too incidental and there was no clean break with the conventional system of direct funding. And probably the times were not ripe for that kind of initiative.[19]

There was of course much resistance among research institutes to the idea of applied research contracts. Instead of acknowledging

[19] More recently some French-funded development projects in Africa have been pioneering contractual research again, this time with small private research units, which were usually also set up with French support as well (and often with staff from previous development projects). I will come to them in the final chapter.

the need for clear contracts between research and its clients, the research institutes designed various manoeuvres to circumvent the painful measures that would be required. Assisted by their partner institutions from the industrialised countries (in fact their consultants) they came up with the idea to organise *platforms, forums* and *partnerships* to bring all and sundry together around the table in the best bureaucratic tradition. They hoped that, by some magic, each of the *stakeholders* would then start playing the roles they had thus far forsaken, thereby creating an environment where research could flourish. That I think was highly naive, surprisingly naive even after more than 40 years of development aid. I am not saying platforms and forums cannot play a catalytic role. I am sure they can if used wisely at the right time, but they are no substitute for the clear and businesslike relations which are needed between the providers of research services and their clients. That may sound like anathema to the defenders of research as an autonomous activity unaffected by the day's fashions. Autonomy has indeed been proven over and over again to be most productive for pure and perhaps to some extent for strategic research and long-term technology development, but applied research is a different matter, for reasons which I do not need to repeat again.

In the early 2000s the word *partnership* in particular was popping up in all those places where actors had failed to make straightforward relationships (with the attendant duties) work. It is this kind of mystifying term which had become the trademark of trouble-evading development bureaucrats and their allies in the NGOs and Universities' international development departments. Partnership is all right to characterise collaboration among actors who have complementary or adjacent tasks (I am also slipping into the jargon), but the relation between applied research and its beneficiaries is, or should be, a completely different affair: research provides services and produces 'goods' and information for the benefit of the agricultural producers, that is what the research institutes have been created for. So, where is the partnership? The term also suggests that the parties are free to enter or not enter into a relationship, but how can applied research *not* engage in a relationship with the users of its results? In Africa they can, but that is a pathological situation. So, partnership between them is nonsense. Why is it so difficult to accept that research is the provider of research services and the farmers are the clients who may be happy or unhappy with (or indifferent about) the services research offers to them? The reason is that acceptance would imply that research has to adopt a much humbler attitude than they are used to and actually start listening to the clients instead of the editors of international journals, at the price of being kicked out if the client is not satisfied. The problem, however, is that the agricultural producers in the underdeveloped world, in particular in Africa, have no power at all over research, at least not the kind of real power which would allow them to demand adequate services. That kind of power only comes with control over the money. I therefore think that the way to go is to fund most of applied research through research contracts with farmer organisations, development projects and NGOs. And if those organisations do not have the means to pay for the contracts, donors and governments can step in to provide them, not directly to research this time, but to its clients enabling them to 'purchase' research from the best provider.

Long-term research with uncertain outcome on the other hand, including technology development (which is outside the loop of Figure 8-1) and basic crop breeding, should continue to be funded from public means, otherwise it will not be carried out at all. The question is, however, how much of that kind of research poor countries can afford or should engage in while they are poor. I will not venture into that discussion, though, except for one small anecdote, to conclude this chapter.

In 1973 I visited the south of Spain to prepare for a student tour of the area, the closest proxy of a subtropical climate available in Europe. At that time the Spanish government was mounting a massive campaign to rehabilitate the ailing olive industry, which involved, among other things, rejuvenating old orchards and fashioning new trees into three stems. I went around a number of olive farms with an extension agent who was taking part in the olive *blitz*. He was a very motivated young man, strongly convinced of the messages he was bringing to the olive growers. Being an academic at the time I started wondering what the messages were based on and whether there was any back-up research going on at a research station somewhere. When I asked, the extension man replied there wasn't, nor did he think any was needed. Spain was too much in a hurry to catch up economically with the rest of Europe and could not afford to wait for the outcome of long-term research before taking action. And anyway, enough was known from research done elsewhere to design a sensible production system which could be introduced with confidence to the olive producers. Basic research could wait. That sounds like the attitude of FSR *avant la lettre*, except that the messages were handed down in a rather authoritarian way, more like the green revolution methods in India and Indonesia. But the vision about the role of research and extension was that they were there to serve the interest of the farmers in the first place. If African research and extension had really adopted that kind of attitude, instead of paying lip service to it, they might not have been mired in stagnation the way they were at the turn of the century.

Chapter 9. The Modelling Sorcerers and Their Apprentices[1]

9.1 The promise

FSR was not of course the only novel research concept of the second half of the twentieth century. Another one was the use of computer power to simulate the growth of whole organisms and the behaviour of complex systems under different circumstances by a technique called systems or computer modelling. We are now going to look at this.

Computer modelling in biology and agriculture is almost as old as computers themselves but it only made its appearance in the tropics around the time FSR reached its climax, in the mid-1980s. Around that time the modellers, who had dazzled their colleagues in the industrialised countries with the predictive power of their computer models, or at least the semblance of such power, started looking for ways to bring their blessings to the needy folks in the developing countries, as well as scouting for new money and employment for their own offspring.

Although the claims made by the advocates of crop modelling about its potential have often been excessive, there is no doubt that the power of the computer offered exciting new possibilities to push the frontier of crop science beyond the limits imposed by the human mind. Some even thought that by converting all the existing knowledge about plant processes into computer simulation models, much of conventional empirical research would become superfluous because its results would now be predictable. At a less exalted level it was expected that computer modelling would make experimental research better focused by concentrating on those processes which the models would tell us were not well understood.

If the potential of modelling was indeed as impressive as its practitioners claimed or hoped, it should also have relevance for tropical agriculture, since the biological, chemical and physical processes are

---

[1] The characterisation is due to Dirk Zoebl, a tropical agronomist and a keen analyst of agronomy's feats, fads and fashions.

the same, only the conditions under which they take place differ. But has computer modelling in fact lived up to those claims even in the industrialised countries? That is the first question to answer, before its relevance for tropical agriculture can be judged. I will therefore indulge in a rather lengthy survey of modelling in agriculture as it developed in the industrialised world. Apart from being necessary to understand what modelling is about, it will also be a welcome break from some of the rather messy stuff we have had to deal with in the previous chapters, FSR in particular. My survey is heavily biased towards the Dutch school, but that does not really matter, since several of the principal modellers were Dutch, and developments elsewhere, especially in the USA, have been very similar.

9.2 What is a model?

A model is a representation of an object or a process by something else which has analogous properties. A drawing of a plough is a model of a plough: every part of the drawing corresponds with a part of the real thing. The drawing is useful to explain what a plough does and what the functions of its parts are, but it is not very useful for scientific study, because you cannot manipulate it to see how the real plough might react to different conditions, for example when the soil is wet or dry, soft or hard, light or heavy. Another simple model of a very different kind is a graph which shows the response of a crop to the application of some growth substrate, like nitrogen If you want to know how much yield you get from applying a certain amount of N, you plug that amount into the graph and look up the yield. This model can predict the response to applied N, but only for a crop which is grown under exactly the same conditions as those in the trial from which the graph was put together. So, it is only moderately useful.

There are various ways to characterise models, the most important being *stochastic* or *deterministic* models, *physical* as opposed to *numerical or mathematical* models, *static* versus *dynamic* models and *empirical* versus *process* models. We will briefly look at each category before going into more detail about modelling in crop science.

The drawing of a plough is an example of a *physical* model, while the graph for nutrient response is a *numerical or mathematical* model. Physical models were quite common before the computer era. I remember a set-up in one of the Wageningen laboratories in the early 1970s which simulated unsaturated water flow in a soil by a system of tubes, tube clips, valves, capillaries and suction pumps. They stood for the driving forces and resistances to water flow in the soil. Going beyond the drawing of a plough, this physical analogue could be manipulated to study the effect of things like drought and soil compaction by changing the settings of the valves and pressures. All the important soil properties had an analogue in the tube system. In the agricultural sciences physical analogue models have been almost completely replaced by mathematical models, substituting the tubes and valves by equations.

Models can also be categorised as either *static* or *dynamic*. A graph for a crop's response to N is typically a static model. It only shows the final outcome of the physical and biological processes of nutrient release and transport in the soil, uptake by the roots and usage by the plants. The underlying processes may not be completely understood, but their outcome usually has a particular form: yield increases linearly with increasing N applications as long as the amounts are relatively small and then levels off as more is applied until a maximum is reached, after which there is no further increase. This pattern is quite robust – it usually describes the response to fertiliser quite well. But, even though we may be confident that next time the pattern will be similar, the precise shape of the response curve will differ with location and year. So the model is not very useful, until we know how the response curve's parameters vary with soil type, the concentration of other nutrients, the climate, etc. One way to increase the power of the model is carrying out many experiments and drawing a collection of graphs which together represent crop response to N in different locations and soils. The set of curves taken together can then be said to constitute a static model for the response of a crop to N.

An example of a dynamic model is the physical analogue for unsaturated water flow in a soil. It starts with some initial conditions, for example, a completely dry soil, and then a certain amount of water is put in from the top or the bottom or laterally which starts flowing through the system, as it would in a real field. You can now measure the flow at different points and hope things will be similar in the real world. The changes over time, which we hope mimic similar changes in the soil, make the model dynamic.

However, that kind of analogue device has largely been replaced by a mathematical representation consisting of equations and parameters to calculate moisture contents and flow by going through the equations repeatedly with small time steps.

A most important distinction is that between *empirical* and *process* models. The nutrient response graph is not only static, it is also empirical, in that it represents the outcome of one or more experiments. Predictions made with this model assume that the same or similar results will be obtained in the future under similar conditions, but it tells us nothing about the mechanisms that lead to this outcome. In a process model, on the other hand, the processes which result in a certain phenomenon, say carbon assimilation by a crop canopy, are explicitly accounted for in the model, which would calculate light interception and photosynthesis by individual leaves in a canopy and sum them to the photosynthesis by the entire canopy. The processes, which happen at the next lower level, however, like the conversion of radiation, water and carbon dioxide into assimilates, are usually represented by empirical relationships. So a model which is process-based at one level may still be empirical at another level. It was the modellers' ambition to eventually replace the empirical relationships by summary equations which themselves are the outcome of modelling the processes further down.

Finally, we have the distinction between *stochastic* and a *deterministic* models. The former is one which contains an element of probability while the latter does not and generates only one particular outcome. Some modellers have tried to introduce stochastic elements into crop models, but I will leave that aside. Later on we will meet an example of stochastic modelling of rainfall events, an eminently stochastic phenomenon.

These classifications may seem a little artificial, but they are helpful to clarify modelling concepts. The crop models at the end of the twentieth century were all numerical and run on digital computers. Most of them were dynamic process models, which generated phenomena at the whole crop level by stepping through equations for the processes which take place one level lower down. The equations, properly linked up to account for interaction between different processes and sub-processes, constituted a process model, which is what crop models are now usually understood to be.

9.3 Elegance and simplicity

Crop modelling started with C.T. de Wit, the most accomplished Dutch agriculturist of the second half of the twentieth century who almost single-handedly reformulated crop science in quantitative terms. De Wit was born in 1924 and trained, in his own words, as an ordinary agronomist. But he had a strong inclination for exactness and his way of thinking was that of a physicist. When he graduated agronomy was a largely empirical discipline and his first line of attack was to provide a theoretical foundation for some well-known but incompletely understood phenomena. The first one was the response of crops to nutrients applied to the soil. There were semi-empirical laws of more or less general validity which had been found to describe such responses reasonably well, like Blackman's law of the minimum and Mitscherlich's law of diminishing returns. These laws contain an analytical element which accounts for a certain constant structure of the responses observed in the field. For example, the diminishing returns from successive amounts of nutrients in Mitscherlich's law is a logical necessity, whereas the exact shape of the response curve is governed by the values of its parameters which have to be found empirically.

The second phenomenon de Wit tackled was water used by a crop and how it was related to yield. Agronomists of the day used a very simple empirical 'law' which postulated that a crop's biomass yield was proportional to the amount of water it transpired. The reverse, the amount of water transpired per unit of biomass yield, was called the crop's transpiration ratio. If this law was valid, you could predict a crop's biomass production by multiplying the available amount of water by its transpiration ratio.

Finally there was the response to planting density. At very wide spacing, biomass yield will increase in proportion to the planting density, but as the density is further increased the increments become smaller and smaller until the yield finally levels off to a maximum. The resulting so-called saturation curve can also be classified as a semi-empirical law, similar to that of Mitscherlich.[2]

---

[2] The shapes of the Mitscherlich and the density response curve are broadly similar but the underlying equations are different.

These were useful laws but you still had to carry out trials to find out what the transpiration ratio would be under specific conditions, or how quickly maximum yield would be reached with increasing density and what that maximum yield would be.

De Wit first looked at the response of crops to fertiliser for his Ph.D. thesis, which was published in 1953. He did not do any trials himself (he once said, in an unguarded moment perhaps, that he had no patience for that), but dissected those carried out by others. His key idea for analysing crop response to fertiliser was to break it down in two parts: how much the crop will yield when different amounts of nutrient are taken up, and how much is taken up when different amounts are applied to the soil. De Wit connected those processes through three graphs arranged in three quadrants. Figure 9-1 shows an example for N and rice yield published more than 20 years later by Herman van Keulen in Indonesia. The second quadrant, at the upper left, shows the conventional graph for rice yield at different N applications, reflected about the ordinate. The only thing you have to do to draw that graph is weigh the grain at harvest and

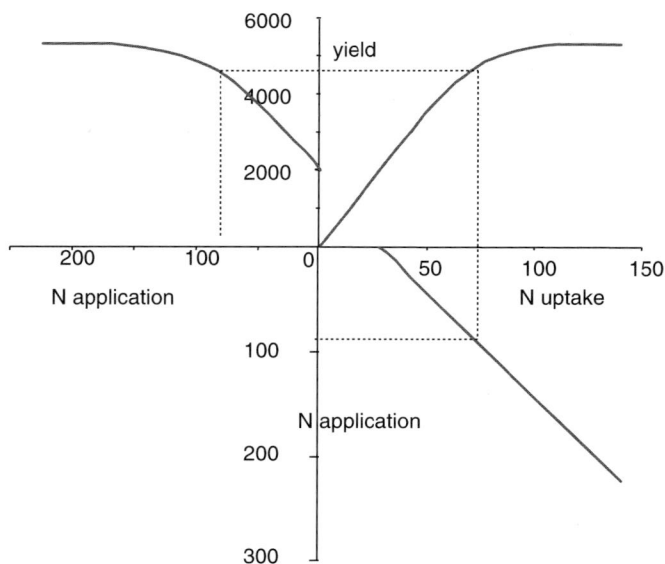

Figure 9-1. Nitrogen application, uptake and yield. (Adapted from van Keulen, 1977.)

plot the weights against the amount of N applied to the soil.[3] It does not say anything about N uptake, only about the grain yield that results from applying a particular amount of N to the soil. If you want to know how much N was taken up you have to digest the plants in the laboratory and extract and measure the N they contain. After doing that, you can also draw the graph in the first quadrant, at the upper right, which shows how much N the crop had actually taken up at each yield level. From these two graphs you can now construct the one in the fourth quadrant (bottom right) which is the amount of N taken up at different rates of N applied to the soil. The dashed lines show how that is done: read the yield for a given application rate in the second quadrant, look up N uptake corresponding with that yield in the first quadrant and finally plot the relation between N uptake and application rate (read on the lower *y*-axis).

By breaking down fertiliser response into an uptake component (fourth quadrant) and a utilisation component (first quadrant) each of them can be studied separately. That looks like a simple and obvious way of analysing fertiliser response, except that nobody had thought of it before. Let us look at the two component processes more closely. I will use two studies, one for potatoes in the Netherlands, published by Jan Vos in 1997 and again the one for rice in south-east Asia by Herman van Keulen in 1977. Figure 9-2 shows the relation between N uptake by potatoes and their biomass yield (that is the process in the first quadrant of Figure 9-1) as measured by Jan Vos. In spite of differences in soil and weather conditions between the years when the data were collected, the pattern was always practically the same. Yield first increased more or less linearly with N uptake and then levelled off to a maximum.[4] Even the maxima were not much different across the years. With this graph you can predict fairly accurately by how much dry matter yield of potato will increase for each additional unit of N taken up by the crop. That is a surprisingly robust response pattern, but isn't that because the trials were carried out under ideal conditions in a location where year-to-year variations are modest?

---

[3] The measured data themselves are not shown, only the curves which fitted the data best.

[4] The initial linearity does not quite follow from the data, because there were practically no measurements at very low uptake. But when uptake is zero, yield must also be zero, so the graph has to pass through the origin and the linear interpolation between 0 and the data points is reasonable.

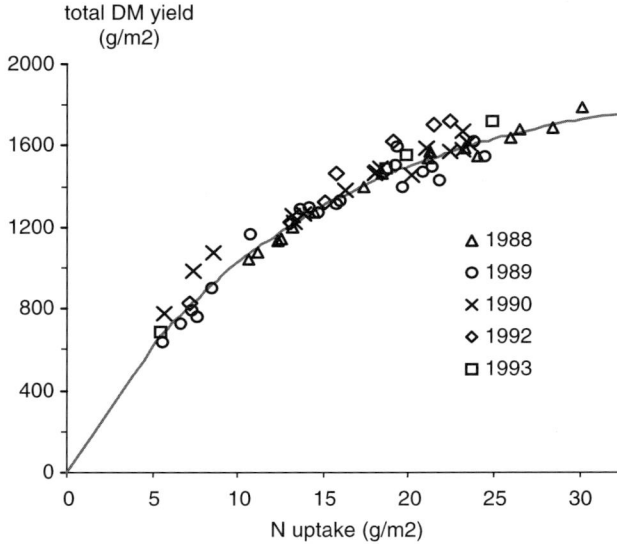

Figure 9-2. Nitrogen uptake and potato dry matter yield. (From J. Vos, 1997.)

For an answer, let us look at data from an entirely different environment. Herman van Keulen analysed a large number of rice trials in Asia in the 1970s and he also came to the conclusion that the initial slope of the uptake–yield curve, that is the yield increase per unit of N when yield was low, was almost always the same in that about 14 kg of N taken up produced 1,000 kg of grain, whatever the growing conditions. But what happened at higher-uptake rates? Yield cannot go on increasing linearly for long. As more N is taken up, other factors become limiting, for instance there may not be enough phosphorus or some other nutrient and then yield will no longer increase in proportion to the amounts of N taken up. And even if there is enough of everything, yield will eventually be limited by solar radiation. In the case of potato the maximum yield was always more or less the same, because other nutrients and water were adequate and solar radiation must have been similar (except perhaps in 1992 when the maximum seems a little higher). In van Keulen's rice data, however, there were large differences in maximum yield which may have been caused by a variety of things, like water management or soil type or weather conditions, but which in most cases little or nothing was reported about. Crop modelling, however, had already advanced enough in 1977 to estimate the theoretical maximum yield a crop could attain if solar radiation were the only limiting factor, which is what van Keulen did. And whenever the theoretical yield ceiling, set by solar radiation, was not reached he knew there had to be other factors which were depressing yield. Explaining the difference between the theoretical maximum and the actual yield has come to be known as yield-gap analysis.

The constant initial slope of the yield–uptake response curve appeared to represent a robust law of the kind scientists are always looking for. If the law is violated and a rice crop at a low yield level takes up much more N than 14 kg to produce 1,000 kg of grain you know something must be wrong. Perhaps there had been a pest attack at flowering or during seed set, or the N was applied too late and ended up mainly in the straw. The level of the yield ceiling is also interesting. If a gap remains between the theoretical and the actual ceiling yields, even when we think all conditions are optimum, we know we are missing some other necessary factor and can start looking for it.

So much for the conversion of nutrients taken up by the plants into biomass or grain. Next we look at the fourth quadrant of Figure 9-1 for the other component process: the uptake of N by the plants as a function of the amount applied to the soil. That is not a simple thing at all. At one side the plants pull at the nutrients in the soil solution around the roots and take them up actively across a concentration barrier, at the other side the soil solution is replenished from the soil's own stock and from nutrients applied in one form or other. The graph does not account for those processes, it simply shows how much nutrient was taken up at each application rate. Two interesting things can be gleaned from the curve, though: the first is how much the soil itself supplied when nothing was applied. That is read at the point where the curve (actually a straight line in this case) intersects with the x-axis. And second, you can see how much of the applied nutrients actually ended up in the plant, about 50% in the example. That is very important, because it says something about the efficiency of fertilisation or, equally important with today's concern about pollution: how much was lost into the groundwater or the air. By looking at many such curves you get a feel for good or acceptable efficiency. Van Keulen's analysis of published rice data showed efficiencies ranging from 10–70%, while in Jan Vos' potato data it was mostly in the range of 50–80%. If the efficiency were much below 50% you would start wondering whether something was wrong with the soil or with the way the fertiliser was applied.

These combination graphs definitely qualify as a model, but is this an empirical or a process model? It is the former, of course, because it simply breaks down the empirical relationship between nutrient application and yield in two equally empirical components, leaving the underlying processes still unaccounted for. Since de Wit's Ph.D. thesis, however, process elements have been brought in. As I have mentioned already, van Keulen was able to calculate ceiling yields as a function of solar radiation (thanks to another major contribution by de Wit, by the way, as we will see presently) and thereby added a process element to the graphical analysis of nutrient efficiency. Without that, the model was purely empirical, though much more informative and amenable to analysis than the simple relationship between N application and yield. A genuine process model would have to simulate plant growth and the nutrient transport and uptake processes themselves, rather than analysing their final outcome. That was going to be the road taken by crop simulation, as we will see later on.

After his Ph.D. thesis de Wit spent a few years in Burma, where he became interested in crop yield when water is limited. That of course is also a key issue in agriculture. I have already mentioned the 'transpiration ratio' which agronomists used to like a lot. If that were a reliable parameter and you knew approximately how much water would be available you would be able to predict crop yield by simply multiplying that amount by the crop's transpiration ratio. The famous (agricultural) physicist H.L. Penman did not agree. He argued that when there is plenty of water, transpiration is a passive process determined by radiation, temperature and air humidity, as if the plant were simply a collection of conduction tubes and sponges. He went on to say that crop growth can show enormous variation and that 'there is no reason to suppose that a plant must transpire a fixed amount of water to produce a given quantity of dry matter'. Hence, he argued, a fixed ratio between transpiration and crop yield would make no sense. Penman's concept of transpiration *under non-limiting water supply* as a physical process has stood the test of time, but de Wit showed that his argument against the transpiration ratio did not stand up to scrutiny. While accepting that Penman was right when there was abundant water supply, he argued that there should

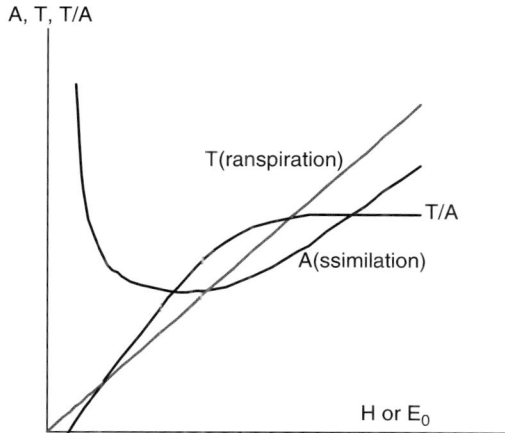

Figure 9-3. Relation between transpiration, assimilation and their ratio in dependence of radiation. (After de Wit, 1958.)

still be a relationship between transpiration and yield when water was limiting. The argument is interesting, because it foreshadows the course crop modelling was going to take. First he looked at how crop transpiration and assimilation (photosynthesis) react to solar radiation. Figure 9-3, which I took from de Wit's paper, published in 1958, shows that Transpiration (T) increases steadily with radiation (H) while assimilation (A) levels off at some point. As a result their ratio (T/A) must be more or less constant at intermediate radiation levels but increases at higher radiation. The former applies for temperate climates with frequently overcast conditions, the latter for semi-arid climates with a lot of sunshine.

What does that mean for the transpiration coefficient, i.e. the ratio between transpiration and biomass production, T/P? The transpiration coefficient is only one step removed from T/A because biomass yield of a crop is proportional to its assimilation, so we may substitute P for A. At low to medium radiation (as in temperate climates), both A (or P) and T will increase linearly with radiation, so their ratio will be more or less constant. Hence:

$$P = k \cdot T \text{ for temperature climates}$$

The parameter k is of course the inverse of the old transpiration coefficient.

For semi-arid climates, however, while transpiration will still be proportional to radiation, assimilation is not. Hence, T/P should increase in proportion to radiation (see Figure 9-3). According to Penman the evaporation of an open water surface, E_0, is proportional to radiation H, so there is no need to measure solar radiation itself, you can just use E_0, which is very simple to measure. Estimates of E_0 are published by all meteorological services. The following relation is then found between production and transpiration:

$$P = k \frac{T}{E_0} \text{ for semi-arid climates.}^{5}$$

Next de Wit analysed a large amount of experimental data, from the US Great Plains, India and the Netherlands, to test this model and found that it was quite robust and fitted most of the data quite well. Since there are climates which are intermediate between those of the Great Plains and the Netherlands, he postulated the following general relationship:

$$P = k \frac{T}{E_0^x}, \text{ whereby } x = 0 \text{ for the Netherlands and } x = 1 \text{ for the}$$
Great plains of the USA. No data were found for intermediate situations and I am not aware of anyone who has looked for them since.

The analysis is considered overly simplistic today, but the implication stands that in semi-arid areas there is luxury transpiration, because the extra radiation increases transpiration but not assimilation. In other words, whereas potential yields are high in climates with a lot of sunshine, more water is needed per unit of biomass than in those with less sunshine. So water is used more efficiently in climates where potential yield is lower. Dirk Zoebl has repeatedly drawn attention to this implication.

Was this a process model? Look again at the flow of the argument. First, biomass production and transpiration were shown to be simple functions of solar radiation. Next the two functions

---

[5] Strictly speaking, this is only valid in the absence of water stress. However, de Wit argued that during drought T and P are relatively small so that their ratio over the entire growing period will not be much affected, unless a large part of the dry matter is formed during drought.

were combined to form two equally simple relationships between transpiration and biomass production, one for temperate and one for semi-arid climates. So I think we may conclude that the model qualified as a process model, except that the constants k which occurred in the functions, were empirical. They did not follow from the process analyses but had to be measured at the crop level. So, the model was process- based and generated simple relationships between biomass production and transpiration but the unknown parameters had to be found by measurements at the crop level. And, of course, it was static.

Next de Wit moved on to competition in plant communities, and published a paper in 1960 which was to dominate research in that area for many years. The term 'competition' sums up the processes by which the growth of plants is affected by the presence of their neighbours, whether of the same species or variety or a different one. I have already given an account of de Wit's competition model in Chapter 3 and Appendix 2, so I can be short here. When two species with similar growth cycles grow in a mixture they are thought to compete for occupation of the same 'space' and their 'relative crowding coefficients' determine their ability to occupy space at the expense of the other. That is all. With this simple concept de Wit could account for a large amount of experimental results on mixed cropping as well as on plant spacing[6] by one and the same formalism. In a further extension of the model he looked at mixtures of plants which only partially compete for the same space, for example a cereal and a legume, where the legume fixes its own N and therefore may compete less for soil N. Or one of the species may mature earlier and then cede its space to the other. That would mean that the combined space available to the two species is larger than the actual area occupied by them and their combined yield would be more than when they were grown in separate plots. Not surprisingly, studies on mixed cropping have been popular among tropical agronomists and de Wit provided a solid scientific underpinning of the biological advantages of mixed cropping, which had often been suspected but never really proven before him.

---

[6] Spacing effects for a single species were treated as competition of a species with space which is empty instead of occupied by another species.

What kind of model was de Wit's competition model? It did not account directly for the processes by which plants compete with each other during their growth cycle, but it used a kind of proxy for those processes in the form of the species' ability to 'crowd' for space, which seemed to capture their operation quite adequately. The (single) competition parameter occurring in the model, the relative crowding coefficient, however, could only be found by actually carrying out mixed-species trials. In other words, this was a mixed model again with a process part, the 'crowding for space' principle, and an empirical parameter, the relative crowding coefficient, which had to be measured for particular species under specific conditions. And again it was static: it accounted only for the outcome of the competition processes, not for their change over time.

There was a clear progression in these three pieces of work. The analysis of fertiliser response still used an essentially classic, empirical approach, but the studies on water use and on plant competition went one step further in that they derived functional relationships from the underlying processes which were used to predict the outcome at the crop level. But not quite, in each case there was a missing parameter which was needed to make the model complete and there was no way to obtain it except by actually growing the crop and estimating it from the final outcome. Nevertheless, reducing complex growth phenomena to simple process-based relationships, was no small achievement. And the parameters the models needed could of course be found the way agronomists had always done it, by simply carrying out many trials under many different conditions, each time estimating the missing parameter and making a table for all those values. Other people could then look up what the value was likely to be under their circumstances and plug that value into the model. That was all right for ordinary agronomists but for someone with the mind of a physicist, having to grow a crop to measure an essential parameter for its model was obviously a major annoyance. De Wit tried for a while to find a more basic way to estimate the competition parameters, but without much success. He then turned his attention to a phenomenon which could be entirely built up from the process level: canopy photosynthesis.

9.4 Enter computing power

9.4.1 A static model for canopy assimilation

While studying crop water use de Wit had already developed a
formula for the amount of radiation intercepted by the leaves in a
canopy, by which he could make a crude estimate of potential
canopy photosynthesis for different latitudes and dates. There were
some clever simplifications there, otherwise it would have been
impossible to capture the process in a single analytical function.
A few years later the first computer arrived in Wageningen and he
started all over again. This time he worked out the distribution of
leaf inclinations for different types of canopy,[7] calculated how much
radiation the leaves in different inclination classes would receive at
different times of day and how much photosynthesis that would
cause. The whole process was then translated into the Fortran computer language and run on the University IBM 1620 machine. That
was the first real process-based crop simulation model, or rather a
simulation model for a most important component of crop growth:
the production of carbohydrates by the leaves. Two years later, in
1967, a group of American scientists, led by W.G. Duncan, published a very similar model.

Canopy photosynthesis is just one component of crop growth,
although a most important one, and being able to calculate its
potential rate does not mean that you can predict crop growth.
The photosynthesis model could only calculate photosynthesis
when given the structure of the canopy and the photosynthetic
response curve of individual leaves as an input. It said nothing
about how the assimilates were used and converted into new plant
tissue and how the canopy structure came into being. Though a
genuine process model, it was still static. Obviously, the modellers
were not going to stop there. What they were after were real
dynamic models which could simulate crop growth processes from
seed to maturity.

---

[7] For example planophile (tending to horizontal), erectophile (no explanation
needed), spherical (with the same distribution as the surface elements of a sphere).

9.4.2 *Dynamic simulation of crop growth*

During the 1960s computing power increased tremendously and new programming tools were developed by the Forrester group in the USA to simulate industrial processes 'dynamically'. That means that you start from a given initial situation and then calculate what happens in the next hour or day or so and how that changes the state of the system. With the new state the calculations are repeated for the next time step and so forth. It was argued that biological systems, although much more complex, resembled industrial processes and that the same simulation methods would be relevant for both. In the late 1960s, de Wit in Holland and several researchers in the USA started tackling plant and crop growth processes using dynamic simulation methods. This must sound rather abstract to people who are not familiar with the concept, so I will explain what dynamic simulation looks like in the case of a crop, using the simplified diagram of Figure 9-4.

If you want to simulate growth you have to start from something, an initial state, a seed for instance, or a young seedling, with its first leaflets, stem and roots. For simplicity let us take the latter and assume that the seed reserves have been exhausted, so the plants now have to manufacture their own carbohydrates and take up water and nutrients from the soil; they have become autotrophic as the term goes. The model calculates how much assimilates are produced by the leaflets, how much plant tissue will be manufactured from them and how much nutrients the roots must take up to match that. The model then updates the weights of the plant parts and starts all over again. That is essentially how it works.

Figure 9-4 is a generic diagram and therefore contains little detail. For a real crop the shaded part must be filled with equations

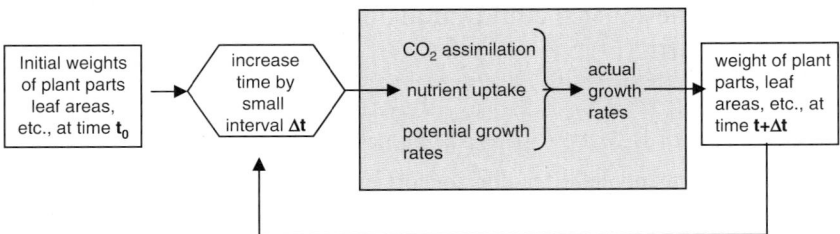

Figure 9-4. The concept of dynamic crop simulation

and procedures which are specific for that crop. An early example was ELCROS (for Elementary Crop growth Simulator), the first whole plant simulation model which de Wit and his collaborators built after completion of the canopy photosynthesis model. The growth of reproductive organs was not simulated by ELCROS. The idea was to look at vegetative growth first and, if that worked well, move on to more complex phenomena like flowering and the growth of cobs or fruits.

(a) The elementary crop growth simulator

The ELCROS model for maize in the vegetative growth phase was first presented at a meeting held at the University of Nottingham, UK, in 1968, by Brouwer and de Wit. It is worth looking at in detail, because it is an excellent example of good physiological modelling. Some of the sound physiological concepts used by ELCROS were lost in most later models, both in the Netherlands and elsewhere, with dire consequences. The two papers which were published about the model contained fairly detailed information about the way some of the processes were simulated, but the actual computer code has been lost.[8] That is a sad example of the volatility of electronic media which many people have warned of. De Wit for one became very adamant that the computer code of any model should be published in print, but apparently he was not yet as firm at that time. I will therefore have to do some guessing about the way some processes were handled by ELCROS.

In order to grow the plants need enough water to maintain their turgidity, and growth substrates, that is carbohydrates and minerals from which new biomass is made. ELCROS assumed a situation where there was no shortage of plant nutrients at any time, so the model was essentially a carbohydrate balance model, modulated by the crop's moisture status. Consider a maize crop at a particular time between planting and harvest. The state of the crop at that (or any) moment can be represented by the state of its parts: the weight and length of the stem, the areas and weights of each of the leaves, length and weight of the roots, and so forth. During the next time interval all kinds of things happen and at the end of the interval the state of all the plant parts will have changed. The model keeps track

---

[8] I have checked with practically everybody who was associated with ELCROS at the time but nobody could find the computer code.

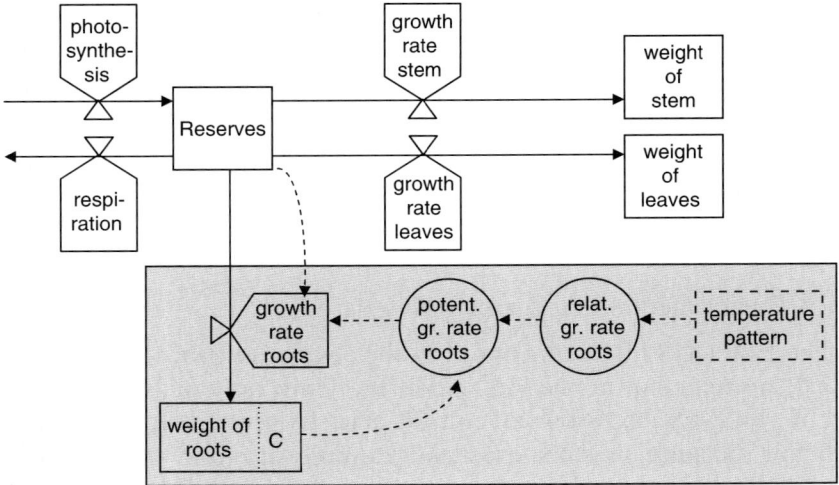

Figure 9-5. Relational diagram of part of the ELCROS model. (After de Wit et al., 1970.)

of all those changes, such as the increase in weight and area of the leaves, the increase in length and weight of the stem, the appearance of new leaves, etc.

The ELCROS flow chart is shown in Figure 9-5, and is quite simple. Rectangles stand for quantities, like weight of tissue or carbohydrate reserves; valves are process rates like photosynthesis and growth rates of organs; circles are 'auxiliary variables'; solid arrows are material flows and dashed arrows are 'information'. Look at the five rate valves first. The one at the top left corner is the assimilation (photosynthesis) rate which supplies the substrate which the plants use in largest quantities, carbohydrates. The one below it is the rate at which carbohydrates are used for respiration, to fuel the plants' growth and maintenance processes. The other three are the rates at which carbohydrates are built into new plant tissue. Supply and consumption are not coupled instantly, they pass through a kind of buffer, the 'Reserves' box at the upper left.[9] That is not just a modelling trick, it is real as can be appreciated from the fact that growth does not stop abruptly when plants are put in the dark.

---

[9] In reality reserves are not in fixed location, that was just a modelling convenience.

Most people will believe that without questioning, but if you do not then carry out a simple test with a young bean plant, put it in the cellar and regularly measure the length of a growing leaf. The plant's pool of reserves buffers it against short-term fluctuations and since ELCROS used simulation time steps of one twentieth of a day, the reserve pool played a crucial role.

How did ELCROS match carbohydrate supply and consumption? That of course is a crucial question. I think the way it is handled by a model determines whether the model is good or not. And ELCROS handled it very elegantly, by first calculating carbohydrate production and demand separately and then trimming down demand until it matched supply. That is less simple than it sounds, though. It is clear what assimilate supply means and how it can be calculated, but what is the demand? It must somehow be related to the growing capacity of the plants' tissues: the greater that capacity the stronger will be its pull on the assimilates. ELCROS made the reasonable assumption that assimilate demand by an organ is proportional to its potential growth rate. So the question becomes: how does one calculate an organ's potential growth rate? Let us see how it was done for leaf growth. Reliable simulation of leaf growth is essential for any crop model, because growth of a crop hinges on the growth of its leaves. First, it consumes a large part of the assimilates, and second, it determines the increase in the crop's assimilate production. Hence, inaccuracy in the simulation of leaf growth will feed forward, thereby driving the whole simulation off course. ELCROS had an elegant (though in the end inadequate) routine to simulate leaf growth[10] which is explained in Box 9-1.

Leaf growth has turned out to be particularly intractable for modellers and has rarely been handled adequately, either in Wageningen or elsewhere. Many later models simply gave up and required the user to supply measured leaf growth data instead of simulating them.

Having good routines to simulate potential growth of the plants' organs is very important but it is not enough. There are growing tissues all over the plants and they all pull at the assimilates with a strength which is assumed proportional to their potential growth

---

[10] The publications about ELCROS do not say so explicitly, but the fact that simulation of leaf growth was dropped from ELCROS' successor, BACROS, suggests there were problems.

Box 9-1. ELCROS: matching assimilate supply and demand[11]

Simulating leaf growth

Detailed measurements had shown that the youngest visible maize leaves were still fully capable of growth while the fraction of growing tissue of older leaves decreased quickly with leaf number. ELCROS kept track of each leaf's actively growing fraction and at each time step calculated its potential growth rate by multiplying its current weight by that fraction and by the temperature-dependent RGR. When water uptake cannot keep up with transpiration the leaves lose water and their turgor pressure (the water pressure inside the cells) goes down and so does their growth rate. ELCROS therefore adjusted the leaves' potential rate by a multiplication factor which was a function of the plants' root-to-shoot ratio. That was a critical function which affected the plants' functional balance between roots and shoot as described in Box 9-2.

The initial weights of successive leaves were not simulated, the model simply assigned measured weights to the young leaves as a so-called forcing function. That can only be an ad hoc solution: the initial weights are themselves a result of the plants' previous growth history. I will come back to that most important issue later on.

The supply side: potential assimilate production

Potential assimilate production was calculated by de Wit's canopy assimilation model which is represented by the valve at the upper left in Figure 9-5. ELCROS kept track of the assimilating leaf area which was needed for the calculations. The assimilates continuously replenished the reserve pool from which the plant obtained assimilates for respiration and growth. That is the supply side.

The demand side: potential growth rates

Demand for assimilates was mediated by the crop's potential growth rate, which is the sum of the potential rates of its organs. It is a kind of suction force, called 'sink strength'. The shaded part of Figure 9-5 shows how the model calculated the potential growth rate of the roots, growth rate being defined as current weight multiplied by a Relative Growth Rate (RGR). In the primordial stage of a tissue all cells take part in growth and RGR is constant, but with time part of the tissue matures and stops growing. The model kept track of the actively growing root tissue, the 'C' in the box at the bottom, and calculated their potential growth rate by multiplying their weight by the temperature-dependent potential RGR. The procedures for (stem and) leaf growth were similar (they are not shown in the diagram to keep it readable).

---

[11] For a modern review of supply and demand and how they are matched, see Vos and Marcelis, 2007.

Matching supply and demand

The next step was to match assimilate demand and supply. That is not simple, since growth is distributed over the crop in a non-uniform way and distance from the source of assimilates will play an important role The Nottingham paper said that 'the organ which will be most successful in obtaining its requirements is that which is nearest to the source', but little was known about the mechanisms involved, nor has it much improved since. Let us see how the model handled it.

Access to the assimilate pool

Supply and demand were not directly linked in ELCROS. The assimilates would go into the reserve pool, from where they were extracted by the growing tissues. At each time step the model checked the state of the reserves (the dashed arrow starting at the reserve box) and allowed the leaves to grow at their potential rate when the reserves were in excess of 4% of the plants' total dry weight. Growth was reduced linearly below 4%. How about the roots? In response to a question Brouwer said: 'The leaves take materials from the reserves first, followed by the roots about 3 hours later, which means that the roots are growing optimally at only a 2 per cent higher level of reserves than the leaves.' I am not sure I understand this but I think it meant that after serving leaf growth the model assigned assimilates to the roots in proportion to their demand and to the status of the reserve pool, but the resulting root growth would only occur 3 h later. The only way to tell would be by looking at the computer code, which is no longer there (except somewhere in some people's drawers).

Functional balance

Brouwer had shown in the early 1960s that there is a characteristic ratio between the weight of roots and shoots, depending on the growing conditions. When the ratio is disturbed, for example by excising some of the roots or clipping some of the leaves, the plants will try to re-establish it. Brouwer argued that this is mediated by the turgor status of the leaves. Excising part of the roots will lead to lower turgor and therefore lower leaf growth potential, which in turn results in more assimilates going to the roots. When the leaves are trimmed less assimilates are produced but since at the same time turgor of the remaining leaves increases, relatively more assimilates go to the leaves. The turgor-dependent RGR in combination with preferential access by the leaves to the assimilates could also correctly (at least qualitatively) simulate the effects of other treatments. Consider, for example, what happens when radiation decreases. Lower radiation results in decreasing assimilation and, because of the leaves' preferential access to the reserves, in relatively more leaf growth. The leaf/root ratio will then increase until it reaches the new equilibrium value. The opposite happens in case of drought, which the reader should be able to put together himself. This is one of the best examples I know of a model generating phenomena at one level by simulating the workings of a process one level lower down.

rates. But some tissues are closer to the source than others, so there is likely to be another factor, something like 'resistances' in the pathway between the source of the assimilates and the growing organs, which will affect the amounts which eventually arrive at the growing centres. The combination of the tissues' potential growth rates and the pattern of resistances to substrate flow results in the so-called partitioning of assimilates over the growing parts. The way ELCROS handled that was also not entirely satisfactory but it was real plant physiology, contrary to the gimmicks used in most later models. If you are interested, Box 9-1 also explains how supply and demand were matched.

One of the appealing features of ELCROS was the way it predicted the well-documented functional equilibrium between roots and shoots and the shift in that equilibrium which will occur when conditions change. For example, when there is drought (real) plants will react by making more roots and when there is less radiation they will make relatively more leaves. Box 9-1 also explains how the ELCROS logic automatically generated this, at least qualitatively. It turned out, however, that root growth simulated by the model was smaller than that of real field-grown crops, so there must still have been something wrong with the procedure for the allocation of assimilate. Perhaps the roots pulled harder than was thought and so the relative priority given by the model to the leaves was too strong. The authors were thinking in that direction. When answering a question from the audience at the Nottingham Symposium, Brouwer said: 'As yet this [assumption that leaves take material from the reserves first] is just a substitute method but we badly need quantitative data on translocation.'

Was the ELCROS model supply-driven or demand-driven? In other words, did the amount of carbohydrates produced by the crop (the supply) determine how much it would grow, or was crop growth in the first place a function of growth potential of the plants' organs (the demand)? Or is that a meaningless question, as I think? We can only say that the model, like the real plant, had a mechanism to dynamically match carbohydrate production and growth rate.

Even though the physiological concepts underlying ELCROS were sound, it soon became clear that better translocation and leaf growth data would be needed for the model to yield satisfactory results. But the data were simply not there at the time and a considerable amount of physiological research would be needed, which the modellers could not wait for. So they devised ad hoc solutions to fix the problems. In principle there was nothing wrong with that. One of the roles modellers saw for themselves was to point out gaps in the existing knowledge and ask the basic sciences to fill them. That is what Brouwer had done in his comments at the Nottingham conference. De Wit therefore assembled a group of young scientists around him to study the key growth processes in more detail than he had done himself and improve the simulation routines. He also tried to get plant physiologists from other departments to work on the quantification of physiological processes such as leaf growth. De Wit's group would not wait until that work had yielded results, however, and they replaced ELCROS' assimilate partitioning and leaf growth algorithms by some, presumably temporary, patchwork which improved the simulation results, but fatally damaged the heart of the model. The result was the Basic Crop Simulator (BACROS), the successor to ELCROS.

(b) The basic crop simulator

BACROS was published in 1978, 10 years after ELCROS. De Wit's group had refined the routines for canopy assimilation and transpiration and calculated better parameters for the conversion of assimilates into plant material and for the energy needed for the maintenance of the cell apparatus.[12] Not much progress had been made, however, in analysing leaf growth and assimilate allocation, so ad hoc schemes had to be used. Like its predecessor, the new model only simulated potential dry matter growth of a (maize) crop in the vegetative stage. For growth partitioning BACROS distinguished two types of tissues: stems and leaves lumped together, and roots and allocated assimilates according to a scheme based on the plants' relative water content. Below 80% of maximum water content all assimilates went to the roots and above that the fraction allocated to the shoot would increase with increasing relative water content.[13] The crop's leaf area at different times was also needed, of course, in order to calculate its assimilation, but growth of leaf area

---

[12] The amounts of assimilates needed for the synthesis of various plant compounds and for maintenance had been calculated by Frits Penning de Vries from the biochemical pathways.

[13] Assimilates needed for maintenance of cellular structures (maintenance respiration), however, were discounted first, before applying these fractions.

was no longer simulated. Instead, the model used leaf growth data measured in field experiments, which were simply put in as a so-called forcing function.

The ELCROS concept of potential growth as a suction force on the assimilates, the demand side in the supply–demand equation, had been eliminated. Assimilates were allocated according to a fixed scheme without reference to the growth potential of the plant's organs: the model had become entirely supply-driven. That of course is not how real plants work and using a predefined resource partitioning scheme is bound to lead to trouble, particularly when dealing with indeterminate species which have a great ability to change the balance of growth between different organs (Box 9-2 explains what determinate and indeterminate plants are).

The authors were of course well aware of the model's limitations and they stressed[14] that '[their] aim [had been] to evaluate the simulation of the dry matter accumulation process only' and that 'further development [should] not so much [be] directed towards the improvement of the present program, but towards the simulation of morphogenesis'. And morphogenesis is how a plant's structures develop, how leaves, shoots and roots grow and how their growth is

---

Box 9-2. Determinate and indeterminate plants

Determinate plants like cereals, pineapple and cauliflower make a definitive switch from vegetative to reproductive growth. Their terminal growing points convert into flowers and no more stem and leaf tissue can be formed. By the time their fruits mature the leaves have stopped functioning and the plant dies (or it regrows from new buds at its basis, like bananas). Some species which do not have a terminal inflorescence nevertheless stop making vegetative tissue as the load of fruits increases, like soybeans and some varieties of cowpeas.

Indeterminate species, even though they will also give priority to the growth of reproductive structures, always allow vegetative growth to continue. Groundnuts and tomatoes are indeterminate.

Some species like cotton, cowpeas and pigeon peas have both determinate and indeterminate varieties. As the number and size of fruiting points increase growth of vegetative tissue may grind to a halt. Cotton may even be determinate under some conditions (a hot humid climate and fertile soil) and indeterminate under others. So these characteristics are not as unambiguous as all that. For species with an uncertain classification the rather indeterminate term *semi-determinate* has been coined.

---

[14] de Wit et al., 1978.

influenced by the plant's growth history and by its carbohydrate partitioning strategies, all of them things which were almost entirely missing in BACROS. They were confident, however, that BACROS was a reliable tool for the simulation of crop photosynthesis and transpiration and for the conversion of assimilates into plant constituents. And rightly so, because those processes were to a large extent species-neutral and had been thoroughly studied under well-defined conditions. But not much progress had been made in clarifying and modelling plant morphogenesis. A plant is a complex organism where the growth of every structure is related to that of every other structure, constrained by an inbuilt flexible blueprint or template. The template makes sure that the plants follow broadly the same growth pattern so that nobody will mistake a soybean for a tomato plant. And there are all kinds of feedback and feedforward mechanisms: what happens in one part has a direct effect on what happens elsewhere and what happens today will have an effect on what will or can happen next week. Leaf growth in particular is complex and potential growth of a particular leaf is not just a property of that leaf itself but also depends on the plant's history. That is because a leaf originates from a growing point, the size of which harbours the history of past growth. De Wit of course was keenly aware of all this. One of the members of his original team was a plant physiologist who worked on leaf growth in the 1970s but left after some time, never to be replaced for some reason. Later, a Japanese post-doctoral fellow at the department, T. Horie, carried out a detailed study of the dynamics of cucumber leaf growth, intended no doubt as another step towards the reintroduction of a real plant into the model, instead of a collection of leaves, stems and roots.

9.5 The fabulous cotton plant

My own modelling work, with de Wit as Ph.D. thesis supervisor, was also about morphogenesis, with cotton. Cotton is a beautiful plant with a very regular structure and its growth pattern is ideal to explain what is meant by a growth template and how it can be built into a crop model. Figure 9-6 shows the essential features of a cotton plant in the flowering stage. Like most dicotyledons, the plant starts by unfolding two leaflets which are already present in the seed, soon

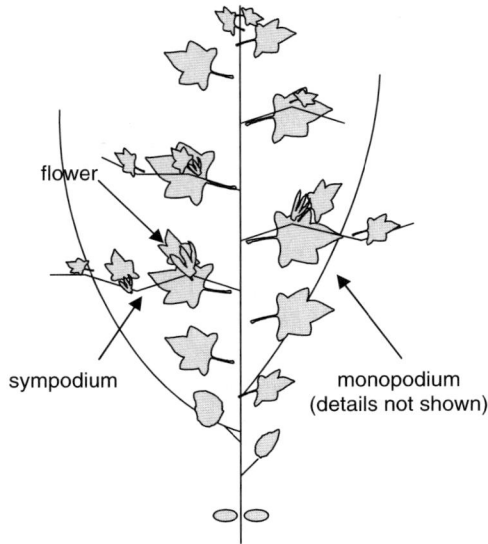

Figure 9-6. Schematic structure of a cotton plant

followed by the first 'true' leaf formed by the plantlet's growing point. Each leaf has a bud in its axil which may start growing after a while and form a branch. The first few branches, called 'monopodia', are vegetative: they are rough replicas of the main stem. Not all the lower buds on the main stem form a monopodium and in a very densely planted crop only a few plants may carry monopodia at all. From the fifth or sixth main stem leaf onwards, however, every axillary bud becomes a generative branch, called a 'sympodium', which will carry the flowers and cotton bolls. A sympodium is a special type of branch which does not grow in the same way as the vegetative branches do. The bud in the axil of a main stem leaf which is going to develop into a sympodium first splits off a small leaf-like structure called a prophyll, followed by a true leaf and then the bud terminates in a flower bud, called a square. The branch is extended from a bud in the axil of the true leaf, which again forms a prophyll, a true leaf and a flower bud and so forth. So the sympodium is in fact a succession of one-sided ramifications which gives it a typical zigzag structure, as shown in the figure. The botanical name for that kind of zigzag branch structure, which occurs in many species, is also 'sympodium'. A genuine plant model should of course generate that regular growth pattern but that is not enough. While maintaining the crop's overall growth pattern, characteristic for cotton, it must at the same time mimic the way the crop adapts to constraints like water stress, low light intensity or a high plant density. How the plant does that is a question crop physiologists must answer before a realistic cotton model can be built. In order to see how that could work the course of a cotton plant's development must be analysed in more detail. That is done in Appendix 7 which you may leave unread if you feel it is too much for you.

So, a realistic cotton model, or any crop model for that matter, must be driven first of all by the plant's innate growth template which is flexible enough to allow the plant to adapt itself to different and varying growing conditions, but always within the limits imposed by the template. Plant species are very different in the degree of flexibility of their template. Maize for example is particularly rigid while cotton is quite flexible: the kind of cotton plant you get under warm, humid and fertile conditions is quite different from one grown in a semi-arid environment with cool nights and low soil fertility, yet both will remain recognisably a cotton plant.

The importance of the plant growth template cannot be overstressed. If it is not adequately represented a model cannot be expected to simulate the real plant's behaviour, especially when its adaptation mechanisms are called upon. For instance when insects remove half of the cotton squares, or drought severely reduces vegetative growth.

9.6 How the plant disappeared from the models

The disappearance of the plant from the simulation models was not an exclusively Dutch phenomenon; the same thing happened in some influential crop modelling centres on the other side of the Atlantic. In Appendix 8 you can find a lot of detail about several important crop models to make your own assessment. Here, I will just briefly summarise how it all happened.

During the heyday of crop modelling there was great enthusiasm among its practitioners to build real physiology into the models which did justice to the plants' beautiful regulatory mechanisms. That enthusiasm transpires in the almost euphoric tone of some of the papers which were published in the 1960s and 1970s. At the same time the modellers were aware, or shortly became so, that structural plant growth was poorly understood and hence poorly accounted for in the models. Many physiological studies were therefore carried out on the growth of plant organs, like those by Horie on cucumber leaves and by myself on cotton and those by Don Baker, John Hesketh and Jim Jones in the USA on maize, cotton and other crop species, with the aim of strengthening the models' plant content. Some of the work was rather cursory and superficial; the researchers being clearly in a hurry to get the necessary data and carry on with modelling.

The most successful and influential studies were those on plant phenology and the results of that work have found their way into all the major modelling families. Box 9-3 explains briefly what phenology means and why it is important, while Appendix 8 has all the details. Another breakthrough was the finding that plants can store quite large amounts of carbohydrates in their stems, up to 20% of their weight, which can be remobilised during maximum grain growth. That is obviously a useful feature for the crop. When the grain in a heavy cereal crop grows at its maximum rate, assimilation cannot keep pace and will be supplemented from the stem reserves. If a model lacked that mechanism it could never adequately simulate grain filling and yield.

Understanding and simulating plant phenology and internal carbohydrate storage and remobilisation were significant breakthroughs and they showed what good collaboration between modellers

---

Box 9-3. What is phenology?

Phenology is the study the plants' growth stages and how their succession and duration are influenced by temperature and day length. The phenology of rice for example describes when the plants start tillering, when the apex converts into a panicle primordium and when the panicle appears. All these transitions are influenced by temperature and day length and there are important differences between varieties. Contrary to the rather gimmicky growth partitioning factors, phenology was fairly adequately accounted for by most models of the late twentieth century as a genetic trait, modulated by temperature and day length. It is a key property which a model must capture or it will stray completely off course. For example, if it got the one or more of the transitions between growth stages wrong the model may think that there is no panicle yet, whereas the real plant is already busy filling the grain. So adequate simulation of phenology is very important, especially for plants with sharp transitions between growth stages like cereals.

and plant physiologists could achieve. But the other aspects of morphogenesis and assimilate partitioning, which make real plants into self-regulating machinery, was never adequately accounted for. Practically all models continued to be assemblies of process modules stitched together in an ad hoc manner so that they would 'predict' plant growth under a particular set of conditions, without any guarantee that the predictions would come out right under another. And since in most cases they did not, the stitching had to be changed again. That was called parameterisation. It means for example that you change the proportions of assimilates which will go to different organs, until the model reproduces the crop you see in the field. If you do that often enough and for many different conditions, you can make partitioning tables for each set of conditions and tell the users which one to use when. In the meantime the model has degenerated into a complicated way of curve fitting which can be done more efficiently and with more transparency by the kind of simple statistical techniques which dynamic simulation was supposed to replace. Parameterisation of the models for each new set of conditions became the occupation of many second-generation modellers who did not want to acknowledge that the processes themselves should be the object of study if real progress was to be made.

Why did things go astray like that? I think one of the main reasons was that the plant physiologists gradually lost their pre-eminence in crop modelling, their role taken over by software developers and by engineers with no taste for the painstaking process studies with real plants which were needed to take modelling forward. In Holland the people de Wit collaborated with early on had all been plant physiologists: R. Brouwer (functional balance), W. van Dobben (mixed cropping) and F. Penning de Vries (assimilate conversion) and later on scientists who worked on leaf growth and morphogenesis: H. Lof, J. Bensink, G.A. Pieters, T. Horie and Jan Goudriaan. Some quite complicated papers resulted from that work, but very little found its way into simulation models. Meanwhile the modellers carried on improving the routines for better understood phenomena dominated by physical, biochemical and soil processes while turning a blind eye on the rudimentary state of the routines simulating structural growth. They were criticised by the plant physiologists for the superficiality of the plant physiology in their models, while the modellers expressed their impatience with the physiologists, endless studies. In the end they stopped talking to each other.

In the USA something similar happened, although the influence of plant scientists perhaps remained a little stronger. One of the most influential modellers and certainly the one with most endurance, Jim Jones at the University of Florida, was a mechanical engineer by training. But he collaborated for decades with the physiologist Ken Boote, who contributed a lot to modelling, in particular in phenology and functional balance between root and shoot growth. In cotton, modelling was dominated for many years by Don Baker, an engineer-type of scientist who collaborated for some time in the 1970s with John D. Hesketh, a crop physiologist. Hesketh did cutting edge research on photosynthesis and leaf growth and their collaboration resulted in GOSSYM, the most successful cotton model. By the early 1980s, however, Hesketh had dropped away from the scene. I remember a rather contemptuous remark by Baker to the effect that Hesketh had never written a single line of code. Apparently, writing computer code had gradually become more important than understanding the processes which the code was supposed to simulate. Another (British) plant physiologist, Basil Acock, made a brave attempt to bring plant physiology back into cotton modelling when he joined Baker's group in Mississippi, but he seems to have lost the battle in the end. A few years later he moved on to head the modelling group of the USDA Agricultural Research Service modelling group where he continued working on GOSSYM at first and then on its successor, the Cotton Production Model, of which Appendix 8 contains a short characterisation. I think it came closest to what I had in mind as a morphogenesis-based model. Its testing and further development ended in 2000 when the US government stopped the funding. In general, however, morphogenesis in the major US models remained surprisingly weak, with the exception of phenology.

Once a machine has been put together, however poorly, it becomes difficult to convince the engineers of the need to disassemble it and re-engineer its parts or the whole thing. In real life a poor machine will not stand up to the stress of the market place and the company will go broke if the design is not quickly replaced by a more successful one. In research, however, especially the publicly funded kind, it does not work like that. A faulty design often will die only when its designers retire or when some brilliant youngsters come around and produce radically new ideas and concepts, after which the old ones are quietly dumped or forgotten. There are signs that something like that may now be happening and that new ideas are being hatched which may in due time bring crop simulation to the level of sophistication (which is not the same as complicatedness) which its originators had in mind. I will return to that later in this chapter.

In spite of their defects, some models have worked surprisingly well, though. Even deficient models can do a reasonably good job when used to simulate dense, uniform crops grown under uniform conditions with high inputs of fertiliser and few disturbances. That is how field and glasshouse crops are grown in industrialised countries, sugarcane and oil palm in well-run plantations in tropical ones and rice in intensively managed paddies with high inputs in Asia. Under such conditions few of the plant's adaptive mechanisms are called upon, which is the reason why crop modelling has been quite successful. The things that matter most under those circumstances are carbohydrate production, water and nutrient uptake, transpiration and, yes, phenology, all of which are well-understood processes. It is like flying an airplane at cruising altitude under ideal weather conditions. Or, as de Wit said in 1982, somewhat cryptically:

"Fools rush in where wise men fear to tread. Much of this rushing in simulation in biology is done by agronomists, perhaps because they are fools, but maybe because they deal with systems in which the technical aspects overrule more and more the biological aspects."

In other words, technology-dominated production has fooled the agronomists into believing that their models were adequate representations of the biological systems, whose inner workings were in fact largely hidden from sight.

Smallholder agriculture in tropical countries is mostly not like that, with the exception again of intensive irrigated rice production in Asia with short-straw varieties and a lot of fertiliser and pest control. Not surprisingly, that is where modelling for tropical agriculture has been most successful. The question is, though, whether you really need all that complexity to simulate what is essentially potential crop production, because that is what you get when everything is ideal and radiation is the main limiting factor. In any case, let us now examine whether twentieth-century crop modelling has produced something which has turned out to be useful for the general practitioner, whether in industrialised or tropical countries or both.

9.7 Has crop modelling been useful for non-modellers?

The hope of the modelling pioneers and the claim of their successors was that modelling was going to be useful for everyone, scientists, practitioners and farmers alike. Has that promise been fulfilled? Well, as usual there is not an unqualified answer. Modelling has been very useful in industrialised countries for certain purposes, like pest and disease forecasting and climate control in greenhouses. Modelling of crop growth processes, the main topic of this chapter, has certainly helped to sharpen research by showing the gaps in our knowledge and some of it has also found its way into higher-level models which simulate global climate change or global food production. But its success has been much smaller when it comes to developing tools for practical agriculture. The models have hardly left the researchers' laboratories, except to be sent to researchers in other laboratories. Even so, after deflating the overblown claims and peeling off the ornamental layers applied over time, a few things do remain which may yet turn out to be of lasting value for the general practitioner, in spite of the modellers' failure to convert their methods into practical tools. There is a sense of waste here, because most of what has turned out to be potentially useful for practical application was already available by 1975.

My programme for the remainder of this chapter is to examine areas where modelling has made significant, or at least potentially significant contributions for tropical agriculture and which are visible to the eye of the uninitiated.

The first success was the calculation of potential assimilation and potential crop yield as a function of solar radiation. That is comparatively simple in principle, and I think practical agronomists should be able to do it, but most are not. The second area is the model-assisted improvement of nutrient-use efficiency, which is particularly important for poor countries where increases in the price of fertiliser have put that commodity out of the reach of many farmers. Then there are various model-based decision support tools, the most ambitious being the Decision Support System for Agrotechnology Transfer, DSSAT, pioneered at the University of Florida. I hesitate to call that a genuine success, since it has hardly reached the rank-and-file agronomists, but at least it shows what form the practical application of modelling might take in the future. And finally there is plant breeding, where modellers have made many unproven claims about their ability to tell breeders what varietal traits they should be looking for, but where new thinking may yet lead to success.

9.7.1 Potential assimilation, biomass production and crop yield

One outstanding contribution of lasting value has been the calculation of potential assimilation by a leaf canopy. That in itself is not particularly interesting for practical agronomists, who are more interested in potential crop yield. But if you want to calculate potential yield, you will need some estimate of potential assimilation first. In 1955 it was not possible to calculate that, but today you would expect that any literate agronomist would be able to do it, or at least to look it up in a textbook. In fact most do not. I will therefore spend some time explaining, broadly, how it can be done. Those who want to see the actual calculations should also read Appendix 9 for the finer details. It is even interesting for the general reader, who will find it revealing to see how complex things like potential assimilation and biomass production are eventually broken down into some fairly simple equations.

(a) Potential assimilation

When we talk about 'potential' assimilation, it is usually implied that solar radiation is the only limiting factor and that the crop's leaf canopy is completely closed and intercepts practically all of it. That sets an absolute upper limit, because there is nothing you can control about the incidence of solar radiation on the canopy, at least not in the upward direction. Temperature will have an effect as well, but much smaller, nor can anything be done about that either except in the glasshouse. So potential assimilation rate becomes a kind of ecological property of a particular location, at a particular time much like potential evapotranspiration.[15]

---

[15] For potential evapotranspiration the standard vegetation is the well-known dense well-watered short grass cover. For potential assimilation a canopy with LAI 5 and a 'spherical' structure (see footnote 7) is taken as the standard.

There are many detailed models which calculate potential assimilation, such as the one by de Wit, which I discussed earlier in this
chapter. All of them consist essentially of four components:

- The amount of photosynthetically active radiation (PAR) as a
  function of location and time of the year
- The three-dimensional structure of the leaf canopy from which
  the amount of radiation intercepted by the leaves at different
  times of the day can be calculated
- The response pattern of a leaf to different levels of irradiance,
  that is how much assimilation will take place when a particular
  amount of radiation is intercepted by a leaf
- A routine to add up the contributions of each leaf

Comprehensive assimilation models are pretty complex, especially
their geometric part, but there are also simpler approximate methods. Not even those have found their way into the agronomists'
toolbox, though, in the way statistical tools and procedures for
calculating potential evapotranspiration have. Modellers have not
been the best of communicators, not even with their own kind.

There were some notable exceptions to this statement. Goudriaan
and van Laar published a paper in 1978 which explained how dummies could calculate potential assimilation, even with a program-
mable calculator, using a relatively simple approximate model. The
calculation method is described in the first part of Appendix 9.[16]
Interestingly, its results were quite close to those of the comprehensive model which the authors used as a benchmark and whose predictions they considered as accurate. That was precisely one of the
tasks the modellers had set for themselves in the early years: put
everything you know about a process in a detailed simulation model,
and then capture its essence in much simpler approximate models
which turn out essentially the same predictions.

Ten years later, at a 1988 modelling symposium, Thomas Sinclair
of the University of Florida came up with an even more drastic
simplification, based on the well-known observation of 'a linear
relationship . . . between accumulated crop biomass and intercepted

---

[16] The paper also contained tabulated values, so it was not even necessary to do the
calculations, the results for a particular location and time of the year could just be
looked up in the tables. An extract from the tables is also given in Appendix 9.

solar radiation'.[17] Since roughly the same amount of assimilates is needed to produce a unit of biomass, assimilation per unit of intercepted radiation should also be more or less constant for a particular crop. The ratio of a crop's assimilation and intercepted radiation is called its radiation use efficiency (RUE). If the RUE and the amount of intercepted radiation by a canopy are known gross assimilation can be calculated from the definition of RUE:

$$\text{gross assimilation rate} = \text{incident radiation} \times \text{fraction intercepted} \times \text{RUE}$$

('gross' means before any of the assimilates are used for growth or consumed for maintenance). The clue is to find a value for RUE for some easy case which may then be applied to other cases. Sinclair showed a simple way to do that. Next, in order to calculate gross assimilation you must know the incident radiation for a particular location and date and how much of that is intercepted. Measured solar radiation data can be obtained from most meteorological stations, or you can use standard astronomical equations or meteorological tables, combined with information on cloud cover. The remaining factor, the fraction of radiation which is intercepted, depends on the type of canopy and on sun angles, but usually a decent approximation can be obtained by using a simple exponential extinction function. The procedures are explained more fully in Appendix 9.

To give you an idea about the orders of magnitude, I have assembled Table 9-1 with some key figures for six important crops. Column 2 contains the crops' RUEs, as calculated in Appendix 9, while column 3 shows the (calculated) amount of PAR on 15 July at 10° N with 40% cloud cover (10.48 MJ/m²/day). Potential assimilation by a crop (column 4) is then found by multiplying daily radiation by the crop's RUE and converting the result to kg/ha (the figures in the other two columns will be explained later). Note that assimilation is given as kg CO_2, the reason being that laboratory or field measurements record CO_2 uptake. If you want carbohydrate production, multiply the data by 30/44, the molecular weight ratio of (CH_2O) and CO_2.[18] Potential assimilation was calculated here

---

[17] As observed by several authors as early as the 1960s and generalised by Monteith in 1977.

[18] The figures must be treated with caution, as explained in Appendix 9. Just consider them as approximate.

Table 9-1. Daily radiation, potential assimilation and biomass production by some important crops at 10° N, on 15 July, with 40% cloud cover

Crop	RUE, g C O_2/ MJ PAR	PAR MJ/m^2/ day	Potential assimilation kg CO_2/ha/day	Carbohydrate conversion kg biomass/kg CO_2	Potential biomass production kg/ha/day
Rice	5.95	10.48	602	0.37	224
Maize	6.73		681	0.35	239
Cassava	5.74		581	0.37	216
Soybean	5.97		604	0.29	176
Groundnut	6.81		689	0.27	187
Cotton	5.80		587	0.32	183

with Sinclair's method, which gives you very similar figures as Goudriaan-van Laar, for latitudes between 0° and 40° and between mid-March and mid-September (and the mirror period on the southern hemisphere) as long as the sky conditions are not too bright and not too overcast and the species' assimilation capacity is low to moderate (see Appendix 9). That is a lot of ifs, of course, but for the rough estimates we are looking for, Sinclair's method works reasonably well.

(b) Potential biomass production

Now let us see how we get from potential assimilation to potential biomass production. Assimilates are the plant's primary raw material, but not all of it is converted into biomass. The plant has to keep its physiological machinery in working order by a process called maintenance respiration, which consumes assimilates. It has first priority, otherwise the machinery would break down, so the assimilates for maintenance respiration are subtracted from gross assimilation first. Maintenance respiration is highest in metabolically active biomass, and the more active the tissue the more maintenance it will need. For our rough calculation method we will simply assume that a crop's metabolic activity and its maintenance respiration are proportional to its assimilation rate. Published data show that maintenance respiration may consume between 15% and 30% of gross assimilation[19] and I will stay on the conservative side

---

[19] The figures are from Penning de Vries et al., 1989.

by assuming it to be 25% for annual crops. Temperature of course has an effect but I will ignore that. The remaining 75% is converted into plant substances and the next question is how much of assimilates a plant needs to make 1 g of biomass.

In order to answer that question F. Penning de Vries from de Wit's group studied the plants' biochemical pathways and calculated how much glucose was needed for the synthesis of each of the major compounds. There are quite large differences: making starch is cheap: less than 1.3 g of glucose is needed per gram, because starch itself is also a carbohydrate, but for energy-rich fatty compounds about 3.2 g is used and for protein and lignin something in between. Penning de Vries' figures are used in practically all detailed crop models. Then we need to know the chemical composition of the biomass, because it makes a lot of difference whether you deal with a cereal which consists mostly of carbohydrates and proteins, or with an oil producing crop which is energetically more expensive to make. Once the overall composition is known it can be calculated how much CO_2 must be assimilated to make a gram of plant biomass.[20] Appendix 9 explains in detail how it is done and column 5 of Table 9-1 above shows the results for the important crops considered there.

We are now ready to calculate potential biomass production. That is straightforward, look at the examples in Table 9-1 again: multiply the species' potential assimilation rate (column 4) by the amount of biomass produced per unit of CO_2 (column 5). This gives us the potential biomass production (column 6). We can do the same thing for any geographical location, time of the year and species, if we have the necessary data, as explained in Appendix 9.

Interestingly, as early as 1930 Boysen Jensen had estimated, from field observations, that maximum daily biomass production in summer in northern Europe was around 200 kg/ha/day of dry matter (quoted by de Wit in a 1968 paper), quite close to what today's simulation models predict as the maximum you can expect. That does not mean that all this modelling gymnastics has been trivial, though. Since the model predictions are close to what was measured experimentally more than 75 years ago and many times since then, we feel confident that the models can make reliable predictions for

---

[20] I am also ignoring the fact that the plants' composition changes with age.

any area at all. And that was precisely one of the things models were meant for.

(c) Potential crop yield

With the problem of potential biomass production solved quite satisfactorily, we now come to the difficult part: how to estimate potential crop yield? That is the kind of thing simulation models deal with, but practical agronomists who are unfamiliar with such models would still like to have some idea about potential production, without having to learn crop modelling first. So let us see how far we can get with a much cruder attack. We start from potential biomass production and make some rather sweeping simplifications which may look a little irresponsible, but which are in fact not much worse than a lot of things which are hidden from sight in many models' computer code. I think it is quite entertaining to see how far you can get with the 'back of the cigar box' method I am going to propose.

The approach will be to add up the crop's biomass production rate throughout its life cycle and then multiply that by the harvest index to obtain potential yield. I have already explained how biomass production by a closed canopy under ideal conditions is calculated (that is, by definition, potential biomass production), so now we need the fraction of the radiation that is intercepted at different times to calculate the crop's life time biomass production. In an annual crop it takes time for the canopy to close and at the end of the season its leaves start ageing and drop off. The trick I used and which is explained in more detail in Appendix 9, is to replace the crop's canopy by an equivalent one, which is fully closed throughout a considerably shorter time than the crop's actual life cycle. For a tropical maize crop, for example, I estimated (see the Appendix) that the duration of the equivalent full canopy crop would be 80 days instead of the actual 120 days. Table 9-1 now tells us that daily growth of biomass at 10° N around the middle of July with 40% cloudiness would be about 240 kg/ha. Radiation varies only little during the growing season, so we estimate that the crop's total biomass yield will be around $80 \times 240 \approx 19,200$ kg/ha. Part of that will be below ground, in the roots, say 15%, leaving about 16,300 in the tops. How much maize yield would that be? Modern maize varieties have a harvest index close to 50%, so this dry matter yield would be equivalent to 8,150 kg dry matter, that is about

9.5 t of maize grain with 14% moisture content. And that is indeed the kind of maximum yield which is obtained in that environment.[21]

For fast growing, short-cycle crops like maize and soybeans an analysis along these lines should be all right, but it is more difficult to find the 'full canopy equivalent' for a crop like cassava. In Africa cassava may stay in the field for up to 2 years and lose some or most of its leaves in the dry season. In such cases, rather than making all kinds of assumptions about seasonal canopy development it is better to find real field measurements to keep the whole thing realistic. The more complex the situation and the more assumptions have to be made, the more the exercise will degenerate into what de Wit called 'koekenbakkerij'.[22] Even in the relatively simple case of maize a change in the senescence period from 30 to 20 days would result in a 600 kg higher simulated grain yield. So take the predictions with a grain of salt. What counts is the order of magnitude.

9.7.2 Nitrogen use efficiency

Another area where modelling has achieved something of practical value was the optimisation of N nutrition of field crops. I am not really familiar with the most recent developments here, so you have to content yourself with a rather superficial, perhaps outdated and certainly mainly Dutch treatment of the subject.

The question of nutrient efficiency is one of great importance for at least two reasons. First and foremost there is the question of economics. The more of the applied nutrients that end up in the harvested produce the better it is for the farmer's balance sheet. Then there is the issue of pollution: the larger the share of the nutrients taken up by the crops, the less will end up in the environment as pollutants. It therefore comes as no surprise that plant nutrition has always been a major concern of farmers and scientists. And modellers have tried to make their own contributions, as we have already seen at several occasions. There are many complex models

---

[21] There is another method which avoids assumptions about the harvest index. It calculates assimilation during grain filling and adds a certain amount of assimilates remobilised from vegetative tissue. An example for rice is given in Section 9.7.4.

[22] After a lecture delivered in the early 1970s by a diligent scientist who had done a similar calculation for rice. 'Koekenbakkerij' means a bungler's job, literally 'cookie baking'.

which simulate the soil and plant processes involved in the release, uptake and utilisation of nutrients, in particular N. Practically all of them are essentially research tools of which the common agriculturist is blissfully unaware. Things have been changing, though, because there has been pressure on the modellers to convert their sophisticated models into something useful for practical applications, for example in precision farming. That is a much-hyped area of investigation where methods and equipment are being developed for very precise management and fine-tuning the application of inputs such as fertiliser to the small-scale variations in soil and crop conditions in the field. Obviously, some kind of decision mechanism or model must prescribe the action according to what is registered by the equipment's sensors, such as how much N to drop in a particular location. Precision farming has been pioneered in the USA and it is not exactly a major concern for most developing countries, except perhaps, theoretically at least, for high-input paddy rice growing. But the optimisation of fertiliser use in accordance with the overall fertility conditions in the farmer's fields certainly is highly relevant for the tropical farmer.

My example is therefore about the use of crop modelling to optimise fertiliser-N application to paddy rice, not by adjusting it to a field's micro-variation, but by finding the best possible distribution of applied N over the crop's life cycle for a particular location. That sounds cryptic, but in a moment the meaning should become clear. The work I have in mind was carried out in the context of the Simulation and Systems Analysis for Rice Production (SARP) Project, funded by the Netherlands government, about which I shall have a few more words to say later. The project set up an Asian research network which fed the (mainly Dutch) scientists who were sitting in the centre of the web (at IRRI in the Philippines) with a lot of data, allowing them to validate their simulation models. One of the project's research topics was the optimisation of N application through the use of a simulation program, combined with an optimisation routine. Don't worry, I will explain. Optimisation of N application was thought to be important because the amounts applied by paddy farmers in some areas were becoming too heavy, fertiliser prices were soaring and expensive wastage was likely unless the application pattern (through time) closely followed the crop's needs. The N-optimisation program called MANAGE-N was developed under the leadership of Hein ten Berge. It repeatedly ran a relatively simple simulation program, ORYZA-0, each time with a different application pattern to find the pattern giving the highest yield for the same N-fertiliser dose. ORYZA-0 is briefly described in Appendix 8. It captured only those processes, which were essential for the crop's N uptake pattern in a semi-empirical way, because, as ten Berge said in a key publication in 1997: 'The complexity of explanatory models . . . limits their application.' At each time step the growth section of the model calculated biomass growth as a function of solar radiation and leaf N-content. Then it worked out the crop's N-demand and the potential uptake from the soil and from the fertiliser and equated the largest of the two with actual N uptake. Finally the new biomass and the N actually taken up was allocated to the different tissues including the leaves. That is broadly how it worked. Real epicureans are referred to Appendix 8 for more details. The ORYZA-0 model was rather primitive, as the authors would have been the first to admit, but it seemed to work, as long as the model was fed with the appropriate locally measured parameter values. The gain from the best application pattern selected with MANAGE-N, compared with recommended practices in two sites in India and China was quite small, though – between 5% and 10%. Not surprisingly, the Indian and Chinese agronomists had done a good job without modelling, but it might be different in areas with less experienced or less skilful researchers.

MANAGE-N was an interesting example of a special purpose model with the ambition to become a practical tool, but without ever reaching the stage where it could actually be used by the general practitioner, rather like the potential biomass models of the previous section.

9.7.3 DSSAT, a model-based decision support system

Meanwhile the modellers in the USA had not been sitting idle, of course. I think some of them were actually keener than their Dutch counterparts to convert their models into practical tools for the general practitioner (though not necessarily more successful). Especially the group around Jim Jones at the University of Florida, Gainesville worked hard to package crop models, especially but not exclusively those of the CERES family, together with soil and climate routines into what they called a Decision Support System for Agrotechnology Transfer, DSSAT. The group originated in Jones' agricultural engineering department, so tool building came naturally to them. The package was designed so that non-modellers should be able to operate it without knowing exactly what went on inside but trusting that the instrument and its parts had been well designed.

In order to understand what DSSAT in its original configuration could do, consider the yield of maize in different years and locations. It is important for planners, extension agents, politicians, aid workers, and maybe even for farmers to be able to predict long-term average yield and how yield is likely to vary over the years. You could of course give them the model and tell them to run it for the conditions they have in mind, but that would be about as useful as a climatologist telling them to study their historical rainfall records if they want to know how likely it is that August will be dry. Climatology has of course the longest tradition of popularisation, which is not surprising because we live with the weather every day and it has managed to translate its data and the output of its models into something which is comprehensible for mere mortals. Consult any atlas of some sophistication and you will find the fruits of their efforts in the form of easily readable climatic charts, or switch on the television for the news. The DSSAT team did something comparable. The idea was to plot the yields predicted by their crop models and their variation into geographical maps. They therefore had to generate time series of arbitrary length of (imaginary) rainfall events for a location, say for two hundred or a thousand years, by a technique called stochastic rainfall modelling. The parameters for the stochastic model were derived from the actual rainfall records, so the time series, although they had never occurred like that, would still have the same long-term pattern as the actual rainfall at that location. Then they ran CERES-Maize for each of those years and plotted the outcome in the form of maps of the maize mean yield and its variation, as the climatologists do for rainfall. In the late 1980s when DSSAT was built it took a lot of computer time to generate rainfall series and run CERES-Maize a thousand times to draw 'potential crop yield' maps, but on today's computers it takes no time to speak of.

DSSAT did not stop at mapping potential crop yields, of course. The ambition of its creators was much greater. They wanted the package to develop into a powerful tool which could do many more things, such as predicting the effect of soil and crop management practices, designing effective pest management strategies and predicting the effect of in-field variation in precision farming. And, yes, optimising N-fertilisation, although I doubt that DSSAT had the power to screen any number of fertilisation schemes in an efficient way, the way MANAGE-N did. DSSAT's intended clientele were not just other modellers, the group wanted to break down the barriers separating them from the ordinary agriculturist and extension worker, by making the package as user-friendly as possible. Without much success so far, however, as Jones et al. admitted in a readable essay on their work published in 2003: 'A [more] difficult issue is, however, the gap that exists between [cropping systems models] and their application for decision support at the farm level.' The paper then made a weak attempt to explain why that was so and pleaded for more research to remove some of the impediments. The use of modelling tools by non-modellers, or rather its absence, is of course a very important issue which I will return to at the end of this chapter.

9.7.4 Telling breeders what to breed?

The last example of a potentially successful application is modelling in support of plant breeding. Crop scientists had always nurtured the idea in the back of their minds that they might one day help rationalise plant breeding by telling the breeders what traits to select for. That idea only started to evolve from a vague illusion into a potential reality when crop physiology was quantified, first through the largely tautological growth analysis and then through crop modelling. In 1968, well before the expansion of simulation modelling, an Australian agronomist and breeder, C.M. Donald, had argued for designing the best possible plant type for a particular set of conditions. He called that an *ideotype* and gave as an example that a cereal variety for high-density high-input cropping should be low tillering. Ideotype design was an appealing idea which stuck and turned up regularly in scientific papers, especially by crop physiologists, but did not have much direct impact on breeding practices. For one thing, because from the start scientists disagreed about what the ideotype for a particular set of conditions should be. And successful ideotypes were mostly defined *after* the breeders had produced them, rarely before, like the combination of traits brought together in the short-straw wheat and rice varieties of the green revolution. I think most breeders saw the design of ideotypes as a rather frivolous pastime of crop physiologists with questionable results on which they were not going to waste their precious time. And anyway, what they themselves had always been doing – selecting the best possible combination of traits – was not that ideotype breeding in the first place, so what was the big deal here? Then the modellers entered into the game and reasoned that optimum plant types simply could not be designed without the help of simulation models because of the complexity of the growth processes involved and their interactions. In the early years, when the models were still rudimentary, there was little scope to apply them to breeding issues, except perhaps for the search for plant specimens with a high leaf assimilation rate. But that turned out to be a disappointment when scaled up to the crop level. With time, however, the models became more complex and covered an increasing number of processes with parameters specific to the crop being modelled. We have already met several of them: length of the juvenile phase, parameters characterising tillering habit in cereals, assimilate partitioning factors, specific leaf area, and many others. Many of these parameters had originally been slipped into the models as ad hoc entities, until such time as plant physiologists would come up with something better. They were poorly understood and their values had to be measured every time the model was applied in a new area. In fact, these parameters were just expressions of our ignorance and it was rather naive to think that they had a direct genetic basis. Nevertheless, some crop modellers liked to think they did represent something in the real world and termed them 'genetic coefficients'. Some of them did have a genetic basis, especially those which were related with phenology, as we saw a while ago. But many others, especially those related to morphogenesis, were just fanciful. That was especially the case with assimilate partitioning factors which were introduced because the earlier attempts to build the plant's functional balance into the models had been unsuccessful. Or with things like leaf sizes, specific leaf area and stem girth which were often handled as forcing functions. Even more importantly, none of the models had a 'memory' which would relate present to past growth, for instance of successive leaves, or link the size of a fruit with that of the stem meristem from which it originated. Memory of past growth is such a crucial trait of real plants that models will become rigid and fail to mimic the plant's natural resilience if it is not accounted for.

The weakness of crop models in morphogenesis had not gone entirely unnoticed, though. I have already mentioned Horie and de Wit's work on leaf growth of cucumber and my own on cotton morphogenesis. And in the USA Piper and Weiss showed, in 1990, that CERES-Maize would stray rather seriously off course if one tried to simulate the effect of a reduction in plant stand or leaf area during the season. The model simply did not possess the adaptation mechanisms which plants have (or else the modelling was incorrect).

That kind of inadequacy did not show up in the beginning when the models were tested with crops grown under good management in different environments or with different varieties, because each time the model was 'fine-tuned', which means that the values of several empirical parameters were adjusted so that the results would come out right. But when a range of varieties were grown in a range of environments, which is what breeders will do in order to measure what they call Genotype × Environment (G×E) interaction, simulation would break down. It took time for the modellers to acknowledge that failure. When Professor Wang of Zhejiang University and I reviewed the SARP project in 1995–1996 we ventured the conclusion that attempts to simulate GxE interaction would not work with the models of the day, but that conclusion was not at all obvious from the project's own writings. I do not mean the project scientists were hiding the truth, they were simply not stating it clearly, a little like market vendors wanting to sell their goods. And some of the models' results were indeed of excellent quality. For example, manage-N allowed the users to fine-tune N application in paddy and obtain the best possible yield for a given total N-dose. And the more comprehensive rice model ORYZA-1, in spite of its rudimentary representation of structural growth, was capable of making meaningful predictions about the kind of rice plant that would be needed to break through the yield barrier which had been immovable since the 1970s.

How could that be? Well, the reason was that the absolute maximum yield is obtained when all conditions are ideal, water and nutrients are plentiful and the crop functions as a comparatively simple machinery in a steady-state mode. Martin Kropff et al. in the SARP project at IRRI carried out calculations to see whether it would be possible to breed a rice variety which could yield up to 15 t/ha in the dry season in the Philippines and what such a variety would look like. The maximum yield attainable at the time (1993) was 'only' 10 t and the variety that was going to break through that ceiling had already been named the NPT, or VHYV by the IRRI breeders, as you will remember from Chapter 7. The modelling group carried out the calculations in two ways (see Kropff et al., 1994) – the simple way, using rules of thumb and approximate figures, much like the potential yield calculations of Section 9.7.1 above, and the formal way, using the SARP model ORYZA-1. The simple method is shown in Box 9-4. It says that the new variety should have:

- Thirty-seven days of effective grain filling (the publication says 38)
- Large panicles, with 60,000 spikelets per m^2
- Maintenance of full canopy assimilation throughout the grain-filling period

No detailed modelling here, the only things which were needed were the results of a summary model for biomass production and some good physiological research about the remobilisation of stem assimilates. What additional requirements for the hypothetical new plant type did the formal model contribute? In fact, very little. It

---

Box 9-4. What would a VHYV of paddy look like?

Suppose the new rice variety had to be able to produce a maximum yield of 15 t under (practically cloudless) dry season conditions in the Philippines, that is about 13 t of grain dry matter. The maximum assimilation rate of a healthy fully closed canopy with optimum N-content will be around 300 kg/ha, as you can work out from Appendix 9. If the full 13 t of grain dry matter had to be accumulated from assimilation during the grain growth phase alone, that would require almost 45 days of *effective* grain growth duration (that is the number of days if the sigmoid growth curve is approximated by a linear one). That kind of plant type simply does not exist. However, cereals have developed a mechanism that allows them to grow panicles which are larger than the plant can supply from simultaneous assimilation: they remobilise 'non-structural' assimilates accumulated in the vegetative tissue. This remobilisation can supply up to 2 t of grain dry matter, so simultaneous assimilation only had to supply for 11 t, i.e. 37 days of effective grain growth. That was considerably longer than the duration of grain growth in existing varieties. Also, the number of grains needed to store that kind of yield was much larger than that of existing varieties. At a 1,000-grain weight of 25 g, 60,000 grain (or spikelets)/m^2 were needed. Finally, the canopy had to stay green and fully active throughout the grain-filling period.

said, for instance, that there should be a minimum of unproductive tillers and that the panicles should be placed lower down in the canopy, to allow more assimilation by the now more fully exposed leaves. Those were good ideas but they did not really require detailed modelling either. Even more unsettling was the fact that all these conclusions could have been drawn with the tools available 20 years earlier, and most had in fact been drawn, for example in a paper which de Wit et al. contributed to a book on plant breeding published in 1979. Anyway, by the late 1990s the low panicle trait was now being incorporated in the NPT and modellers and rice breeders were slowly converging on the same ideas about the ideal plant type or ideotype to break through the yield ceiling.

By the early 2000s a new wind was beginning to blow among crop modellers. The Florida and Wageningen groups published a joint paper (Boote et al., 2001), where they were refreshingly modest about the capacity of contemporary models to account for the obviously existing differences among crop varieties in their response to particular environments: 'We believe there is an inadequate degree of genotypic specificity in present crop growth models [. . . because, among other things] most lack the ability to describe the subtle complexities associated with genotypic differences within species.' After 35 years some introspection was taking place and new ideas were beginning to stir. Or perhaps it were old ideas which were reasserting themselves.

9.8 New trends are emerging

In 1996 Xinyou Yin, a former member of the SARP team, defended his Ph.D. thesis on rice phenology in Wageningen and then joined the faculty of the C.T. de Wit Graduate School for Production Ecology. As I have argued earlier, the representation of phenology had been one of the successes in dynamic crop modelling, but Yin and his colleagues (which included some of the old hands from the early days of modelling in Wageningen, like Jan Goudriaan and the omnipresent Gon van Laar) realised that there was an urgent need to come up with something equally reliable for structural growth, instead of the customary parameterisation which in de Wit's words bordered on a 'complicated way of curve fitting'. That new insight was stimulated in no small measure by the rapid advances in genomics research which forced the modellers to think more clearly about the genetic basis of the plant properties they had built into their models. They argued that, in order to be useful to breeders, a model should only contain varietal traits which could be expected to link directly with a small number of genes and which could be measured directly on growing plants. Most parameters for phenology fell in that category, but tabulated factors for carbohydrate partitioning obviously did not. The new insights and results from new physiological studies were laid down in a new crop simulation model, GECROS (for Genotype-by-Environment interaction on CROp growth Simulator), published by Yin and van Laar (the same) in 2005. Partitioning of carbohydrates (and N) was of course pivotal to plant growth, so what kind of mechanism was steering that and how should it be modelled? Yin and his group first of all went back to the functional balance concept which was thought to control the partitioning between roots and shoots. The balancing mechanism itself was thought to be shared by all plant species, which would only differ in its parameters. Then came the partitioning within the shoot. Yin and van Laar wrote about this as follows: 'It is assumed that the strengths of growing organs as sinks [. . .] determine the carbon partitioning.' Well, well, doesn't all that sound somewhat familiar? Yes, these were exactly the same principles which had been built into ELCROS more than 35 years earlier – and then abandoned. But in spite of its sound principles GECROS still treated the crop as an amorphous set of roots, stems, leaves and reproductive organs, with no apparent structure linking them up, apart from the functional balance and the sink strengths. And, as I have argued repeatedly, it is precisely that structure which distinguishes one plant species from the other and even to some extent one variety from another within a species.

In the years between ELCROS and GECROS, modelling had become dominated by engineering with clumsy tools based on ad hoc assumptions and the fact, well known to the early modellers, that the plant was not well represented in the models had gradually been forgotten. Weiss (quoted by Yin and van Laar) said in 2003: 'Crop growth simulation models have hardly been further developed over the last two decades when most work was devoted to applications.' I think de Wit foresaw the danger when he commented in his academic acceptance speech in 1968: 'We can imagine the sorry state mechanics would be in if Newton, when observing a falling apple, would have had a computer at his disposal: a model, however successful in mimicking reality, is no substitute for a theory in the natural sciences.'

It is just a matter of time, however, before the modellers find out that they will have to start paying more serious attention to the real plant, in particular to its morphogenetic template as the steering principle which governs its growth. I have no doubt that is where the most direct link will be found with genetic factors distinguishing one variety from the other and where the modellers can realise their hope to be listened to by the breeders, because they will have something important to tell them.

9.9 Does tropical agronomy need modelling?

Towards the end of de Wit's career, in the mid-1980s, his apprentices decided that their tools were good enough and that they should now go out and help the ignorant scientists in the Third World. It was not the first time they ventured out there, but the earlier projects were mainly meant to extend the modelling methods into a more exotic environment, rather than to educate the local scientists. This time around, however, education was the primary aim, although the further development of the modelling tools came as a welcome bonus. So they talked the Ministry for International Cooperation into the need for modern computer-aided research to be introduced in agricultural research institutes in Asia. Something similar was done by Jones' group in Florida who went to Africa on the same mission. I am less familiar with those efforts, so I will restrict myself to the Wageningen expedition. The project to bring the blessings of modelling to Asian research was called SARP about which I have written earlier.

The Ministry of course wanted the modellers to state development-relevant objectives to justify the considerable amount of money that would be needed, so the project's goal was formulated as follows (the quotes are from the project's official documentation):

To build research capacity in systems analysis and crop simulation at national agricultural research centres and Universities.

Yes, OK, but why is that useful?

To integrate knowledge about biological and physical processes, thereby developing 'systems thinking', stimulating interdisciplinary research and increasing
research efficiency.

Oh, yes, 'interdisciplinary' is good and 'increased efficiency' also.
And 'systems thinking' sounded holistic which was also considered
good in the 1980s when the project started. But how exactly would
computer modelling help in increasing research efficiency?

Well, that was obvious. Models could explore a wide range of
management practices to find the best combination for optimum
resource use. They could be used as 'what if' tools to explore the
implications of changes in management. They could predict the
performance of varieties and fertiliser in different environments
and they would help design plant ideotypes to improve the effectiveness of breeding. As a result, there would be much less need
for expensive and error-prone experiments and those that were
still needed would be lean and well-focused. In short, models were
going to be a blessing for researchers, institute treasurers and
farmers alike.

There was of course little evidence from the record in the
industrialised countries to substantiate these claims. But the conviction on the part of the modellers reinforced by the eloquence of
their discourse convinced the aid bureaucrats, so they could go
ahead, for the next 11 years. De Wit's prestige was no doubt one of
the reasons why the project was approved, although I know he himself was not so convinced that it was a good idea. But as a good sor-
cerer he probably felt that the apprentices should now have a chance
to go out and try on their own. By the end of the SARP project the
crop models, although they had been much refined, had not much
advanced towards the fulfilment of their promises. There had been
some notable successes for sure, as I have shown above, but I do not
think they were in proportion to the vast amount of money spent
on the project. That in itself would not be a disaster, nor would the
waste be comparable with the waste in development aid in general,
as we will see in the next chapter. And several Asian scientists whom
I met later said they all had had a good time and learned a lot. The
question is, however, whether what they learned was in any way
going to be useful for them and especially whether it would help
their institutes to rationalise crop research. I doubt that very much.
As I have argued earlier, modelling did not even come near to

accomplishing its self-declared role as the great purifier of experimental research, because of the deficiency of the models and the failure to convert their useful parts into practical tools. And I do not think that scientists in developing countries, should be bothered with such speculative business as building process models for complex organisms, and certainly not those who are expected to carry out applied research in the service of smallholder farmers in their countries.

IITA also ventured into systems modelling in the late 1980s when the institute hired a scientist from the University of Florida school, Shrikant Jagtap. Jagtap doubled as the Institute's climatologist which was fortunate for him when it turned out that there was really not much a crop model like CERES-Maize, which he introduced at IITA, could accomplish. He launched into GIS instead and when he left, I think in 1998, he returned to Jim Jones' group in Florida. It was not for lack of potential topics that modelling did not take off, though. I have shown in previous chapters that there were important issues around alley cropping, one of IITA's major technologies, where simple modelling could have made a contribution and, indeed, helped to focus the experimental research. To my knowledge nobody picked up the challenge and the technology more or less died a natural death, at least at IITA it did.

Perhaps in a few years from now or more likely in decades, when some key processes in the growth of plants have become better understood crop modelling may still fulfil its promises. Meanwhile it would be good if the modellers could start to convert their sophisticated programs into something which will attract general practitioners rather than repelling them by their austere and frightening computer code. Once models have been developed to the point where they can become genuinely useful for practical applications and converted into user-friendly tools it is the time for developing country agronomists to get involved with them in a serious manner.[23]

---

[23] This is not a plea to keep the scientists in developing countries out of the loop. I am talking about agronomists and other scientists involved in applied research. If a country decides to have an institution for basic research they can of course opt to join the model development effort.

able to the open, in order to understand the

Chapter 10. Donors, Experts and Consultants

10.1 Development aid: A short and mainly African history

In the previous chapters I have talked a lot about agricultural research but only little about agricultural development. I will redress the balance a little here, by first picturing a background of post-independence development in Africa in general followed by an account of agricultural development on that continent.

Development aid was born from genuine concern about the backwardness of most former colonies and other countries which, though not colonised, suffered from the same underdevelopment syndrome. Aid was meant to help them build modern state apparatus, infrastructure, industry and agriculture and teach their people the skills to run them, in order to disentangle themselves from the poverty trap. There were of course other less honourable motives as well but I will leave those aside. Aid was seen as a short cut to development, which should allow the young countries to bypass the age-long development processes which the industrialised nations had gone through.

I think we can assert without risk of contradiction that, in spite of good intentions on the whole, the record of twentieth-century development has been rather dismal, especially in Africa. If the intentions of aid were honourable and the funds spent considerable, why have the results been so pathetic? Many factors have been responsible for that, some of them technical and some historical, social and political. I think among the most noxious were impatience and a barely disguised sense of superiority, combined with an often shocking lack of competence on the part of the aid providers. It is a well-known fact that the lower esteem an aid recipient is held in and the smaller the aid provider's awareness of his own short-comings, the more likely he is to interfere instead of assist. And that is what most development aid has been: interference where catalysis was needed. The name for that game was development help,

assistance, aid or cooperation, or whatever euphemism was
invented for something that at best was benevolent paternalism and
at worst, imposition of unsuitable development models by self-declared
western experts.

Euphemisms were symptomatic for twentieth-century development
philosophy and the hypocrisy of its practitioners, starting with the
term 'developing countries' itself. It is hilarious to read the following impatient remark in a famous book by René Dumont, the first
edition of which was published in 1962: 'the large majority of what
are called, disingenuously,[1] "developing countries" are stagnating,
mainly for reasons of a defective socio-economic structure". These
lines could have been written with equal or more justice in 2000,
after four decades of attempting to change that socio-economic
structure, and we can now add that the international money which
has been thrown at those countries since the 1960s has made things
worse, as Dumont also clearly foresaw. In spite of his fame in some
circles, Dumont was considered by many as a maverick, best to be
ignored, but many of his lucid, often blunt but always sympathetic
observations and analyses have been vindicated by history. External
interference has prevented the locals from making their own mistakes and suffering the consequences, which could have led to
home-grown creativity and perhaps bring about home-grown development. Instead, donor money has corrupted the minds of two
generations of African public servants because it could be got with
a minimum of effort, provided one played along in the donor-
driven development game. Let us take a quick walk through the history of development aid and see what evidence there is to support
this harsh judgement.

10.1.1 Great needs, high expectations

People are prepared to suffer for grandiose ideas and postpone the
satisfaction of more immediate, materialistic wishes, as long as
they believe in their future fulfilment. When the excitement about
the newly gained independence was over, the new governments
therefore had to start thinking about how to build up local production capacity to supply their populations with consumable

---

[1] In 'L'Afrique noire est mal partie'. The word Dumont used was 'mensongèrement'.

goods and earn foreign exchange by exporting agricultural and industrial produce. The former colonisers and the world's super powers, the USA and the Soviet Union and later on also Japan, were going to help them, motivated by a variety of interests, in addition to genuine concern.

A few basic ideas were shared by all aid-giving countries. They all agreed that it would be good for everybody if the new countries' economies would quickly enter into sustained growth at a rate which considerably exceeded their own in order to catch up and that they should be helped, both financially and technically to do so. But there was less consensus about how that was to happen, in what order and at what rate, what should be the role of government and the private sector and what kind of aid donors should provide. In the 1960s there were two opposing views, with equal strength of conviction, about the role of government in the quest for productive development. These views coincided with the world's division into two great political blocks, the capitalist and the socialist/communist, also known as the first and second world. The capitalist view was of course that government should abstain from productive activities and just create an environment conducive to private enterprise. In the socialist world view, governments were the prime actors, not just as facilitators but as owners and managers of key industries, in the most extreme case all industry, including agriculture. Many young African countries were attracted to socialism, from conviction or simply because private industry was practically non-existent and state capitalism seemed to promise a quick fix. The USSR and the communist countries in Asia, which were trying to industrialise themselves out of backwardness were gleaming examples and many African countries wanted to follow their example. Even countries which favoured capitalist ideas opted for a fair measure of state intervention during the transition to a capitalist future. And the emerging aid bureaucracies in many donor countries, even the capitalist ones, were also sympathetic to a strong role of government in the countries they supported. After all, aid was almost entirely financed by public money and it was therefore handled by government bureaucracies, on both sides, not to mention the fact that many of the new aid officials, especially those in Europe, had definite interventionist inclinations. I will largely ignore the few African countries which associated themselves with the Soviet Union and its allies (for one thing, because I have had only limited experience with them) and talk mainly about those countries which sided with 'the west' (later augmented by Japan), adopted its development views, accepted its funds and welcomed its experts.

In the early years there was also much debate about whether a country should first develop industry rather than agriculture, or the other way around. Many were convinced that industrial development should come first and that it would eventually pull along agriculture. They included Ester Boserup whom we have already met and my agricultural economics teacher in Wageningen, Professor Joosten. But others, for example, René Dumont and Clifford Geertz, argued that historically surpluses had first been generated by agriculture, which were then extracted to help embryonic industries develop. This had happened both in Europe and in Japan and it would probably have to happen again in the poor tropical countries. There is a large body of academic literature on the subject but I wonder how much influence that really had. In practice the choices were quite unsystematic and ad hoc, now setting up a project for industrial development, now for agriculture.

There was enough to do to keep everyone busy anyway, from digging wells for drinking water to building steel factories and from creating village infrastructure to building electric power plants. And since there was no indigenous private sector to speak of, the state took on those tasks, assisted by the donors. All those things had to be planned and as long as there were no local planning experts, the donor countries sent in their own. The foreign experts usually had as little experience with that kind of planning as the locals did, but that did not really matter. The most urgent needs were so obvious that nobody could fail to see them, with or without planning experience: schools, many schools to quickly raise the literacy level of the population and start turning out skilled workers, rural clinics, roads of course and power plants. In the beginning the new countries did not have the people who could build those things either, but the donor countries would send experts to help them and meanwhile train the locals to take over. Many useful things were done and a lot of white elephants appeared as well, like the huge empty grain silos in the Dar es Salaam port, dairy plants without milk, a new capital city with a monstrous cathedral in the middle of nowhere and all the other well-known examples, but that assumedly was inevitable. And in those early years the nonsensical was easily eclipsed by the useful.

If the expensive failures and nonsensical wastage had only been a matter of money it would not have been so bad, provided everyone had learned from them. But they gradually corrupted the minds of the receiving governments and their bureaucrats; and not only theirs, also those of the donor bureaucracies and their consultants. Corruption is a relatively slow erosive process. In the beginning one is shocked by the ease with which large sums of money are spent with little or no measurable impact, but soon one tends to ignore the waste and only look at the positive, like the employment provided by the project, however temporary (even counting the experts' domestic personnel), and the training of many people, even though it may not be clear whether the skills they have learned will ever be useful outside the project. In the world of business, wasteful behaviour will soon be penalised by the loss of competitiveness and eventually by bankruptcy, but there were no such mechanisms in the world of government-sponsored investment and development projects. The results would only become visible after the ministers and the bureaucrats who made the decisions had long been gone. And governments rarely went bankrupt, not even those of developing countries, because they would eventually be bailed out again by some consortium cf donors, who shared responsibility for the waste.

In the 1960s and 1970s, the world's major economic fault line had been that between capitalism and socialism/communism, but in the early 1990s most socialist economies collapsed, after which capitalism affirmed its superiority as an economic model, practically sweeping away its competitor. Market liberalism then became the norm and was imposed by the World Bank and the IMF on developing countries who wanted their loans, in particular in Africa, even though the major free marketeers refused to fully apply its principles to their own economies.

(a) Projects

After this bird's-eye view of development concepts and principles, let us look more closely at the nuts and bolts of development aid in the countries associated with the capitalist and social democratic world. In the first two decades projects were aid's defining feature, or, in today's development speak, its major modality. The term 'project' suggests an activity with a limited time span and a well-defined output, like building a bridge or a factory, and those were

indeed among the things donors liked to invest in. They provided
concrete proof that aid was contributing something tangible, if not
useful. Gradually the meaning of the term 'project' broadened and
became also associated with activities, whose output was not as
clearly circumscribed or whose lifespan might turn out to be anything but short, like setting up an efficient civil service or extension
organisation. If there was a limitation to those projects' lifespan it
was related to the donor's patience rather than to the nature of the
tasks at hand.

There were roughly four kinds of projects. The most straightforward were those which came closest to the dictionary definition of
what a project is: 'a piece of work that needs skill, effort and careful planning over a period of time' (according to the *Longman
Dictionary of Contemporary English*, 1989). They were concerned
with building infrastructure: roads, bridges, new factories, power
plants and such things, perhaps technically complex but institu-
tionally simple tasks, or so it seemed. Once completed the works
were turned over to the locals, who had meanwhile been trained to
run them. The second group of projects were concerned with the
development of government bureaucracy, public institutions and
services, like healthcare or agricultural extension and research.
Most newly independent states inherited a working bureaucracy
from their previous colonial masters, but the ambitious new tasks
of nation building and rapid economic development required considerable expansion. And even though some countries like Ghana,
Nigeria and Ivory Coast also inherited healthy monetary reserves
and profitable enterprises and plantations, their development needs
were beyond their means and donors had to step in to replenish the
resources and assist the bureaucracies to grow and take on new
tasks. Then there were the small-scale community development
projects which were set up to help mostly rural communities with
everything that might improve their livelihood. They bloomed for a
short while in the 1960s, especially in Asia and then withered away
because the slightness of their achievements did not match the
weight of their ambitions. But their ideas would not die and some
of them reappeared in the participatory development movement of
the 1990s.

And finally, there were the rural development projects, which will
concern us most here. They would work in a particular area,
because it was poor, or because it had great potential or for some

other reason, and they dealt with everything economic. The large rural development projects of the early post-independence period were quite costly affairs. They were like self-contained microcosms, with their own personnel, materials and equipment, like army divisions which can operate as autonomous units because they have everything from infantry to artillery to tanks. Rural development projects would have a fleet of vehicles, tractors and lorries, they would have a budget to build stores, buy seed and fertiliser for distribution or sale to farmers. They could help farmers set up a cooperative and provide a credit fund, organise market access, and many other things which were needed to stimulate the rural economy. The idea was that after a number of years the project activities would be taken over by local organisations which the project itself had helped to bring about. At the head of the project there would be a project manager, the analogue of a military divisional general, invariably an expatriate, assisted by a staff of experts. They were also expatriates in the early years, until that was no longer acceptable to the donor or the recipients or both, because the country now had enough experts of its own. But by the time that happened the era of self-contained projects had also come to an end and aid had entered into a new mode as we will see presently.

(b) Aid money, easy money

Bilateral aid, that is aid from individual donor countries as opposed to international organisations, has mostly been in the form of grants. It has been particularly easy to get, because bilateral donors have often been overly eager to provide it, measuring their bureaucrats' effectiveness by their spending prowess. Their counterparts at the receiving end were quick to conclude that the money apparently meant little to the other side and that they could play off one donor against another, as the latter appeared to be competing for scarce spending slots, rather than the aid recipients having to compete for scarce funds. In fact, they did not have to do anything, except informing a donor of their general interest in being financially supported. And not always even that. The donor would call in its consultants to *identify* a fundable cause, carry out a *feasibility study* and *formulate* a project and then hire other consultants to carry it out. After a few years yet another group of consultants would be sent in to *evaluate* the project. The recipient country would of course participate in the identification, preparation and evaluation of the projects. They would appoint their own experts, at first middle echelon officials from the ministry and later on consultants from local firms which started to spring up everywhere in the 1990s. They would all be paid by the donor, it being understood that the recipient country had no money. In the beginning the participation by the locals often amounted to little more than ritual, especially in Africa, until by the end of the last century the locals had learned the tricks by example, set up their own consultancy bureaux and became almost as proficient in the artful ways of consultants as the expatriates were.

One of the worst things for poor countries to do is rely on foreign aid to get out of the poverty trap. It corrupts the minds of their leadership and prevents them from taking things in their own hands. Aid dependency paralyses a country's initiative. Its politicians and bureaucrats let the donors decide what is good for them and, if things go badly, let the same donors bail them out again, in international aid's merry-go-round. One might think that aid money in the form of loans like those provided by the World Bank would be better than grants, because the former would be less likely to be spent irresponsibly, but in practice the difference was only marginal. The loan hangover should have come when the loans had to be paid back, but by that time the donors, feeling guilty about making loans available for ill-conceived programmes (which they usually had conceived themselves in the first place), could often be talked into rescheduling or even scrapping the debt, at least those of really poor countries, or not so poor ones whose coffers had been emptied by their own leaders.

Things were made worse by the fragmentation of aid. There were numerous donors and aid agencies and all of them pursued their own pet projects, within the overall paradigm of the day, with little concern for coordination and without being forced to subordinate themselves to the interest of the country they were trying to help. Most of the time the central authority was simply not capable or willing to put order to this chaos. Foreign aid just seemed to happen and obeyed its own logic and there was little one could do except making sure one got a large slice of the cake.

(c) Volatile experts and restless bureaucrats

Another bad feature was excessive reliance on expatriate project personnel. In the first decades there were simply not enough qualified local people around to spend the considerable amounts of aid money responsibly, so the donors sent in their own to get things moving. The intentions were mainly honourable and the experts' presence would only be temporary, or so we told ourselves, but little by little aid addiction crept in on both sides. It is easy to see now that the countries which have fared best are those which refused to be pushed around by donors and would not accept consultant-driven aid, like South Korea, India, China and Malaysia. In less assertive countries, which included practically all the African ones, large number of experts were brought in to provide the skills that were thought to be lacking. In the early years that meant practically all imaginable skills, in agronomy, civil engineering, extension, secondary school teaching, indeed anything. The experts were expected to transfer those skills to their local counterparts who would be able to continue on their own after a few years. But things were not as simple as that. The technical skills could be transferred all right, but it was less easy to create the institutional environment where they could be effectively applied. As long as the projects were there with their expatriates and a lot of resources they could create islands of efficiency where things were done as they were at home, but such projects would not survive after the expatriates and the resources were gone.[2] One solution was to stay on indefinitely or almost so. There have actually been projects which continued for 20 or more years, with each next extension being justified by the argument that stopping now would jeopardise the previous investments. More commonly the donor would get tired of the project after a few years, close it down and start something new elsewhere, hoping (usually in vain) that things would work out better this time around. And, more significantly, aid gradually started to gravitate away from technical interventions to institution building, a domain which largely nondescript aid officials and consultants could more easily appropriate as their own. I will come to that in due course.

Who were those experts who thronged the development projects and the aid agencies? In the early years many of them were former colonial officers but the number of these was small relative to the tasks at hand and there was a great sense of urgency. A new army

---

[2] Of an entirely different type were the voluntary services modelled after US President John F. Kennedy's Peace Corps. Volunteers were placed with local organisations with no special facilities and for a local salary. Veterans of these services have been an important source of recruits for aid projects.

of aid workers had to be raised to handle the growing flood of aid money which was going to help the poor countries skip a few phases on their way to prosperity. Western universities churned out increasing numbers of graduates who had specialised in all kinds of development-related studies. They were mostly motivated by genuine concern for the poor combined with youthful hubris about what they could do. And there were also people with more conventional training and career interests for whom a few years stint in a developing country was attractive because of the combination of adventure and broadening one's horizon. Some western employers even appreciated that, provided it had not lasted too long.

Once the new experts had arrived at their posts in the projects, their hubris was rarely tempered, nor were their skills significantly improved by their supervisors, as they would have been in more mature trades. The experienced people who could have moulded them into effective aid workers mostly worked in the new aid departments rather than in the field, so many young experts were just turned loose to carry out unfamiliar tasks in unfamiliar environments and had to learn on their own. They would have a supervisor all right but in the development branches of the United Nations, the largest single employer, the occupancy of senior positions was based less on skill than on equitable allocation of positions across member countries. And the new consultancy firms that specialised in development were not much better. Their field supervisors' distinction rarely amounted to more than a few years age advantage over the supervised and the corresponding amount of experience, whipped up to an impressive volume in their biodata.

With time the new experts developed entirely new skills which had more to do with their own survival than with the technicalities of development. The most needed survival skill was flexibility, that is the ability to spot new trends and fashions and quickly adapt one's professional profile and curriculumvitae to the exigencies of the times, and often one's attitude to the whims of an insecure or arrogant supervisor. Soon it turned out that bringing in real experts from industrialised countries was mostly an illusion. They were too busy at home, had no appetite for working in developing countries, or did not want to jeopardise their careers. Even established western consultancy firms, when launching into aid projects, set up special departments to carry out those projects. They might be able to draw on their home departments for highly specialised staff who could occasionally be sent out short-term on a specific assignment, but most of their staff would be exclusively involved in development projects. Meanwhile, the development trade had drifted away from the original idea of transferring resources and especially skills from the industrialised countries to the underdeveloped ones, by experienced people from the former who were to teach the ignorant ones in the latter. That is how Tsar Peter the Great had done it in the seventeenth century when he lured ship builders from Holland to St Petersburg to help set up modern shipyards there. In the twentieth century the roles were reversed. It was not the locals who solicited technical assistance from the west, although it was often made to look that way, but western bureaucrats who formulated development objectives and plans and provided the means and the people to carry them out. International aid was, in today's parlance, supply-driven from the start.

The new experts were also highly mobile. Projects would be set up for periods of 3 or 4 years and even though they were usually extended a few times, the experts' shelf life often did not exceed a single project phase. The aid bureaucracies and the consultancy firms were busy building their staff strengths and would move junior staff around to diversify their experience and 'build up their CV'. That improved their suitability for future projects and the range of projects a consultancy firm could successfully bid for, but it did not result in solid training on the job.

The impermanence of the consultants was matched by that of the aid bureaucrats. That had to do with their having been part of the diplomatic service. In the early years aid was managed by remote control from the donor capitals, but as the volume of aid increased, the embassies were given more and more responsibilities. Technical specialists were recruited for the embassies' development desks and given latitude to make their own decisions about the projects to undertake, as long as they stayed within their government's general policies. So new projects usually carried the imprint of the officers in charge. You would expect that those officers would stay at the same embassy long enough to see most of their projects through to the end, but the foreign ministry did not encourage that, nor were most aid officials interested in protracted stays in any one country. And after its creator had left a project might languish in the shadow of the new officer's own projects or be completely overhauled, because every officer wanted to leave his own imprint.

That rather frivolous, highly personalised and sometimes plainly irresponsible conduct of the aid business contributed in no small way to the lack of lasting impact of many development projects. I will give some striking examples of that later on.

As the money flow and the number of projects increased so did the demand for people who could formulate and evaluate them. This generated a new breed of experts, not in any particular discipline, but in project identification, formulation and evaluation itself. They usually had spent a few years in the field and then went back home because of the children's education, their partner going crazy from not being able to have a job, or simply because they did not much like living in the tropics after all. Some of these people ended up in the international cooperation ministries, but when those ranks were filled some of them started offering their services as experts in project formulation and evaluation. With time they set up bureaux specialising in that kind of thing. At the receiving end something similar happened some years later. Erstwhile bureaucrats and ministers who had observed the consultants' manoeuvres while in office started offering their own consultancy services. At first they charged much lower rates than the foreign consultants, but once it became fashionable and was considered proper for donors to hire local consultants, their fees started to converge. These local consultants' skills had not increased miraculously by the sole act of exchanging a bureaucrat's chair for a consultant's, nor was the quality of their services guaranteed by their knowledge of the local conditions. But since the skilful execution of consultancy rituals gradually became more important than the content of its output, that hardly mattered.

(d) What about accountability?

With so much money flowing into development aid, surely the donors must have kept a constant watch over the way it was spent. In other words, was the development industry kept accountable for the large sums entrusted to them? In the colonial era, when most people in the West knew little more about life in the colonies than what they heard from their missionaries, aid was charity and accountability for its use was straightforward. When a missionary was about to go out for the first time he would climb the pulpit in his home parish, preach about the needs of the people in the poor country he was going to and leave with a healthy sum of money to build a church, or a school or clinic. He would send regular contact letters telling the parishioners about the good things he was doing and occasionally asking for more money to buy a motorcycle or some other useful item to make his work easier and more efficient. And when he came home on furlough, after 5 or more years, he would mount the pulpit again to refresh the parish's memory and re-establish its support by explaining what he had been doing and how much larger the needs had become. So the link between the aid financiers and the implementers was direct and effective, in spite of the distance and the slowness of the communication channels and the people trusted that the money was well spent, because it was spent by their own sons and daughters.

Things changed of course when aid became government business and increasingly large amounts of aid money were provided from donor governments' regular budgets. Their citizens were no longer free to give or not to give money for aid because it was taken out of their taxes, and the stories about abuse which started to be heard added concern to alienation. With time, however, they stopped worrying and accepted aid as inevitable and something best to be ignored, like many other things government did with their tax money. Aid became the business of self-declared development experts, a paradoxical title, because how can one be an expert in something as elusive as development? The experts designed their own criteria to measure whether what they did was successful in order to convince other experts who were involved in the same ritual. The rest of the world sometimes asked what they were doing but they were quickly put off by a jargon which could only be understood by the initiated. In the end even normally critical non-aid politicians would shrug their shoulders or make vitriolic comments, but only few of them dared to draw the conclusion that the holy cow might as well be slaughtered.

Meanwhile, the general public had no idea what was going on, except that a lot of money was spent for unclear purposes and with unclear results. They were told that mistakes had indeed been made in the past which were now being corrected, but that at the same time the aid money, even though it looked much, was actually not enough, considering the magnitude of the needs. And that message was effectively driven home by the ugly images of war and starvation broadcast by the media in an increasingly interlinked world.

(e) Surely, the World Bank must have done a better job?

The single largest donor was the World Bank, although 'donor' is not really the right term for a bank lending rather than giving away money. But terminology is often used quite loosely in development aid. The money spent by individual bilateral donors was peanuts compared with the loans from the World Bank and its sister institutions, the Asian and African Development Banks and the International Monetary Fund. But they were loans and you would think that the development banks had all kinds of checks and safeguards to ensure that the money was spent for profitable purposes, because that is the only guarantee that the interest and eventually the principal will be paid in time. You would therefore also expect that development activities financed with borrowed money were less soft than bilateral projects funded by grants. After all, the money had to be paid back, so the government would not use it for ill-considered projects with a high risk of failure. That, however, was an illusion.

First of all, a country would rarely approach the World Bank on its own initiative with a worked-out loan proposal, as a client of a commercial bank would. The country would just make a request for financial assistance (or be prodded to make one) and the Bank did the rest. Once the request had been submitted and the country was considered eligible[3] for a loan the World Bank circus got under way and all kinds of missions would be flown in to work out the details of the programmes. Apart from providing loans the Bank also used to tell governments how they should spend them. So, the Bank employees did most of the things which in normal life a borrower would have to do himself to convince a bank of his serious and bankable disposition. Appearances were of course kept up that normal banking procedures were being followed. Government's own officials were formally in charge and all the agreements had to get parliamentary approval. But the Bank would set up an office in the capital from where they stage-managed the whole thing, assisted by some local staff seconded by the country's government. The Bank

---

[3] Today the conditions to qualify for a loan are stricter than they once were and usually include reform of public finances and adoption of liberal trade principles. If you are really poor you can get a loan on soft conditions from a special branch of the Bank, the International Development Association (IDA).

was very good at getting the best people for those local staff positions
who then got a big boost in salary in order to make them more
effective. That was also an illusion. The local officials who until
then had to survive on a meagre salary all of a sudden found themselves with a several times higher pay plus all kinds of perks, which
allowed them to get the best possible medical care and send their
children to the best schools. Apart from their personal stake in the
goings on, they were also proud to be associated with so prestigious
an institution as the World Bank and very soon they would start
speaking the Bank's language, unless they were exceptionally independent and incorruptible people who kept their distance and took
care of their country's interest instead of their own. Those are rare
traits anywhere and under these conditions exceptional strength of
character would have been needed to really stand up to Bank staff,
who usually combined a strong personality with the arrogance
nurtured by power, an effective mix when the ideas are good and a
deadly one when they are not.

Another reason why World Bank loans in Africa only superfi-
cially resembled those of a real bank was that they were often used
to bring overstaffed, inefficient government services back on their
feet. However urgent institutional reform may be, the question for
a bank is not whether reform is needed, but whether spending loan
money on it is good investment. The bank economists could probably show with their clever but unproven models that it was, but
anybody in his right mind would think that investing large amounts
of borrowed money in government bureaucracies is the last thing
an African country would want to do. The main thing those loans
did was adding to the countries' indebtedness without producing
the monetary benefits which would have been needed to service the
debt. In the real world having many large defaulting customers on
its books would discipline a bank into more cautious lending. In the
case of the World Bank nothing of the sort would happen. If a
project failed and the borrowing country defaulted on the loan, the
World Bank would not loose a penny, because in the end the
indebted country would be bailed out by the rich countries and the
Bank would still cash in, in spite of the error of judgement it made
when awarding the loan. And once an indebted country had been
bailed out the whole show could start all over again.

So, in spite of the semblance of sound banking principles, there
was as little accountability in the Bank's loan operations as in other

forms of development aid; nor did simple market forces lead to its collapse which would have occurred with any normal bank with a loan record as bad as the World Bank's. Instead of being exposed to the harsh and healthy mechanisms operating in the real world, the Bank would regularly carry out the kind of ritualistic evaluations which were commonplace in the world of development aid. These evaluations almost invariably showed that, in spite of minor hitches, the programme was on track and that the money spent could not fail to result in the expected benefits. And by the time they were proved wrong, nobody would remember who had been responsible.

The Bank was not entirely devoid of sense, though. Whereas the periodic field assessments of its loan programmes were little more than expensive trivialities, the Bank did have a mechanism by which it regularly adjusted its policies. Whenever the criticism of its policies became too loud to be ignored they would commission a penetrating study by some famous or clever individuals with the gift of the pen. In the late 1960s prestigious international committees were appointed like the Pearson and the Tinbergen and Peterson Committees to analyse the achievements of the 'first development decade' and design strategies for the next. In the years that followed, Bank employees or sympathetic outsiders continued to produce a steady flow of high-level studies, policy reports and development strategies which explained why the previous policies had not worked and the next were going to do much better.

By the late twentieth century, however, the world was going through fundamental changes which were going to have profound effects on development assistance. History seemed to have decided in favour of liberal democracy, market economy and free trade as the way to go and the World Bank made their adoption a condition for approving loans to countries in need. The changing times were also reflected in a major study commissioned by the Bank which analysed the effectiveness of aid in the previous 40 years and charted the way forward, the so-called Dollar paper (Dollar is the name of a person). The study, published in 1998, concluded that a lot of money had been wasted, because it had been thrown at poor countries with bad policies and that donors, including the World Bank itself, had measured their effectiveness by the amount of money they were able to allocate, rather than by its impact. That showed that the Bank was capable of self-criticism, although its repentance came too late to repair the financial and moral damage its ill-conceived loans had caused. But what about the future, how was the Bank going to avoid the wastage which had been so typical of the past? The new principle that henceforth money was going to be lent to deserving countries only sounded good, but the question remained what kind of projects the World Bank was going to promote and whether they were going to have a significant impact this time. The signs were not promising. The Dollar paper crowed that institutional reform was the way to go for poor countries and that investing loan money in reform could not fail to pay off. But that conclusion was contradicted by the paper's own observations, which showed that significant development had only occurred in countries which had adopted sensible policies by their own volition and carried them out with great and sustained efforts of its people, not with World Bank loans and expatriate consultants; nor did those countries' policies necessarily involve the kind of market liberalisation the Bank advocated. The examples had been there for everyone to see: South Korea in the 1950s and 1960s, China in the 1970s and 1980s and India, Malaysia and to a lesser extent Indonesia and Thailand in the 1980s and 1990s, all of them in Asia. Why these countries seemed to have done most things right while others had not remains a question which has not been fully answered, but a major cause is certainly one of cultural differences, the mention of which had long been considered as indecent. Rather than acknowledging these differences and analysing their implications, development thinkers and donors have preferred to indulge in apologetics and the recital of the mantra that all people are essentially equal, which may be true but is not really relevant. René Dumont knew all that already in 1962 and he expressed his views loud and clear, without being listened to.

10.1.2 Development aid transformed

The general public, unaware of what exactly was happening, but increasingly exposed through the media to the often shocking realities of the poor countries, gradually started to feel intuitively that a lot of money, their money, was being wasted. That eroded the public's support for development aid and made them wonder whether they had been fooled all those years into believing that something useful was being done with their money. In the 1990s the gulf between aid supporters and detractors had become almost unbridgeable. The latter's most radical faction argued that development aid had better be stopped entirely, because it was doing no good at all and the countries which had received least in fact had done best. Another faction, led by the American aid establishment, had put their hope in the private sector and non-government institutions as the future flag-bearers of development. The majority of the aid officials, however, especially in European and international agencies, argued that aid provided through governments could still be salvaged, provided the recipient governments would adopt the reform measures proposed by the World Bank and aid itself was put on an entirely different footing.

First, projects run by expatriates were made the scapegoats for the failures of foreign aid and even the use of the word 'project' became almost taboo in the early 1990s. It stood for artificiality, lack of integration and technical hobby-ism. It was argued, correctly I think, that the large number of expatriates had themselves been part of the problem. They would set up their own offices, impose their home-grown work ethics and efficiency and leave the locals out of breath from trying to keep up. That had prevented the latter from taking things in their own hands. So out went most expatriates. Not all of them, though: most donors still wanted to keep an eye on the money and push things along here and there, so they would appoint coordinators or advisors in key positions to do that. But what if the locals did not live up to their new responsibilities and put the project (by now called a programme) in jeopardy? The logical thing to do would have been to withdraw because there is no point in flogging a dead horse. But development aid was ruled by a different kind of logic, the logic of bureaucracies which had to spend a large amount of money one way or another. Simply withdrawing from a project, or, at a different level, even from a non-performing country would not help their cause. It would reflect badly on the aid administrator's and his consultants' competence and harm their careers. So what did they do instead? They tried to achieve by remote control of their projects what they had failed to accomplish by direct management. For that purpose clever methodologies were conceived, for programme planning, management and evaluation, which the locals had to adopt and which could not fail to guide them along the right path. The remaining experts were going to teach them these tools instead of doing everything themselves. That was rather naive perhaps, but not dishonest. Most aid experts were truly eager, not to say desperate to improve the impact of what they did.

Another presumed cause of the failure of aid was that the old projects had been rather authoritarian affairs. The aid workers remorsefully acknowledged that they had decided what people needed rather than listening to what they said they wanted. Things were going to be done differently this time around and the beneficiaries would henceforth be consulted and their active cooperation solicited. That was the 'participatory approach' to development, which we have already met and which descended directly from FSR, although its adherents would be loath to admit it. Another equally sound and simple new concept which emerged a little later was decentralised government, which implied that power and responsibilities should be brought as close as possible to the people. The two together, transfer of power to the local level and participation by the governed in their own governance, were going to be the two columns of the new development thrust, and a complete departure from the authoritarian style of government which had been the rule in Africa.

(a) Decentralised participatory development

During the early 1990s the tandem concepts of decentralisation and participatory development became mainstream among donors who now put pressure on the governments of the poor countries to transfer responsibilities to the local level, reduce corruption and make government more responsive to the needs of its subjects. All these things were summed up in the new term of 'good governance' and its adoption now became a condition for receiving aid. Like most previous development concepts, good governance was very much a donor-driven idea, but at least it made a lot of sense. And it could be applied immediately in ongoing programmes: the 'Area-Based Programmes', which had become very popular among donors in the 1990s. They were integrated area development programmes, the direct successors to the earlier rural development projects but with even broader objectives: building roads, schools and clinics, improving agriculture, stimulating private entrepreneurships, strengthening the local government administration and the democratic process. But their approach was quite different, and more akin to that of the 'community development projects' of the 1960s: they practised, or at least advocated people-centred development, as the latter did. And they were directly subordinate to the local government rather than being run by aid workers. That, however, also exposed them to the inefficiencies of the local governments, a major one being lack of autonomy.

African regimes have been notorious for their unwillingness to hand responsibilities down the government hierarchy, leaving little or no room for local initiative. That paralysed the local government machinery which would only move when given instructions from the top. It severely hampered the effectiveness of the early area-based programmes and induced their expatriate officers to devise all kinds of clever ways to defeat the system rather than trying to improve its effectiveness, which would most likely have been in vain anyway. But the adoption of good governance changed all that drastically. One of its key principles was decentralisation and the local governments, severely lacking experience in making their own decisions, had to be helped assume their new responsibilities. The area-based programmes eagerly took on that task and sent local government staff for training in programme management, organised planning workshops, called in local and foreign consultants to teach participatory methodology, and shored up the local governments' finances, because without money all the rest would be meaningless.

At the same time resident technical assistance was reduced to the minimum. In the old days all kinds of technical experts had populated the projects as well, but that was no longer acceptable, since there were supposed to be enough well-trained locals and if they were not as effective they had to learn by doing. There usually remained a single programme adviser who worked as a trouble shooter wherever needed, helping to prepare the annual plan and budget here, ensuring that the contractors built schools or roads of reasonable quality there, as jacks of all trades. But quite powerful jacks of all trades, because they were also expected to keep an eye on the way the donor money was spent, even when the donor funds were (at least in theory) integrated into the local government budget. When there was an occasional need for specialised skills which were not locally available, someone could be brought in from outside on a short-term basis. The programmes budgeted money for that, but when the locals saw how expensive those short-term experts were, they would rather build another school or equip the government offices or send one of their own abroad. That was also a consequence of the devolution of responsibility. Development workers who wanted to work in the new programmes declared themselves institutional development experts and adjusted their CVs accordingly. That was not a difficult thing to do for people with many years of project experience, because whatever they had done would invariably have involved some institutional development. Even I could have written a quite convincing profile for myself as an institutional development expert.

The new-style development programmes were now going to address the people's real needs instead of those imagined by the aid workers and since people's needs were many and diverse, so would be the new participatory programmes' objectives. Instead of just starting to dig community wells or improve dairy production, because somebody had decided that was important, the programmes would now start off with objectives cast in the broadest terms, such as 'enhancing the people's well-being by improving government's capacity for service delivery'. The strategies to get there were not defined precisely either. In fact, detailed programming had been discredited by the disappointing achievements of the earlier projects. It was difficult to foretell what would turn out to be the real problems and opportunities and in particular the local people's perception of those. Planning was therefore replaced by a 'process approach',[4] which meant that the programme would grope its way forward and formulate more precise objectives and activities as it went along.

Interesting things happened as a result of decentralisation, which had not quite been foreseen. For example, when it was left to the local councils in Ethiopia, Tanzania and Uganda to decide how to spend their money some of them, would allocate little or nothing to agriculture, because they felt it had just been a waste of money in the past. That was encouraging because it showed that people will speak their mind if they are really put in charge. Of course, rash decisions like that may be quite disruptive, but when the local councils start to see the implications of their decisions, for example that they will have to lay off most of the extension personnel, they will think again. That is how people will gradually learn that their new

---

[4] I think the term was invented by one of the gurus of Dutch aid, Wim Zevenbergen.

power also brings new responsibilities. New donor programmes were designed to assist the learning processes which were denominated 'good governance' support programmes.

These were quite sensible trends but they begged the question of what the residual value of the trimmed down area-based programmes really was. They trained local personnel and made an annual financial injection which allowed the local government to build or repair roads, improve and build schools and set up a few more dispensaries. But couldn't that also be done if the same amount of money simply came from the central government? Well, not quite of course: a support programme with one or two dedicated advisers could do a lot of good. Getting the money straight from the donor avoided a lot of red tape, it would arrive in time and there was no risk that part of it got lost along the way. But was that what the donor really wanted, simply circumventing the administrative inefficiencies of the recipient country? Should those inefficiencies not be tackled directly?

(b) The donors withdraw to the capital cities

Once the area-based programmes had reached the point where that kind of questions was being asked it looked as if their fate was sealed. For why should one district get aid while another did not and why should a donor decide on that? And if there was no longer need for technical assistance, why give aid at the local level at all? Would it not be better to help government set up proper planning and control processes and improve the skills of the people who were to carry out the plans? If the plans were good and the local government was capable, why not simply provide budget support and keep an eye on how the money was being spent? And that is what started to happen in the late 1990s. Donors gradually withdrew from local development and started to provide money through the central budget in support of the government's regular so-called sectoral programmes, like those for health and education.

The World Bank had already been operating in that way for some time through its 'Sector Investment Programs', but now a new label was conceived which suggested that there was more to it than just money. The term was Sector-Wide Approaches, or SWAp, accompanied by the usual theoretical noise which I will not bother you with. The SWAp idea caught on quickly with bilateral donors, tired of their feel-good but inconsequential local development programmes.

The World Bank and the bilaterals would form consortia, usually dominated by the former, to support particular sectors, like health, education or agriculture and push the government to come up with sector development plans. Or they called in their own consultants to do it for them. Once the plans carried the Consortium's approval each member would pledge money and every year or so they would put together a multi-donor review team which would go out and see how things were working out in the field.

The sector approach was something of a revolution in aid philosophy. First, donors would no longer involve themselves in the day-to-day business of development programmes, they would rather provide funds to a government to do it, while looking over their shoulder to see whether they did as agreed. Second, it appeared that the integrated, holistic approach, so popular among development thinkers in the 1970s and 1980s, was being dropped in favour of rather old-fashioned compartmentalised sector support. The sectors which a particular donor would support, though in theory suggested by the recipient government, in actual fact was decided mostly by the donor itself. And agriculture became one of the least popular sectors, because the aid bureaucrats and politicians had become fed up with decades of unfulfilled promises. The Netherlands for one, a staunch supporter of agriculture in the past, had dropped the sector entirely in most of its aided countries by the early twenty-first century. Those were ominous trends. Although investments in education and health were necessary to generate a well-educated and healthy population, the countries would remain dependent on foreign aid unless there was a steep increase in agricultural and industrial productivity. But instead of trying to understand why there had been so little progress in agriculture and then try to do better, many donors simply turned their back on it. This was more surprising as several Asian countries had pulled themselves out of stagnation by the spectacular yield increases of the green revolution.

There was also a contradiction between the support provided by donors for decentralised government on one hand and for vertically organised sectors on the other. Support to individual sectors seemed to consolidate the separation between them, whereas decentralisation was expected to bring about better coordination between the sectors by making them work together at the local level. For example, the health department could take care of health education in the schools and stem the HIV/Aids pandemic by making the schoolchildren aware of the dangers. And the agricultural department could help set up and run school farms, which would serve the dual purpose of producing some food for the schoolchildren and teaching them good agricultural methods. But by pouring new money into sectoral funnels, the donors actually strengthened the vertical separation between sectors, thereby further consolidating the separate empires. The donors were of course aware of the danger but they argued that support to the sectors should not prevent them from working together where most of the action was: at the basis. The local government therefore had to be capable of breaking through the barriers separating the sectors and make them cooperate in an integrated manner for the benefit of the people. And in order to do that devolution of power was needed from the central government to the lower echelons as well as a drastic change in the mentality of the civil servants. The donors were keen to help the process along as well, in spite of their commitment to sector support, so they stretched the definition of the term 'sector' a little and declared 'good governance' or even 'decentralisation' sectors in their own right. The area-based programmes could have been good testing grounds for the new ideas at the grass roots but many of them had been phased out so most donors had lost their forward positions in the field, nor did they want to go back there. They preferred to stay where they were and send their embassy staff and consultants to have an occasional look at what was happening.

The choice for the sector-wide approach in the early 2000s also marked a shift in power in the donor agencies and the embassies from the technicians to the diplomats. Hard-core diplomats had always looked with a certain suspicion at the development workers with their half-baked ideas, immature methods and rather dismal achievements. I think many of them felt that the embassies' development advisers, who belonged to the same league as the development workers and often had been recruited from their ranks, should not be entrusted with the responsibility for their embassies' development programmes. And that was precisely what had happened in the 1980s and 1990s when many of them had been hired and given a diplomatic rank which real diplomats could only hope to attain after some years of dedicated service. The adoption of the sector-wide approaches provided an opportunity for the diplomats to regain control because it lifted development cooperation to the level of central policy making, where diplomatic skills carried more weight than technical competence. It marked the imminent defeat of the technocrats at the hands of the diplomats, for which they only had themselves to blame, after decades of spending vast amounts of aid money and producing very little to show for it.

(c) . . . and substitute incantation for control

So, by the early twenty-first century the trend was for donors to withdraw from direct involvement in development activities and transfer the responsibility for spending aid money to the beneficiaries. That looked like a mature idea and a simple thing to do too, but of course it was not. Donors soon started worrying whether the local government officers actually had the skills to spend the money responsibly and since in their hearts they thought they had not, they had to think of ways to push them in the right direction. Since they no longer had direct control there were only two options. The first was to simply let things run their own course, accept that the way the money was spent would not be perfect but monitor what happened and advise the government on how to improve little by little. That was the SWAp way, at least in theory. Obviously, it made the donors not a little nervous to lose practically all controls and perhaps in the end to be blamed for the inevitable wastage. A second option was therefore to set up a remote control system by giving detailed prescriptions to the central and local governments and their bureaucrats on how to plan, execute and monitor their programmes and account for the use of the money. Fortunately, a host of control methods had already been invented by field workers when they had been forced earlier on to hand over the management of their projects and programmes to the locals. Those methods were now going to be scaled up and extended to all levels of government.

ZOPP

Let us start from the beginning: the design of a development programme. A programme must of course have objectives. In the early years when the donors were still firmly in control, they were spelled out in detail before the programme or project started, mainly by the donor himself. From the late 1980s, however, when the process approach became popular, particularly in Dutch aid and later in that of other countries as well, objectives were going to be formulated in only the most general terms, such as 'improving people's livelihood by strengthening government services and stimulating productive development'. The programme would start by improving the local government offices and other facilities, purchase equipment and vehicles, send some people for training, and so on. Once the programme was well under way, a process would be set in motion to work out more detailed objectives. That surely was a sound principle, but also a little scary, especially in later years when the donor's representatives were no longer in control. In practice it was therefore applied only half-heartedly. Instead of letting the beneficiaries work things out in their own way, *methodologies* were conceived which told the locals how they should design, plan and run the programme. The method of choice was the famous *ZOPP* approach invented by the GTZ, which means *Objective-Oriented Programme Planning*.[5] I will not bother you with the details, but in a nutshell it involved bringing together everyone with an interest in the programme, the stakeholders, putting them through a few *workshops*, preferably far from their home base, and not releasing them until they had agreed on what the programme should be involved in for the next few years. Applying the ZOPP methodology required special skills of course, which the locals were not supposed to possess. They were the business of consultants who had meanwhile declared themselves experts in *participatory planning* and would be contracted to *facilitate* the entire process on the spot. All kinds of tools were used, ranging from sensible to infantile. The first workshop would usually start with *idea cards* which everyone was invited to stick on the wall. A drafting committee or the facilitator him/herself would then group the ideas around themes and build the *problem tree*, which would show what the major problems were according to the participants and the causes underlying those problems. You will remember what a problem tree looks like from Chapter 8. The workshop would also feature *focus group* discussions around specific topics, and probably a few more methods taken from communication science, which I will leave aside. As you can probably imagine, the process would yield a comprehensive but unworkable wish list which then had to be ranked and converted into more specific objectives which the programme was to attain. How successfully that was done indeed depended to a large extent

---

[5] The German acronym ZOPP stands for *Ziel-Orientierte Projekt Planung*.

on the skills of the facilitator. I have observed many participatory planning processes and even participated in a few and I know how easily things can get derailed by a combination of incompetent facilitators and stubborn participants. If everyone was to be maximally satisfied by the choice of priorities, the outcome would be an equitable but unworkable collection of pet projects contributed by the various stakeholders. And the participants were usually quite adamant to get their interests taken care of, because they feared to get flak from their superiors back home if they were not.

In the next step of the process the objectives had to be converted into a programme with activities and outputs. For that purpose another tool was invented: the Logical Framework.

Logical framework, monitoring and evaluation

Although I personally find Logical Frameworks unappetising, the method was actually an excellent idea, until it also fell victim to development's main affliction – ritualisation. Look at the standard logframe Table 10-1,[6] or rather the logframe *matrix*, because developers do not like simple words. The first column states the programme's objectives, the expected outputs and the activities. Simple and logical, for sure, but not very exciting. The real strength of the logframe was in the next two columns. The second column contains *objectively verifiable indicators* for each of the objectives and outputs, which means the kind of things one would look for as proof that the objectives were indeed attained and that the outputs were realised. The next column ('means of verification') shows how those indicators were going to be measured. A logframe was just that, a frame, which had to be accompanied by a precise description of the things to be measured, how that would be done and how the information would be reported and, most importantly, used. That was known as *Monitoring and Evaluation* (M&E).

Project cycle management

ZOPP, logframes and M&E, augmented with some additional ornamentation, formed a kind of management toolset which came to be known under the name of *Project Cycle Management*. If properly carried out it would impose a healthy discipline upon the

---

[6] Adapted from the IFAD Guide for Project M&E.

Table 10-1. Typical specimen of a logframe matrix

Narrative summary (intervention logic)	Objectively verifiable indicators	Means of verification	Assumptions and risks
Goal			
Long-term goal towards which the project makes a contribution	Indicators at goal level – high-level impacts	How necessary information will be gathered	For long-term sustainability of the project
Purpose/objectives			
The immediate purposes, observable changes that should occur as a result of the project	Indicators for each purpose – lower-level impact and outcome indicators	How necessary information will be gathered	Assumptions in moving from purposes to goal
Outputs			
The results that must be attained for the purposes to be achieved	Indicators for each output	How necessary information will be gathered	Assumptions in moving from outputs to purposes
Activities			
The actions that are required for delivery of the outputs	Needed inputs		Assumptions in moving from inputs to outputs

programme officers because they would no longer be able to avoid the hard question as to whether they were achieving what they had promised to. And if they were not, they would have to work harder or adjust their objectives and make them more realistic. In any case, things would now be out in the open for everyone to see what was happening. Consultants were expected to master Project Cycle Management and instruct the locals in the field how to use its tools and how to write reports using prescribed formats. Donors liked all this very much, especially the larger ones like the European Union and the United Nations development branches, because they hoped that the transparency of the reports' structure would be matched by the significance of its contents. Let us see how that hope also turned into an illusion.

As I said, the logframes, intended as icons of clarity, were particularly sensitive to ritualisation. A symptom that the whole thing was becoming ritualised was that the requirement of well-formedness gradually superseded content. As long as there was a logframe at all with the right labels on the rows and columns, that would go a long way to satisfy the requirements of the aid bureaucracies, especially those of the large donors. I am exaggerating a little here, but not much. Consider objectives like 'the average cassava yield in the area will be increased by 25% in the next 4 years' or 'the school attendance rate of girls will be increased from 35% to 70%' in the next 4 years. These were not invented by me, they were taken from really existing logframes. They are perfectly well-formed statements, but anyone who is not completely ignorant knows them to be ridiculous claims anywhere in Africa. Yet they were quite common in programme logframes, nor did the donors who funded such programmes seem to be particularly disturbed by them.

The trivialisation of the logframes was matched by that of the M&E process which accompanied them. M&E is really quite a simple concept. It means that one continuously watches what is being achieved and makes timely corrections if things are not going well. In reality that rarely happened. Like the other parts of Project Cycle Management, M&E was blown up to something that seemed beyond the powers of mere mortals, especially those in underdeveloped countries. That provided an easy excuse for programme implementers not to do any monitoring and evaluation at all and argued that external consultants had to be called in to do it for them. This obviously defeats the whole idea.

Further ritualisation: SMART indicators, SWOT analysis, DAC criteria and best practices

Of course donors and their consultants were not entirely stupid. They soon became aware that the new methodological rituals were not yielding the results they had expected, so more tools were added, in the vain hope that they would improve the quality of the outcome. Take the logframe analysis for example. It soon turned out that the objectives and indicators, instead of emerging as a result of a proper analytical process, were often dictated by expediency or political desirability. The intellectual leadership therefore concluded that the guidelines were too loose and allowed the practitioners too much latitude to indulge in their usual sloppiness. Their solution was to define more precisely what constituted meaningful objectives and measurable indicators. In the established tradition of development methodology the criteria which they had to satisfy were couched in a new catchy acronym: they had to be *SMART*, which means *Specific, Measurable, Achievable, Realistic* and *Time-bound* and the practitioners now had to analyse each of their objectives and indicators for SMARTness. Those are sensible requirements of course, but the fact that they needed to be stated at all foretold their likely fate: they soon became meaningless incantations along with the rest.

Something similar happened with M&E. Instead of being part of the daily routine, M&E developed into a speciality for which external consultants could be hired. They would carry out a survey, write an unread report and return home. In practice it meant that hardly any meaningful monitoring was done and by the time external programme evaluators arrived, there was very little concrete information to base their evaluation on and the evaluators had to collect the information themselves. This was of course impossible, so they used the nearest substitute: anecdotal evaluation through superficial interviews and casual observations. And instead of raising hell and making sure that proper M&E data were collected next time, the donors would conclude that all this could not be helped and that it was better to accept that consultants were going to do the job if it were to be done at all. The anecdotal methods used by those consultants would then have to be upgraded to a formal status, however, which was done in the usual way by adorning them with pseudo-scientific trivia dressed up in further new acronyms: *SWOT analysis* and *DAC criteria*.

SWOT stands for Strengths, Weaknesses, Opportunities and Threats. No doubt it was a good idea to analyse development programmes from those perspectives, which between them cover just about everything of importance for a programme. Good evaluators would quite naturally, even if not explicitly, do a SWOT analysis, but for not-so-good evaluators it helps to have a formal method and to know that certain criteria have to be satisfied, for example that each of S, W, O and T must be treated separately in the report. It also allowed the donor to quickly scan through the report and see at a glance if the consultant had covered all the SWOT points and what he thought about the programme. But it did not by itself guarantee the quality of the analysis.

Then there were the DAC[7] criteria. They dealt with Efficiency, Effectiveness, Impact, Relevance and Sustainability. Brilliant! If you could indeed quantify those things, you would have a really solid basis to assess a programme. It did not take long for methodology to smother the DAC criteria either. It explained what the different criteria meant, for instance what the difference was between effectiveness and efficiency – not so easy, come to think of it – and how the criteria could be scored and tabulated and the scores added up, thereby adding numerical prestige to the semblance of seriousness. But the problem was of course in those scores, what they meant and how you assigned them. In fact, they would be rather arbitrary without a real in-depth analysis or an effective year-around internal monitoring process. But SWOT analysis and DAC criteria were precisely meant as a substitute for such real data, which completed another vicious circle.

As a final safety measure, in case all these analyses would not yield the desired result, donors introduced the 'Best Practices' concept. That was a collection of lessons from previous projects, things that should be avoided to prevent failure or should be done to ensure success. Donors and consultants started to compile lists of best practices which they would then let loose on the next project they were going to design or evaluate.

At the end of the day, nothing had been gained by these methodological embellishments, quite the contrary. What used to be just messy a few decades earlier had now become both messy and obscure. The results were entirely trivial, but the process gave donors the feeling that they had objective criteria to measure the effect of the programmes they sponsored. Especially the large international institutions like the World Bank and the European Commission became very fond of logframes and SWOT and DAC analyses and they elevated them to the status of required methods for their programmes. For their own staff, monitoring the programmes in their portfolios thereby became much easier indeed and if they were really busy otherwise, they could just scan the consultants' reports for logframes and SWOT and DAC scores which told them all they really needed to know.

---

[7] DAC is the Development Assistance Committee of the European Union.

Project Cycle Management with its methodological frills had
now become another jargon-studded pseudo-discipline which could
only be mastered after extensive training. It was quickly appropriated by versatile consultants and allowed them to survive in an era
when donors no longer wanted to involve themselves with the
everyday chores of development. As a survival strategy that was a
wise thing for international as well as local consultants to do: all the
donors, especially the larger ones, had jumped on Project Cycle
Management because it gave them the illusion that they could
gently steer development processes from a distance with a minimum
of effort. The development world had now reached the stage where
processes and methodology had completely replaced content.

*The apotheosis: Poverty reduction strategy and comprehensive
livelihood framework*

The final stage in the withdrawal of the donors from the turmoil in
the field to the Olympian heights of policies and strategies occurred
around the turn of the century. The World Bank and its associated
bilateral donors had formally adopted poverty reduction as one of
the key principles of its development policy. Henceforth they
demanded that a country prepare a *Poverty Reduction Strategy* as
the starting point for all its development plans, and of course they
gave them a helping hand to formulate the strategy. Around the
same time the *Sustainable Livelihood Framework* emerged, an analytical tool of an eminently holistic nature which I described briefly
in Chapter 8. Of course, the World Bank was quick to spot this brilliant idea and started to promote it in the context of the elaboration
of the Poverty Reduction Strategy which a country had to formulate to qualify for further bank loans. However, the term
Sustainable Livelihood Framework was not all-encompassing
enough for them, so they coined the term *Comprehensive
Development Framework* which, in the Bank's own words[8]: 'empha-
sizes the interdependence of all elements of development – social,
structural, human, governance, environmental, economic, and
financial [resulting in] a holistic long-term strategy'. The art of
designing development strategies had now advanced to the stage of
generality and holism where the analytical documents could be

---

[8] On its 2004 website *www.worldbank.org*. The 2006 website reads slightly differently
but the essence was the same.

written practically without any specific knowledge of or reference to a particular country.

10.1.3 Enter the NGOs

The donors' withdrawal from direct involvement in development projects and the shift to sector support were dramatic changes which, as a matter of fact, made a lot of sense. They promised genuine transfer of responsibility to the aid beneficiaries, even though the donors and their consultants did attempt to maintain a measure of control by stage-managing development programmes through methodological rituals. Another important change was the withdrawal of the governments of the developing countries from things that could be done more efficiently by the private sector and by civil society. That move was also inspired by the donors who were going through the same manoeuvres at home. But in Africa civil society was even less developed than government, so how do you pass on tasks to something that hardly exists? That is where the international NGOs came in. There was a large variety of them and some of them had already been around for a long time, doing all sorts of mostly useful things. They had been set up by churches and charities, for disaster relief and for refugee care, but also for less dramatic things like non-formal education or organic farming or the promotion of cooperatives. In fact, there were NGOs for almost anything you could think of. They were more than willing to fill the gaps left by the withdrawal of government and the donors were happy to provide the money. The international NGOs formed associations with local ones which started to spring up everywhere, nourished by an increasing stream of aid funds. The Americans were the first to channel a significant part of their aid through NGOs and other bilateral donors followed suit. Soon the international NGOs started operating much like consultancy firms used to do, bidding for contracts from donor governments and embassies to carry out relief work, but also for local development projects. And that is how such projects came back in through the back door, often complete with expatriate personnel and all. For donors the international NGOs and their local associates were convenient vehicles for delivering aid at the local government and community level. They felt they needed little supervision because, deservedly or not, most of them had a reputation for sincerity and honesty. National and local governments were less enthusiastic though. They often resented the NGOs' independent and sometimes arrogant attitude and their lack of respect for the government's authority, and tried to get them under some sort of control. But that would have jeopardised an arrangement which was quite convenient for the donors, and who therefore argued that good governance needed a strong civil society and that the international and local NGOs would help bring that about. It was clear that the NGOs were there to stay.

NGOs also became popular among the people of the donor countries. They saw a steady stream of disasters on television and wondered whether there had been any effect of regular aid at all. Many people, especially those who had no direct stake in development aid, started wondering whether all conventional aid had not been a big mistake. Anyone who has worked in Africa knows that the first question he would be asked back home was whether aid really made sense and whether 'they' were learning anything at all. Those were fair questions which made us a little uneasy and should have made us think harder about what we were doing, but they hardly did. We simply went on doing the same things in the same or slightly different ways and building our careers on increasingly irrelevant skills. At the same time a lot of people were seen rushing to the disaster scenes and appearing to do good work. These were the new missionaries, many of them working for NGOs, and because of their visibility and the instant effect they had by providing shelter and taking care of the sick and hungry they had no difficulty in raising funds from the public. So, by the beginning of the twenty-first century the taxpayer and many domestic politicians in donor countries had become weary of regular development aid and were much more inclined to lavish money on highly visible NGOs, especially when their cause was taken up by popular public figures such as artists and pop singers, who made the conventional aid workers and bureaucrats look aloof and callous as well as wasteful in comparison. Some successful business people even started to set up private projects during their holidays. And when disaster hit an area where they used to go or might conceivably have gone for holidays, or where their foster or adopted children had been born, they very generously contributed relief money to be spent by highly visible NGOs. Development aid had come full circle.

There were, however, reasons for concern about the excessive growth of NGOs, their sometimes dubious competence and the absence of coordination among them. More and more funds were being channelled through them, both from donor governments' aid budgets and from funds raised through jamborees around disasters, which were spent with little real accountability. At the same time governments receiving aid had come under increased scrutiny about the way they spent aid money, which of course they resented more as the international expatriate-dominated NGOs got more of it and were controlled less. They saw that, not unreasonably, as lack of confidence on the part of the donors (which is not to say that that lack of confidence was unjustified).

10.1.4 So, What did we learn from all this?

Obviously, disaster relief can be no substitute for development aid, nor will it have a lasting effect unless it puts a country back on its feet and allows it to resume a development course which had been cruelly interrupted by events beyond its control. But in many cases, especially in Africa, there had been little or no real economic development, either before or after disasters, or even in their absence. So, even though disaster relief was a humanitarian necessity, it hardly helped development. Development will only take place if the countries themselves take things into their own hands. And that had not really happened, nor were there favourable signs of it happening soon.

A fundamental question was therefore whether donors should not have abstained entirely from interference, whether through direct or remote control, and simply let the locals do their own thing, rather than sending their bureaucrats and experts to mess things up and get away with it because of the power of money. Clearly, it would have been better if the rich countries had simply assisted the poor but deserving ones financially and adopted trade measures which favoured their products. That opinion had been defended by a minority since the early days of development aid, but now it seemed that more attention was being paid, but more from fatigue than from real conviction.

By the early twenty-first century, however, the rich countries were rudely woken up from their slumber by loud protests from a new generation of westerners about the shamefulness of allowing an African child to die from starvation every 3 seconds. Such simple but well-packaged messages were broadcast by people of international fame, including pop singers and film stars, who were genuinely upset by what they saw, but just as condescending as their bureaucratic counterparts had been before them. They could not be ignored by the politicians, who bowed their heads remorsefully and promised to do something about it. Their response consisted of a new round of perfunctory rhetoric which culminated in a call for a 'Marshall Plan' for Africa, made by the British government in early 2005. The Plan, which attracted a lot of transient international attention, implied debt relief, doubling aid and removal of barriers to import of agricultural produce from developing countries. Some of the ideas made a lot of sense too. Debt relief was only reasonable, since a lot of debt had resulted from ill-conceived projects invented by the donors' own bureaucrats, especially those of the World Bank. And making access to western markets easier made as much sense as it did 40 years earlier, when nobody had listened. But doubling the aid budgets was a facile but not necessarily a good idea. The Dollar paper had already shown that aid only made sense if the recipient country had 'good policies' and a 'good institutional environment'[9] and the aid-giving countries had some trouble finding countries which satisfied those criteria. And the ones that did qualify often could not absorb the large amounts of money thrown at them and spend it in a responsible way. René Dumont had already said all these things in 1962, with the visionary's passion which had since been replaced by the World Bank's and the IMF's cold analyses. And he had added that even if all the political and economic requirements were fulfilled, development would still remain an elusive goal unless it was carried by dedicated and honest people, from the smallest villages all the way up to the top of the government. Such grandiose visions, romantic dreams of the 1960s, sound perhaps a little pathetic, but I wonder whether the haughty diagnoses offered by the World Bank and the IMF could ever be a substitute for inspiration. However that may be, the Dollar paper sounded a much needed note of caution about the wastefulness of indiscriminate spending. It was doubtful, however, whether the donors were going to heed the advice. After all, a lot of money had to be spent somehow and there was more to come if all countries

---

[9] Six years later the World Bank's sister institute, the IMF, concluded that aid had had no effect at all, irrespective of the recipient countries' policies. This was symptomatic for the intellectual chaos among those who were expected to lead the development debate, and lent an entirely new dimension to the term 'dismal science'.

were going to respect the 0.7% norm which had been set by the
United Nations in 1970 and was reconfirmed in the Marshall Plan
for Africa 35 years later. But a more or less embracing and coherent vision about development was nowhere in sight, least of all for
Africa, except that the continent should adopt western free-market
policies, and make sure that everybody was properly educated, free
from hunger and healthy. So the question was how all that aid
money was going to be spent without deepening the moral erosion
caused by the smaller sums of the past. The bilateral donors,
instead of starting a real debate about development concepts,
priorities and approaches, chose to hand even more power to large
multilateral institutions, the World Bank in the first place, which
was very convenient for them because they could now simply transfer the unspent balance of their lavish aid budgets to the international agencies. And the aid ministers told their voters that those
agencies were much more capable of guiding development than
were the numerous fragmented bilateral donors with their ineffective and wasteful programmes, including those inherited from their
predecessors.

   These were important issues which loomed large at the end of the
twentieth century and which needed to be dealt with if progress was
to be made in helping the poorest countries pull themselves out
of stagnation. Agriculture was just one of many areas of human
endeavour which had to be lifted to a higher level of productivity, but
it was obviously a very important one for the poor countries. Progress
in agricultural production was highly dependent on progress in other
sectors of the economy and conditioned by the social, the institutional and not to forget the cultural environment. Development aid
at large, if it had any effect at all, would therefore always have a direct
or indirect effect on agriculture as the main occupation of a majority
or a large minority of people. That was the reason for writing this
short history of development aid, as background for the history of
aid to agricultural development, to which I will now turn.

10.2 Agricultural development

By the end of the twentieth century agriculture had lost much of its
glamour. How different it had been in the early post-independence
years when we had all been excited about the tropical farmers who

we thought were their societies' backbone and their countries' hope. They formed an unlimited reservoir of talent and possessed the vitality which was needed to pull their societies out of underdevelopment. What had happened to them and why were those early promises not fulfilled? I do not pretend to know the whole answer, but I do have a few notions about what went wrong, and even about what can still be done to find the road out of today's stagnation. It was written from the perception of an agricultural scientist turned consultant, that modern brand of peaceful mercenaries which has been one of development aid's defining features.

10.2.1 Early settler-based schemes

We have already met the early and largely unsuccessful attempts to convert peasant farming into an efficient market-oriented sector using all the modern techniques, materials and forms of organisation available at the time. The most striking examples were the Farm Settlements in the anglophone countries and the Unités Expérimentales in Sénégal. What was striking about them was the enthusiasm of their proponents and, with hindsight, the naiveté of their ideas. The Farm Settlements disappeared, as far as I have been able to trace, with hardly any published analysis of their failure. The Unités Expérimentales, led by scientists and therefore more likely to publish their findings, have fared a little better in that their history was documented in detail in what one of its early workers called the UE Bible. It was published in 1986, after 20 years of operation (Benoît-Cattin, 1986).

Besides the rather small Farm Settlements, there were other, much larger settler-based schemes, whose scale, though not the achievements they realised, were proportional to their conceivers' confidence, not to say hubris. The best known were the Tanganyika groundnut scheme, an unmitigated failure, the Gezira-irrigated cotton scheme in Sudan, a relative success, and the Office du Niger-irrigated rice scheme in Mali, which has weathered many storms and has survived until today in reasonably good shape. They were all set up during the colonial era. Each of these schemes has been extensively documented. The failure of the settler-based Tanganyika groundnut scheme was particularly shameful because it was due to wrong technical assumptions. On the other hand, if it had been technically in order it would probably still have failed after independence by incompetent management, as did so many schemes of a lesser size in Africa, so its early failure probably prevented a long and even more costly agony. The other two schemes survived because they were essentially based on sound technical principles and they were managed as large enterprises with settlers from outside the area as tenants. By the end of the 1980s, however, the inefficiencies inherent in these large government-owned schemes, where tenants were little more than cheap labour had resulted in de facto loss making. And those losses, as you may have guessed, were also mostly taken over by international donors. I do not know the current conditions in the Gezira scheme, but the Office du Niger was put on an entirely different footing during the 1990s when the settlers were emancipated and became genuine tenants. Later in this chapter I will come back to the 'Office', which I think is one of the few real success stories in agricultural development in sub-Saharan Africa.

10.2.2 Agricultural development projects

The brief period after independence when modern settler-based farming schemes were popular came to an end when it became clear that peasant farmers could not just be catapulted into the industrial age. And even if they had been successful, they would in any case only have been a drop in the ocean. Large projects were therefore undertaken in the 1960s and 1970s to improve the productivity of indigenous farming, like the ADPs in Nigeria and the Sociétés de Développement in the francophone countries. In the beginning their intentions were as ambitious as those of the Farm Settlements and the Unités Expérimentales: transform peasant farming from subsistence production with primitive means into market-oriented production with modern methods. The ADPs brought in heavy equipment to remove the vegetation and make the land suitable for mechanised tillage. They provided tractor rental services, organised fertiliser and seed supply and marketing and set up or strengthened the agricultural extension services to bring the good news to the farmers. And they established close links with agricultural research to supply them with new materials and information. But peasant farmers would not be transformed into modern producers just like that, especially when de-stumping the land followed by mechanised

farming turned once reasonably fertile land into a decapitated
skeleton in a few years time, unless all kinds of protective measures
were taken, such as growing a protective cover between the crops,
ploughing and planting along the contours, planting hedgerows or
practising minimum tillage. All those measures were indeed proposed,
but the more complicated things became, the less likely farmers
were to adopt them. Nor was there any guarantee that the new
system would be any more productive than the traditional one
could have been with some simple improvements such as better
varieties and fertiliser application.

Cash crops or food crops?

The ADPs paid little or no attention to cash crops. That was not
surprising, even though it was unfortunate. At the time of independence many African countries had flourishing cash crop industries: cocoa, oil palm, rubber, sugarcane, coffee, cotton and quite a
few other crops of lesser importance. Oil palm, rubber and sugarcane were mostly produced in large foreign-owned commercial
estates with salaried labour: I am not concerned with those here.
They managed to weather many a political and economic storm
and some have survived until the present day. My story deals mainly
with smallholder agriculture, so I will look at the crops which were
mainly produced by small producers: cocoa, coffee and cotton. In
the colonial days efficient organisations had been built around these
crops which took care of everything that was needed by the growers: planting material, fertiliser, pesticides, processing, marketing
and technical advice. And, whatever can be said about the colonial
regimes, they generally did treat the farmers fairly, or at least in an
uncorrupt manner. Ghana, the first former African colony to gain
independence, started life as a nation with healthy national
accounts due to the earnings from the cocoa industry. In most or all
anglophone countries the industrial crop sectors gradually faded
away after independence or they fell prey to greedy and corrupted
bureaucrats. By the end of the 1970s and sometimes even earlier,
many of the smallholder cash crop industries in the anglophone
countries in West and Central Africa had collapsed by neglect,
incompetence or embezzlement and more likely by a combination
of all three. It takes much less time to destroy a healthy industry
than to build one and once the cash crop infrastructure was gone it
became almost impossible to put the whole thing back on its feet,

even with donor assistance. The experienced specialists on both sides were simply no longer there. It is no wonder that the ADPs turned a blind eye to the cash crops and concentrated on the food crops instead. And besides, cash crops were shunned by left-leaning aid politicians and bureaucrats as politically incorrect, associated in the minds of many with colonial exploitation. I think neglect of potentially lucrative crops has been one of the big mistakes made in the post-colonial era.

In the francophone countries the organisations built around the cash crops before independence were kept alive through tight links with institutes in France which continued to send managers and technical specialists, while taking care of marketing for mutual benefit. And when the new governments wanted to increase food production as well, it was understandable, though not necessarily wise, that they looked to the well-organised cash crop organisations for help. Improving the food crop sector was a macroeconomic as well as a social necessity for a country, but it was not obvious how a commercially oriented cash crop organisation could recover the cost involved in helping farmers improve their food crop production. Having that kind of responsibility dumped in its lap almost killed the Malian Cotton Development Company (CMDT), until it returned to its core business in the early twenty-first century. So, the cash crop industry was not neglected in the francophone countries, but the mistake made there was the continuing dominance of the French. When they finally started loosening their grip in the early twenty-first century, the locals had to take on responsibilities which they had been denied before, in spite of appearances to the contrary.

The ADPs and similar projects in the anglophone countries, though they left the languishing cash crop industries untouched, did some good for smallholder agriculture. In the West African savannah, for example, they boosted maize production, a crop which hitherto had only been grown in backyards as we have already seen. The ADPs were organisational anomalies, however, dominated by expatriates and funded with a lot of easy money. Once it became clear that their ambitions could not be realised and that agriculture had to be transformed by small incremental steps instead of by quantum jumps, the ADPs in their original configuration became largely irrelevant, although in Nigeria they plodded on for many years with national money.

10.2.3 Area-based programmes and agriculture

The ADPs were succeeded by a new generation of development programmes with much broader development objectives, which came to be known as *Area-Based Programmes*. Agricultural development was just one, although an important component of those programmes. In most countries the agricultural extension service had survived on little more than the money to pay for staff salaries, but it was still there and could be put back on its feet. That was one of the things the new area-based programmes wanted to do to improve food production. Besides the programme coordinator, they would therefore hire an agricultural expert, who in theory was seconded to, but in fact tried to control the local extension service. The programmes wanted to help the extension service do a better job which, if done adequately, would have meant laying off poor performers, hiring qualified people in their place, and providing them with well-equipped office space and means of transport. That could have been handled by the area-based programmes at the local government level, but it would have created jealousy and unacceptable inequality among neighbouring districts. And reforming the extension service, especially firing staff, was very tricky and could only be done by the central government. That is where the World Bank came in. The Bank had been a major sponsor of agricultural development and the ADPs had been their creation, so when the latter went out of fashion they had to find new targets capable of absorbing money at a rate commensurate with the Bank's resources. Obviously, the national extension services were fitting targets. They were full of personnel and small in facilities and skills and upgrading them was going to be expensive and time consuming.

Extension services had been inherited from the colonial times and, as happened with many services in Africa, their staff numbers had steadily increased whereby competence was not an important criterion for being hired, and the point was eventually reached where little money remained for meaningful work once the salaries had been paid. Obviously, if the service was to be reformed, staff quality had first to be improved, which meant firing the patently incompetent and hiring new people with better qualifications. Then the offices needed to be refurbished and means of transport had to be purchased. The World Bank loan provided the money for separation allowances and for training, office furniture, typewriters and later on for computers, motorbikes, vehicles, the lot. So far things were simple, and only painful for those who had lost their jobs. The next question was how the new organisation could reach as many farmers as possible, delivering meaningful messages. That is where one of the most influential extension methods of the second half of the last century came in, the Training and Visit system (T&V).

10.2.4 New extension methods: T&V

The T&V method was developed by Daniel Benor, an Israeli extensionist with the World Bank who first introduced the method in India in the late 1960s and later practically everywhere else where the Bank was financing the agricultural sector. It was a highly structured top-down method, based on the extension-knows-all principle. Figure 10-1 shows the organogram of an extension service using the

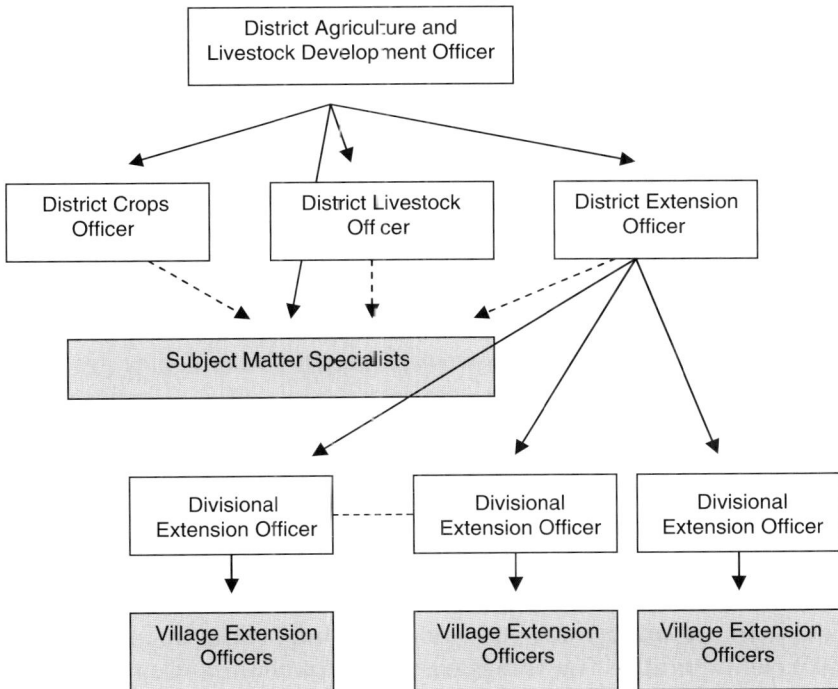

Figure 10-1. Specimen of an organogramme for the extension service under the T&V system

T&V system, at the lowest level of comprehensive government, say, a district. There were many variations on this theme but they all came down to the same essence. This one is from Tanzania. Most important were the village extension workers who supposedly were in daily contact with the farmers, and the subject matter specialists, who were the repositories of technical knowledge and who fed the field workers with the necessary information. There was of course also an administrative superstructure above the district level with supervisory tasks, which often appeared to mean chivvying the lower ranks with meaningless requests and field visits. Let us just ignore them.

In order to reach as many farmers as possible the village extension staff worked through contact farmers whom they visited regularly and who were themselves expected to pass the information on to their colleagues. That was the 'visit' part of T&V. The 'training' part involved two types of periodic training sessions. In the so-called fortnightly training sessions the contact farmers would come together with the village extension officers and the subject matter specialists to be lectured to about what was important at that particular time in the annual farming cycle and to be taken to demonstration plots where new technology was shown. The other type consisted of Monthly Technology Review Meetings, where extension personnel themselves were educated about the latest technological novelties by scientists from the research stations and the universities. The process was clearly unidirectional: the researchers told the extension officers about new technology, who passed on the message to the contact farmers who did the same to their neighbours. That had worked very well during the green revolution days in India, because the messages were simple and effective: grow the new short straw rice and wheat varieties, apply a lot of fertiliser and control the weeds and pests. And, what was perhaps even more important, the seed, fertiliser and pesticides were actually there and in time. By using all those things the farmers could triple their crop yields. The mini-green revolution in the West African savannah was not as spectacular, but a similar kind of technology did make it possible to grow maize with yields which were unheard of in those areas, if the season was right and the rains came in time and in sufficient quantity. But things were not as well organised as in Asia and the input supply chain broke down easily with sometimes drastic effects. And the crop yields were much more sensitive to the vagaries of the weather than the yield of rice in Asia which was grown under irrigation. Apart from the relatively successful maize technology for the savannah, however, there was very little in the way of modern production methods to boost food crop production, especially for the forest zone as we have seen before. So the extension services had little to offer except some simple prescriptions, most of which ranged from trivial through nonsensical to plainly harmful. For example, the omnipresent message that crops should be planted in lines, which did not make sense when everything was done manually, or that it should be planted at a higher density, which had no effect unless fertiliser was applied also or could even be harmful when other crops were grown in mixture with the maize. In any case, the extension workers themselves did not bother to apply their messages in their own field, a fact which the farmers could not fail to notice. The regular T&V meetings nevertheless had to take place and since there were hardly any meaningful new practices to teach they tended to degenerate into events with very little practical relevance. And everyone started to complain bitterly that T&V took too much time and was too expensive. Even in the savannah it was like that. Once the farmers had understood that they could grow an excellent maize crop with the new varieties and fertiliser there was little else the extension officer could teach them. For the other food crops like sorghum, groundnuts and cowpeas the extension service had even less to offer. For cotton the problems were mainly organisational rather than technical and as long as the organisational environment was poor that limited the effect of everything else.

All this shows that pumping more resources into extension would make no sense unless they had something meaningful to extend. The World Bank experts, however, thought that the extension service had to be equipped first, while anybody in his right mind could see that they were just going to drive around spreading senseless talk about the need to adopt 'modern technology'. A significant part of the indebtedness of many African countries has been due to money being wasted on the rehabilitation of extension services without a message, which the countries were unable to keep running once the World Bank loan had been spent.

Not all Bank officials were mindless project pushers, of course. Some of them did listen from time to time to sensible individuals inside as well as outside the organisation, who told them that things were not well. Thus they became aware that, in spite of claims to the contrary, research had actually very little to offer and that the large extension organisations which were supposed to transmit research findings to the farmers resembled hollow barrels producing mainly random noise. The reader should be able by now to guess the Bank's response to these observations. Indeed, they offered their financial assistance to help the research institutes improve their performance. That meant that the national governments had to take further big loans to strengthen their research institutes.

10.2.5 Could research help?

As we have seen in earlier chapters, the research institutes had not been sitting idle waiting for the World Bank to come and rescue them. As early as the 1970s, while development workers were busy promoting their ineffectual technologies borrowed from temperate agriculture, some eminent researchers had argued, and even proved in some cases, that something different was needed to bring agriculture forward, something which fitted better into the farmers' existing production systems. They designed clever methodologies to find out what the farmers really needed and called it FSR, or Farming System Approach to Research or Research with a Farming Systems Perspective. We have already gone through all that in great detail. The ideas were picked up by some donor agencies with a knack for spotting promising new trends, in particular the Ford and Rockefeller Foundations and the Canadian International Development Research Centre (IDRC). They started sponsoring FSR in several national research centres as well as at some of the international ones. That launched a movement which reached its heyday in the mid-1980s when many bilateral donors had jumped onto the FSR bandwagon and sponsored FSR teams in national research institutes, attracted by the promise of almost assured success. The promise was that the new approach could not fail to come up with new technologies which farmers would be sure to adopt, contrary to the earlier ones which they rejected. Box 10-1 gives a simple example of how that would work. Adoptable technologies were of course the lifeblood of extension – without them they could not be the driving force of agricultural development, which they were intended to be. That was the situation in the mid-1980s, when it became clear that the large World Bank-funded ADPs were failing for lack of adoptable technologies.

Box 10-1. A very brief history of cowpea breeding

In northern Nigeria creeping cowpeas are traditionally sown into cereals, both for their grain and for hay to feed to the animals. In the 1980s IAR and IITA introduced erect, early maturing varieties, but that was a very different animal. It was extremely sensitive to insects, much like cotton, had to be grown as a sole crop, and failed rather miserably. After wasting many years trying to promote the crop with chemical control, breeders finally accepted the facts of life and started to breed 'dual purpose' creeping varieties, to be grown as intercrops with maize or sorghum, much like the farmers did. That worked much better of course and the new varieties did find their way into the system, which shows that the FSR workers were right: that something that fits the farmers' needs will find its way into their fields without much effort.

The Bank had always been aware of the importance of technology and they were already making important contributions to the international institutes' budgets. Now they started thinking about giving assistance to national research institutions as well in order that they could try out new technology with farmers first and the ADPs would not waste their time with things which in the end might not work at all. That was precisely what FSR was all about, so in 1984 the Bank commissioned a study by Dr. Norman Simmonds, a plant breeder well known for his work on bananas, to see whether this FSR held any promise. I remember him coming to IITA in 1984 and talking to most of us in the Farming Systems Programme. I do not think he was very interested in our brand of FSR (which was essentially OFRs), although he did give endorsement to the Bank's providing some money for national institutes to set up their own OFR teams and work with the ADPs. His report talked much about 'New Farming Systems Development' (NFSD) instead, which most of the work done by IITA's Farming Systems Programme was supposed to contribute to, but it did not give a clue as to what he thought these new farming systems might look like, or whether their adoption by peasant farmers would be at all possible. He simply advised the World Bank to encourage research in that direction:

With regard to the interests of the World Bank, I suggest that . . . it should encourage the CG institutes . . . to think hard about NFSD, and to take perennial plants[10] seriously. . . . . The Bank should . . . also be prepared to be bold and try to construct and exploit one or more types of NFSD in the wet tropics. [This would] provide . . .

---

[10] Tree crops and 'auxiliary species' such as the hedges used in alley cropping.

valuable guidance on how to 'do' NFSD and short-circuit what may otherwise prove to be an intolerably prolonged research process. There is a real need for bold thinking coupled with resources and who better to provide them than the World Bank?

Well, yes, true, but not very helpful. The new farming systems remained what they had been before, an idea and probably an unworkable one, because new farming systems were very unlikely to be adopted except by compulsion.

In any case, Simmond's report did convince the World Bank that they should pay more attention to agricultural research by national institutes because they had to become efficient players in the technology development, testing and dissemination process. The Bank therefore started providing seed money to set up FSR teams in national institutes which worked in areas where the Bank was also financing ADPs. In that way they hoped that the former might come up with new technologies which would contribute to the latter's success. And since the national institutes were not considered capable of starting FSR on their own, international research centres like IITA were often contracted to provide technical assistance. That was in the heydays of FSR, the mid-1980s, when many donor agencies were putting their hope and investing their funds in these convincing looking ideas. In the early 1990s it became clear, however, that FSR was not going to deliver on its promises. The teams usually remained quite isolated even within their own institutes and hardly had any effect on the attitudes of the hard core scientists who should have been co-opted to do their part in designing better, adoptable technologies. After several years of surveys and on-farm trials the approach which once looked so promising had little more to show for itself than conventional research had, and probably less.

In the end, FSR was almost blown off the scene by its successor, the participatory movement, which held that FSR had only pretended to listen to farmers, and that research had to become really participatory and client-oriented, not just some small FSR teams, who were not taken seriously by their colleagues, but the whole of research. The Bank, desperate to get something useful out of research to match their heavy investments in extension jumped at it. They hired a new batch of people from among the believers in participation, recruited ISNAR's services to help national institutes reform their programmes in a participatory direction and provided another round of big loans for the new programmes to be carried out, this time in participatory style. But people do not become participatory overnight any more than they had become genuine farming systems researchers in the past. The Bank was in a hurry, however, so a few corners had to be cut to convert research in a participatory and client-oriented direction as soon as possible. So they started promoting a variety of gimmicks which I have already discussed extensively in Chapter 8 to give the farmers themselves a major say in what research was going to do. In the true and tested bureaucratic tradition that meant setting up all kinds of mixed committees and boards with scientists, government officials and farmers who would scrutinise the institutes' research proposals and even make proposals of their own. But farmers in Africa are hardly or not at all organised, and the committees were soon populated by the usual token farmers, more skilled in political and donor games than inclined to be the *portes-parole* of the farmers they were supposed to represent. At the end of the day, research remained what it had been before: some of it useful, in particular plant breeding and biological pest control, and most of it trivial and inconsequential. The new World Bank loans were consumed to build or recreate another empty shell, until the money was finished, the country's debt further swelled and the Bank had to start looking around for the next grandiose idea. In actual fact it did not wait for the next idea, but quietly started to pull out of support for agricultural research and extension altogether.

10.3 Three donor-assisted programmes

To conclude this chapter I would like to give three examples of donor-assisted agricultural programmes. Two of them failed, I would say unnecessarily, and one succeeded, and if you want a common factor underlying the two failures I think it was instability, not just of the recipients' institutions, but just as much of those of the donor. The example of the successful programme shows what could be accomplished by the dedicated and sustained efforts of all concerned, and how fragile the accomplishments were even then.

10.3.1 FSR in Nyankpala, Ghana

This story starts way back in the early 1980s. At the time German international aid, like that of most bilateral donors, was strongly

biased in favour of agricultural development. Many of their
projects were carried out by the German Institute for Technical
Cooperation GTZ[11], a semi-autonomous organisation which operated more or less as the technical branch of the aid ministry. GTZ
developed an interest in FSR in the early 1980s and wanted its
African projects to learn and apply OFR. Kurt Steiner, who was in
charge of the 'mixed cropping project' at GTZ headquarters, was
made responsible for training the field staff in the African projects
in OFR methods. Steiner had been associated with IITA for some
time and he requested us to come and organise a training workshop
with them. Since Germany was one of IITA's major donors, management approved and we, relative novices ourselves, thought it was
a great opportunity to try out our ideas. The workshop was held in
November 1984 and hosted by the research station in Nyankpala,
in the northern Ghanaian savannah, which was one of the institutes
supported by Germany. We came with three people from IITA and
I also invited Neil Fisher from Samaru in Northern Nigeria,
because of his experience in the West African savannah and because
he was an excellent agronomist. Research at Nyankpala at the time
was quite conventional, dominated by a large number of German
scientists who mainly carried out their research on the station.
Perhaps the quality of the work was good, Neil Fisher thought so
I believe, but the institute was insular and introvert and most
researchers were hardly interested in OFR. As usual we all enjoyed
the workshop and we learned a lot about farming practices in the
Ghanaian savannah. I think we also came up with some good ideas
about the kind of OFR the institute might undertake after the
workshop. Although there were signs that the things we were
talking about was not considered serious research by the
Nyankpala institute, we hoped that the three staff who participated
in the workshop, two agricultural economists and one agronomist,
would start their own on-farm work and, by the force of the antic-
ipated results, would pull the others out of the ivory tower into the
real world.

   We co-organised a similar workshop with GTZ for francophone
countries in Togo the next year, but in the years that followed there
was no further contact between IITA and Nyankpala, because Ghana

---

[11] GTZ stands for Gesellschaft für Technische Zusammenarbeit.

was not among the countries where IITA was most active, nor were there any further requests for our assistance. I think the FSR ideas never really caught on, either among the Ghanaian or the German scientists at Nyankpala.

Then in 1997, several years after leaving IITA, I received a request from GTZ to carry out a consultancy at the Nyankpala research station. The terms of reference explained that the institute wanted to reorient its research towards the real farm and that its staff had to be trained in methods of on-farm experimentation, in particular the design and analysis of on-farm trials. That came as a great surprise, 13 years after the same institute with the same donor had set the first steps on the FSR road, assisted by the same scientist (me), though nobody remembered, neither the scientists at the station nor the people at headquarters. It was nice to be back at Nyankpala, but the signs of the institute's decline were apparent everywhere. There were no coherent ideas about the institute's future and the best (and youngest) scientists were looking for opportunities elsewhere. The donor seemed to have given up long since and tried to close down the show as elegantly as possible. The amount of money which had been spent over close to twenty years must have been enormous and the campus still showed how comfortable the working and living conditions had been in earlier days, but there was little else to show for it. The support project was now in its final phase and someone, probably an evaluation team, had recommended that the institute should still make an effort to embark upon OFR even at this late stage. Several on-farm tests were already running, some quite good ones, carried out by clever young scientists who were very enthusiastic and eager to learn new things – they were a pleasure to work with.

My first assignment in 1997 went very well. We worked on the nuts and bolts of on-farm experiments and I demonstrated how they should be analysed with examples I had brought along. I would come back a second time, in 1998, and meanwhile the scientists were to compile their own trial results so that we could work on them and see how their analysis could be improved. That was less easy. Some scientists who had participated enthusiastically the first time (in 1997) now started dragging their feet, because their materials were not ready yet, or they were otherwise occupied, or for whatever reason. Since my time was short I got pretty annoyed and there was an open clash at some point, but the conflict was resolved gracefully in the end. It was clear that the consultancy had passed the stage of non-committal pastime and that some serious work had to be done now. In the real world, this could just have been the first movements in a longer process, but the Nyankpala project was about to be closed down. The institute's management saw the importance of what we were doing, or at least claimed to, and they came up with the idea to find money elsewhere to continue my advisory role. Probably they never tried or they did not put in enough effort, but I never heard from them or GTZ again. And that was the end of what could have developed into a fertile collaboration, if it had been sustained for some years or, better still, made a permanent arrangement, backed up with a modest amount of money for equipment, computers and software.

10.3.2 Soil science and FSR in Tanzania

The second story is about soil fertility research in Tanzania and starts in the 1960s. At that time the FAO launched an ambitious programme in many African countries to demonstrate the blessings of fertiliser, the 'FAO Fertiliser Programme', which plodded along for more than 30 years, with frequent changes in its avowed mission.

In the beginning the programme's objectives were simple: lay out a large number of simple demonstration plots all over the continent to show to farmers the effect of a moderate dose of fertiliser on their major crops. From the fertiliser responses measured in the demonstration plots and the results of the soil analyses which were also carried out, correlations could be established between soil analysis and fertiliser response for different soils, as was done in the industrialised countries.[12] I assume that after some years that kind of analysis was indeed carried out and published, but if you search for the results today in a particular country, you will probably find nothing. In Tanzania I know you do not, because I tried several times. I would be surprised if it were different elsewhere in Africa.

By the early 1990s the fertiliser programme, which had been cosmetically renamed 'Plant Nutrition Programme', had become completely fossilised in spite of its incorporation over the years of

---

[12] Chapter 2 explains how those correlations are carried out and why they are important.

all kinds of fashionable ideas, like farming systems and participatory approaches. While still being carried out by FAO, it was now funded by the Netherlands Embassy and I participated in the two final reviews on their behalf, in 1995 and 1997. Responsibility for the reviews rested with FAO's own internal review department, however, a very undesirable arrangement but common in UN organisations and in the World Bank as well. Nevertheless our reports made it clear, at least for those who could read, that the purpose of the programme had become completely blurred, it worked practically in isolation, its trials did not make any meaningful contribution to knowledge, and no effort was made to publish the little that might still be useful. In the first review, carried out in 1995, we recommended that the remaining 2 years should be spent on salvaging the legacy of decades of field work, which was stored in a container in the office yard, and bring it into a form which would allow others to use it. Some half-hearted attempts were made to do that but when we came back to do the end-of-programme review in 1997 it was not even remotely ready. If the matter were to rest there, the results would be lost forever and most of the efforts and a lot of money wasted.

Simultaneously with the FAO Fertiliser Programme the Netherlands had been funding two other agricultural research programmes in Tanzania. One was based at the national soil research institute in Mlingano, near Tanga, where at its high point up to five Dutch researchers were helping the institute to set up a national soil service. The other was a FSR programme with a mixed FSR team of Tanzanian and Dutch scientists, based at the Ukiriguru Research Institute of old fame, near Lake Victoria. You would expect that these programmes had cooperated closely with the FAO programme, because soil fertility was a primary concern of both, but there had been absolutely no coordination, all of them worked practically in isolation from the others, in spite of them being supported by the same donor. The only plausible explanation I can think of was that each of the programmes was technically supported by a different foreign institute – the Fertiliser Programme by FAO's own soil fertility department, the National Soil Service by a soils institute in Wageningen and the FSR programme by the Royal Tropical Institute in Amsterdam, better known as KIT. The Mlingano programme closed down in 1994 or thereabouts and it was rumoured that it had left behind an extensive computerised soil and soil fertility database. When we started looking for it during the first review of the FAO programme it turned out to be either non-existent or inaccessible, or it had been taken to Wageningen by the former programme scientists. In any case, it was not available in a form which would have made it useful for the country.

The FSR programme in Lake Zone was reformulated in 1996, 6 months before the FAO Plant Nutrition programme ended. It had made some interesting contributions to methodology but, like most FSR programmes, it was fairly weak on technology. A good map of the farming systems in the Zone had been prepared, but the problem with such maps was that they often ended their lives as sophisticated wallpaper, rather than as a research or extension tool. In order to be useful, the map would have to be linked with information about the soils associated with each farming system and their likely response to nutrients, so that specific recommendations could be formulated for each system. Soil fertility was one of the major concerns of the FSR programme and the review team, which I chaired, therefore saw a great opportunity for the programme to exploit the results obtained by the FAO and the Mlingano soils projects. The two soils programmes had ended; however, we recommended that the task of exploring their soil fertility data be made a part of the next phase of the FSR Programme. The idea was to prepare maps of expected fertiliser responses and then explain to FSR researchers and extension workers how to formulate recommendations for fertility management. That I think was a concrete, well-defined and manageable objective. But the times were not favourable for that kind of thing and the Embassy was only marginally interested. Only if we would wrap it in the participatory rituals which had by then become obligatory were they going to consider it. We acceded to this because that was our only chance, and proposed that the programme would also develop a participatory 'counselling method for Integrated Plant Nutrient Management (IPNM)' to take the results directly to the intended beneficiaries (Box 10-2 gives some more detail).

And, just as we feared at the time, the premature emphasis on delivery methods prevented the main objective from being achieved – salvaging decades of field work by mapping likely nutrient responses for different zones of the country. There were two reasons why that was predictable. First, you cannot develop a counselling method for things that do not exist. Nutrient response patterns first had to be extracted from the chaos left by the two earlier soils

Box 10-2. Integrated plant nutrient management

Integrated plant nutrient management (IPNM), also known as integrated soil fertility management (ISFM) is one of those modern concepts which go by the name 'holistic'. Yet, the idea is very appropriate and simple. It says that recommendations to farmers for fertility management must take a lot of things into consideration: the farmers' crops, the kind of soils they are grown in, how the crops respond to applied nutrients, whether the farmers practise fallow, what they do with the crop residues, whether they have cattle which produce manure and many other things. That is needed to calculate a farm's nutrient budget and counsel the farmers on how to make the best use of the scarce nutrients they have access to. A number of things must be known quite precisely, otherwise the method quickly degenerates into a trivial play with generalities. In particular, you must have a good idea about the likely response to nutrients of the crops when grown in the soils of the area. That kind of information was supposed to be available in Tanzania from decades of soil and soil fertility research and the IPNM project was expected to unlock it for different kinds of users. It started to do that, but the project came to an end before the work was finished and the material was left in the unfinished state which had become typical of many donor-sponsored projects.

There have been other attempts to develop IPNM methods, such as the work by Toon Defoer and collaborators on farmer counselling for nutrient management in Mali, in another KIT-led FSR project. The difference with the Tanzanian example was that it dealt primarily with 'methodology'. It remains to be seen whether that kind of work has enough substance to resist the erosive effects of time, but I am not optimistic.

programmes. In the fanciful world of development aid, however, content carried less weight than the packaging so the counselling method had to be developed almost simultaneously with nutrient response mapping. Second, both the response maps and the counselling method had to be completed in 3 years, an impossible task for one expatriate and a few loosely associated national scientists. Anyway, the Royal Tropical Institute sent a seasoned soil scientist, Wietze Veldkamp, to do the job. He struggled for 3 years and produced quite a respectable output, but by the time the project ran out, the job was still not finished. Veldkamp left behind a voluminous report describing the agro-ecological maps which he had put together, the data-sets that went with them and descriptions of fertility management strategies for some sample locations. But the material was still far from operational – using it in its present form was beyond the capacity of the average scientist, let alone the extension officers who were to be the end users. A neat little booklet, (Ley et al., 2002) summed up the situation as follows:

[W]hile the outputs are available electronically and in hard copy, they are not user-friendly. . . the question of how and when the information can be used . . . by larger groups . . . still needs to be addressed.

Towards the end of the 1990s the Netherlands Embassy decided that agriculture was no longer a sector they wished to support, which also meant the end of support for agricultural research. That was the final death blow for several decades of soil fertility research, which had consumed many millions of dollars, occupied scores of expatriate and national scientists, but of which the results were likely to evaporate in a few years time. Compare that with the work on soils, soil fertility and land use carried out in south-western Nigeria before independence, which I mentioned earlier. In 1962 Smyth and Montgomery brought together the results of decades of colonial research on soils, soil fertility and land use in a hardcover book which has stood the test of time and even today remains the best source of information about the area.

By the late 1990s the FAO and the World Bank had taken an interest in integrated soil fertility management (ISFM) and the two teamed up, not to continue the work which had already been done, such as that of the earlier projects in Tanzania, but to start all over again. Their new brilliant initiative was called the Soil Fertility Initiative (SFI). I will abstain from going into that, but I fear that after a few years it will suffer the same fate as the earlier FAO Fertiliser/Plant Nutrition programme and disappear without trace, to be replaced by the next fashionable acronym.

10.3.3 And a successful case: the office du Niger

Not everything has been as bleak as the previous examples suggest, though. When asked for a really successful agricultural development programme in Africa, the one that comes to my mind is the large irrigated rice scheme in the floodplain of the river Niger in Mali, the Office du Niger.[13] The scheme was set up by the French

---

[13] The name 'Office du Niger' refers both to the irrigated scheme itself and to its managing authority.

colonial government in the 1930s for the production of irrigated cotton, groundnuts and rice, initially by settlers from the densely populated Mossi plateau in Upper Volta (now Burkina Faso). In the 1940s many of the original settlers returned home and were replaced by new ones from areas adjacent to the scheme. Cotton growing soon proved a failure and for decades the settlers produced mainly rice with pathetic yields of between 1 and 2 t/ha. The early post-independence years saw the usual experiments with socialist models involving collective farms, government-created cooperatives, production targets, government-controlled marketing boards and all their attendant abuses. But contrary to many other development schemes, the Office du Niger weathered all storms and survived, even though for a long time its production barely exceeded subsistence level.

In the early 1980s a liberalisation process got underway which was going to lead to a spectacular rejuvenation of the scheme, an increase in the area under regular irrigation from 25,000 to almost 65,000 ha, tripling of rice yields and expansion of lucrative off-season vegetable production. If I had not seen the scheme with my own eyes in the 1990s and early 2000s I would not have believed it. There have been very few successes to report from Africa, so one would like to know what had been behind this remarkable story, whether it was likely to be durable and whether something similar could happen elsewhere. An interesting book (Bonneval et al., 2002) about the scheme was published by CIRAD, which presents a convincing, though somewhat partisan picture, but the essential factors which could explain the success tend to get obscured by an overdose of detail contributed by the many authors. Let me elaborate on the recent history of the scheme to see whether a straightforward explanation can be found.

The overhaul of the scheme which led to the spectacular increase in productivity started in the early 1980s, at a time when the scheme was at the brink of collapse. The rescue operation was carried out with the help of (in fact by) three foreign donors – French and Dutch bilateral aid and the World Bank. They allocated considerable funds for the rehabilitation of the irrigation infrastructure, the roads and the villages, as well as for credit facilities, equipment, even fertiliser delivered in kind. The ideas were mainly contributed by the French and the Dutch, whose views were of course very different. But since there was enough room for everyone, they were assigned different parts of the scheme to bring their ideas into practice, while the World Bank took care of the diversion dam in the Niger river and the primary channels. The French approach was highly technical, with detailed designs of the secondary and tertiary irrigation channels, clever water distribution structures and mechanised field levelling to obtain a uniform water layer for the paddy plants to grow in. Furthermore, they introduced and actually imposed transplanting of rice seedlings, instead of broadcasting the seed, as was the custom in the scheme. The guiding principle of the Dutch approach was to push for a high level of self-management by the farmers, who thus far had been little more than farm labourers ordered around by the *Office* officials, and to make them responsible for everything that went on within their fields and farms. They also introduced small threshing and milling units which could be owned and operated by village groups or even by individuals and which replaced the large units managed by the *Office* (Figure 10-2). That tallied with the liberalisation of the rice trade which the government had decided to enact. When I visited the *Office* for the first time in the 1990s the small threshing and milling units had completely

Figure 10-2. Small rice threshing unit in operation in the Office du Niger, Mali. (Reproduced by permission of Dr Frans Geilfus and CDP, Utrecht.)

replaced the old big machinery. Both projects helped to put the extension service back on its feet, set up credit facilities, stimulate farmer organisations and provide training facilities. There was one highly artificial element in the whole thing: the projects, especially the Dutch one, were run by large numbers of expatriates and the local staff hired by them. That was becoming unfashionable at the time, but the intellectual originator and driving force behind the project in the Embassy and later at the Ministry, Chris van Vugt, had the necessary authority to pull it off and he personally chose the people to staff the project. At one time there were 18 expatriates in the Dutch project alone. Even today those people, though scattered over the globe, treasure their memories of a time when they belonged to that band of pioneers which jealous outsiders referred to as the 'van Vugt mafia'.

The amazing thing about the *Office* rehabilitation projects, compared with most expatriate-dominated projects, was that their achievements survived the onslaught of time after the projects had ended[14] In 2004 *average* paddy yield stood at 4.5–5 t/ha, with some farmers getting as much as 8–10 t, which are Asian yield figures. The infrastructure continued to function, rice was being threshed, milled and sold by the farmers themselves and the equipment was maintained by a network of small private workshops. Why did foreign aid result in apparently enduring development here, whereas it didn't in so many (one could say most, if not all) other places? I am not sure I fully understand, nor have I heard anybody else explaining it convincingly, but I will try.

I think the first cause can be found in the unique character of an irrigation scheme, which can only function if the infrastructure to deliver the water to the fields is kept in working order. That requires a tight organisation, skilled operators and enough money. Large organisations in Africa have been notoriously inefficient, but as long as the effects of sloppy management have not been too dramatic they have usually got away with it. Dysfunction or collapse of a diversion dam or the main irrigation channels, however, cause unmitigated disasters, which in the case of the Office du Niger were not allowed to happen, because the government and the donors

---

[14] Well, they never really ended. Both the French and the Dutch continued their support until this day, although at a much lower level than in the heydays of the 1980s and 1990s.

always stepped in when the danger signs went up. Perhaps
paradoxically, that has created a stable organisation with a strong
esprit de corps and pride of belonging on the part of the scheme's
managing authority. Not surprisingly, the Office du Niger was seen
as a state in the state with its high degree of independence, organisational as well as financial. Financial independence originally
came from having control over the paddy trade, and since the
reforms of the 1980s from irrigation fees levied on the water users.
And international donors could be called in to finance expensive
rehabilitation works wherever decline had set in.

I think we can agree that stability is a necessary condition for
development, but it is not a sufficient one. Thus the Office du Niger,
in spite of having a stable management, for decades hovered close
to the subsistence level, with revenues barely enough to cover the
salaries of an inflated work force and the most essential operational
costs. In the late 1970s the government issued invitations to donors
to help rehabilitate the scheme, to which French and Dutch bilateral aid and the World Bank responded. If they had only refurbished the physical infrastructure it would eventually have slipped
back into decline without a lasting effect. But they went far beyond
that and brought in two innovations which set the scheme on a
course to genuine development. One very successful innovation,
contributed by the French project was the intensification of paddy
production by the new system of transplanting rice seedlings in the
Asian way, instead of the traditional broadcast sowing, combined
with adequate fertiliser application. A rice breeding station was set
up with Dutch funding, which started to crank out new high-yielding
varieties adapted to the conditions of the scheme. The other innovation, or rather innovative complex, contributed by the Dutch
project, was to transfer as much responsibility as possible to the
farmers and their newly formed organisations. Especially the introduction of small-scale rice threshing and milling machines, in
combination with liberalisation of the rice trade, caused nothing
less than a revolution, because the farmers and their village cooperatives now controlled the entire rice production chain, up to and
including sales of the final product. In a short time the large *Office*-
owned machines and mills went entirely out of business. With the
loss of control over the rice trade the *Office* could no longer
provide seed and fertiliser on credit, the cost of which used to be
recovered through the harvest. The farmers now had to purchase

their own seed and fertiliser, for which credit schemes were set up. The best elements of French and Dutch approaches, although experimented within separate corners, eventually found their way into the other camp.

OK, but once again, what were the essential causes of the success of the rehabilitation projects? What I have said so far does not really qualify as an answer, it is a narrative rather than an analysis. I doubt that an answer which satisfies scientific criteria is possible, though, so you will have to make do with my opinion instead. I think the main causes for the success were the following:

1. The necessary physical conditions were created for high productivity – well-functioning infrastructure, excellent rice-growing technology and readily available inputs, in particular seed and fertiliser.
2. The farmers and their organisations were given responsibility for everything they could handle themselves, with the parastatal *Office* taking care of what went beyond, especially the infrastructure.
3. The donor-funded projects supported the emergence of a network of private construction, maintenance and repair shops for the farmers' equipment: farm implements, threshers and rice mills; the repair shops were viable because there was a critical amount of demand for their services.
4. The *Office* parastatal continued to have access to assured revenues through water fees levied on the users.
5. The level of corruption was drastically reduced by temporarily putting expatriates in key positions assisted by local project staff under their control.

Although the projects would not have had the success they did in the absence of any one of these, I think the second cause carried most weight. That may sound rather obvious to those who have not been associated with development aid, but the development experts have taken a long time to come to the conclusion that, whatever their shortcomings, farmers should be made responsible for running their own affairs, rather than being treated like children who must be taken by the hand.

The third point in the above list is especially intriguing. While equipment introduced by projects, such as small tractors (and large ones in former Marxist countries), have invariably fallen apart in a short while and now litter the continent, especially in savannah areas, the mechanics at the African countryside manage to maintain a considerable fleet of cars, motor cycles and bicycles, without any projects coming to their aid. If there are large enough numbers of a highly valued piece of equipment around, such as the omnipresent motorcycles, or animal-drawn ploughs in the African savannah for that matter, there will be a huge demand for their maintenance and the mechanics are sure to meet the challenge. But a few tractors with implements purchased by some projects are not enough to genuinely interest them, of course.

The last factor, i.e. the role played by expatriates, was very politically incorrect, as the term goes. As I have said before, early development projects used to justify their expatriate staff by the demonstration effect they would have on the local officials, but the opposite was often the case – the locals would learn nothing and just sit waiting until the foreigners were gone. In the case of the Dutch project in the *Office*, however, things were done just right. Whether it was by chance or by design I do not know, but after taking time to build up the project's capacity and set farmer empowerment in motion (excuse the term, but that is what it was) they turned their attention to the *Office* authority, knowing full well that without them their efforts could not survive. And over the next 10 years or so they gradually incorporated all the activities not handled by the farmers into the *Office* departments, and gradually gained their allegiance.

One final point, I wonder whether all the factors responsible for the scheme's success could have come together in such a powerful mix, if the environment had been less favourable. I mean the combination of one dominant, eminently marketable crop, grown under exceptionally stable and uniform conditions, by a migrant community. In any case, the five points will, *mutatis mutandis,* be valid under any conditions although some of them will be hard to achieve when the circumstances are less favourable. The sheer fact that there are so few success stories to relate testifies to that.

I cannot close this success story without saying a few words about its sequel. By the early twenty-first century the activities sponsored by French and Dutch bilateral aid had much diminished and responsibility for the sustenance of their achievements had mostly passed into the hands of the grass roots organisations created by them and those of the *Office* authority. There was concern, however, whether they were really ready to continue on their own and there were signs that they might not be. Meanwhile, the Malian government had decided that public and semi-public institutions, such as the *Office*, should further reduce their involvement in economic activities, including agricultural extension and the supply of inputs, to be replaced by farmers' own organisations and the private sector. And government had the ambition to dramatically increase the area under irrigation, from the present 60,000–200,000 ha in 20 years,[15] surely a rather megalomaniac but physically possible target. Donors were invited again to step in and finance these plans, but none of them fancied doing that in a major way. They therefore commissioned a study to develop a strategic plan for the next 20 years, which the government would then use to negotiate assistance with a consortium of donors. That was in the early 2000s and a strategic plan was indeed published in 2004, but I do not think anything much had moved at the time of writing these lines (late 2006). The two traditional donors were maintaining some presence in the *Office*, especially the French, trying to further strengthen private organisations, but the massive efforts needed to implement the plan were not in sight, nor had donors committed the necessary funds. It will be extremely interesting to watch further developments in the *Office*, that treasure in the floodplains of the river Niger, which should not be allowed to perish. It will be a test case for Mali's capacity to manage its own affairs and for donors' willingness to provide meaningful assistance, but only to the extent it is really needed and without taking over responsibility.

---

[15] Part of the extension would be for sugarcane, which is being grown in the scheme as an estate crop.

Chapter 11. Can African Farming Be Improved? (And Can Agronomists Help?)

I must confess that when I started to write this final chapter I was not sure what I was going to say. It had to be something of a grand finale, of course, looking into the future, armed with the lessons from the past and bringing the materials from the earlier chapters to their full fruition. I had been jotting down ideas, but taken together they did not form a coherent vision. And perhaps they could not have, because I am in the habit of vacillating between enthusiasm about the African farmer and the good things he is capable of doing, and despondency about the little that has been accomplished and the slow but steady demise of smallholder farming, while the farmers' sons are trying to escape en masse to the west. That does not augur well for a thoughtful synthesis which can help chart the course ahead, but I will try anyway.

It cannot be denied that agriculture in many parts of Africa at the turn of the century presented a depressing picture, having lost its earlier stability and failing to find a way out of stagnation and into the ranks of the optimistic farmers of more successful countries. Almost half a century of development aid has not helped a bit either, with only very few exceptions; nor do we seem to understand why things are as bad as they are. And until we do understand, the solution will continue to elude us. Agriculture is of course embedded in society as a whole and its backwardness is symptomatic for what is wrong with that society. It is therefore unlikely that farming can be significantly improved unless the societal ills are cured first. In spite of this interwovenness I will try to understand the factors which are more directly associated with agriculture itself and see how far we can get in curing its defects.

What have we learned in almost 50 years of attempts to modernise agriculture and 30 years of FSR? I think the lessons can be grouped into four categories. The first one is about the farmer himself, his motivation and skills and his ability to break out of the poor productivity syndrome. The second is about extension methods which can help farmers improve their skills. Third, there is the need for innovative technologies and what research can do to generate them. And finally, we must ask the question to what extent agriculture can progress in a deficient social and economic environment. Although I am aware that the importance of the first three issues pales before the last one, I will mainly stick to the former and only hint at the latter.

11.1 The African farmer and his potential for change

What do we really know about the African farmer, his skills, his potential and his ambitions? And are we justified to talk about 'the farmer' or are there so many different ones that it is meaningless to lump them all together? The anthropological literature of the nineteenth and early twentieth century with its meticulous observations was more informative about the human side of agriculture of those times than most of what has been produced by the FSR movement and its successors has been about that of ours. FSR assumed that it was not all that difficult to understand farmers' conditions and that for the rest it was simply a matter of logical analysis to find appropriate technology which would help farming forward. But the poor record of agricultural development in Africa has effectively refuted these claims. It is doubtful, however, that old fashioned anthropological research would be useful today to understand why agriculture has stagnated and perhaps find remedies. Agricultural communities are in a constant flux and the glue of traditions holding them together is rapidly being dissolved, so the findings from long-winded studies will be outdated before they have even been formulated. So we have to make do with what we have learned from 30 years of FSR to formulate more effective approaches for agricultural development.

11.1.1 A brief profile of the African farmer

In spite of much rhetoric suggesting otherwise, FSR has left the real African farmer mostly invisible. He has been an object of study, a partner in and a client of participatory research and his constraints and technological needs have presumably been the start and end point of FSR, but who is this farmer and what are his skills? In earlier chapters I have pictured him[1] as a skilful and rational operator who has managed to extract a living from agriculture under difficult

---

[1] As always, 'him' is shorthand for 'him and her'.

conditions, with very limited means and in the face of many constraints. The question today is what is his capacity for change and what is preventing him from becoming an efficient producer, who is equipped to face today's and tomorrow's challenges.

(a) Forest and savannah farmers

First I would like to say a few words about the difference between a (West) African forest farmer and his counterpart in the savannah. The forest farmer is not really a farmer in the sense of someone who effectively controls and manipulates the environment for his own productive needs. The traditional forest dweller would slip into the bush to collect what he wanted or to clear a patch of forest and quickly extract a crop before it closed down again upon the fruits of his toils. It would have been technically impossible to do otherwise. That makes it difficult for him to consider a permanent form of land use, even if that were technically feasible today. And that remains doubtful as we have seen, except where the multispecies forest is replaced by a mono species one, in the form of a cocoa, rubber or oil palm plantation, all of them tree crops which have been grown successfully by smallholders for many decades.

The savannah farmer on the other hand is a real manipulator of the physical environment. He adapts his crop choice to the soil conditions in his fields, carefully staggers planting and sowing to minimise the chances of crop failure due to the vagaries of the weather, uses various smart tricks to correct soil nutrient deficiencies, like *écobuage* or planting legumes right after a grass fallow. In many areas he has also adopted animal traction and he will go out of his way to get fertiliser in order to grow nitrogen-hungry but productive crops such as maize.

That does not mean that the forest farmer is any less skilful than his counterpart in the savannah. What it does mean is that development pathways which take into account the restrictions imposed by the environment and exploit the farmers' specific skills have to be very different in the two areas.

(b) The African farmer as a rational operator

In principle then, the traditional African farmer is a rational operator. The mere fact that he has survived in his harsh environment testifies to that. Just one simple example to show that today's farmers

continue to make rational choices and that agronomists had better
work from that premise even if superficial analysis would suggest
otherwise.

The story is this. As you may remember from Chapter 8, IITA
carried out OFR for several years in some villages in south-western
Nigeria. One of the things we wanted to know was whether it would
be beneficial for farmers to apply fertiliser to their maize, which
they rarely did. Our observations showed that with the local varieties only about half of them would make a profit from applying
fertiliser, the rest would lose money. And since they could get by
without fertiliser and still get a reasonable yield, only very few of
them would take the risk. So I think not using fertiliser was a
rational choice. Compare that with the northern savannah area,
where maize growing had expanded tremendously since the 1970s.
The response to fertiliser was impressive whereas the crop would
yield next to nothing without it. So if you wanted to grow maize at
all you had to apply fertiliser. That is why farmers were seen queu-
ing up in front of the fertiliser stores as soon as the government and
the traders had put their act together (which was usually too late).
I have given many more examples in the earlier chapters which leave
no doubt that the farmers' production practices have been based on
rational choices and continue to be so.

(c) Yesterday's skills and today's challenges

This may all sound like politically correct babble, when you look at
the sorry state African smallholder farming is in today. Perhaps its
historical development was steered by rational choices, but surely
its present state in many areas shows that something has gone badly
wrong. Apparently, the correction mechanisms which allowed
farming to adapt gradually to changing conditions did not work
well when the changes were rapid. Box 11-1 describes examples of
how things can get out of hand when abrupt changes take place.

In today's world the way farmers respond to new challenges and
opportunities and especially how quickly they do, decides whether
they are successful and eventually whether they will survive as
farmers. That is how it has been in industrialised countries and so
it will be in Africa. Innovation in itself is not a recent phenomenon,
only the rate at which today's farmers must innovate is new. In
Europe, for instance, agriculture slowly transformed by the adoption
of more intensive livestock keeping, made possible by new fodder

---

Box 11-1. Tree loggers, farmers and the environment

Farmers in much of humid Africa have developed land use systems which cause minimal disturbance. The way they clear a patch of forest will allow the vegetation to quickly re-establish itself and their crop combinations provide full canopy cover until the next fallow starts taking over. In some areas it does not seem to work like that, however. Sometimes it is difficult to understand why that is so. In north-western Cameroun, for instance, some of the hills were covered with thick forest while others had been completely denuded by farming. Perhaps the combination of soil properties and the mid-altitude climate did not allow the forest to come back once it had been removed. For other areas where forest was entirely gone, I have heard anthropologists explain that the area had been colonised by people from the savannah who continued to use farming methods which were unsuitable for the forest. I would like here to talk about two examples where the causes of environmental degradation were a little more evident.

In Bendel State in south-central Nigeria and in the south-east of Congo–Brazzaville, where the natural vegetation would normally be a tropical forest, the landscape in some places looks every bit a savannah. There are very few trees remaining and the stumps from which secondary forest could re-grow are mostly gone. The destruction has not been caused primarily by intensive land use, though, – the population density in both areas is quite low. So what happened? The primary culprits were the tree loggers who removed the valuable timber, pushed down most other trees in the process of evacuating it and once they were done, they were succeeded by farmers who finished the job and then planted their plantain bananas. In Congo the loggers had long been gone, but in Bendel State in Nigeria their chain saws could still be heard in the distance in the early 1990s. I think the alienation caused by the logging led farmers to forget their traditional methods and put fire to the remaining trees and stumps. And once they had used up the fertility of the forest topsoil, there was no way that a secondary forest fallow could establish itself, by the combination of thorough clearing and a chemically and physically poor soil. The result was a shrub vegetation, dominated for some time by *Chromolaena*, eventually followed by spear grass, *Imperata cylindrica* Demanding crops like plantains and cowpeas would disappear from the system, followed by the maize, leaving their place to cassava. It will be clear what that will do to the nutritional quality of the people's diet.

---

crops such as clover, lupins and turnips. More intensive livestock keeping in turn yielded more animal manure which boosted crop yields, in some areas reinforced by the use of town refuse and the contents of urban latrines. Similarly, the peasant class in Africa has absorbed many innovations over the centuries at a leisurely pace, which has also profoundly transformed agriculture – novel crops such as cassava and maize were introduced (few people are aware today that they are of foreign origin), a smallholder cocoa industry was established in the west African forest areas and animal traction was widely adopted in the savannah.

(d) Agricultural evolution and involution

What happened in western Europe when the system started to crack at the seams, for instance when there was not enough farm land for all the young men to establish themselves as farmers? In medieval times there was always 'wasteland' which could be cleared, while the growing towns also started to absorb excess rural population and employ them as craftsmen or labourers. Meanwhile, the productivity of the land increased slowly, but when it could not keep pace with the growing urban population, overseas trade provided the balance. The Industrial Revolution marked a period of accelerated changes in the agricultural production process. The rapidly growing industries could absorb large number of poor landless people from the rural areas, while the remaining farmers increased their efficiency by the use of equipment, manufactured by the new industries and by the use of chemical fertiliser. That set in motion a process which was going to lead to today's industrialised agriculture with its very high productivity and the attendant abuses. While less than a century ago farming in many European areas was not all that different from that in the African savannah today, it is hard to find any similarity between West European and African farmers now, except that they are known by the same name.

In some tropical areas an entirely different process took place, which Clifford Geertz called 'involution': ever increasing labour inputs by which an ever larger number of people could be fed from the same land. Geertz' celebrated examples were the irrigated paddy fields on the islands of Java, Bali and Lombok in Indonesia. The intensive land use in the periodically inundated lands of Bangladesh, which I described briefly in Appendix 5, also falls in this category. In Africa there are several examples of a similar process where very intensive land use allowed high population densities to develop under extremely harsh conditions, such as the hillside farming in the Mandara mountains in North Cameroun, the Machakos area in Kenya, the Pays Somba in Mali and the Atacora region in Bénin. In the African examples we are dealing with very traditional societies who are thought to have retired into the hills centuries ago to escape from submission by dominant nomadic tribes. One might expect that the mountain dwellers, when migrating into the plains in search of new fertile land, would adapt their meticulous system to the new conditions and continue to practise their intensive farming methods. But the opposite has been the case, as Westphal (1981) has described in his book on indigenous agriculture in Cameroun. Once they settled in the plain they quickly adopted the extensive land use system of the locals. And that indeed made sense, because its *labour* productivity was much higher than that of the hills they had left. The old system certainly had aesthetic appeal as well as effectively conserving a fragile environment and it continues to attract western tourists in search of natural and cultural purity. But it is very hard work simply to stay alive and as soon as there is a chance the people will do what we would all do under such conditions, opt out and start a less strenuous life. What a relief! Rational, for sure, but in a sense also retrogressive, because it means abandoning a highly sophisticated land use system for a much simpler one. Intensive hillside farming in Africa is technically interesting and perhaps even contains useful lessons about how fragile environments can be protected, but it is the end point of a process of involution, not a system which can easily make the transition to the twenty-first century.

Agricultural involution has been the exception rather than the rule in Africa, however. In most areas there had been a slow but steady historical process of innovation in historical times by the uptake of new crops and, in the savannah, the adoption of animal traction, until in recent times increasing population density forced farmers to intensify by shortening the fallow, without significantly changing their farming methods. Also, the mutually beneficial interaction between agriculture and industry, which has been so important in the industrialised countries, has been practically absent in Africa and the population overflow from rural areas has mainly resulted in a parasitic urban underclass without much hope for a better life.

(e) New kinds of farming are emerging

There have been a few favourable and hopeful exceptions, though. Visit the large cities of Accra, Bouaké or Bamako and you will note an interesting phenomenon which is called urban, peri-urban or city farming – mainly young fellows growing vegetables in neatly laid out, well-maintained and regularly watered beds, or producing ornamental plants and seedlings of fruit trees, all for the urban market. This is not a new phenomenon. I do not know how long it has been going on, but G.J.H. Grubben (1975) published a dissertation more than 30 years ago describing the production of the leafy vegetable *amaranth* in city gardens in Dahomey (now Bénin Republic). Another equally interesting development has been the emergence of peri-urban dairy production, with mainly small units of just a few dairy animals, and medium to large-scale egg and broiler production close to the cities. These are all quite dynamic sectors, which can and do respond to new opportunities. The customers are the more affluent part of the urban population and the fortunes of the small producers go up and down with those of their relatively few customers.

This brings back the question whether an entirely new type of farm will eventually emerge in Africa or whether some of the current peasant farmers are likely to evolve into the dynamic entrepreneurs Africa needs, as it happened in modern times in Western Europe. You will remember earlier comments, in particular Dick Lowe's, that there will not be an unbroken line connecting the old farmer with the new one. That is certainly true for the peri-urban gardener, whose father or grandfather may have been a farmer, but who initially went to town to do something completely different and, when that did not work out, picked up intensive vegetable production instead. And the peri-urban dairy and poultry farmers are usually bureaucrats or professionals who have invested money they earned otherwise, much as Lowe predicted.

When I worked in Cameroun on my first African job (where I also first met Lowe), I did not yet believe that things would happen that way. I once made the students of the agricultural college where I worked calculate the profit that farmers around the capital city of Yaoundé could make from intensive production of high value food crops such as plantain bananas, yams and vegetables. The outcome was quite spectacular, but nobody was doing it. In fact a lot of the city's supplies were hauled from far away, from the fertile hills of Western Province, populated by the dynamic Bamilekes. Ethnic differences are obviously a tricky topic, especially in Africa, but it is not unthinkable that some tribes simply make better farmers, for whatever reasons, and that some of them may indeed evolve directly into effective market producers. But I am now inclined to believe that there may often times be no direct road leading from subsistence farming to commercial agricultural production and that a new breed of farmers is needed to really bring agricultural production forward.

11.1.2 Can a bad farmer be converted into a good one?

Let us return to the present-day farmer again. No two farmers are the same, in Africa anymore than anywhere else. Some of them are successful producers while others are not. I have tried to find factors that make the difference between good and bad farmers in Chapter 6, without much success, and I concluded that one hardly gets beyond the trivial conclusion that good farmers do most things well most of the time. That is not very helpful for an extension agent or a development worker. They may be able to tell farmers which varieties are available and which fertiliser formulation would be most suitable for their soil,[2] but the ambition of agricultural extension is more than that. They want to counsel farmers on how to become better producers. Take the simple case again of fertiliser in the West African forest zone again. It turned out that it was not profitable to apply fertiliser to maize in a field where the overall yield, with or without fertiliser, was low due to a variety of causes. The extension worker would want to translate that into practical recommendation: 'if you do such and such your yield level will be high enough to justify the use of fertiliser and you will then get an even better yield'. But that was precisely the problem. First we did not know precisely what 'such and such' would be in the coming season, and second we could not rely on farmers' previous success to be sure of the same this year. That makes the task of extension rather miserable, as we know it often is.

In the Netherlands Vinus Zachariasse of the Landbouw-Economisch Instituut (LEI)[3] thought he could do better than that for Dutch farmers. During his two-year study in the late 1960s his ranking of farmers according to their crop yields varied little between years, but in a follow-up study 8 years later some of the lowest producers had moved up the scale considerably while some of the highest producers had moved down (Meijer et al., 1979). If

---

[2] In Africa even that goes beyond what most extension workers can do.
[3] Institute for Agricultural Economics.

that were all that had happened nothing would have been gained because the overall yield had also hardly budged. But there were indications (that is what scientists will say when they believe something they cannot quite prove) that several farmers had picked up the lessons from the original study and improved their efficiency even though their yield had not changed much, which led the authors to conclude that 'it ... appears meaningful to continue supporting the farmers vigorously in their attempts to improve their entrepreneurial and technical skills'.

Apart from studies like Zachariasse's, the LEI institute operated (and continues to operate) a Farm Accountancy Network, whereby a large sample of farmers kept detailed accounts for many years. LEI's objectives were scientific data gathering and analysis rather than extension, but the participating farmers benefited by getting detailed feedback about their performance every year. It somewhat resembled the 'Conseil de Gestion' method practised in francophone Africa, which is also a tool for management counselling (that is what the words mean). Both methods have the same appeal in that they deal directly with the day-to-day workings of the farm, the farmer can learn to look systematically at the implications of what he is doing and, in the case of the Conseil de Gestion, the extensionist can use the lessons learned by one farmer to help the other. Both of them are optimistic tools – they assume that extension can indeed help farmers to do better by learning from their more successful peers.

Well, maybe they can, but it will only be true if there is something worth copying and farmers do indeed have the ability to emulate the example. In the industrialised countries the less talented farmers have long since been shaken out of the profession and modernisation has been led by an innovative and aggressive minority who have pulled along those who could follow, leaving the weak ones to perish. The tragedy of African agriculture, in particular in the forest zone, is the absence of such a successful vanguard which could spearhead its transformation without which the struggle against underdevelopment and malnutrition cannot be won. I will leave those larger questions aside for the moment and first look more closely at the extension methods the profession has come up with to enhance the skills of the agricultural producers and lift their production to a higher level.

11.2 Extension: organisation, methods and education

11.2.1 New trends in extension

The most interesting extension methodologies introduced into Africa in the past few decades were the Farmer Field Schools and the 'Conseil de Gestion' or Farm Management Counselling. We have already met both of them in earlier chapters. They treat the farmers as adult individuals with their own mind and experience, rather than as ignoramuses who have to be preached to or ordered around to do what the extension officer tells them. At least that is the theory. If applied as intended the methods are a significant departure from the good old top-down approaches. Their adoption has been timid, however, and the old reflexes remain as strong as ever, but at least there is hope for something better, because taking the farmers seriously is the first condition to achieve something with them. The other major development has been the emergence of private extension bureaus. They hold promise for the future in the eyes of many and may eventually replace the large monolithic and highly ineffectual public extension services which are now getting starved of funds, because donors have finally given up on them.

The crumbling of sclerotic extension services and the emergence of new organisational models and methods have been hopeful developments, which may yet change agricultural extension for the better and eventually help pull agriculture out its downward spiral. But before looking further into the African situation I like to make a brief excursion into agricultural extension in the Netherlands, where radical changes have also taken place in the last two decades, to see whether there are any lessons for Africa there.

(a) Extension in the Netherlands

In the early twentieth century the structure of Dutch agriculture was simple, transparent and stable. There were dairy farms, arable crop farms and mixed farms with both dairy animals and arable crops. Each type had its own geographical distribution and the assortment of crops was rock solid. The extension agents knew their farmers, and they had enough knowledge about their animals and crops to give them useful advice. Most of their knowledge, once acquired, remained valid for a lifetime, and what novelties came out of research or blew across from other countries could easily be absorbed at a leisurely pace. There were of course considerable differences in skills among farmers, but even the less skilful ones could still make a living from agriculture, because they were cushioned against the challenges of the wider economy by the production of their own food.

In the 1950s things started to change. Farming families became increasingly integrated into the overall economy and subsistence farming, still common in many areas early in the century, became a thing of the past. Increased need for monetary income forced farmers to become efficient entrepreneurs and look for more profitable crops and better production techniques. Small inefficient producers dropped out or died without a successor, allowing others to take over their land and increase their farm size. Over the years new crops, new management methods and even new animals appeared and sometimes disappeared again in the farmers' continuous struggle to generate an acceptable income, comparable with those outside agriculture. These continuous changes put a heavy strain on the farmers' skills, not just technical, although that remained as important as ever, but also managerial and financial. The demand for external advisory services therefore changed as well. Farmers who wanted to launch into new ventures often had to incur considerable investments which had to have an acceptable profitability if the bank were to provide the money as a loan. The demand for financial and farm management advice therefore increased, for one thing because the banks would demand a solid business plan. As a result, the services requested from the national extension services became more and more of an economic nature and the extension agents would help the farmer write business plans rather than stalking through the crops and advising them on how to keep the weeds or insects under control.

In the 1990s, the national extension service was privatised, but long before that farmer demand for personal technical advice had dwindled and when the newly privatised extension agency started to charge fees it virtually disappeared. Extension now dealt mainly with farm management issues and its agents participated increasingly in group-based extension activities, either for a fee or with government subsidy. Technical advice was more and more provided by commercial firms in the seed, animal feed and 'crop hygiene' business or by cooperatives, and financed by them from their profits, or by companies specialising in 'crop hygiene' who would conclude contracts with farmers for pest, disease and weed control.

The continuously changing agricultural sector called for a new kind of learning where all actors put their heads together, to learn from each others' experiences and grope their way into an uncertain future. That resulted in an interesting development, and one which is highly relevant for tropical farmers as well – the emergence of study clubs, in which farmers, extension personnel and research institutions participated. They would be organised around specific crops or such issues as integrated pest management to reduce pesticide use, or mineral bookkeeping, imposed by the government to roll back pollution by excessive organic and inorganic fertilisation. This shift towards group-based approaches reflected the self-confidence of a new generation of farmers as well as being an effective way of dealing with the increased sophistication of farming and learning from each other's experiences.

In summary, the key changes which had taken shape in agricultural extension in the Netherlands by the end of the century have been a shift from advice to counselling, from technical content to economic and management analysis, from individual to group-based approaches, and from government to private (or privatised national) extension.

(b) Towards private extension in Africa?

The world has become a small place and similar changes as in the Netherlands have been taking place in agricultural extension in Africa, under very different conditions. Large amounts of funds, mainly from donors, had been pumped into national extension services, practically without any real effect. During the last decade, therefore, initiatives got underway to create private extension units, in anticipation of the ultimate collapse of government extension. In particular the French have been involved in such initiatives in West Africa by setting up small bureaus, usually staffed by motivated young people. The question is of course whether they will be viable. The smallholder farmer cannot afford to pay for their services, and if they could they would not necessarily be convinced that the investment will pay off. So, another solution has to be found. For crops like cotton things are comparatively simple. Where the cotton trade is dominated by one company, as is often the case in francophone Africa, the company can create its own (cotton) extension service and recover the cost of running the service from its profits. In the end it is still the farmer who pays, of course, but if the service is any good, he will benefit too. Other examples where such arrangements could work are cocoa, oil palm and coffee, but I am not aware of successful examples here, perhaps because the trade in these crops is highly fragmented. An alternative would be the creation of co-operatives with their own extension service, but 'cooperative' remains a dirty word in Africa because of its history of nepotism, embezzlement and other forms of corruption.

For crops which are even less 'vertically integrated', in particular food crops, the funding of private extension services is an almost unsolvable problem. Several highly artificial arrangements are in use, which are typical for the fairy tale world of development coop-eration, such as donor-funded development projects which hire extension bureaus after first setting them up themselves. Non-governmental organisations have also ventured into agricultural extension, sometimes with an ideological flavour such as biological farming, or with the idea to access the market for fair trade prod-ucts in the industrialised countries. Many of them can hardly be considered private extension, though, bearing more resemblance with missionary stations which dabble in agriculture with money provided by the faithful back home.

An interesting example of the beginnings of privatisation comes from the Office du Niger, the large irrigated rice scheme in the flood plain of the river Niger in Mali, which I described in Chapter 10. Up until the late 1970s the *Office*'s managing authority controlled almost everything, from the irrigation infrastructure, through the supply of seed and fertiliser to the processing and sale of the rice. Since the 1980s new legislation has gradually transferred many of the *Office*'s traditional tasks to the farmers themselves, their organ-isations and the private sector, with a lot of support from donors. The idea was that the *Office* would eventually be only responsible for the infrastructure and the delivery of irrigation water up to the secondary channels, where users' organisations would take over. Extension was one of the last services to be privatised, an extremely difficult task since it implied that the farmers would now have to pay one way or the other. Unless government would regard exten-sion as something it would still have to pay for, even if it were deliv-ered by private organisations. At the time of writing these issues were yet to be sorted out. Meanwhile, a French project had set up small advisory bureaus staffed by young graduates and former extension officers laid off by the *Office*. They were trained in various skills, in particular *Conseil de Gestion* methods, by another facility which was also created by the same project, called URDOC,[4] and run by two dynamic individuals, Paul Kleene, whom we have met before, and Yacouba Coullibaly. URDOC also conducted applied research and tried to formulate solutions to important problems farmers were facing, which would then be passed on to the extension bureaus. The bureaus were set up as private entities and the farmers would have to pay for their services, either directly or indirectly. It was obviously an illusion to expect them to hire the bureaus directly and pay the bill, so the next best would be for farmer organisations to hire their services, and since they hardly existed either, they had to be created also. And so they were, supplied with donor money again to contract the extension bureaus. That was seen as a temporary measure for sure, but still, the whole thing started off again as a donor-initiated, donor-funded and donor-led operation.

Even so, the arrangement certainly had logic on its side and the French, who probably pioneered it elsewhere also, liked it very much. In the early 2000s they set up similar operations in Mali-Sud, the centre of cotton growing. It was no coincidence that this was also one of the country's economically most endowed areas and therefore had the best chances for privatisation of agricultural services to succeed, simply because there was money around. Whether that was indeed sufficient for success was not at all clear at the time of writing, but at least some of the results reported by URDOC looked very promising. Large public extension and research institutions in Africa were moribund and here at least were the beginnings of an alternative.

In spite of their donor origin these were promising initiatives. They might attract fresh minds and raise youthful enthusiasm to replace the large anaemic research and extension institutions which all but their own *salarymen* had given up on. They were frowned

---

[4] URDOC stands for Unité de Recherche-Développement Observatoire du Changement, or unit for development-oriented research and monitoring of change.

upon by those institutions, who felt they were being left in the cold by the donors. And indeed they were, because the donors had grown tired of throwing money at powerless and uninspired bureaucracies which were never going to deliver. But setting up a few dynamic extension and research teams, well supplied with donor funds is one thing, putting them on their own feet and extending them across an entire region or country is something else.

The World Bank of course quickly spotted the promise and also started to experiment with contracts between local governments and private research and extension bureaus or privatised former public institutions. That, I think, was an unsettling development. One would like the World Bank to stay away from such embryonic ideas for as long as possible, until they have gone through their childhood ailments and reached a level of maturity where they could survive the onslaught of mediocrity.

I find it hard to look into, let alone predict the future of agricultural extension in Africa, but I do believe that there is a future for relatively small autonomous units, which either specialise in money spinners such as cotton, cocoa or paddy, or in farm counselling and group extension approaches. But the challenge will be to find ways to increase their density or, in developmental new-speak, to *mainstream* the approach, as well as to pay for their work if their bills cannot be paid by the individual farmers. And development organisations should not again fall into the trap of taking the lead, but rather let thousand flowers bloom and just play the role of facilitators and financiers at a distance, coming in only when needed and, even more importantly, explicitly requested.

11.2.2 Farmer organisations

Creating new extension units and ensuring that they stay alive is easier said than done. So far, the donors and the foreign consultants have taken the initiatives, recruited the staff, developed the programmes, organised the clients and paid the bills. You might therefore feel that the whole thing may as well be relegated to the rubbish heap of bad ideas. But no, the ideas look too good this time, in spite of their dubious origin. It would surely have been much better if the initiatives had been taken by the locals and the fact that they have not may still undo them in the end, but let us hope that the ideas will at least be carried out with a minimum interference by expatriates.

I am not entirely confident on that score, though, because there is not much initiative left among African agriculturists after decades of spoon-feeding (not my qualification, but one by Bede Okigbo, a Nigerian if ever there was one) on easy donor money. Perhaps the initiative can come from people who have been less affected by the mental corruption caused by development aid. I mean the farmers themselves and their organisations. The large majority of African farmers are barely integrated into the wider economy, however, and they lack the ability and the drive to organise themselves, so salvation may have to come from a new breed of farmers who are now beginning to appear on the stage. They are the cotton, cocoa and paddy growers, the peri-urban vegetable and fruit producers and the dairy and poultry farmers who have the skills and the means to create professional organisations, define their need for advice and raise the money to pay for it. They may transform themselves into a well-organised dynamic farming class and tug along their less dynamic brethren that sounds like a plea for cooperatives, and indeed it is, in spite of the dismal record of the government-imposed organisations that went by that name. Genuine farmer cooperatives, can be a blessing in that they can take up activities which are far beyond the reach of individual farmers: collective bargaining and purchase of inputs, central processing if needed and collective marketing of quality products, and yes, farm advisory services. Much the way things have happened practically every-where in the industrialised countries. Donor money, made available sparingly, can help to cushion such organisations in their fledgling stage against the impacts of heavy weather. That, I think, is more meaningful than the so-called platforms of stakeholders which were the donor craze in the early 2000s. Platforms could play a role in the future, but only if farmers are represented as serious, well-organised and strong partners, that is after farmers have organised themselves first, instead of donors and their field aides doing it for them.

11.2.3 The new extension methods

Let us now talk about the two extension methods, or rather extension tools, which I see as the most promising additions to the arsenal made in the past few decades – *Conseil de Gestion* and Farmer Field Schools.

I have already talked extensively about *Conseil de Gestion* in Chapter 8, so I can be brief here. The *Conseil* is a counselling

method, whereby the extension worker together with the farmer or
group of farmers analyses farmers' management decisions and how
their management may be changed for higher productivity or profitability. The method can also be applied to more specific issues,
such as fertility management, which may involve the analysis of
nutrient flows, how they can be better exploited and how losses can
be minimised.

The *Farmer Field School* idea was pioneered by the FAO in
South-east Asia in the 1980s, as a method of familiarising farmers
with IPM. The history and successes of IPM make another interesting story which I touched upon in Chapter 7. It was first developed for cotton in California in the 1960s and vigorously promoted
for irrigated rice in Asia since the early 1970s, when it was observed
that the Green Revolution was leading to abuse of highly poisonous
chemicals for pest control. It is perhaps hard to imagine today, but
in Java, with its very dense pattern of rice paddies and settlements,
even aerial spraying was practised for a while, at the height of the
Green Revolution in the 1970s. IPM is less simple than walking
through the field with a sprayer on your back. It involves many different components, which vary with the pests, the ecological conditions, the farmers' means and their knowledge, the chemicals
available (the use of chemical control is not entirely excluded), and
so forth. So, there are no simple prescriptions which can be easily
handed down to farmers. FAO therefore came up with the idea of
bringing together groups of farmers, say once a week, and study
with them the life cycles of the major pests, what could be done to
break their cycles so as to reduce the damage, or how their natural
enemies could be helped to be more effective. The teacher or
moderator would of course bring in technical knowledge and experiences from elsewhere. That was a really appealing idea, which
spread like bush fire throughout Asia, and which is claimed to have
had an enormous impact on the use of pesticides and has boosted
rice yields at a small cost to the farmers. How real the successes
have been I do not know. There is a lot of partisan literature, but an
objective history of the Farmer Field Schools (and of IPM in developing countries for that matter) is yet to be written, as far as I am
aware. In any case, it was soon realised that the concept could also
help to address other complex issues, such as the management of
fertility on the farm and it is now popping up in many places.
Farmer Field Schools and *Conseil de Gestion* have also started to

exchange elements and the two methods perhaps can be seen as two variants of the same idea, the main difference being one of intensity.

Both approaches also have similar limitations. The first one is precisely that they are very intensive and time-consuming and can only reach a limited number of farmers. In Africa that is a particularly serious handicap, because farms are usually small, the number of farmers is correspondingly large and there are few skilled extension workers who can adequately deliver the methods. The other limitation is that, without sound technical options which farmers can choose from, the methods tend to degenerate into empty ritual, as has so often happened with methodological innovation in Africa. I am not implying that nothing can be improved in the African farm without new technology. As I have argued earlier, there are considerable differences in skill and performance among farmers and the study clubs and field schools are good for less successful farmers to learn from the more successful ones. But that is not enough. Farmers must have the feeling that there is more to it than just peeping into each other's affairs or listening to an extension worker who tells them that they could do a little better with their small means. There must be some real novelty there, a new way of doing things, or a new technology, otherwise they will quickly lose interest. Farming Systems researchers have often neglected the need for new technology, with the extremists among us even arguing that indigenous knowledge was as good as or better than the technologies invented by the scientists. When looking at the history of European farming, however, you can hardly doubt the importance of new technology as a driving force for progress. And, to stay closer to home, the best results with farmer field schools were obtained precisely when applied to the promotion of integrated pest management, a highly technical subject. Extension methods are just delivery techniques which only work when there is something worth delivering. But once that is the case it does make a lot of difference whether it is done in an authoritarian or in a participatory way.

11.2.4 Agricultural education

A country's intellectual prowess and its economic performance are directly linked to the quality of its educational system, and most African countries do poorly indeed on that score. I am mainly

concerned with agricultural education here, not about general education, but the quality of the one is intimately connected with that
of the other, so I must say a few words about general education
first. The failure of most African governments to provide quality
services to its citizens is most painfully felt in education where the
flower of the nation is prepared to build a prosperous nation, a task
in which their elders have largely failed. In recent times many
donors have opted out of agriculture and moved into increased support for health and education. We can all agree that quality education and good health will greatly help a country to find a way out
of backwardness, so the choice looks logical, but I suspect that
some of the reasons are less noble and have to do with the politicians' wish to disentangle themselves from the disappointing
productive sectors and get involved in less opaque business. But
external support for health and education will only make the countries more dependent on donor funds, unless the productive sectors
start to generate significant local resources. Many bilaterals now
prefer to leave that to the large international organisations including the World Bank, however, and you know by now how I think
about that. Donors' loss of interest in agriculture has also affected
agricultural training, which in many countries is in a pathetic condition, nor will there be any improvement unless their own politicians, educators and in particular emergent farmer leaders start to
take cognizance of the primary importance of good agricultural
training.

It starts right from the primary and secondary schools, where the
interest in agricultural production is usually minimal or nil, even in
rural areas where farming is the occupation of majority of the children's parents. Everyone in the west has seen pathetic scenes of dark
shabby classrooms with bedraggled children crammed into wobbly
desks, or sitting on the floor, and the teacher lamenting to the inter-
viewer about the absence of facilities for proper teaching, which is
very sad indeed, but at the same time you wonder whether with a
little initiative of their own and the help of the children and their
parents they could not at least do something to improve the situation. Like for instance setting up a small school farm to supplement
the school's meagre means and perhaps the children's diet as well,
while at the same time teaching the children a few things about biology and agriculture and perhaps even demonstrating some new
farming methods. That would also render agriculture more

respectable in the eyes of the children, instead of something they should run away from as quickly as they can. I am certainly not the inventor of school farms – it has been tried many times and usually with very little enduring success. Apparently the idea is mainly appealing to the development worker who comes in from the outside and will leave again after a while, not for the people who have to do it. Development workers sometimes come with vague notions about nineteenth-century European village teachers, rooted in their society and devoting their lives to the uplifting of the rural people, children and parents alike. And that the African village teacher should do likewise. It cannot be denied that a good portion of idealism would be very beneficial, the question is how that will ever happen if there is nothing in the example set by the country's leadership to suggest that idealism has any value. In the villages in south-western Nigeria where we did our OFR for many years the teachers at the secondary school would arrive in the course of Monday from Ibadan, where they kept their families, and left again on Friday or even on Thursday. And even on the days in between you could often see the students sitting in the classroom, disciplined but idle, without any apparent teaching going on. The school's leadership was not at all sensitive to the idea of a school farm or making agriculture part of the curriculum, because 'no science teacher had been appointed by government', which shows how little sense it makes to start something like that at the local level, unless it is vigorously promoted from the top.

What about the education of the agriculturists themselves, the extension workers and the other professionals, who will be expected to bring innovation to the farmers' doorsteps, stimulate their initiative and assist their organisations to improve their performance, to the benefit of their members. What they learn at their training institutions and in the universities should prepare them for that role as well as instilling respect for the farmers they are going to work for. Some of that is happening here and there, but I think agricultural teaching has barely been touched by the many new ideas which have been developed over the last three or four decades. The bulk of it remains very conventional, with agronomy organised around individual crops, plant spacing, planting dates, fertiliser rates, chemical pest and weed control, mechanisation, all those things which are considered modern but which the real farmers (or the extension workers themselves) do not practise. And when they start working they will try to make farmers do what they have learned and get frustrated very soon when they find out they will not listen.

Good agricultural education is one of the keys to the urgently needed transformation of agriculture. It should breed a new kind of agriculturists who understand what the real issues are, who are impatient to go out and help those farmers who have the skills and mentality to progress, and yet who are realistic enough to understand the many blockages and realise that real progress will take a lot of time and perseverance. Among all the things African governments and their donors should have done, agriculture teaching has been among the most neglected, except in the first few years after independence, when many training schools and agricultural faculties were created, although with quite conventional, western kind of curricula. If there is one field where the seeds of change can yet be sown it is in agricultural education, provided there is a consistent model of what that education should look like. Hardly any serious attention is being paid to it today.

11.2.5 Are regulatory measures needed?

To conclude this section I want to devote a few words to the dangerous issue of regulations and restrictions which may have to be imposed on farmers for their own benefit and that of their fellow citizens, present and future. Farmers must first of all take care of their own survival and they cannot be expected to be much concerned about the wider environment or the future productivity of the land. That is so in the industrialised countries as much as it is in Africa. European governments have therefore always considered it their duty to impose restrictions on the way farmers use their land. It started with the medieval three-course rotation[5] imposed in continental Europe, whereby a field had to be left fallow once every 3 years to restore its fertility. In the last century a rule was enacted in the Netherlands that potatoes could only be grown once every 3 years on the same land, to control cyst nematodes. Another example, which I find particularly interesting, is the compulsory rotation with grassland for farmers on the lighter soils in one of the Dutch

---

[5] I believe this is the English term for what is called 'drieslagstelsel' in Dutch and 'Dreifelderwirtschaft' in German.

young polders. The reason was that these soils needed a high organic matter content to maintain good structure. The implication was that on the lighter soil mixed farms were the rule and the tenants had also mostly been recruited from areas where the soils were light and mixed farming was the rule. When the rotation rules were relaxed in the 1970s many farmers chose to sell their animals, plough up the grassland and specialise in field crops. But they soon discovered the wisdom of the earlier rules and started to use other means to maintain the organic matter content of their soil, voluntarily this time, such as growing green manure and swapping land with dairy farmers.

Colonial governments also had the habit of trying to control farmers' farming practices if they thought that was needed for the common good. In the Dutch East Indies, for example, in areas where two successive paddy crops could be grown, the authorities would withhold irrigation water for the preparation of seedling nurseries until egg laying by the moths of the white stem borer was over. I will come back to this intriguing method a little later. In several hilly countries in Africa, such as Rwanda and western Cameroun, farmers had to plant their crops along the contours rather than up and down the slope, in order to reduce erosion. The rule was often resented, because making ridges along the contours was quite laborious and the practice was abandoned at the first opportunity, without even a suitable, less laborious alternative to replace it.

These are just a few examples of serious threats where a responsible government must step in to regulate farmers' behaviour. Widespread erosion in hilly lands is certainly such a threat, the effect of which can be seen everywhere, but African governments usually lack the vision, the will, or the capacity to seriously tackle such issues. And if they do, the whole thing may quickly be marred by the endemic disorders of corruption, including extortion by officials. Donors and idealists have come up with all kinds of ideas for the even more spectacular destruction of the tropical forests by loggers and farmer settlers, which cannot be stopped either, unless national governments face up to their responsibilities. In Africa there are no signs they will do so any time soon, perhaps not before most of the forests are gone, the way it probably happened in Holland ('Holland' means woodland I am told, but there is very little of it around today). Clever ideas, such as establishing concentric circles of different types of land use around forests have been tried, but they are far too complex and fall victim to the disintegrating power of bad government.

I do not know what the solution to these larger challenges is. But I do know that the problems must be addressed at an international level and that the usual politically correct approach with its excessive reverence for incompetent African governments is not going to help. The scope of the challenges far exceeds the realm of agriculture and so do the actions needed to confront them. But agriculturists at least can start making their own contribution, by making an inventory of the most pressing problems and bringing the technical solutions which are already there to the attention of the decision makers. I know it sounds awfully soft considering the seriousness of the issues, but we must start from somewhere.

11.3 Can agronomic research help agriculture?

I must now come down to my own level again and consider a most important question for agronomists: what can agronomy do today to help African agriculture to advance? Before attempting to answer that question I will indulge in the classification and characterisation of different types of tropical agriculturists, something which I said in Chapter 1 that I could not do then. That is not only an amusing pastime, it will also serve as a summary of the things agronomists have done in the past which will help to understand what they may be able to do (or should stay away from) in the future.

11.3.1 Types of agronomist

(a) The geographer-historian

We owe a lot of our knowledge about the history of tropical farming to keen observers of local customs and agricultural practices from the colonial era. Many of them would not even be classified as agronomists or even agriculturists. We have met an example in Chapter 3 in the person of G. Tessmann, an anthropologist who worked in Cameroun in the early 1900s. Another, even earlier example was the pharmacist G.E. Rumphius who compiled an invaluable inventory (selling today at $5,000) of useful plants on the island of

Ambon in the Dutch East Indies in the eighteenth century. A similar, more comprehensive monograph for all of the East Indies was written by K. Heyne, an economic botanist, in the 1920s. He was the inspiration for a present day botanist and geographer, Egbert Westphal (1975), who worked in Ethiopia and Cameroun and most recently headed the publication of a multivolume inventory of useful plants in Asia, modelled after Heyne's monograph. This list is heavily skewed towards Dutch research, but I am sure similar examples can be given for other countries, about which I must shamefully admit ignorance. There is even a discipline, called ethnobotany, which studies the use of plants for nutritional and medicinal purposes in traditional societies. Examples closer to our own times are the thorough work on soils, soil fertility and land use by Nye and Greenland and by Smyth and Montgomery, both carried out shortly before the high time of nation building in Africa. And we have seen the studies on traditional agriculture and its likely future by Esther Boserup and by Clifford Geertz, both social scientists. I have lamented earlier and will do so here again, that those studies belonged to an era when people could afford the time to build a complete scientific edifice rather than just a series of journal papers which few practitioners will read and which posterity will have difficulty in assembling into a whole. Today, while we pay lip-service to the need for knowledge-based development, real knowledge gathering of this kind has become rare in tropical agriculture. Instead, we have indulged in a sterile play of concepts and processes and forgotten what really counts for development. But when real knowledge is needed we have to run back to books written in more thoughtful times to find real information, even if some of their relevance has worn off.

(b) The hard scientist

The hard scientist is primarily interested in the exciting biological and physical issues involved in agricultural production. At some point in his professional career he (there have indeed been few she's here) may regret his choice of agronomy and specialise in a technical sub-discipline, or he may dig into a particular aspect of crop production such as mixed cropping, where endless mildly interesting questions are waiting to be solved. For an agronomist from an industrialised country the decision to work in a tropical country instead of staying at home may be motivated by many things, but most commonly there is some romanticism involved. And there is also the greater likelihood to make a splash because of so many unresolved issues.

We have already met several examples of accomplished hard scientists: Bremer (1928) and Jeswiet, the sugar cane cytologist and breeder from the Dutch East Indies, Kang and Lal, soil scientists, and Akobundu, weed scientist from IITA. And of course several generations of plant breeders, and workers in biological pest control, especially (but not only) at the international research institutes. And crop modellers from industrialised countries, who have sometimes made forays into tropical agriculture as well. The work of all these people is indispensable as the source of new insights, methods and materials to drive development, provided there are others who can carry on where research ends and application begins. And that is precisely where things go wrong. At IITA, for example, there were few direct links with our farmer clientele, so the hard core scientists had to imagine a virtual user for their results, which in many IITA publications (and with suitable changes in those of other international institutes) was referred to as 'An African Farmer'. Some scientists were genuinely interested in the real African farmer, but mostly in a rather abstract way, not by the frequent and intensive contacts with farmers or extension agents or both, which are needed to continually confront their ideas with the reality of the real farm, nor were they alone to be blamed for their ivory-tower attitude. Applied researchers can only be really effective if their work is flanked by that of others who pick up the new ideas and technologies, transfer them to the farm and bring back the news about their performance. We have seen that FSR was meant precisely to fill the gap between research and farmers.

In the absence of an effective delivery system there may still be exceptionally good and robust technologies which find their way to the farm, either because they are so appealing to farmers that they can spread unaided, such as very good crop varieties, or their spread is entirely independent of the farmers' own action, like the predators of the cassava mealy bug which have been released by IITA and national institutes in many African countries. But that kind of technology is the holy grail of many hard scientists, and finding one the good luck of only a very few.

(c) The action researcher

That brings us to the action researcher, who goes out and works with the people for whom the products of research are meant, tests new technologies or methods in a real-life situation or even invents such along the way. Many action researchers started out as hard scientists and, perhaps out of impatience with the lack of impact of their work (or because of their lack of scientific proficiency), converted themselves into action researchers. FSR was a form of action research. It attracted many bright and highly motivated people, as well as quite a number of hopeless cases who were tempted by the promise of easy success, attainable with a minimum of skill. But that is only natural when a new fashion emerges and attracts a lot of attention and new money – it happens in all walks of life. The emergence of FSR happened in response to the very real problem of the disconnectedness of the development of innovations and their use, especially in Africa. FSR was not the only form of action research in agriculture. There have always been agronomists who felt happiest when working with farmers and arguing with them about the things to do and the innovations to use to get out of the subsistence trap. They would test all kinds of technologies with farmers without sticking a label (such as FSR) on what they were doing. One example in IITA was George Wilson, who tried for many years but in the end gave up, rather frustrated I think, both by the lack of impact and by the lack of recognition by the hard-core scientists who found his work (and that of the FSR scientists for that matter) rather second rate.

In Nigeria and elsewhere the kind of thing Wilson did came to be known as 'research-extension linkage', for which the research institutes appointed special officers (called RELO, for Research-Extension Linkage Officer) and facilities in the early 1980s. I think the idea was picked up by the World Bank at some point, it must have been, since I met RELOs on both sides of the continent. But instead of being manned by the most dynamic and outward-looking people the institutes had, it became the dumping place for disposable ones. So the whole thing got nowhere. FSR teams and RELOs existed side by side for years with essentially the same mandate, competing for scarce resources – another example of the near-impossibility in Africa to scrap an institutional layer and its personnel once it has been created. This is true not only in Africa, although probably to a more extreme degree there.

(d) The educator

The agronomist-educator is the twin brother of the action
researcher. It is his passion to mould complex things into practical
language and explain them to farmers, agriculturists and politicians. Good educators must have an excellent grasp of technology
and they can bring across to their students the enthusiasm which
motivates them. I have met many born educators over the years and
they can become quite a nuisance, unless their skills find a fertile
and dynamic environment to thrive in. The first educator I met after
I graduated was a Frenchman whose name I forgot, who came to
visit us in Indonesia when I worked there on my first job, in cotton.
This man would immediately dive into any cotton field he found
along his path, dragging the farmer along, analyse a plant's history
on his knees from little clues in its present make-up, and lecture
about what had gone wrong and what the farmer should pay more
attention to next time. I learned more about cotton from spending
one afternoon with this man than from a year's plodding along on
my own. In fact, he was a breeder by training and there are indeed
many plant breeders like him, who become educators the moment
they go to the field. But they need a well-structured working environment for their skills, otherwise they may become unguided mis-
siles, generators of varieties which never leave the station, and the
subject of much repeated anecdotes.

Most educators are of course found in the extension profession,
where their skills and motivation are potentially most useful. I have
seen many a good educator withers away in the suffocating environment of the extension service, his initiative killed by resentful
superiors whose major skill consisted in preventing the boat from
being rocked. But if he is in a position of responsibility he can pull
along an entire organisation, which is good if his ideas are relevant,
and very harmful if they are not. I think the case of Benor, the
father of the Training and Visit system of extension, has been an
example of the former in Asia and of the latter in Africa.

(e) The consultant

There is quite a significant number of tropical agronomists who do
not belong to any of the above categories. They are the people with
no specialist skills who are mainly found among today's development consultants. If you want to be malicious you could place me

in that category, although I like to think of myself as one who has been searching for an anchoring place without really wanting to find one. I also flatter myself by thinking that that has at least allowed me to write this book.

As we have seen in the previous chapter, the consultants' most important trait is their chameleonic ability – adapting quickly to the exigencies of the times and the rapidly changing fashions of development aid. By the early 2000s many had declared themselves experts in institutional development. That kind of expert was in great demand, because of the retreat of the donors from the untidiness of the field into the more relaxed and familiar environment of offices in the national capitals, where they could lecture the government about the need to reform their institutions and then call in their self-declared experts to drive home the new ideas. I think that the first symptom that an aid-receiving country is making progress is when they start to kick out the foreign institutional experts and take things into their own hands.

11.3.2 What can agronomic research contribute?

Now I must try to answer the question posed at the beginning of these paragraphs: what can agronomy do today to help African agriculture advance? I have argued in this book that agronomy has contributed very little of direct practical value to the African smallholder during the last 40 years or so. Many of us have therefore converted into institutional experts and become involved in development projects in that capacity, because there seemed to be more urgent problems than the availability of agricultural technology which stood in the way of progress. But in the last analysis, it will still be technological innovation which drives progress, once a number of societal ills have been cured and socio-economic constraints removed. Therefore, agronomists must rethink how they can apply their skills in the service of the agricultural development in Africa. I will talk no further about things which have occupied such a prominent place in this book – agronomy's indulgence in endless station research, detached from the reality of the field, the failure of extension to reach the farmer, the phoney development theories and methodologies, the donors' failed attempts at remote control, the linguistic pollution, the dominance of careerism over idealism. In the remainder of this final chapter I will deal mainly with the possible future role of agronomic technology and a little about the research to generate such technology and then close with a brief review of the larger issues, far beyond the competence of agronomy, which in the end will determine agriculture's success or failure.

(a) Colonial agronomy, a technological treasure trove

I will start by moving back in the history of tropical agronomy again, to see whether there is anything in the way of technology which remains valid today. I have found the achievements of early researchers and development workers fascinating, as you will have noticed in several previous chapters, a fascination which has grown with time. That must be a natural thing when old age is an approaching reality, rather than a distant notion. C.T. de Wit once told me that, after his retirement, he wanted to be a historian of agriculture. I was quite surprised by that at the time, but no longer. There is an amazing quantity of research results from the early part of the twentieth century, much of it now forgotten, which is highly relevant for today's tropical agriculture. If this had been properly studied, a lot of recent work would have been unnecessary. I am not going to present an elaborate survey of historical technologies, though, just mention a few particularly intriguing examples.

First, a lot of research was done early in the twentieth century on plant species which Ferwerda, the professor of tropical agriculture in Wageningen, called 'auxiliary crops': cover crops, green manure, shade trees, shrubs for slope protection, etc. Part of that work was done for plantation crops, of course, in the search for species which could keep weeds under control without becoming a nuisance themselves, or fix nitrogen and stimulate soil life, but the colonial departments of agriculture soon realised that smallholders could also benefit from the same technologies. The technical journals of the first half of the twentieth century are replete with long-term studies on the effects of these species on soil properties and crop yields. As a result of this work, the use of auxiliary crops became a matter of routine in plantations, but adoption by smallholder farmers has been almost zero. In the early post-colonial years there was an upsurge of interest in green manure crops which were touted as the cure for many ills, but dropped from sight again when the middle-class farm model of that period turned out to be a dead end. In the cyclical pattern of memory-free development fashions a rebound occurred in the 1990s when a new generation of development workers, convinced that biological farming was the way to go, became interested in green manure crops again. Fashion-conscious agronomists in institutes like IITA also started another round of largely redundant experiments, trotting out information that had been available for close to a century. I am not saying that the renewed interest in green manure crops made no sense. In fact, it was now more appropriate than 60 years earlier, because by now the soil conditions in many areas had definitely deteriorated and farmers were actually now seen sometimes to adopt cover crops themselves, for instance in Bénin Republic where they used *mucuna* for speargrass control. But there was no need at all to carry out an elaborate research programme about green manure all over again. All the information that was needed was already there, it was just a matter of picking it up and going straight to the farm. I do not have to repeat that whole argument again.

I want to give just one example of another line of work done by colonial agronomists, on pest control in the Dutch East Indies (or pre-independence Indonesia), a quite spectacular example, I think. It is about a potentially devastating pest of paddy rice in the drier areas, white stem borers, and how they were controlled in an age when killing them chemically had not yet become the agronomist's first reflex. The white stem borer, the caterpillar of a moth (it is the moths that are white, not the caterpillars), feeds on the shoots of the rice plant by boring through the leaf sheaths and eating the soft tissue inside the whorl. The biology of the white borer had been studied since the 1920s in order to find clues for its control. It was found that the borer population increased rapidly during the monsoon while there was rice in the field, but whether they reached harmful levels depended on where they started. If the population fell enough during the dry period when there was no rice in the field, they would remain below the economic threshold, in spite of their rapid increase during the previous paddy season. *Ergo*, the key to their control was to make sure that the population was as low as possible when the new paddy season started. In the early 1930s the researchers came up with an effective method which involved manipulation of the planting time so as to prevent the moths laying eggs in the new seedling nurseries when they came out of their dormancy with the first rains. Box 11-2 gives some more details of the method. An extensive paper was written about the method by P. van der Goot (1948), which was published posthumously after the

---

Box 11-2. Controlling the white stem borer by crop scheduling

---

The white stem borer was endemic in areas with a pronounced dry season. Each year its population would build up rapidly in the course of the west monsoon (the wet season) when there was abundant paddy and in the dry season the mature larvae would hibernate in the stubble. Early in the next west monsoon enough irrigation water would be released for farmers to prepare their seedling nurseries, but the first rains would also trigger the emergence of the borers from dormancy. The first flight of moths emerging from the stubble with the first rains (the 'stubble flight') would deposit their egg clusters on the young seedlings in the nurseries. Thus, the way to drastically reduce the population was to prevent these moths from finding rice plants to lay their eggs on, and the control method therefore was simply to withhold irrigation water until the stubble flights were over. That was simple, because the colonial government had almost full control over the irrigation system. The method proved to be very effective, although farmers disliked the delay in planting, but that price was considered worth paying.

Second World War, in the agronomic journal *Landbouw*, a rich source of information about pre-war research and extension for smallholder agriculture in the East Indies. Another posthumous paper by the same van der Goot, published in 1951, dealt with an equally devastating pest, the paddy rat, and came up with another interesting crop scheduling technique to keep the rats out of new rice plantings. The common denominator of these biological control methods, as they would be called today, was 'crop scheduling'. It was also its major challenge. The colonial government with its mild regimentation measures, succeeded in imposing prescribed planting times, but after independence such measures were dropped because they were tarnished by their colonial origin. And the incorruptible state apparatus needed to implement them was rapidly substituted anyway by one which sets less store by such things.

In due time these environment-friendly methods were forgotten when the appearance of powerful insecticides made them obsolete. Eventually it was realised that they caused an inordinate amount of collateral damage by disturbing the ecological balance, so the work on biological control measures was resumed. Most researchers were unaware of all the work that had been done and which lay hidden in the old journals, much of it written in incomprehensible languages like Dutch and German. I think it is about time that somebody or some institute takes up the task of bringing the material buried in journals and scientific reports from colonial times to the surface again. It would cost an infinitesimal sum, compared with the vast amounts of money which have been spent on ill-conceived ideas in the last 40 years. I am not sure whether 10 or 15 years ago a proposal to do that would have stood a chance to be funded, unless the initiative had come directly from a developing country, but today there appears to be more interest in this kind of work as long as it is undertaken jointly with an institution in such a country. It would be an exciting and very useful undertaking indeed.

(b) Post-independence technology

So much for early technologies, which have fallen in disuse, become obsolete or have simply been forgotten, but which may have unexpected relevance today or at some later time. But there are also more recent, equally interesting technologies, which lie waiting to be used by a new generation of farmers in search of more profitable ways to run their farms. I have gone through several of those in Chapter 7, and tried to explain why their adoption has so far been disappointing. There are many more technologies which scientists have called 'promising', usually meaning that they did well at their research stations, and sometimes in their on-farm tests, but which farmers have not chosen to use I will not go into detail about them, just describe some of them very briefly in Box 11-3. If you are interested, it is easy to find more information.

So there does not seem to be a shortage of technologies, just a lack of adoption. If there is one thing the poor adoption record of all those good-looking technologies shows, it is that the real farmer, struggling along in his antiquated ways, is a very different person from the imaginary farmer in the mind of the scientists. Until such time as the agronomists become aware of this, they will just go around in circles, repeat what has been done several times before and invent new technologies for which there is no client either. That was the original observation which triggered the rise of the FSR movement, so we have now come full circle. That is a sobering comment on our efforts to bring the real farmer closer to the ideal one. But we should not give up, only try to learn from our failures and determine not to repeat them all over again.

Box 11-3. Some more promising technologies

I will just randomly list another few more or less promising technologies here, with a short explanation for each of them.

Hybrid Oil Palms for West Africa's 'Agroforesters'
Remember the natural oil palm stands in food crop fields in south-western Nigeria and Bénin Republic. By gradually replacing the tall, low yielding palms by modern dwarfish hybrids oil yield may be tripled at the cost of buying seedlings. Farmers must be aware, however, that letting the hybrid palms regenerate from dropped seeds, as is normally done, will give disastrous results.

Taungya for Reforestation Areas
Taungya is a land use system from Thailand, whereby farmers are allowed to grow crops for a few years between tree seedlings which have been planted for reforestation. The trees benefit more from maintenance by the farmers than they suffer from competition by the crops they plant between them. This system has been practiced in forest reserves in Nigeria but has largely been abandoned.

The Heifer International Model
Heifer International (HI) is an American charity which has been very successful in promoting small-scale dairy farming in developing countries. They provide heifers (first calf animals) to farmers who have built a proper permanent shed and grow sufficient fodder crops for stable feeding, according to a management model developed by HI. The farmer repays in kind by passing on the animal's first female calf to another beneficiary.

Direct Seeding of Rice
Transplanting rice has been a major innovation in African wet rice growing, but more recently there has been a return to direct seeding in Asia. It is less work and allows three annual crops in some areas, but it can only be done if the paddy fields are perfectly levelled and farmers apply herbicides to control weeds.

Peri-Urban Agri/Horticulture
There is a lot of simple technology available for intensive vegetable production which can be applied by small-scale producers around or inside urban centres. It is especially interesting when irrigation water is available in the dry season from streams of wells.

Composting, Use of Town Refuse
Good recipes for compost making with various materials abound in the literature. The use of town refuse can also be developed much more than it is at present.

(c) Technology repertories and users' guides

If there is no shortage of good production technologies, both of recent and remote origin, there is definitely a problem of documentation and information. It is amazing how little many of today's tropical agronomists know about the historical record of technology development, even in their own institutes, and which of those technologies remain valid today. The gurus of FSR knew this, of course, and they therefore made the preparation of an inventory of 'available technologies' part of their methodology. In retrospect, that was naive. Such inventories should not be taken lightly and cannot be done simply as part of a diagnostic study. They require a major effort, particularly getting at the older results, which will often turn out to be the most interesting. In the last 12 years I have carried out many evaluations of donor-sponsored research and extension programmes all over Africa and beyond, and one recurrent theme has been precisely that of available technologies. I have not made myself popular by requesting, a little maliciously perhaps, for summaries of the many technologies which the scientists claimed to be available, but which they said were not adopted for all kinds of reasons, other than that they were trivial or even non-existent. When pushing hard enough you almost invariably find that there is very little 'on the shelf', apart from some crop varieties, which many times turn out to have been grown only at the research station. Or the scientists are blissfully ignorant about what even their own institute has produced in years past. I am not implying that nobody has shared that concern, though. Willem Heemskerk of the Tropical Institute (KIT), for one, set in motion institute-wide inventories of available technologies when he worked in Zambia and Tanzania, and actually published them, but his example is a rare one.

On a positive note, I think that any research institute should keep an up-to-date record or repository of all the technologies they have 'on the shelf', with all the details needed by on-farm researchers and extension agents to assess their suitability and subject them to on-farm tests. And the repository should not just contain technologies generated by the institute itself but also include relevant ones from elsewhere. I have started to call for that kind of documentation when at IITA and noticed how difficult it was even there to get at the details you needed to test technology on-farm. It was all there of course, but not in an accessible form for potential users. I thought at the time that it would be an exciting idea to bring together all those technologies the institute had generated and publish them in the form of a technological sourcebook or users' guide. The then Director General, Ermond Hartmans, agreed and told the head of documentation, Jack Keyser, to set the work in motion, but soon after the latter left under a cloud, and that was the end of it. I kept trying, with too little energy perhaps, and found out how difficult it was to get scientists to put together a user guide even for their own technologies – because, you know, it was not really ready yet and needed a few more years of work (or something to that effect). Or maybe they had already published their stuff in a scientific journal and had moved on to something else. There should be a firm institutional rule which says that the work is not finished until the technology has been documented in a form which makes the technology directly applicable by the on-farm researchers or extension agents. In the US land grant universities that was accomplished by putting research and extension agronomists together in the same institution and in IITA to some extent by having FSR workers work alongside the research scientists. But the radius of influence of a few FSR scientists was small, so in order to reach the wider research and development community which IITA was expected to serve, an IITA technology sourcebook would have been a real asset.

It never happened. And the FSR workers in the national institutes who could have made use of technology guides most urgently, in fact rarely seemed to feel the need. Many of them just carried on 'testing' very ordinary and often trivial technologies such as line planting, while everything was done manually, timely weeding, as if farmers did not know that was important fertiliser, which farmers would have used anyway if it had been profitable. I have gone through all that several times. In order to have significant impact on smallholder farming, genuinely innovative technology would have been needed. It is the life blood of dynamic agriculture and once that is realised, the demand for precise information on new technologies will rise. The people who work with the farmers should tell the research scientists to stop their ritual dance of inconsequential and mostly redundant research, properly document the technologies they and their predecessors have already produced and start a really meaningful research programme which addresses real issues and builds on the achievements of the past.

(d) What kind of research is needed?

The question is, with all those technologies already available, is there a real need for new research? Look at strategic or innovative research, called 'upstream' by FSR's word magician Michael Collinson. I mean the kind of research that develops new technologies such as alley cropping or new crop varieties, rather than just testing them. That cannot be casually dismissed, because it is one of the sources from where real innovation springs. But how much capacity is in a poor African country for doing really innovative research and how much of it can a country afford? It was precisely in response to that kind of question that the Ford and Rockefeller Foundations started the network of international agricultural centres in the 1970s, which were going to do the necessary things which the countries themselves could not. Nevertheless, however poor a country, it will resent being entirely dependent on foreign institutions for strategic research. Several institutes in West Africa have chosen the middle road, by associating themselves closely with international research centres and making some significant contributions of their own in that context, especially in plant breeding. But how many institutes would IITA have been able to cooperate with closely, without jeopardising its own research? I think a developing country's research leadership should not devote a major part of its scarce human and financial resources to strategic research, which is unlikely to produce much of value, as the record of the last four decades clearly shows. It is better to obtain results and materials from better endowed institutes elsewhere and work from there. In any case, the bulk of agricultural research in Africa has been of the applied type, dealing with crop management, fertiliser, pest and weed control, and things of that sort, and even the products of such research have rarely found their way to the farmers. That is the defect which FSR was going to fix, so far with rather disappointing results as we have seen.

Does that mean that the whole agricultural research circus might as well be scrapped and that development workers should just scout around for technology generated by the few, mainly international institutes which have the capacity and can afford the cost to carry out significant research? Perhaps a poor country should concentrate on development rather than spending scarce resources on research which donors are no longer willing to fund. That is an appealing thought (except for the research workers themselves, probably), except that somebody will still have to do the scouting for relevant technologies which have been generated elsewhere, such as 'auxiliary crops' or novel tools, and put them through some tests before passing them on to the farmers. In other words, there will still be a need for applied research facilities, but much leaner and very different from what is there today, with a strong accent on modern information technology. The research facility should also include a basic capacity in variety selection, which can at least test crop varieties obtained from elsewhere under the local conditions.

But, most importantly, there has to be capacity to go out to the farm and work with the farmers to test innovative technology, of whatever origin. That is what used to be called Farming Systems or On-Farm (Client-Oriented, Participatory) Research.

(e) What about FSR?

So, there will still be a role for FSR in the future? Surely, the movement's professed attitude towards the farmers should play a prominent role in the future, even though we have made many mistakes in its practical implementation. I have brought many of them to the fore in this book and I like to re-emphasise some of them again. One sad mistake of FSR has been to forget that agricultural research, and FSR in particular, should be the hand-maiden of development. Another one is that we got carried away with our data collection, pretending that a lot of diagnosis would help technology adoption, which it did not – quite the contrary, it was a waste of time. And finally, we let ourselves be fooled by the excellent performance of our technologies in farmers' fields as long as we were present, and did not look back at what happened after we had left. If we had done that sincerely, we would have seen that all was not right and learned some lessons about the reasons for our technologies failure. In spite of all this, the attitude to applied research advocated by FSR is the best we have and should be part of any future strategy to finally set agricultural development in motion.

FSR has attracted both visionaries and quacks and some of the latter built mini-empires in their institutes, financed by hopeful donors who in the end lost confidence, and let the empires crumble.

Again, setting up a separate FSR team has turned out to be a bad idea which should not be repeated. But if the FSR ideas themselves remain valid they should become an integral part of the philosophy of agricultural research, carried by a new generation of workers whose primary concern is the interest of the African smallholder or his successor, rather than the favour of the development financiers.

(f) Privatised research?

But how about the institutional side of agricultural research? Much of what I have said about the bankruptcy of large public extension organisations in Africa also applies to the research institutions. As long as they were kept on their feet with a lot of foreign money their failure was screened off from the public view, but now that the donors seem to have given up they are starting to disintegrate as well. One good thing about this is that it will force the long overdue reflection on the future of national agricultural research, what remains necessary and what can be dispensed with.

A country should not deprive itself entirely of a capacity for strategic research, so there will remain room for some lean and smart, publicly funded, units to do that, possibly incorporated into the university system. Furthermore, there have to be some facilities for on-station screening of essential technology including crop varieties. Most emphasis should be on OFR, however, and the challenge will be to choose the most effective institutional set-up to carry it out. The large and expensive networks of research stations, dating from colonial times or set up with donor funds afterwards, have failed miserably to play that role, nor can they be maintained in the future from their countries' meagre means anyway. Their generally poor achievement record also does not provide any valid reasons why they should.

A reform process must be set in motion which decides what should be the government's task and what is better left to the private sector, whether private research bureaus could take over some of the research traditionally carried out by public institutions and how their services would be paid for. That will involve tough decisions which have been called for in the past, but have never been taken. Apart from some ornamental changes I have seen few signs that a basic reform process is actually taking

place in Africa's public research systems, or that there is even
thinking and discussion going on about it – with most of the
donor money gone, the institutes are gradually fading away.
Meanwhile, new initiatives are beginning to take shape outside
the old research establishment, such as the creation of private
research entities, the first hopeful development in decades. Let us
hope the donors will be wise this time and abstain from direct
support beyond what is needed to get them over the initial hur-
dles. Good private research will best be helped by research contracts which they will be able to acquire once they have shown to
be competent and can deliver the goods.

(g) Is there need for technical assistance?

The creation of private entities for applied research may bring
new inspiration and dynamism where there has mainly been
paralysis and decay. But for the time being, these fledgling institutions are still as dependent on foreign money and ideas as were
the large research monoliths of the past. The ISNAR was
intended to help national institutes reformulate their research
strategies, but its achievements have been minimal, perhaps
because they have been too much influenced by the World Bank's
nostrums and lost their objectivity in the process. In any case,
they were downgraded a few years ago into a branch of the
International Food Policy Research Institute (IFPRI), and have
now reached the ultimate state of insignificance, unlikely to contribute to the search for new institutional models for national
research.
   One could of course adopt a laissez-faire attitude and let things
sort themselves out, with intellectual starvation and eventually
brain death of the research community as the most likely outcome.
That might be consistent with neo-liberal thinking, but most
donors would not like to see things going that far. They should
therefore be ready to offer some help to a new generation of
researchers who want to establish themselves in private research
units and can bring youthful enthusiasm to the task. Their survival
will depend in the first place on a steady flow of research contracts
from a variety of 'clients': their own government, donors, NGOs
and in the future hopefully farmer organisations. And second, they
need to establish an intellectual lifeline by somehow linking up with
similar groups and getting access to information which is not available

at their level. As long as a research entity is supported by a donor that is not difficult, the external link is usually provided by a research institution from the donor country which has probably set up the whole operation. But by the time the unit is weaned from that kind of support, another source is needed from which to feed itself with ideas and materials.

The danger of intellectual starvation has lurked around the corner ever since donor enthusiasm to support agricultural research started to cool in the early 1990s, when privatisation of research was still far from people's minds. In the early 1990s Georg Weber, Mark Versteeg and I at IITA concluded that the institute should step in and provide serious technical support and on-the-job training to national research institutions, rather than just using them as multilocational testing sites. We therefore proposed to set up what we called SPARCs, for 'Support Groups for Adaptive Research Co-operation' (a clumsy name I admit, but the acronym sounds OK), as an alternative for ISNAR, where we thought heads were too far up in the clouds. We prepared an extensive proposal to start with one SPARC in West Africa, to be extended to other parts of the continent if successful, but IITA was thinking in a different, 'upstream' direction and wanted to go into biotechnology and genetic engineering instead. So it never happened, which was one of the reasons why Weber and I left the institute. Today one or more SPARC units would be even more relevant than they were 15 years ago, to provide the kind of technical backing which a new generation of researchers, no longer working in large research establishments and spoon-fed by donors, will need and which they will find it increasingly hard to get. There are many exciting things the SPARCs could do, some of which I put into Box 11-4. As distinct from earlier forms of technical assistance, the SPARCs should maintain a certain distance from the research units they support and only intervene at the level and with a frequency desired by those units. And another difference would be that they would not carry out research of their own. The SPARC scientists should already have satisfied their own appetite for active research, built up a solid publication record and be prepared to continue their career in the service of others. And their most important ambition should be to make sure that the fruits of research finally reach those they are meant for: the African smallholder farmer.

> *Box 11-4.* SPARCs: technical support at a distance
>
> SPARCs were meant to promote regional cooperation in applied research with international technical assistance 'at a distance'. They would emphasise good OFR on a solid technology basis and would carry out the following functions:
>
> 1. Promote good participatory OFR methods
> 2. Provide technical support upon request from individual teams
> 3. Assemble technology source books, databases and decision support systems; examples are FAO's Ecocrop (available on-line at http://ecocrop.fao.org/) for choosing suitable crop species for a particular environment and LEXSYS for grain legumes
> 4. Help national teams to prepare technology repertoires for their own area
> 5. Facilitate experience sharing among the teams through exchange and advisory visits and secondment
> 6. Hold regular thematic workshops and meetings

11.4 Are we missing essential elements?

11.4.1 Societal ills

When one observes the trouble many African countries are in at the start of the 2000s, to talk about privatised extension and research must sound not a little frivolous. The continent is in continuous turmoil and even countries which were once shining examples of stability may still spin out of control one day. In most of Africa production has stagnated and the hope that a new class of farmers with a more entrepreneurial attitude would emerge has not materialised. Instead, rural youth have flocked into the urban centres in a mostly vain search for employment, leaving their parents and their less adventurous siblings behind. Instead of a modern enterprise, farming has become the occupation of the losers, those who see no other option than to struggle on and scrape a living from an increasingly impoverished soil. That is a gloomy picture and many international aid donors have tacitly given up hope that farming as it is can ever be pulled out of stagnation by government-implemented and donor-funded development projects.

The emergence of a modern agricultural sector will not be possible unless some very serious ailments of African societies, which are beyond the control of the sector itself, are remedied. An experienced development worker who used to be a friend once said that nothing would be more important for the development of Africa than the Africans stopping to cut each other's throat. When some time later I reminded him of this politically incorrect one-liner (which I very much agreed with) he snapped that he could never have talked in such terms. That typifies the dilemma of people who are sympathetic to Africa, but in a critical way – while being keenly aware of Africa's ills and prepared to say so, they do not easily tolerate others to quote them if they sense even a whiff of negativism on the other side. Africa's societal ills are many indeed, as most people have become aware in the last 10 years, and their effects on agricultural production are incomparably more serious than all the things I have described in this book. This is not the place to elaborate on them, but I would still like to say a few words about the way some of them have affected agriculture.

The most important ill, in particular in southern and eastern Africa, has been the AIDS pandemic, which is wiping out an important part of the active population, including those involved in agriculture. Again, I will not further elaborate here, although the effect is dramatic, as for example Louise Fresco, until recently FAO's Deputy Director General, foresaw a long time ago. I will just mention a few relatively smaller ills, which will take on more weight once the larger ones are gone, as we hope they will one day.

Perhaps the societal ill with the next most damaging effect on African societies in general and on agriculture in particular is the lack of genuine, disinterested and dedicated leadership. Without such leadership, from the top all the way down to the villages, there will be no end to corruption and no effective farmer organisations and cooperatives can be created. And if they are created willy-nilly, they will succumb in their turn to embezzlement by parasitic officials, whether from government or from their own ranks. School teachers and doctors will continue to shun the rural areas where their presence and initiatives should have been the focal points for development, and the increasingly inhospitable rural areas will continue to lose the young generations who should have been their society's lifeblood. There is little or nothing donors and foreigners can do about all that, however idealistic they may be. The remedies must come from the countries' own people and from them alone. The only thing donors can and should do is give a helping hand if they are requested to, and only then. That much we should have learned from five decades of so-called development aid: no amount of aid money and foreign technical and institutional expertise can substitute for the locals' own initiative, whether in agriculture or anywhere else. As long as local leaders, politicians, scientists and citizens let themselves be pushed around by donors and consultants instead of taking things in their own hands, there can never be progress.

11.4.2 A new breed of farmer after all?

Agriculturists cannot solve the larger ills of Africa's societies, but they can make their own relatively modest contributions. I have described and sometimes explained in this book why agriculture has stagnated the way it has, and how agronomists can sometimes help to pull it out of its quagmire. I do not have to repeat that here. But one thing that has been missing is a clear vision about what Africa's future farmer will or should look like. Implicit in the FSR approach, of which I have been a long-time adherent, has been the assumption that the new farmer will be a direct descendant of the traditional subsistence farmer. But doubt has crept in over the years and today I wonder whether the future will not have to be with a new type of farmer after all, one who has seen the world and is open to the challenges and opportunities of modern times. How much simpler the life of an agronomist would be if the African farmer were a middle class farmer in a reasonably well-organised society, who makes rational choices based on scientific evidence, supported by efficient services, whether provided by government, by the farmers' own organisations or by the private sector. His choice of the crops to grow would to a large extent reflect the potential of the environment, with the forest areas specialising in tree crops and the savannah areas growing mainly annual crops, perhaps with small tree orchards in suitable places. Sound soil management practices would protect the soil against Africa's harsh climates in both zones and fertiliser and manure would be applied in a judicious way to maximise their effect. Farming in the forest zone would look very different from what it is today, with perennial crops such as cocoa and oil palm taking place of choice in each farm as the main source of income. Food crops, grown in small intensively managed fields, would be produced mainly for the farmers' own consumption in order not to be too dependent on purchased food. It is not all that difficult to put together convincing and profitable models for such farms, as was done by Nye and Greenland and by René Dumont as early as the 1950s and 1960s and by several others since. Many projects have actually tried to introduce them, all of them failures in the end, because the farms were set up as small artificial islands, which would rapidly be swallowed by the surrounding ocean once the protective barriers were gone. And the peasants who operated them were defenceless against the onslaught.

Perhaps after all the idea of Richard Lowe (1986) will prove valid, that new kinds of farmers will eventually establish themselves, coming from outside agriculture with an entrepreneurial attitude and management skills learned in different trades but with enough affinity with agriculture to absorb sound farming practices. They could form a future vanguard in the modernisation of agriculture and pull the more dynamic of the traditional farmers along. This is not at all an unlikely scenario, considering that some of it is already happening with the intensive vegetable, fruit, poultry and dairy producers around large urban centres.

In the long run this may yet, to quote Smyth and Montgomery again:

set a new standard of living in the rural areas, a general improvement in the lot of the farmer and [a] new regard, by the population at large, of farming as an enviable occupation, to be sought after by young educated school leavers.

Appendix 1. C.T. de Wit's Analysis of Plant Competition

A1.1 Two species competing for the same space

Consider a plot of unit area, say 1 m² or 1 ha or whatever, where two species will be grown as a mixture. The plot is divided into cells of equal size m, so there are altogether m^{-1} cells per unit area. In each cell one seed is planted of one of the two species. Remember this cell size m and note the auxiliary role it plays in the theory. It will return several times later, until it is finally eliminated in the yield–density function that de Wit derived from his theory.

The numbers of seeds of the two species planted in unit area are Z_1 and Z_2, so the areas available to each species are proportional to their seed rates:

$$A_1 : A_2 = mZ_1 : mZ_2 = Z_1 : Z_2$$

$$\text{and } A_1 + A_2 = (Z_1 + Z_2)m = 1 \text{ (i.e. unit area)}$$

It looks as if the cell size m has already disappeared from the equations, but it is still implicit, because the seed rates add up to the total number of cells m^{-1}.

Assume that during the growing season the two species compete for the same space and that the more successful one 'crowds into' the other species' area. The competing ability of a species in a mixture is represented by its 'crowding coefficient' b, and the spaces eventually occupied by them are assumed proportional to their seed rate multiplied by their crowding coefficient[1]:

$$A_1 : A_2 = b_1 Z_1 : b_2 Z_2$$

and, since the species still compete ('crowd') for the same space: $A_1 + A_2 = 1$.

In the above equation b_1 and b_2 may be multiplied by an arbitrary constant, hence the competition relation is completely determined by the *relative* crowding coefficient of species 1 relative to

---

[1] The crowding coefficients are analogous to the activity coefficients which determine the partial vapour pressures of the components of a mixture of liquids.

species 2, i.e. by $k_{12} = b_1 / b_2$. If we further define the *relative* seed rates as

$$z_1 = \frac{Z_1}{Z_1 + Z_2} \text{ and } z_2 = \frac{Z_2}{Z_1 + Z_2} \text{ , then we get}$$

$$A_1 : A_2 = k_{12} z_1 : z_2 \tag{1}$$

and, since the yield of each species is, by de Wit's definition, the product of the area A it eventually acquires and its yield in monoculture M, the yields in mixture equal:

$$Y_1 = A_1 M_1 \text{ and } Y_2 = A_2 M_2 \tag{2}$$

From (1) and (2) we obtain expressions for the yields of the two species as a function of their relative seed rates:

$$Y_1 = \frac{k_{12} z_1}{k_{12} z_1 + z_2} M_1 \text{ and } Y_2 = \frac{z_2}{k_{12} z_1 + z_2} M_2 \tag{3a}$$

and, since $z_2 = 1 - z_1$ and $k_{12} \times k_{21} = 1$ they may also be written as:

$$Y_1 = \frac{k_{12} z_1}{(k_{12} - 1) z_1 + 1} M_1 \text{ and } Y_2 = \frac{k_{21} z_2}{(k_{21} - 1) z_2 + 1} M_2 \tag{3b}$$

that is the yield of each species is a function of its own relative seed rate and monocrop yield and its crowding coefficient relative to that of the other species. I have kept de Wit's original notations which does not result in very lucid equations, but do not get discouraged by that and try to see through them. The rewards for the effort will come later.

In the 1950s van Dobben had carried out a large number of field trials with barley and oats under widely different conditions, comparing the yields of the monocrops and three different mixtures. In all cases the same 'normal' seed rate was used in all treatments; in other words, the total number of cells m^{-1} was always the same. De Wit used van Dobben's data to test his competition model. The parameters of (3) (k_{12}, M_1 and M_2) were estimated from the data by trial and error until yield curves were obtained, which were as close as possible to the measured data. A typical example is shown in Figure A1-1. The model was amazingly successful in describing all the experiments, and showed that barley and oats competed strictly for the same space. In most cases oats had the highest monocrop yield but was the weaker competitor, as in the figure. Only under low pH was its relative crowding higher than that of barley.

Yield (grains /10^6)

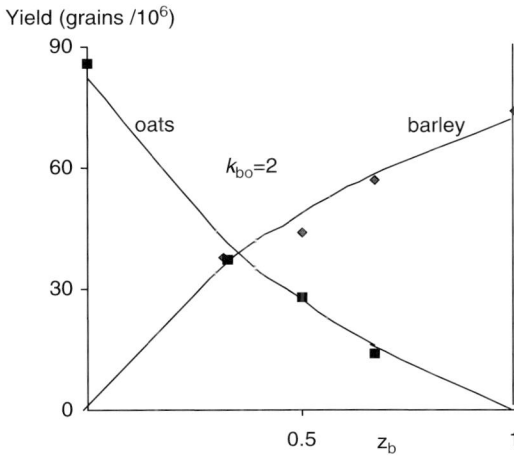

Figure A1-1. Yields in one of van Dobben's oats–barley mixtures plotted against relative seed rate

That is interesting, but a practical person would want to know why farmers would grow the crops in mixture, as they often did at the time, if there was no yield advantage. De Wit's analysis showed that the yields of oats and barley were proportional to the space they occupied, so it made no difference whether they were grown in a mixture or in separate plots with the same total area. There must have been other reasons why they did it. I will discuss those reasons a little further because possible advantages of intercropping were about to become a hot topic among tropical agronomists.

I think the most likely reason why European farmers used to grow barley and oats together was simply that the mixture made an excellent animal feed, so why grow them separately and then go to the trouble of mixing them again? Another reason was that barley was less likely to lodge when it could lean against the stronger oats straw. Neither of these had anything to do with the factors which the model was built to capture.[2] But there were two things that did. The first was the effect of pH. When pH was very low, 4 or less, barley yield decreased steeply, while that of oats did not. The former's relative crowding coefficient also became smaller than 1, so that oats was now the stronger competitor for space. Now suppose you

---

[2] If the monocrop barley had suffered badly from lodging, the mixture should have shown a yield advantage, though, but not because of competition for space.

have a patchy field with large differences in pH. If you plant barley
alone, the crop would also become very patchy with poor barley (and
a lot of weeds), where pH was low. If on the other hand you had
planted the mixture, the oats would take over where the barley did
poorly. So it looks as if the mixture would produce more than two
separate monocrops with the same total area under those conditions. Can that be quantified? De Wit did not do that but I will try
– it is instructive to milk those relatively simple equations a little
further. Consider a field where 50% of the area has a pH below 4,
in randomly scattered locations; so it is not known in advance
where they are, and a barley–oats mixture is sown uniformly over
the entire field. In the spots with low pH, both the yield of barley
and its relative crowding coefficient would be reduced, while oats
yield is not affected. Their yields can be calculated by applying
equations (3) twice for each species:

$$Y_b = \frac{1}{2}\frac{k_{bo}z_b}{k_{bo}z_b + z_o} M_b + \frac{1}{2}\frac{k'_{bo}z_b}{k'_{bo}z_b + z_o} M'_b \text{ and}$$

$$Y_o = \frac{1}{2}\frac{z_o}{k_{bo}z_b + z_o} M_o + \frac{1}{2}\frac{z_o}{k'_{bo}z_b + z_o} M'_o \qquad (4)$$

where the accents indicate parameter values under low pH.

Assume that oats is not affected at all by low pH and that its sole
crop yield is always 120×10^6 grains/ha, the average in van Dobben's
trials. Table A1-1 shows how much various mixtures would yield for
different pH conditions. The first case is for a uniform field with
pH > 4 everywhere and the others for half the area having pH > 4
and half with a pH which is lower (cases 2 and 3) or much lower
than 4 (case 4), randomly distributed over the field.

The yields were calculated with equations (4). Most interesting
are the lines saying 'sole area needed'. That is the sole crop area,
which would be needed to get the same yield as the mixture, which
is a measure of the advantage of growing the crop in a mixture.
Even when the sole crop yield of barley and its relative crowding
coefficient are halved because of low pH, the advantage is quite
small, only around 3%, but it increases to about 10% when these
parameters are only one-fourth of their original values. We will
see later on that advantages of mixing can be expected when two
species do not compete for exactly the same space, but even when
they do, as in this example, some yield advantage can still result
from variable field conditions, provided their effect on one of the
species is strong compared to that on the other.

Table A1-1. Calculated yields of barley and oats (nr of grains/ha $\times 10^6$) and sole area needed for the same yields; for different mixtures and pH levels

		$z_b : z_o$					
		1:2		1:1		2:1	
		Yields (nr. of grains $\times 10^6$)					
k'_{bo}	M'_b	Barley	Oats	Barley	Oats	Barley	Oats
Case 1, uniform field, pH > 4							
1.5	80	34.0	69.0	48.0	48.0	60.2	29.7
Sole area needed		**1.0**		**1.0**		**1.0**	
Case 2, patchy field, medium low pH							
0.75	40	22.4	78.3	32.6	58.3	42.2	38.6
Sole area needed		**1.03**		**1.03**		**1.02**	
Case 3, patchy field, low pH							
0.6	32	20.6	80.8	30.0	61.5	38.9	41.9
Sole area needed		**1.04**		**1.05**		**1.04**	
Case 4, patchy field, very low pH							
0.3	20	8.3	105.4	14.1	94.6	21.5	79.7
Sole area needed		**1.05**		**1.08**		**1.11**	

But there was something else the matter with the barley–oats mixture. Van Dobben's data showed that the 1,000-grain weight of the oats in the mixture was up to 10% higher than in sole oats. We have seen that under uniform conditions there was no yield advantage of the mixture as far as *grain numbers* are concerned, so you might as well plant the crops separately. But the increased *seed weight* is a real bonus from the mixture. The reason why it occurs is that the barley matures before the oats and stops competing for 'space', so that the oats has more space available at the end of the growing season. And since the only thing which still retains some capacity for change is seed weight, that is where the gain is found. This is an example of two species 'crowding for partly the same space', which we will look at more closely later on.

The competition equations used so far are based on equally sized cells occupied by a single seed. That is fine for barley and oats. But how do you handle species of very different plant size, like maize and groundnuts, which have different 'normal' monocrop densities? Define the cell size as the area occupied by one seed or plant of the smaller species, then one plant of the larger species occupies more than one cell, say a multiple c of a cell. This fraction is of course the ratio of the 'normal' monocrop densities of the large and the small species. If we continue to reckon the species' densities in numbers of plants in a mixture, Z_1 and Z_2, the areas they occupy are proportional to Z_1 and cZ_2 and their relative planting densities are:

$$z_1 = \frac{Z_1}{Z_1 + cZ_2} \text{ and } z_2 = \frac{cZ_2}{Z_1 + cZ_2}, \text{ with } Z_1 + cZ_2 = m^{-1} \quad (5)$$

Otherwise, the relationships remain exactly the same as before.

Another way of handling this, and one which has been more common in agronomic research, is to define a mixture component's relative density directly as a fraction of its sole crop density. If the two fractions add up to 1 that is of course the same thing.

A1.2 Crowding for space in monoculture, a yield–density function

Now de Wit took an elegant step from crop mixtures to sole crop densities, helped in no small measure by van Dobben's very meticulous field studies with barley and oats. When the pH became really low,[3] close to 3, the growth of barley was strongly depressed and the crowding coefficient of oats relative to barley increased up to 20. At that point barley yield was so negligible that the competition experiment 'degenerated into a spacing experiment for oats'. So, de Wit argued, there must be a degenerate form of the competition equations (3) which could be used to describe the results of spacing trials. That form was found by treating the effect of plant spacing in monocrops as crowding for space, regarding some of the cells as empty instead of being occupied by a different species. The yield of the monocrop, say oats, as a function of its relative seed rate then has the same form as (3b):

---

[3] Very low pH had been artificially induced by the continuous use of ammonium sulphate fertiliser in a long-term study by the extension service.

$$Y = \frac{k_{oe}z}{(k_{oe}-1)z+1} M \qquad (6)$$

where the relative crowding coefficient k_{oe} is that between oats and 'empty cells' and the relative seed rate z was defined earlier as a fraction of some 'normal' or reference rate m^{-1}. Expression (6) is not really suitable, we want to express yield as a function of density, of the form $Y = f(d)$, whereby density d can vary between zero and infinity. So we are going to try to replace the relative seed rate z by the absolute seed rate or planting density. Call the absolute seed rate s^{-1}, in analogy with m^{-1}. That means that the area allocated to one seed is s. Now take s larger than m, then z, the seed rate relative to the 'normal' rate equals $z = ms^{-1}$, and from expression (6) $Y = \frac{k_{oe}ms^{-1}}{(k_{oe}-1)ms^{-1}+1} M_m$ where M_m is the yield when all the m^{-1} cells are occupied. Call $(k_{oe}-1)m = \beta$ then:

$$Y = \frac{\beta+m}{\beta+s} M_m \qquad (7)$$

Since m was chosen arbitrarily, we may also choose a different cell size as reference, say m', somewhere between m and s and express Y as a function of the corresponding yield $M_{m'}$ (which is the yield when all cells of size m' are occupied) and relative crowding coefficient k', then:

$$Y = \frac{\beta'+m'}{\beta'+s} M_{m'} \text{ with } \beta' = (k'_{oe}-1)m' \qquad (8)$$

Finally, $M_{m'}$ itself can be written as a function of M_m, by substituting m' for s in expression (8):

$$M_{m'} = \frac{\beta+m}{\beta+m'} M_m \qquad (9)$$

And, combining (7)–(9) and rearranging: $\frac{\beta+m'}{\beta+s} = \frac{\beta'+m'}{\beta'+s}$, therefore, $\beta' = \beta$.

This means that (6) applies irrespective of the size of the reference cell. We take m to approach zero and call M_m, the limiting yield at very high density Ω, then we get from (7):

$$Y = \frac{\beta}{\beta+s} \Omega \qquad (10)$$

where s^{-1} (the inverse of area per plant) is planting density. You may not recognise this immediately, but it is the well-known rectangular hyperbola, which has often been found to fit yield–density relations well[4] and was derived here from very simple assumptions. It may be shown that $\beta\Omega$ is the weight per plant at very low density, hence β^{-1} is the number of plants which would be needed to obtain maximum yield if all plants had their maximum possible size.

A1.3 Crowding for partly the same space

In the tropics and especially in Africa, it is common to grow together crops with very different growth habits. Before de Wit's work little quantitative research had been done on such mixtures, although some observers had suspected that there might be a yield advantage. De Wit argued that for mixed stands to show a yield advantage, the species in the mixture should only partially compete for the same space. That may happen if one of the species competes less for an important growth substrate than the other does. In a cereal–legume mixture, for instance, the legume would compete less for nitrogen, because it can fix its own. Or if the periods of maximum growth do not occur at the same time, the early variety surrenders part of its space to the later one. Such mixtures can be analysed in a similar way as those competing for the same space. Consider again the expressions (3), for the yield of two species grown in mixture. What happens when they do not fully compete for the same space? In the most extreme case there is no competition for space at all, and the functions (3) become two independent spacing functions. For a crop mixture where the crops use partly the same space it is 'most plausible' (de Wit's words) that the expressions also hold:

$$Y_1 = \frac{k_{1\,(2e)}z_1}{(k_{1\,(2e)}-1)z_1+1}\, M_1 \text{ and } Y_2 = \frac{k_{2\,(1e)}z_2}{(k_{2\,(1e)}-1)z_2+1}\, M_2 \qquad (11)$$

where $k_{1(2e)}$ stands for the crowding coefficient of species 1 relative to species 2 and 'empty space'. The assumption that competition is

---

[4] The usual form of the hyperbolic yield-density function is: $\frac{1}{Y} = a + bN$, where N is the plant density. If you convert (9) to this form, the interpretation of the parameters a and b will be clear.

only partly for the same space implies that the sum of spaces which are virtually occupied by the two species is larger than 1. According to their definition by equation (2), those spaces equal $A_1 = \dfrac{Y_1}{M_1}$ and $A_2 = \dfrac{Y_2}{M_2}$, hence, $\dfrac{k_{1(2e)}z_1}{(k_{1(2e)}-1)z_1+1} + \dfrac{k_{2(1e)}z_2}{(k_{2(1e)}-1)z_2+1} > 1$. It can be shown that this implies that $k_{1(2e)} \times k_{2(1e)} > 1$. The parameters $k_{1(2e)}$ and $k_{2(1e)}$ are estimated by fitting the expressions (11) to the yields of the two crops at different relative densities.

This analysis lacks the transparency of the foregoing, nor has it been used much by other authors, at least not by agronomists. The more common approach has been to simply calculate the Relative Yield Total (RYT) at each relative density, that is the sum of the virtual areas occupied by the mixture components: $RYT = \dfrac{Y_1}{M_1} + \dfrac{Y_2}{M_2}$. If the species compete for the same space, the part of the space which is ceded by one of them is balanced exactly by the extra space acquired by the other, so their relative yields have to add up to 1. If, however, the species do not compete exactly for the same space, *RYT* will be larger than 1. In Chapter 3, I have given an example of analysing a mixed cropping experiment with maize and groundnuts in terms of the RYT.

A1.4 Competition in natural plant populations

Farmers harvest their crops and then plant them again in the next season, so competition in mixed species crops covers only one cycle. In natural plant communities and in permanent pastures, however, things keep changing, because after each cycle some or all species reseed themselves or, if they are perennials, they keep competing for space, until some equilibrium is reached. That is the kind of thing ecologists are interested in and de Wit's work has attracted much attention from them. For good balance I will look at de Wit's analysis of such situations for the relatively simple case of two species growing together and reproducing by natural reseeding or through sods and stolons – the precise reproduction mechanism does not really matter. Examples are mixtures of two kinds of grass or of a grass and a legume like clover. The analysis becomes rather complex, but it is really interesting and worth the trouble.

A1.4.1 *Crowding for the same space*

We first look at two species crowding for the same space. The success of a species in the mixture can be expressed by its relative reproductive rate. If it is larger than unity then its representation in the mixture will increase from season to season. The species' reproductive rates are obtained from (3a) and from the identity $Z_1 + Z_2 = m^{-1}$ as:

$$a_1 = \frac{Y_1}{Z_1} = \frac{mk_{12}}{k_{12}z_1 + z_2} M_1 \text{ and } a_2 = \frac{Y_2}{Z_2} = \frac{m}{k_{12}z_1 + z_2} M_2, \quad \text{and} \quad \text{their}$$

relative reproductive rate:

$$\alpha_{12} = \frac{Y_1 Z_1^{-1}}{Y_2 Z_2^{-1}} = k_{12} \frac{M_1}{M_2} \tag{12}$$

Note that m has been eliminated from the equation again. If the rate is larger than unity, the share of species 1 in the mixture will steadily increase.

De Wit devised an elegant graphical analysis for the changes which will occur with time in natural self-reproducing populations on the basis of this rate. From the left-hand part of (12) we get $\frac{Y_1}{Y_2} = \alpha_{12} \frac{Z_1}{Z_2}$, and taking the logarithms:

$$\log\left(\frac{Y_1}{Y_2}\right) = \log \alpha_{12} + \log\left(\frac{Z_1}{Z_2}\right), \text{ with } \alpha_{12} = k_{12} \frac{M_1}{M_2} \tag{13}$$

If $\frac{Y_1}{Y_2}$ is plotted against $\frac{Z_1}{Z_2}$, both on logarithmic scales, a straight line is obtained with a slope of 45°, as in Figure A1-2. The line goes through the origin if $\alpha_{12} = 1$, which means that the composition of the mixture remains the same. The solid line is obtained for an imaginary case of two grassland species competing for the same space, with $\alpha_{12} = 2.5$. Now look at what would happen in a perennial grassland with these two species only. Assume that at the time we start measuring the density ratio of the mixture is 0.1 (point A). After one season point B the yield ratio will be equal to B and the mixture is 'resown' at the same seed or plant ratio, that is at point C. After another season point D is reached and so forth, until after a number of years, species 2 will have disappeared from the mixture. The closer the α is to 1, the longer it takes, but the mixture will always move in the direction of the species whose α is larger than 1. Hence, there can be no stable equilibrium in a mixture of two species competing for the same space, unless α is 1.

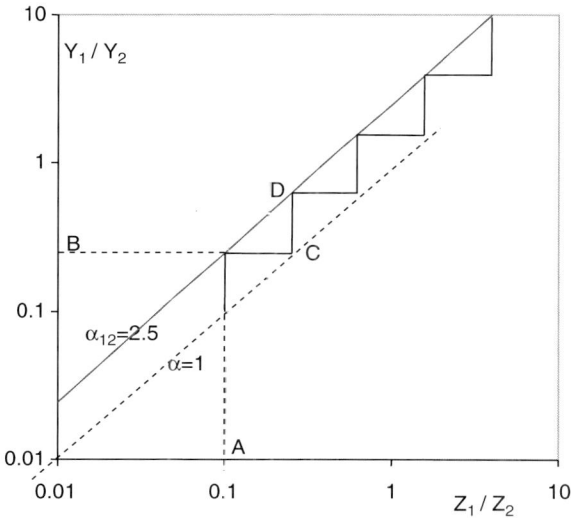

Figure A1-2. Ratio diagram for two species competing for the same space

A1.4.2 *Crowding for partly the same space*

Next consider a mixture of two species which compete for only partly the same space. As an example I will use de Wit's figures for a mixture of two grasses (*Anthoxanthum odoratum* and *Phleum pratense*) grown in containers, to illustrate what may happen. Again, it is rather complex, but it is too late to turn back now. I will also have to give the full story with the real figures and all, otherwise it may not convince you.

The two grasses were planted in different proportions and then allowed to grow for one full year, after which the starting conditions were measured, in numbers of tillers per container. There was one complication though: the initial relative densities of the two species must be defined in such a way that they add up to unity in all combinations, or which is the same thing, that $Z_a + cZ_p$ is constant for all combinations of Z_a and Z_p. Tillers were therefore counted at the end of the first year and the tiller densities of all combinations were found to satisfy the relationship $Z_a + 1.53Z_p = 420$ 'equivalent tillers' per container. Hence, according to expression (5) the relative densities are now defined as:

$$z_a = \frac{Z_a}{Z_a + cZ_p} \text{ and } z_p = \frac{cZ_p}{Z_a + cZ_p},$$
$$\text{with } Z_a + cZ_p = m \text{ and } c = 1.53 \tag{14}$$

and the yields are expected to satisfy equations (11):

$$Y_a = \frac{k_{a(pe)} z_a}{(k_{a(pe)}-1)z_a+1} M_a \text{ and } Y_2 = \frac{k_{p(ae)} z_p}{(k_{p(ae)}-1)z_p+1} M_p \qquad (11)$$

The yields obtained in the trial at different relative densities were plotted in Figure A1-3 and the drawn lines were obtained by fitting (11) to the data. The fitting was done by the same clever trial-and-error method which was also used for Figure A1-1. I will not go into that here. The species did not compete for the same space because of their different growth patterns, their competition indices, estimated in the fitting procedure, being $k_{a(pe)}$ = 2.1 and $k_{p(ae)}$ = 3.4. Estimated sole crop yields M_a and M_p were 1,050 and 400 tillers per container respectively.

We now find an expression again for the relative reproductive rate, from (11) and (14):

$$a_a = \frac{Y_a}{Z_a} = \frac{m k_{a(pe)}}{\{(k_{a(pe)}-1)z_a+1\}} M_a$$

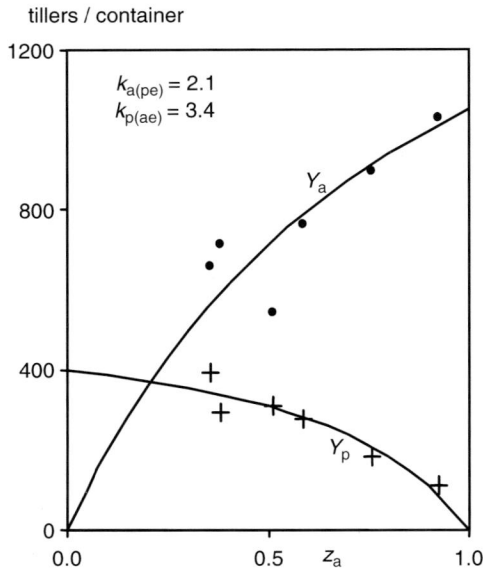

Figure A1-3. Results of a competition experiment with two pasture grasses. (From de Wit, 1960.)

$$\text{and } a_p = \frac{Y_p}{Z_p} = \frac{mck_{p\,(ae)}}{\{(k_{p\,(ae)}-1)z_p+1\}} M_p.$$

$$\text{Hence: } \alpha_{ap} = \frac{Y_a Z_a^{-1}}{Y_p Z_p^{-1}} = \frac{(k_{p\,(ae)}-1)Z_p+1}{(k_{a\,(pe)}-1)Z_a+1} \times \frac{k_{a\,(pe)} M_a}{ck_{p\,(ae)} M_p} \text{, and}$$

$$\log \frac{Y_a}{Y_p} = \log \alpha_{ap} + \log \frac{Z_a}{Z_p}, \text{with } \alpha_{cp} = \frac{(k_{p\,(ae)}-1)z_p+1}{(k_{a\,(pe)}-1)z_a+1} \times$$

$$\frac{k_{a\,(pe)} M_a}{ck_{p\,(ae)} M_p} \tag{15}$$

Note the elimination of m again and the presence of c in the relative reproductive rate. This time the rate is not independent of the population densities and some very interesting patterns emerge. That is best shown by the mixture's ratio diagram of Figure A1-4. The interesting thing is that the curve crosses the diagonal, which is the line for $\alpha = 1$, as in the previous case (Figure A1-2). The arrows

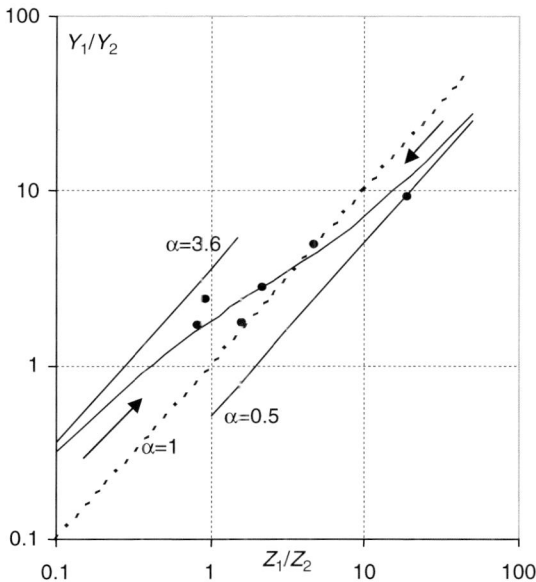

Figure A1-4. Ratio diagram for two grasses competing for partly the same space

indicate the direction in which the composition will change over time and approach equilibrium at the intersection where $\alpha = 1$. The equilibrium value of $\dfrac{Z_1}{Z_2}$ is found by setting $\alpha = 1$ in equation (15) and solving for z_a (remember that $z_b = 1 - z_a$). Substituting the parameter values for this trial it works out at $z_a = 0.715$, which corresponds with $\dfrac{Z_1}{Z_2} = 3.84$. The curve in the diagram approaches the two asymptotes as $\dfrac{Z_1}{Z_2}$ approaches zero or infinity, that is as z_a approaches 0 or 1. The values of α for the asymptotes are calculated from equation (15) as $\alpha_0 = \dfrac{k_{a\,(pe)}\,M_a}{c M_p}$ and $\alpha_\infty = \dfrac{M_a}{c k_{p\,(ae)}\,M_p}$, within this case the numerical values $\alpha_0 = 3.6$ and $\alpha_\infty = 0.5$. Obviously, a mixture will attain an equilibrium if the asymptotes are at opposite sides of the diagonal, otherwise one of the components will eventually disappear. So if we know the competition parameters of a binary mixture, we can calculate whether it will eventually stabilise and at what relative densities.

Appendix 2. Shifting Cultivation and Recurrent Cropping – the Figures

A2.1 A cautionary note

I must start this story on a cautionary, or perhaps I should say an apologetic note. The idea is to develop a rough nutrient budget for shifting cultivation as well as for systems with short fallows, using published data, particularly those collected by Nye and Greenland. Without such a budget it is not possible to make the kind of predictions which agronomists want to make, in particular about the yield levels that can be expected and the nutrient deficiencies which are likely to occur when the transition is made to more intensive land use. At the present state of knowledge, however, it is not possible to put a nutrient budget together without making a lot of assumptions, some of which would be unacceptable in a rigorously scientific text. The hard core of the profession is usually reluctant to do this, with the exception of the modellers, but they tend to shroud their predictions in computer code, unintelligible for the uninitiated. It is my ambition here to put together an intelligible nutrient budget from as much real data as there are, supplemented with assumptions where there are none. The tables sometimes contain figures with a rather preposterous air of exactness, but that is purely computational and has no physical significance. What counts are the orders of magnitude, which I think are quite reasonable.

A2.2 Shifting cultivation

A2.2.1 Why modelling?

As we have seen in Chapter 5, van Beukering's work in the 1930s touched on the essence of shifting cultivation, exploring its limitations and speculating about the possibility of a future, equally stable production system to replace it. He designed the contours of a mixed farm but rather than exploring its theoretical feasibility first, he proposed to try it out in the research station before recommending it to the extension service for transfer to the farmers, in the good old tradition of applied research. That approach has major flaws, the most important one being that a lot of time may be wasted before it turns out that the ideas do not work, or that the farmers did not wait for the research to finish and moved on in a different direction on their own. So you want to cut some corners and explore whether the proposed system is likely to be feasible, before actually carrying out a real life test. In other words, you want to model the system and make informed guesses about its likely performance before going into the trouble of doing many years of research. And if the model tells you it will probably not work, you may drop the idea altogether or change it so that it may still work. Modelling of course has its own pitfalls, one of them being dubious predictions because the data fed into the model are poor, that is the well-known 'rubbish in – rubbish out' adage, or because the processes involved are poorly understood and therefore not adequately accounted for in the model. First of all, hard data are needed about nutrient stocks and nutrient flows in order to calculate with some confidence how long the reserves are likely to last if their extraction rate is increased by more intensive cropping. There are good data for shifting cultivation itself, thanks especially to Nye and Greenland, so it makes sense to first model the nutrient budget for that system and then do the same for what may be called its degenerate form, fallow-based cropping with short fallows. That will show why and how the system is likely to break down and what an alternative system should do to prevent collapse.

A2.2.2 Nutrients in the soil and in the fallow vegetation

The first thing we need to know is the amounts of nutrients cycled through the topsoil and the vegetation under a shifting cultivation system. Nye and Greenland's review of a wide range of field studies showed that 10–12 years was the most common fallow length in forest ecologies where shifting cultivation was practised. Farmers were unlikely to clear older forest because, in Nye and Greenland's words, 'the increased yield to be expected from increased soil fertility is no longer commensurate with the increased labour of clearing more heavily wooded land'. For the budget calculations I will therefore consider a land use system of 10–12 years of fallow and 2 years of cropping. Table A2-1 shows the pattern of nutrient accumulation

Table A2-1. Nutrient build-up in fallow vegetation and topsoil in a 10- to 12-year fallow cycle and after 'very long' fallow (>100 years)

| | Nutrients (kg/ha) present in | | | | | Topsoil (0–30 cm) | | | |
| | Vegetation | | | Organic (humus-) N | | Organic (humus-) P | | Exchangeable K | |
Year	N	P	K	Total	Total increase	Total	Total increase	Total	Total increase
0	–	–	–	3,300–4,600		200–275		200–350	
5	300–500	20–30	200–300	3,400–4,750	100–150	205–285	5–10	205–365	5–15
10–12	**500–700**	**30–40**	**300–400**	**3,500–5,000**	**200–400**	**210–300**	**12–25**	**220–385**	**20–35**
>100	1,200–1,800	100–130	800–1,000	4,500–6,500	1,200–1,900	270–390	70–115	300–500	100–150

in vegetation and topsoil in such a system, based on Nye and Greenland's analyses. You will not find that table in their essay, though. I put it together as a summary of their findings, making some rather sweeping generalisations the authors would not necessarily have agreed with, but I think it captures the essence. The figures show what happens between the beginning of the fallow after cropping has ended (year 0) and 10–12 years later when the land is ready for cropping again. The bottom row shows what the final equilibrium would be if the land were no longer used at all and reverts to real lowland forest. The figures may not convey an obvious message, but their meaning will become clear as we go along.

The humus content of a soil, which is in equilibrium with a mature lowland forest (the bottom row, '>100 years') plays a pivotal role in the calculations. Equilibrium is attained when the build-up of new humus by conversion of decomposing litter and decaying roots is matched by the simultaneously occurring humus breakdown. Nye and Greenland estimated that the soil in equilibrium would contain between 54,000 and 78,000 kg of humus-C,[1] with an average of 66,000 kg. C-content is usually reported in the literature as percentage of carbon in the top 20 cm and, since humus is almost completely concentrated there, 66,000 kg is equivalent to 2.24% C, which agronomists will recognise as a humus-rich soil.[2] The ratio between C and N in stable humus is usually around 12 and if we adopt that figure as a standard for the soil under humid forest, the estimated range of organic-N content works out at 4,500–6,500 kg/ha (the bottom row of Table A2-1) with an average of 5,500 kg. In a land use system where 10–12 years of forest fallow alternate with a few years of cropping C-content at the end of the fallow will eventually settle down at 75–80% of the maximum. That is shown in the highlighted row of the table. The ratio between P and C was found to be about 1:200 in P-deficient soils (i.e. most soils in the humid and sub-humid tropics), hence a humus-P content of 270–390 kg/ha at mature forest equilibrium. The N and P locked up

---

[1] Humus is chemically complicated and its composition is variable, so most authors prefer to report soil carbon rather than humus content. Humus content is usually assumed to be 1.78 times the organic-C content.

[2] The volume of 20 cm topsoil over 1 ha is 2×10^6 dm^3 and weighs 2.94×10^6 kg (bulk density of 1.47). Hence, 66,000 kg of humus-C corresponds with $(66 \times 10^3/2.94 \times 10^6) \times 100 = 2.24\%$ organic-C content.

in the humus only becomes available for plant growth when it is set free by humus breakdown.

Humus, also known as soil organic matter, is a very interesting material with an essential role for soil fertility, especially in the humid tropics. Apart from being the main store of nitrogen and phosphorus, it has properties similar to clay particles. Like clay, it binds or 'adsorbs' positively charged ions ('cations') such as K, Ca and Mg, all of them essential plant nutrients. They are 'exchangeable' in that they can be released into the soil solution from where they are taken up by the plant roots. The clay particles and the humus together form what is called the soil's 'adsorption complex'. The reason why this role of the humus is so important in the tropics is that the binding ability of the clays in most tropical soils is weak, much weaker than those in the temperate zone clays.

According to Table A2-1 the amounts of nutrients stored in the fallow vegetation are quite large. Where did they come from? Practically all the nitrogen has been fixed over the years by free-living and symbiotic bacteria which bind atmospheric nitrogen, but the other nutrients must have been found somewhere in the soil profile. Since the upper part of the soil is depleted of nutrients by the end of each crop cycle due to removal by the crops and leaching, the fallow plants must obtain a lot of nutrients from lower down. Their roots scavenge a large volume of soil and work as nutrient pumps, thereby recovering most of the nutrients which were washed down during the previous cropping period. That is the second important role of the fallow, next to the build-up of humus. If there were no deep-rooting shrubs and trees to pump up nutrients, the whole nutrient cycle would be limited to the top 30–40 cm of the soil and what was washed further down would be out of reach for the crops.

The nitrogen stored in a 10- to 12-year-old fallow vegetation is about twice as much as the N added to the humus over the same period, but most of the former is lost into the air when the vegetation is burned, so the nitrogen available for the crops will not be as abundant as it seems. The phosphorus accumulated in the vegetation is also about twice as much as the amount added to the soil in the form of humus-P, but contrary to N, most of the P in the vegetation is set free by burning and will be available to the crops. With time, however, part of it is converted into less soluble forms.

The K in the vegetation is more than ten times as much as the amount that was added to the topsoil and after burning, most of it will be found in the ashes. It is then washed into the soil by the rains, where it is partly adsorbed to the clay–humus adsorption complex and becomes available for uptake by the crops. Part of the K is washed further down into the profile and may be lost, unless it is recaptured by the roots of trees and shrubs.

After the vegetation has been cut down and burned, the soil conditions are ideal for cropping, only free nitrogen is in short supply. But the soil humus content is high and it breaks down at a rate of 3% per year, releasing the nitrogen on which the crops feed. The humus also releases P, which supplements what comes from the burned vegetation. K is extracted from the adsorption complex (the clay and the humus), enriched with fresh K from the ashes. So, a good crop will be obtained during the first year or two after clearing the fallow. Then decline will set in. Cations are removed by the crops and depleted by leaching, the availability of free phosphorus decreases due to conversion into less soluble forms, weed problems are on the rise and the effort to obtain another crop increases steeply. So, the shifting cultivator will leave the land and look for a new plot to clear. After he is gone,the vegetation will re-establish itself from the remaining stumps and roots and from new seedlings and the recovering vegetation and the soil humus start acting in tandem again as a nutrient accumulator, until the next cropping period.

A2.2.3 The nutrient budget for crop production

That is all very interesting and it explains why farmers do as they do, but we want to go a step further and look at the actual quantities of nutrients moving around in the system. That should tell us whether shifting cultivation is indeed as stable as it seems and why that is so. And, even more importantly, we hope it can help us to predict what will happen if the rules for shifting cultivation are no longer obeyed. That was the question asked 75 years ago by van Beukering, and it remains as relevant as ever, or even more so.

Crop yield under shifting cultivation is constrained by many things: the amounts of available nutrients, shadiness of the fields, and weeds, pests and diseases. It is not obvious a priori which of these is the primary limiting factor, but by looking at the available nutrients and calculating how much yield they would permit, we can perhaps decide whether nutrient availability is likely to be the main limitation for crop yield. And if it is not, something else must be limiting them. Nye and Greenland went about it the other way around. They gathered information about the crop yields obtained under shifting cultivation, calculated the amounts of nutrients the crops would have taken up and compared that with what was actually available, to see whether the two matched. That, I think, is not very satisfactory because it reverses the order of what happens: the soil has a certain amount of nutrients available and the crop yield is the result of what they take up, not the other way around. I therefore used the authors' data to redo the analysis in that order.

Consider a semi-deciduous forest[3] area with two rainy seasons where farmers leave the land under fallow for 10–12 years and then grow one crop of maize, cassava and plantains, all planted at the same time. The maize will be harvested after 100–120 days, the cassava remains in the field for up to 2 years and the plantains for 3 years. In actual fact, by the end of the second year, the new fallow is already on its way in and the third year, when only the plantains remain, is really part of the next fallow, so we will do the nutrient analysis for 2 years of cropping only. What kind of yields can be expected from the amounts of nutrients available to the crops? In order to answer that question, let us follow the nutrient status of the soil, starting after the vegetation has been burned and the land has been readied for cropping.

(a) Gross nutrient availability

I first estimated the gross amounts of nutrients available to the crops. Unless stated otherwise, I have followed Nye and Greenland in using 30 cm as the reference soil depth where field crops obtain practically all the nutrients they need. The estimates are shown in Table A2-2. How did I arrive at those figures?

Nitrogen. From Table A2-1 we know that after 10–12 years of fallow there will be 3,500–5,000 kg/ha of N in the humus, 3% of which is released annually by humus decomposition, that is about 100–150 kg/ha/year, or 200–300 kg for a 2-year cropping period. However, there is a problem with the timing At the beginning of the rainy season there is a so-called nitrogen flush, due to the rapid

---

[3] Semi-deciduous means that part of the tree species drop their leaves during the dry season.

Table A2-2. Estimated 'gross' nutrient availability after 10–12 years of fallow and threshold contents of P and K, in kg/ha for the soil's top 30 cm

	N	P	K
Initial balance	–	22–40	220–385
Released/added in year 1	100–150	32–35	250
Released in year 2	100–150	6–9	–
Total available over 2 years[1]	200–300	60–84	470–635
Threshold, kg/ha	–	20	170

[1] Ignoring immobilisation and leaching losses.

build-up of microbial activity in the soil. A newly planted crop cannot absorb all that nitrogen, so supply and demand are out of phase and an important part of the flush will be lost by leaching, in the form of nitrate. I will come back to that in a moment.

Potassium. For potassium (K), Nye and Greenland estimated that an average of 250 kg/ha is added to the soil by the ashes of the burned vegetation, 30% less than what was there before burning. The rest is in unburned wood and some is volatilised. Together with the K adsorbed to the clay–humus complex at the end of the fallow (220–385 kg according to Table A2-1) there will therefore be some 470–635 kg/ha of K, which is in principle available to the crops at the start of the cropping period. Not all of this can be taken up freely, though. The less there is, the more difficult it is for the plants to extract it. Soil scientists have therefore established thresholds for exchangeable K,[4] short of which deficiency can be expected. Exchangeable K is measured routinely in soil analytical laboratories and expressed in milli-equivalents (meq) per 100 grams of dry soil. For maize a threshold of 0.15–0.20 meq/100 g, measured in the top 20 cm of the soil is often used. When exchangeable K is below the threshold and *other nutrients are optimally available*, crops can be expected to respond to K-fertiliser, and more so as K-content is farther removed from the threshold. Later on we will see that under shifting cultivation the condition *other nutrients being optimally available* is not satisfied because crop yield is actually constrained by available nitrogen, so there will be less demand for K than would be the case if nitrogen were abundant. K will therefore only limit

[4] That is ionic K, in equilibrium between the adsorption complex and the soil solution.

crop growth when there is much less than the official threshold quantity. It is worth explaining that a little further, because thresholds will show up again later, when we look at the nutrient balance under alley cropping, in Chapter 7 and Appendix 4.

Consider the graph in Figure A2-1. It shows, schematically of course, how maize yield depends on the soil's exchangeable K content. The top curve applies when all other nutrients are optimally available so that the soil scientists' K-threshold will hold. When exchangeable-K is above 0.2 meq/100 g, maize yield is almost at its maximum: the plants can obtain all the K they need almost without effort. Applying more K to the soil makes no difference: there is no response in yield to applied K. When exchangeable-K is below 0.15, however, the plants have to pull harder to bring in the K and the lower the content the greater the effort. If K is now applied to the soil the yield will go up and the more steeply so as the soil's K-content is lower.

The bottom curve shows the relationship between exchangeable K and yield when another nutrient constrains growth, for example, nitrogen. The relationship between K and yield is now dominated by N-deficiency and a K-effect can only be expected when exchangeable K-content is much lower than the 0.15 meq threshold.

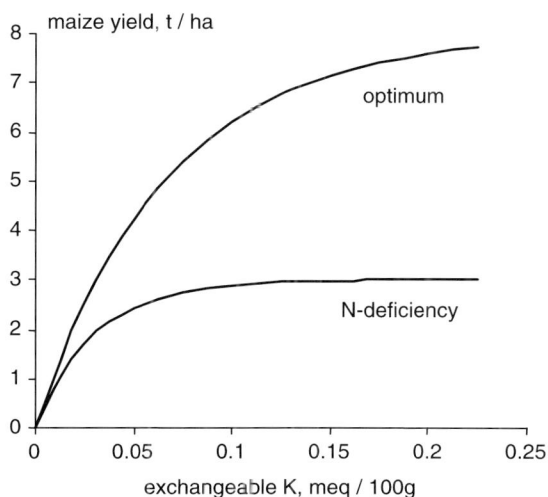

Figure A2-1. Maize yield response to exchangeable K in the soil, at optimum and low soil N-content

We will see later that crop growth under shifting cultivation will be limited primarily by nitrogen and I will therefore use a somewhat arbitrary threshold of 0.10 meq/100 g taken over the top 30 cm of the soil. Everything in excess of that will be considered as 'readily available'. In order to convert meq/100 g to kg/ha, the former must be multiplied by 1716,[5] so a threshold of 10 meq is equivalent to an exchangeable K content of about 170 kg/ha.

Phosphorus. To start with, there will always be a certain amount of available P in the soil after a fallow period, which will be estimated in Appendix 4 (on alley cropping) as ranging from 22 to 40 kg/ha for different soil types. According to Table A2-1 the fallow vegetation contains 30–40 kg of P, but after burning, not all of that is recovered, because some is locked up in unburned wood and some is volatilised. Nye and Greenland reckoned that an average of 26 kg/ha is added from the burned vegetation. Finally, 6–9 kg is released annually by decomposing humus (1/17 of the nitrogen or 1/200 of humus-C). So the amount of new P added to the topsoil in the first year is estimated at some 32–35 kg and another 6–9 kg in year 2, totalling 38–44 kg over 2 years. Hence the total amount of P coming available over 2 years will be 60–84 kg/ha. As with K, there is also a threshold for available P, below which deficiency can be expected. 'Availability' is less clearly defined for P than it is for K, however, and the amount that is measured depends on the analytical method used, but a figure usually adopted as a threshold for maize grown in a tropical area is 10 ppm (parts per million), assuming again that other nutrients are optimally available. When N is severely limiting as in our case the threshold will be considerably lower, as explained for K, and I adopted a value of 4.5 ppm. Taken over 30 cm of soil depth, that is equivalent to 20 kg of P per ha.

So much for the gross nutrient contents of Table A2-2. The figures on their own will mean little to most people, until they are put in the context of crop yield. In order to do that, we must know a few more things: in particular how much the crops take up per unit of biomass and at what time, how much will be lost over time by leaching, denitrification and immobilisation and how much will be left by the time the crops need it.

---

[5] 1 meq of K weighs 39×10^{-6} kg (39 is the atomic weight of K) and the dry weight of 1 ha soil of 30 cm depth equals 4.4×10^6 kg, hence, 1 meq per 100 g over 30 cm is equivalent to $39 \times 4.4 \times 10 = 1,716$ kg K.

(b) Nutrient demand of the crops

The amounts of nutrients, which are taken up per unit of new bio-mass by maize, cassava and plantains are shown in Table A2-3. The figures are similar to those used by Nye and Greenland but I added the amounts in the stems, leaves and roots, because we will assume that the crop residues are left in the field, so that their nutrients will become available again later on. I also lowered the N-figure for maize, because the authors took theirs from a US feedstuff table, which is too high: when maize is given a lot of fertiliser-N, much more N is stored in the grain per unit of weight than when N is in short supply.

Table A2-3. Nutrients taken up to produce 1,000 kg of maize grain (at 'low to moderate' yields), 10 t of cassava tuber and 10 t of plantain bunches

Crop	Yield, kg	kg in the produce			kg in stems, leaves, roots			Total nutrients		
		N	P	K	N	P	K	N	P	K
Maize	1,000	12	2.7	4	6	2	15	18	4.7	19
Cassava	10,000	22	5	45	15	3	35	37	8	80
Plantains	10,000	22	4	55	25	4	100	47	8	155

(c) Uptake and yield of maize

Now let us see how much will be available for the three crops at the time they need it. Table A2-4 has the computational details. The first two lines show the initial nutrient content and release from the burned vegetation, which will be recognised as those in Table A2-2.

Maize is the first crop to mature. It takes 100–120 days from seed to seed and most of the nutrients are taken up during the first 3 months. N and P are released by decomposing humus in propor-tion to the length of the season, hence 60% of the year's release will occur during the maize growing season (line 3). Half of that N is taken to be lost by leaching, dragging along part of the exchange-able K (line 4).[6] Of the initially available P, 10% is assumed to be

---

[6] It is assumed that 30% of the leached N is accompanied by potassium, so each kg of leached N will carry along $0.3 \times 39/14 = 0.836$ kg of K (39 is the molecular weight of K and 14 that of N).

Table A2-4. Nutrient budget for maize grown during the first rainy season after a long fallow

Line		N Min	N Max	P Min	P Max	K Min	K Max
1	**Initial balance**	–	–	**22**	**40**	**220**	**385**
2	Added from burned vegetation	–	–	26	26	250	250
3	Released by humus	63	90	4	5		
4	Leaching, denitrif., immobil.	−31	−45	−5	−7	−26	−38
5	**Available nutrients**	**31**	**45**	**47**	**64**	**444**	**597**
6	Taken up by maize	−31	−45	−8	−12	−33	−48
7	Maize yield, kg/ha	1,750	2,500				
8	**Balance after maize harvest**	**0**	**0**	**38**	**53**	**410**	**550**

immobilised in the same period. I admit those are quite a lot of assumptions, but I think they are reasonable and consistent with research findings. Line 5 of the table shows the balance – the gross amounts of nutrients available to the maize. Remember that these amounts are freely available up to the threshold only, below which, uptake becomes more difficult. By comparing available nutrients with crop demand (Table A2-3), it should be obvious that nitrogen is the limiting nutrient. I have therefore assumed that all the available N will be taken up by the maize (line 6). That is not quite correct, for one thing because the cassava and the plantains are also there and although they will not grow much while the maize is present, some of the nutrients will be taken up by them. I will just ignore that, however, to keep things simple, and assume they will take up all their nutrients after the maize season.[7] Available P and exchangeable K will remain well above their thresholds until all N has been used up. An uptake of 31–45 kg of nitrogen (never mind the air of precision) by the maize would give a yield of 1,750–2,500 kg/ha (line 7). That is indeed the kind of yield a good farmer can get under these conditions, which may make you suspicious, because there are a lot of assumptions in these calculations and I knew beforehand what kind of yield should be expected. But there was no undue tinkering to obtain desired results.

---

[7] I have also ignored the fact that uptake of N will become more difficult as there is less of it.

(d) Nutrient uptake and yield of cassava and plantains

We now continue with the cassava and the plantains. Cassava usually remains in the field up to the end of the second year and by that time the plantains will also have produced their first bunch. Table A2-5 shows the nutrient budget for the two crops. The starting balance (line 1) is of course the same as the closing balance of Table A2-4. The amounts of N and P released by humus decomposition (line 3) were taken to be 40% of the year's total for the second season of year 1 plus 100% of year 2. From here on, things become a little more complicated because the model must account for the decomposition of maize residues and their partial conversion into fresh humus (lines 2 and 4). Box A2-1 explains the details of the calculations. Fifty per cent of the remaining N is assumed to be leached out of the topsoil, while P-immobilisation is put at 20% over 1.5 years (line 5).

How much of the remaining nutrients (line 8) will be taken up by the cassava and the plantains? That is not so straightforward, but I keep telling myself not to complicate things. The plantains are usually planted at a fairly low density, so let us say that each

Table A2-5. Nutrient budget for cassava and plantains grown from the second rainy season of the first year to the end of the second year

Line		N Min	Max	P Min	Max	K Min	Max
1	**Balance after maize harvest**	0	0	38	53	410	550
2	Returned by maize stover	10	15	3	5	26	38
3	Released by humus over 1.5 years	144	206	9	12		
4	New humus	−18	−26	−1	−2		
5	Leaching, immobil	−63	−90	−9	−13	−53	−75
6	**Available for uptake**	73	105	40	56	384	512
7	Taken up by cassava	−53	−76	−11	−16	−115	−164
8	Taken up by plantains	−20	−29	−3	−5	−67	−95
9	Cassava yield, kg/ha	14,346	20,497				
10	Plantain yield, kg/ha	4,304	6,149				
11	**Total uptake**	−73	−105	−15	−21	−181	−259
12	**Balance after cassava**	0	0	25	34	203	253
13	Nutrients locked up in crop residues	16	23	3	4	47	67

Box A2-1. Nutrients released and new humus built from maize residues

1. N and P released by the stover from the first season (line 3 in Table A2-5) is the product of maize yield (Table A2-4, line 6) and N- and P-content of the stems and leaves (Table A2-3)
2. Assume that the dry matter weight of the maize residues equals grain weight. The residues are left on the surface and half will be eaten by termites and other bugs. The rest is converted into humus with a conversion factor from dry matter to humus-C of 0.25 (see Appendix 4).
3. Hence, the weight of fresh humus-C formed from the maize residues is between 219 kg ($1,750 \times 0.5 \times 0.25$) and 313 kg ($2,500 \times 0.25 \times 0.25$)/ha.
4. At C/N and C/P ratios of the humus of 12 and 200 respectively, this fresh humus incorporates 18–26 kg of N and 1–2 kg of P.

time the cassava takes up a packet of nutrients needed to produce 1,000 kg of roots (3.7 N, 0.8 P, 8.0 K, according to Table A2-1), the plantains take up a packet needed for 300 kg of fruit (1.41 N, 0.24 P, 4.65 K). We continue assigning packets until (the readily available part of) one of the nutrients is depleted. I admit it looks somewhat arbitrary again, but let us see what we get. If you work this out you will find that after 14.3–20.5 cassava and plantain nutrient packets, the N will be depleted. The yields from those packets of nutrients would be 14,300–20,500 kg of cassava and 4,300–6,100 kg of plantains, quite realistic yields after 10–12 years of fallow.[8] The cassava plant residues and the banana stems which carried the first bunch will be left in the field (line 14) and will decompose in the course of the following year.

In the third year the situation becomes rather blurred, because the fallow vegetation is now in full development. The plantains gradually lose the battle for nutrients (and sunlight) and the second plantain crop will be smaller than the first. Leaching is also much reduced because the new fallow will take up a lot of N. That is really too much to handle for this simple model, so I did not extend the calculations beyond year 2.

---

[8] I have ignored the fact that the new fallow come in gradually and competes with the crops for nutrients.

A2.2.4 Another cropping period?

Would it be possible or wise to put in another crop cycle before letting the land return to fallow? As far as the shifting cultivator himself is concerned, that is really a hypothetical question, but it has
been studied a lot by scientists, in particular during the 1950s and
1960s when it was thought that the road to development was the
elimination of bush fallow. What does the model predict? If no
nutrients are brought in from outside, the answer would of course
depend on the nutrient status of the soil at the end of the first cycle.
Let us first look at the N and P balances. The N and part of the P
are still locked up inside in the crop residues (line 14 in Table A2-5)
but none of that will become directly available for the crops in
the following season because it will be incorporated into the new
humus formed from the residues. According to the calculations in
Box A2-2, the conversion into humus will cost about 42 kg of N and
2.5 kg of P per 10,000 kg of cassava or plantain. And since the combined yield of cassava and plantains was estimated at 18.7–26.6 t/ha,
about 78–111 kg of N and 4.7–6.7 kg P will eventually be immobilised in fresh humus (never mind the air of precision again, it is
just a rough estimate).[9] That is much more than the residues themselves contain, so it will have to be supplemented by some of the
N and P that is released by humus breakdown in the next
season. Therefore, nitrogen would be even more limiting than in the

---

Box A2-2. New humus from cassava and plantain residues

1. Assume again that the dry matter weight of the cassava and plantain
   residues equals the dry weight of their produce and that the latter's dry
   matter content is 40%.
2. About 50% of the residues is assumed to be eaten by bugs, in particular
   termites, who will release the nutrients again in due time. The conversion
   rate of the remainder from dry matter to humus-C is set at 25%, as before.
3. Hence, the weight of fresh humus-C from the residue of 10,000 kg of
   cassava or plantains equals $10,000 \times 0.4 \times 0.5 \times 0.25 = 500$ kg.
4. At a humus C/N ratio of 12 and a C/P ratio of 200, 500 kg of humus-C
   incorporates 42 kg of N and 2.5 kg of P.

---

[9] The conversion into humus is a complex process with many intermediate steps,
but for our purposes the simplified calculation in the box is adequate. Even so,
uncertainty about the values of the parameters involved is a weak point in the
calculations which has a strong influence on the yield predictions.

first year of the first cycle, reducing maize yield to about 60% of
that of the first cycle. For cassava and plantain, a yield of 70% is
predicted because the soil's available K-content will have fallen
below the threshold.[10] These crops are especially gluttonous for K –
another cropping cycle and their yield will fall to a much lower level
still, because of acute K-deficiency. Available P will also reach a
dangerously low level, but that is hidden by the overriding effect of
K-shortage. If a farmer wants to get the same yields as in the first
cycle, he would have to apply all three elements N, P and K from the
second cycle onwards.

Twentieth-century scientists, hoping to break through nature's
barriers, did a lot of experiments with continuous cropping as we
will see in the following section. The crop yields and nutrient deficiencies they measured in the second and third cycle after fallow,
quoted by Nye and Greenland, were quite similar to what our
model predicts, but the results were variable not to say erratic. That
is understandable, because the buffering effect of the nutrient
legacy of shifting cultivation has now been spent and the differences in the soil's native fertility will show through in the results.

As I said, all this is only of theoretical interest. Shifting cultivators are very unlikely to use the land for another crop cycle, it is too
much work to keep the weeds under control and applying fertiliser
does not come naturally to them.

A2.2.5 The long-term prospects

So much for the feasibility of a second crop cycle. Let us now look
at the more sensible option of a single cropping period alternating
with 12-year fallows and see how long that can last. The easy
answer is 'forever'. That answer is not only easy, it is even true if we
consider the shifting cultivators themselves as part of the system, so
that there are no losses, except some small leakage out of the soil
profile. But what if the nutrients, which leave the field with the
harvest, do not find their way back into the land? In principle, the
answer can still be found, although less easily than in the previous
case. Since the nutrients contained in the produce are now taken
out, we need to know how large the store of P and K is from where

---

[10] That is what further model calculations say, which I will skip.

the losses can be replenished (export of N can be ignored, because it is compensated by fixation of atmospheric N during the fallow).

Finding good figures about nutrient reserves in tropical soils is no simple matter. Nye and Greenland did look at the reserves and at the release of nutrients locked up in undecomposed soil minerals or other tightly bound forms, but they did not get very far. The scarce data showed a wide range of what is called total P- and K-content, that is the amount you get by using aggressive extraction methods, much more aggressive than the plants are capable of. Since Nye and Greenland's days, more information has been gathered on nutrient stocks which I will look at in detail in Chapter 7 and Appendix 4 when dealing with a special case of permanent land use, alley cropping. Here I will just give some summary figures from the Appendix on alley cropping, which allow a rough estimate of how long shifting cultivation can last when only the nutrients in the crop produce are exported from the land.

The figures for nutrient contents of a 'good soil' and a 'poor soil' of Table A2-6 are total stocks of available and moderately bound P and K in a soil profile of 100 cm depth. The crops are assumed to extract their nutrients from the top 30 cm, while the fallow vegetation gets most from between 30 and 100 cm and replenishes the top-soil with the nutrients released by the litter. The crops feed primarily on available and humus-P and exchangeable K, while the fallow plants can also access moderately bound nutrients. The table does not include more tightly bound nutrients of which there may be quite large amounts. In the very long run, part of those will also be set free and very slowly replenish the used up nutrients. I will just ignore those.[11] Finally, we want the soil's humus content to remain

Table A2-6. Long-term nutrient availability for shifting cultivation from free and moderately bound sources (for details, see Appendix 4)

	P			K		
	Non-humus	Available	**Total**	Moderately bound	Exchangeable	**Total**
Poor soil	210	45	**255**	800	400	**1,200**
Good soil	460	70	**530**	2,200	1,100	**3,300**

[11] Their contribution will at least partly offset the increasingly difficult extraction of the more available forms as they become depleted.

stable so the humus-P is part of the fixed assets and does not enter into the overall nutrient balance calculations.

Now consider a land use system where a crop of maize, cassava and plantains alternates with 12-year fallows. The crop produce is assumed to be exported outside the system so their nutrients are lost, but all the residues remain in the field and the nutrients which have been leached or immobilised will eventually be recovered again. With the yields calculated in Tables A2-4 and A2-5, the export per cycle was estimated at 14–19 kg P and 95–136 kg K, with the lower limit applying for a poor soil and the upper limit for a good one. The durability of the system will then be the quotient of the total nutrient stock and the export per cycle. K will be depleted first, after an estimated 13–24 cycles, that is 180–340 years, depending on the soil's intrinsic fertility. That is because cassava and plantains put a heavy drain on the K. Taking out part or all of the residues will of course further reduce the durability. A less demanding combination such as a cereal followed by a legume would cover only 1 year and export 7.5–12 kg P and 15–22 kg K. It would keep going much longer, until the available and moderately bound phosphorus is depleted (after 30–45 cycles, that is a whopping 400–600 years according to the calculations).

A2.3 Recurrent cropping

Let us now look at land use practices with much shorter fallows, called fallow-based or recurrent cropping, the ideal of most agronomists short of permanent land use. What kind of soil changes and what yields can be expected and how long can such a system last? That of course depends on the length of the fallow-crop cycles as well as on the soil's native fertility and the fertility, which may be brought in from outside.

A2.3.1 The nutrient status under recurrent cropping

(a) Equilibrium humus content

As we have seen earlier, humus is a very important factor in the fertility of tropical soils. When the fallow gets shorter, the soil humus content will go down, because more will be broken down than is built up in new humus, until a new equilibrium is reached. Nye and Greenland derived a simple formula which describes how equilibrium humus content depends on the cycle of fallow and cropping. Instead of humus per se, most authors use carbon content and I will follow that convention. Call the equilibrium humus-carbon content of a particular cropping system C_E. We want to express C_E as a fraction of the maximum possible content C_M, which would be attained under a mature forest or after a 'very long' fallow. If the length of the fallow is T_f and the length of the cropping period is T_c, the ratio between C_E and C_M as a function of T_f and T_c equals, according to Nye and Greenland:

$$\frac{C_E}{C_M} = \frac{T_f - 2}{T_f - 2 + T_c}$$

The 2 years, which are subtracted from T_f account for the 'lag period', which the authors postulated for the vegetation to re-establish itself after the cropping has ended.[12] Hence, in a land use system with 2 years cropping and 12 years fallow, the equilibrium humus content will be 83% of the maximum, and with 2 years cropping and 4 years fallow it will be 50%.

When land use changes from one system to another, the humus content starts moving towards the new equilibrium. That will take a long time to reach, barring disasters such as loosing the topsoil to erosion. Consider a field where the humus-C content is 80% of C_M and assume that the land use system is changed abruptly to 2 years cropping and 4 years fallow. That is unlikely to happen in reality, but this is just a thought experiment. The equilibrium humus-C content of the new system will be 50% of C_M, but according to Nye and Greenland's model it would take an amazing 40–50 years for the humus content to decrease from 80% to about 60%, and another 50 years to reach the new equilibrium of 50%. And building up the humus again from 50% to 80% would take just as long. That is shown in Figure A2-2. The curve can be easily generated

---

[12] The simple relationship follows from the assumptions that (i) the decomposition rate of humus during cropping is the same as that during the fallow (after the lag period) and (ii) after a very long fallow the net change in humus content becomes zero.

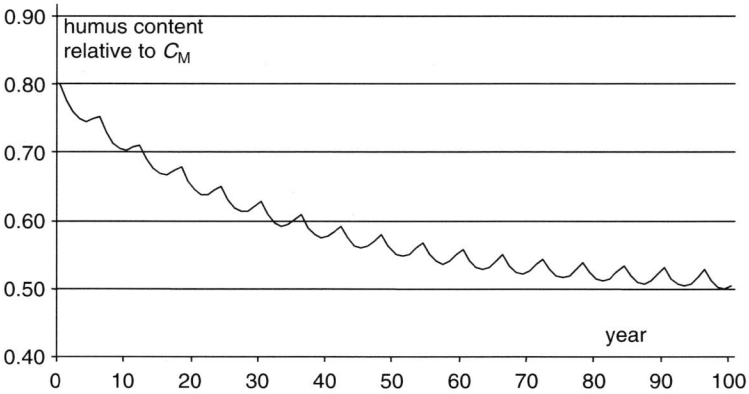

Figure A2-2. Theoretical decline of relative humus content from one equilibrium level (80%) to another (50%) (after Nye and Greenland, 1960)

with a simple spreadsheet procedure, expressing the combination of humus build-up and breakdown by:

$$\Delta C_i = cB_i - dC_{i-1}$$

where C_i = humus-C content in year i
ΔC_i = change in humus-C content in year i
B_i = biomass produced in year i
c = conversion rate of biomass dry matter into humus-C
d = annual humus decomposition rate

(b) Humus C, N and P Content

How much N and P do these relative humus contents represent and how much will be released annually by humus decomposition? Since we have been dealing with relative C-contents so far, we must convert them into quantities of humus-C, in order to calculate the amounts of N and P contained in and released by the humus. And since the relative contents were in respect of C_M, the equilibrium content under a mature forest, we must start from there. We have already put that at 54,000–78,000 kg C/ha in the top 30 cm (Table A2-1). Hence, for soils with a C-content of 80% and 50% of C_M, a C/N ratio of the humus of 12 and a C/P ratio of 200, we find the N- and P-contents shown in Table A2-7.

The figures would apply for well-drained lowland conditions in the wet tropics (never mind the air of precision, I must keep repeating).

Table A2-7. Contents and release of N and P at two equilibrium levels of humus-C

C_E as % of C_M	Content, kg/ha			Release, kg/ha/year	
	C	N	P	N	P
80	43,200–62,400	3,600–5,200	215–310	108–156	6–9
50	27,000–39,000	2,250–3,250	135–195	68–98	4–6

In highland areas and under water-logged conditions, organic matter content may be much higher. For most people including agronomists, the N- and P-contents presented this way probably convey no message at all, but if you have enough endurance their meaning will gradually become clear.

For agronomists, the figures are easier to understand if they are connected with their usual frame of reference. Absolute quantities of C, N and P in the soil are not the kind of figures they are most familiar with. They usually express organic-C content as a percentage of total soil dry weight in the top 20 cm, where practically all the humus is concentrated. That is what a standard soil analysis will report and by looking at hundreds of analysis sheets of known soils, one gets a feel for what the figures really mean. Agronomists use rules of thumb, like 'a soil with a C-content of more than 2% is rich in humus and one with less than 1% is low in humus'. So, for the agronomist's benefit (and to convince myself that the figures make sense), I calculated the carbon, humus and nitrogen percentages in the top 20 cm of soil corresponding with these absolute amounts, assuming that all humus is in the top 20 cm (see footnotes 1 and 2 for the calculations):

C_E as % of C_M	Kg C/ha	Kg N/ha	% C	‰ N	% Humus
100	54,000–78,000	4,500–6,500	1.84–2.66	1.53–2.22	3.28–4.73
80	43,200–62,400	3,600–5,200	1.47–2.13	1.23–1.78	2.62–3.79
50	27,000–39,000	2,250–3,250	0.92–1.33	0.77–1.11	1.64–2.37

(c) Available N, P and K

We are now ready to estimate the available nutrients under a system with 2 years cropping alternating with 4 years fallow, after it has reached the equilibrium humus content of 50%. It would take a long time to reach that point when coming down from shifting cultivation, but eventually it must happen and in some densely populated forest areas in Africa it has. The estimated figures for available

Table A2-8. Estimated 'gross' nutrient availability after 4 years of fallow
and threshold contents of P and K, in kg/ha for the soil's top 30 cm

	N	P	K
Initial balance	–	10–15	150–250
Released/added in year 1	70–100	14–16	100
Released in year 2	70–100	4–6	–
Total available over 2 years[1]	140–200	28–27	250–350
Threshold, kg/ha	–	20	170

[1] Ignoring immobilisation and leaching losses.

N, P and K are shown in Table A2-8. They are 'gross' figures, i.e.
ignoring immobilisation and leaching losses for the moment. I will
explain how I arrived at them.

For nitrogen, things are simple: the only source, at least for
non-leguminous crops, is the humus. Putting the annual decomposition rate at 3% again, which should be in order as long as farmers
practise their usual tillage with little soil disturbance, the annual
release by the organic matter will be about 70–100 kg/ha (Table A2-8),
or 140–200 kg over 2 years.

Estimating the amounts of non-humus P is less straightforward.
The fallow vegetation is an important source of P, but the information on nutrient content of natural vegetation in landuse systems
with short return periods is scarce and shows a lot of variation.
That is because the system is much less buffered than shifting cultivation and because a larger part of the nutrients is shunted out of
the system by a higher cropping intensity and more leaching losses.
The soil must substitute for the losses from its native reserves, which
vary with its native fertility, hence the strong variation in the data.
Reported nutrient contents for a fallow vegetation of around
4 years range roughly from 20% to as much as 80% of that in a
10–12 year fallow[13] (although the figures at the high end may have
been from fields, which were still far removed from the equilibrium). For the model calculations, I assumed an average figure of
40%, so the contribution of P from the vegetation will be about 10
kg. Part of the P-supply is also linked to humus decomposition,
which releases about 4–6 kg/ha over 2 years, hence the total amount
of P released over 2 years was set to 18–22 kg/ ha (Table A2-8).

[13] Data from Nye and Greenland(1960), Sanchez(1976), Ahn(1970) and
Slaats(1995).

Available P present in the topsoil at the end of a 4-year fallow will also be highly variable. I have assumed that at the beginning of the crop-growing period, the available P-content of a 'poor soil' as defined in Appendix 4 would apply, amounting to some 22 kg/ha. I therefore set the content before burning at 10–15 kg/ha. The three contributions (initial content, release from the burned vegetation and humus decomposition) add up to 28–37 kg/ha over 2 years.

For K, things are even more complicated. The uncertainty about the nutrient content of the fallow vegetation of course also affects the K-estimate. I will assume again that the burned vegetation will yield at most 40% of that under shifting cultivation, say 100 kg/ha. Another source of uncertainty is the amount of exchangeable K in the soil at the end of a 4 year fallow. It will certainly be less than what is there after 5 years in a 10–12 year fallow system (see Table A2-1), but how much less? Short of real data, I will just prime the calculations with a value of 150–250 kg/ha. Hence the total for K would be 250–350 kg/ha. These are all rather wild guesses of course. We will take the same threshold contents into account as before, viz. 20 kg/ha for P and 170 kg for K.

A2.3.2 Nutrient budget and crop yields

We are now ready to put together a nutrient budget for a recurrent cropping system with one crop of maize followed by cassava (but planted at the same time), alternating with 4 years of fallow. I dropped plantains because farmers are unlikely to grow them when soil fertility is poor. Plantains are quite fragile and when they do not thrive, they become sensitive to all kinds of disorders, such as nematodes and weevils. Farmers may throw in just a few stands, in spots, which they know are a little more fertile, like the location of a rotten tree trunk, but they are no longer a major part of the crop association.

Another consequence of low fertility is that it is no longer acceptable to assume that leaching of K and N will proceed in tandem. I assumed earlier, on the authority of Nye and Greenland, that would be the case under shifting cultivation (footnote 7), when K in the soil is abundant. But when there is less K, it will be more tightly bound and less easily leached. To account for this I used a simple, though again rather arbitrary trick by putting K-leaching at zero when the soil K-content is less than or equal to 100 kg/ha and letting it increase linearly beyond that, such that 30% of the leached N will be accompanied by K when the soil's K-content is 400 kg/ha.

Finally, there is the effect of low nutrient content on uptake by the crops. In the case of shifting cultivation, available P and exchangeable K would not drop below their thresholds, unless cropping was extended beyond one cropping period. In recurrent cropping, however, K and perhaps P may cross the threshold, and when that happens the model must adjust growth accordingly. I therefore introduced a simple modifier which adjusts growth to below-threshold P and K, whichever is most limiting. The explanatory note with Table A2-9 explains how it works.

Table A2-9. Nutrient budget for maize, cassava and plantains in a 2-year cropping cycle with 4 years fallow

Line		N		P		K	
		Min	Max	Min	Max	Min	Max
1	**Initial balance**			**10**	**15**	**150**	**250**
2	Added from burned vegetation			14	16	100	100
3	Released by humus	41	59	2	4		
4	Leaching, denitrif., immobil.	−20	−29	−3	−3	−14	−26
5	**Available nutrients**	**20**	**29**	**24**	**31**	**236**	**324**
6	Potential (N-limited) maize yield	1,127	1,629				
7	*Maize yield, kg/ha*	*1,115*	*1,629*				
8	Taken up by maize	−20	−29	−5	−8	−21	−31
9	**Balance after maize harvest**	**0**	**0**	**19**	**23**	**215**	**293**
10	Returned by maize stover	7	10	2	3	17	24
11	Released over next 1.5 years	93	134	6	8		
12	New humus	−12	−17	−1	−1		
13	Leaching, denitrif., immobil.	−44	−63	−5	−7	−28	−51
14	**Available nutrients**	**44**	**63**	**21**	**27**	**204**	**266**
15	Potential (N-limited) cassava yield	11,883	17,125				
16	*Cassava yield, kg/ha*	*9,356*	*15,701*				
17	Taken up by cassava	−35	−58	−7	−13	−75	−126
18	**Balance after cassava harvest**	**9**	**5**	**13**	**14**	**129**	**141**

> **Explanatory note on P- and K-limited yields.**
>
> The model first calculates 'potential' N-limited crop yields (lines 6 and 15). If either P or K drops below the threshold, yield (and uptake) are adjusted as follows. Reduction factors corresponding with the actual P- and K-contents are calculated at the start of crop growth, assuming they decreases linearly from unity at the threshold to zero when available P or exchangeable K is depleted (for cassava they will both be unity according to line 13). N-limited yield is then multiplied by the correction factor to get a first estimate of actual uptake and yield. The reduction factors are then recalculated from the average P- and K-contents between the start and the end of the growth and uptake process.

Now let us look at the predictions of Table A2-9. Nitrogen is the most limiting nutrient for maize again and there will now only be enough for a meagre yield of about 1,100–1,600 kg/ha. For cassava, phosphorus is most limiting and the attainable yields are estimated at between 9.5 and 15.5 t/ha. Perhaps surprisingly, in view of its high demand, K is not predicted to be the main limiting factor for cassava. But when you look carefully at Table A2-9, it will be clear that K is following very closely. At the end of the cassava period, both nutrients will have crossed the threshold. Another effect of the shorter fallow is that the soil will become more acid and more phosphorus will be immobilised, so that P-deficiency may show up even earlier than predicted here. Even then cassava may still produce a reasonable yield because it can extract P from immobilised sources where most other crops cannot.

What happens when a second round of crops were grown before letting the land go back to fallow? From the bottom line of Table A2-9 it is clear that both P and K will be now limiting and that fertiliser will be needed to maintain reasonable yields. I will spare you the details, but the model estimates that without fertiliser the yield of maize will only be 650–900 kg and that of cassava 500–5,500 kg/ha. And of course there is the weed problem again, which will be even more severe than in the case of shifting cultivation. So growing another crop without external inputs is really a bad idea, as farmers know very well, of course.

It will be clear from these analyses that the predictions are very dependent on reliable estimates or good data on the P- and K-content of the soil and the fallow vegetation. The burned vegetation is a very important source of P and K. If the vegetation were removed from the field, as scientists sometimes do in their trials, instead of burning it, nothing much would grow without fertiliser, except perhaps cassava, which may still scrape up enough to produce something. Securing the nutrients contained in the vegetation does not necessarily imply that it must be burned, of course, but that is what farmers will usually do, for various reasons. If it is worked into the soil instead, the N will be conserved, the decline in organic matter will be slower and the nutrient losses smaller, but maize yield will be lower because of N-immobilisation by the decomposing biomass. But I think the main reason why farmers burn is really one of convenience – working a massive amount of biomass into the soil is almost impossible for a farmer with only hand tools or light equipment.

A2.3.3 Long-term prospects

Can we estimate the time horizon for recurrent cropping? We can try, as we did for shifting cultivation, but for recurrent cropping, things will be even more speculative, for mainly three reasons:

- The kind of fallow vegetation you get will not only depend on the duration of the fallow, but also on the natural richness of the soil. And the less well developed the fallow, the less successful it will be in tapping the nutrient resources in the subsoil.
- A related issue is the extent to which leached nutrients, in particular K, are recovered by the fallow vegetation.
- Much depends also on where you start: has the system already reached its new equilibrium or is it still changing in that direction?

I will choose a situation where the system of 4 years fallow alternating with 2 years of maize–cassava cropping has attained its

Table A2-10. Long-term nutrient availability for recurrent cropping

	P			K		
	Non-humus	Available	**Total**	Moderately bound	Exchange-able	**Total**
Poor soil	210	30	**240**	800	250	**1,050**
Good soil	460	40	**500**	2,200	400	**2,600**

equilibrium humus content and postulate the same nutrient reserves as in Table A2-6 (for shifting cultivation), except for a lower available P- and exchangeable K-content over 100 cm of soil depth (Table A2-10).

The export by the crop produce will be smaller because the yields are lower, but on the other hand losses of K by deep leaching cannot be neglected. Let us put the latter at 50% of the amount leached out of the topsoil (lines 4 and 13 in Table A2-9). Together the annual export and the leaching losses amount to 8–12 kg P and 67–116 kg K (never mind the suggestion of precision). As before, we estimate the durability of the system as the quotient of the total nutrient stock and the export per cycle. K will be depleted after 16–22 cycles, that is after about 90–130 years.

Appendix 3. Factor Analysis:
An Example

Like any other statistical technique, factor analysis is no substitute for good data and clear thinking. Even so, as long as you do not expect miracles, it is sometimes useful and always enjoyable to those who are not afraid of numbers. The joy is in seeing patterns emerge among variables, without the need for laborious calculations, because they can now be done in a breeze by the computer. But there is often disappointment when the patterns are difficult to interpret. I will demonstrate how it works with data collected from 40 farmer fields in south-western Nigeria in 1988. The crops were maize inter-cropped with cassava, but I will look at only the maize yields. Apart from yield, many other things were recorded such as planting date, shade, termite damage, weeds, plant stand at harvest, etc., and I will explore whether that information can help in understanding the large differences in yield (which varied from 200 to 3.840 kg/ha).

The most commonly used statistical technique for this kind of problem is regression analysis. I will explain in a few words how it works, just in case you are not familiar with it, because it helps to understand factor analysis. The things, which were measured in each filed, the variables, are shown in Table A3-1. There were 17 of them and they may be grouped in three categories as follows:

In 9 out of the original 40 fields, the data were incomplete, so those were not used for this analysis. In the remaining 31 fields, 17 observations had been made; hence, the complete data-set could be represented by an array with 31 rows, one for each field, and 17 columns, one for each variable.

A good way to start is by looking at all possible pairs of variables to see whether they appear to be related. In Figure A3-1 , three such pairs of interest were plotted: yield versus stand at harvest, yield versus weed infestation and clay versus K-content of the soil. The lines drawn here are the linear regression lines, meaning that they are the straight lines which most closely fit the data. The closer the actual data points are to the regression lines, the more strongly the two variables are correlated. That is expressed by the 'r' values in the graphs. They represent the strength of the correlation, which can vary between +1 and −1. Zero means that there is no correlation

Table A3-1. Variables measured in on-farm maize trials in south-west Nigeria, 1988

Maizecrop data	Disorders	Shade, soil
Planting date	Weeds (weighted score)	Shade (score)
Planting density	Termites (nr. of plants)	Silt, clay
Stand at harvest	Rodents (nr. of plants)	pH, organic C, P
Yield	Lodging (nr. of plants)[1]	Ca, Mg, K

[1]Plants fallen over by unknown causes.

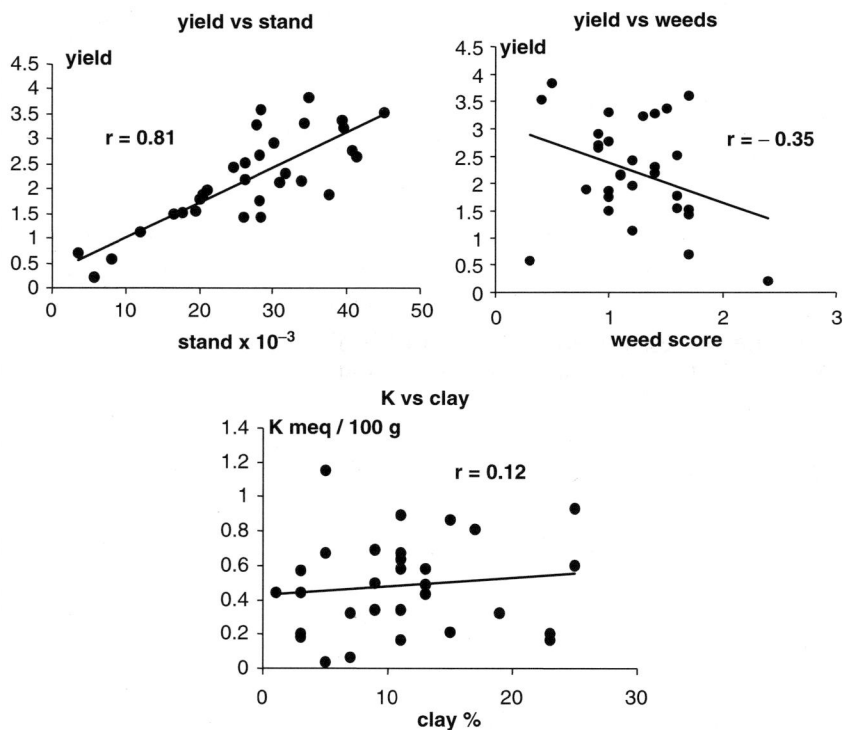

Figure A3-1. Linear regression and correlation for some pairs of variables

at all. Not surprisingly, yield is quite strongly correlated with stand at harvest, because the latter is a 'result variable'. If you knew only the stand, you would already have a fair idea of the yield that can be expected. The association between weediness of the fields and yield is of course negative, the more weeds, the lower the yield, but it is rather weak, and that between clay and potassium content of the soil is weaker yet, and positive.[1] You can easily compute all the correlation coefficients using a spreadsheet computer program or a statistical package and then look at all of them for a first idea about how the variables are related to each other and which are likely to be important.

We are of course most interested in the way yield is affected by the other variables. Their correlations with yield were always fairly weak (except of course for stand at harvest), the highest being for damage by rodents, −0.52, which itself correlated with stand, negatively of course, but not even that strongly (r = −0.35). But we suspect that the combined effect of several of the variables might explain considerably more of the yield differences. Instead of pair-wise regressions you can therefore look at the joint regression of yield on several or all of the variables, by what is called multiple regression. Result variables like 'plant count at harvest' are best excluded from the analysis, because they are really components of yield rather than directly causative (and they just eat up variability when left in). Multiple regression analysis is used a lot in all branches of science and, if done intelligently, it will help, but in the case of on-farm trials, every year the regression equation would look different because of the changing relative importance of the variables. For instance, there may be less rodents or there is more rain and the weeds become more troublesome. No statistical technique can beat that, of course, but multiple regression analysis may make things less rather than more transparent. Factor analysis may give a clearer picture, because, if the data are good, it may reveal important underlying properties, which cannot themselves be measured directly, although their importance will also vary from year to year. This probably all sounds rather cryptic, but I will try to clarify.

The idea of factor analysis is as follows. An object has certain properties which are important, but cannot be measured directly. In psychology, for instance, where factor analysis has been used most, such a property could be 'social intelligence' or 'perseverance' and the art of psychological test design is to find measurable variables, which are related to those underlying properties. In our case a property

---

[1] In a temperate climate this correlation would be much stronger, because the clay minerals' adsorption capacity is much higher there.

could be a field's 'quality of maintenance' (which usually says less about the field than about its owner) or 'overall fertility'. We hope to find properties of that kind, which have something to do with the fields' productivity. You can of course determine a field's productivity by simply growing a crop and measuring the yield, but that only tells you how productive the field was in that particular year, not the reasons why that was so. A number of things may be important for crop yield, like the soil's nitrogen content and the weediness or shadiness of the field, but you hope that there are some broader underlying field properties, which somehow give rise to those measurable variables. If there is such a thing as 'overall fertility', for example, that could be expressed by the soil's organic-C and cation content and its pH. Factor analysis tries to reconstruct such properties (called factors) from the values of many observed variables, which are expected to be linearly related to those properties. I am afraid that is only a little less cryptic, so let us look at the real data. I used Jerry Hintze's powerful yet user-friendly NCSS statistical package[2] for the calculations.

We are searching for a small number of factors which together 'explain' as much as possible of the variation in the original variables over all the fields. If the measured variables were well chosen and the data are good, these factors may capture important underlying field properties. And we hope that some of them will correlate with the productivity of the fields, so we also include 'yield' as a variable (but not plant count, for the reason explained above). The analysis of the maize data showed that most of the variation between fields was taken care of by just five factors, which are given in Table A3-2. The 'factor loadings' of the measured variables are the correlation coefficients between those variables and the new factors. For example, the loading of factor 3 with the soil's silt content was 0.42. If the factor loading for a particular variable is low, that variable has little or nothing to do with the factor. The table shows the factor loadings for each of the five factors when they are larger than 0.40; the rest are ignored.

The results are quite interesting. In a field which scored high for factor 2, there would have been little weeds (or they were well

---

[2] Number Crunching Statistical System; visit www.ncss.com for information on the package.

Table A3-2. Factors, variables and their factor loadings for the same trials as Figure A3-1 and Table A3-1

Factor 1		Factor 2		Factor 3		Factor 4		Factor 5	
Variable	Loading	Variable	Loading	Variable	Loading	Variable	Loading	Variable	Loading
K	0.52	Rodents	−0.82	Termites	0.95	Pl. date	0.47	Shade	0.52
Ca	0.83	Weeds	−0.63	Silt	0.42	pH	0.74	Org. C	0.72
Mg	0.82	Yield	0.76	Yield	−0.44	P	0.88	Clay	0.68

controlled), little rodent damage and good yield. And a field which scored high for factor 3 was affected by termites and tended to have low yield. If we want to attach names to these factors we could say that factor 2 represents the 'quality of maintenance': poor maintenance results in high weed pressure, risk of rodent damage and low yield. Factor 3 measures the field's 'termite infestation': a lot of termites is (weakly) associated with the soil's silt content and results in low yield. The factors which were associated with yield (2 and 3) together explained an important proportion (78% in fact) of the yield differences.

The other three factors are largely descriptive of the soil. For example, factor 5 shows an association of shade, organic-C and clay content. A high value of these variables is often found in cocoa or former cocoa plots, so this factor characterises 'cocoa field conditions' and fields scoring high for factor 5 were probably former cocoa orchards, now being used as cocoa fields. None of these soil factors made an appreciable contribution to yield, though. The reasons why that is so are explained in Chapter 6.

Factor analysis did quite a nice job here, for this particular year, but in other years (and in different fields) things looked quite different. For example, shade played an important role in 1986, but not in 1988, because there were simply fewer shady fields in the sample. And in 1987, there was a very strong effect of planting date, which was caused by a pronounced dry spell in April 1987, after several fields had already been planted. The fields where planting had been postponed did much better than the early planted ones. Incidentally, this shows the wisdom of most farmers' habit of staggering their planting dates. So, every year is different and you cannot simply assume that the conclusions from one year are relevant for another. We can only hope that some factors will turn up regularly and those are the ones the extension people can base recommendations on. Otherwise, a lot of uncertainty remains. Good farmers will be able to anticipate this uncertainty and make the right decisions to minimise the effect of unexpected events as they occur, which is a skill that is difficult to define. The only thing extension and research can do is, find the few general principles, which will always apply, like the importance of good weed management and staggered planting for risk reduction.

There is another indication that the study only partially captured good management: the 'communality' of yield with factors 2 and 3, that is the percentage of the yield variation explained by those factors, was 78%, leaving 23% unexplained. Assuming that 5% of the variation is due to unexplainable causes, whatever they may be, that still leaves almost 20% unaccounted for. Perhaps that is where you also find some of the farmers' intuitive skills which are so difficult to lay your finger on.

Even so, it is moderately useful to keep looking for major yield-determining factors. Studies like this, carried out over a number of years, may find a few robust factors, and that would help the extension service, because they are easy to transfer. But the studies are quite costly and the factors you find may be little more than an experienced agriculturist could have told you right from the start, or what you could have found out yourself by keeping your eyes open. They are unlikely to capture the essence of farmers' skills, which is doing most things right most of the time.

Appendix 4. Nutrient Dynamics of Alley Cropping: A Simple Model

A4.1 Another cautionary note

Before studying this analysis of the nutrient cycles in alley cropping, the reader is reminded of the cautionary note at the beginning of Appendix 2. What is said there about the uncertainties and the sometimes sweeping assumptions in the nutrient budget for shifting cultivation applies a fortiori here for alley cropping. For the rest, I suggest that you just swallow what may seem unpalatable at first and suspend judgement until the final digestion.

A4.2 Nutrient flows and nutrient stocks[1]

In alley cropping, crops are grown in the alleys between permanent hedgerows, usually of a leguminous tree or shrub species. The hedges are regularly pruned until the crops have been harvested after which they are allowed to grow undisturbed, until the next cropping season. The prunings remain in the field and decompose gradually. We want to calculate whether permanent cropping in the alleys is possible in the semi-deciduous forest zone with two rainy seasons in West Africa. That was the environment where the technology was originally developed. First we look at maize grown in the first season, followed by cowpeas in the second. That is not a very common sequence, but those were the test crops in much of IITA's earlier alley cropping research, so at least we have some field data to check the predictions against. Later on we also look at the more common maize–cassava crop mixture.

As befits a scientific treatise (if what follows deserves that name) we start with a hypothesis:

alley cropping, established after a long fallow in the semi-deciduous forest zone can maintain maize and cowpea yields indefinitely without fertiliser and without the need for further fallowing.

---

[1] For the nutrient budgets I have drawn extensively on Kang et al. (1990) and other publications mentioned in the text, sometimes without explicitly referring to them.

The nutrient budget should show whether that is possible, and if not, why not. In practice it is very unlikely that farmers will voluntarily start alley cropping in freshly cleared forest land where soil fertility is high. But the soil conditions under shifting cultivation have been well studied by Nye and Greenland and by taking their data as point of departure, the alley cropping budget will need the smallest, though still significant amount of number acrobatics. If it works well, we can extend the analysis to less favourable and more exacting conditions with some confidence.

We are going to put together a spreadsheet model which calculates crop yield under alley cropping. Four things are needed to do that:

- The amounts of nutrients removed from the field with the crop produce, whereby we will assume that the crop residues stay behind and will eventually release their nutrients into the soil again.
- The soil's stock of nutrients, in particular those which are immediately available or can be extracted in the medium term.
- The ability of the crops and the hedgerows to extract nutrients from the topsoil and the subsoil.
- The mechanisms by which all these components are wired together into an integrated 'system'.

A4.2.1 Nutrient removal by the crops

The first thing the model needs to know is how much nutrients must be taken up to produce a unit of crop biomass. These I put equal to amounts of nutrients found in the different plant parts, which are shown for maize, cowpeas and cassava in Table A4-1.

Table A4-1. Nutrients taken up to produce 1,000 kg of maize and cowpea grain and 10,000 kg of cassava roots, at 'low to moderate' yield levels

Crop	Kg of Produce	Kg in the produce			Kg in stems, leaves, roots			Total nutrients, kg		
		N	P	K	N	P	K	N	P	K
Maize	1,000	12	2.7	4.0	6	2.0	15	**18**	**4.7**	**19**
Cowpeas	1,000	35	3.5	10	20	3.5	17	**55**	**7.0**	**27**
Cassava	10,000	22	5.0	45	15	3.0	35	**37**	**8.0**	**80**

The maize and cassava figures are the same as those used in Appendix 2, while the cowpea data are averages from various published sources. These figures will be used in the nutrient budget.

Our hypothesis stated that alley cropping should maintain maize and cowpea yields indefinitely, so we need to know first how much they would yield approximately in a forest soil with a high humus content immediately after a long fallow. That will be the yardstick for the success of alley cropping. It was shown in Appendix 2 that in the first year after clearing the fallow, maize yield will be constrained by the amount of nitrogen released by the organic matter, so if we know how much nitrogen is released and which part of it will actually be available we can estimate the yield. The calculations are shown in Box A4-1 for a modest humus-C content of 48,000 kg/ha. The maize yield works out at 2,000 kg/ha, provided growth is not limited by other factors. The calculations are the same as those in Appendix 2.

For cowpeas, things are a little more complicated. Its yield cannot be related to a single dominant factor, because cowpea is a leguminous crop, which fixes its own nitrogen, but less so as there is more in the soil, which it can just take up. Another complication is that, yield is depressed when the cowpeas are grown in alleys, because of shading and possibly by competition for moisture. Rather than trying to simulate all those difficult things I have simply postulated ceiling yields: 900 kg/ha without alleys and 500 kg/ha under alley cropping. Both figures were the kind of yields observed in many years of field trials at IITA. The 'actual' yields will be simulated by adjusting growth to the available P and K.

Box A4-1. Humus content and N-limited maize yield after long fallow

1. The soil in a long fallow system will contain 42,000–60,000 kg of organic C per hectare (Appendix 2). For the nutrient budget in alley cropping I will start with a figure of 48,000 kg, which is equivalent to 1.63% organic-C content.
2. With a C/N ratio of the organic matter of 12 and a decomposition rate of 3%, the annual release of N will be: $48,000 \times 0.03 / 12 = 120$ kg/ha.
3. If 60% is released in the first season and 50% is leached that leaves 36 kg for the maize.
4. According to Table A4-1, maize takes up 18 kg N to produce 1,000 kg of grain, so from 36 kg you can get a maximum of 2,000 kg of maize grain, *if there are no other limiting factors.*

If there were no hedgerows, crop yield under continuous cropping would soon start to decline. In Appendix 2 we found, for example, that for a crop combination of maize–cassava–plantain to maintain the same yield level, all three major nutrients N, P and K would have to be supplemented by fertiliser as early as the second cropping cycle after a long fallow. In alley cropping, however, the hedges will pump up K and put it back in the topsoil, unlock P from less accessible sources and maintain the soil's humus content by its leaf litter and prunings. And cowpeas are also much less K-hungry than cassava and plantains. But can the hedges really put enough nutrients in the topsoil to maintain yield, even of the less exacting maize–cowpea sequence? If that were so, we would have done something quite remarkable: producing yields similar to those under shifting cultivation, at one-tenth of the area. But it does not come free: the extra cost is more work and, most importantly, being more alert because you have to keep the hedges trimmed while there are crops growing between them, otherwise you may get much less or even nothing at all when the field gets very shady. Even so, what we hope alley cropping can do still sounds like magic. The nutrient budgets, which I will introduce shortly should show where the magic ends.

A4.2.2 The soil's nutrient stocks

Before we can put the budget together, however, there is a lot of preparatory work to do. In particular the soil's nutrient stocks must be carefully estimated, because their size largely determines the durability of the alley system. In the following paragraphs I will first determine broad ranges of available and not so available nutrients as they occur in West Africa from published sources. Then I will define the nutrient status of a relatively poor soil, as the majority of soils in lowland West Africa are, but one with a high humus content built up through many cycles of shifting cultivation. That will be the benchmark for the first round of model testing.

(a) Humus and nitrogen

Most of the nitrogen on which the crops feed comes from decomposing humus, so if you want to get the same yields year after year, the humus content must remain stable so that the same amount of N will be released. New humus must therefore be built at the same rate as old humus is broken down. Its raw material are the crop residues, which are left in the field and, most importantly, the hedgerow prunings. That biomass must be produced first, however, for which a lot of nitrogen is needed – that is why nitrogen-fixing leguminous species are preferred for the hedges.

How much biomass must roughly be produced to keep the soil's humus content stable? The calculation in Box A4-2 shows that in order to maintain humus-C at 48,000 kg of humus-C, about 5,800 kg of pruning biomass is needed annually if all of it were to come from the hedgerow prunings alone. Experimental data show that the most commonly used species, *Leucena leucocephala,* planted in hedgerows 4 m apart, can easily do that, once it is firmly established and provided other essential nutrients are in sufficient supply. *Leucena* can produce up to a maximum of 9,000 kg of biomass dry matter per hectare per year, as long as there is no shortage of P and K (or other nutrients of course, but I will ignore those). When the budget predicts such shortages it must adjust growth of the crops and the hedgerows accordingly, as we will see later.

Box A4-2. Annual amount of hedgerow prunings needed for a stable humus content

---

1. At an organic-C content of 48,000 kg/ha and an annual decomposition rate of 3% about 1,450 kg is broken down annually. That same amount of fresh organic-carbon must be formed to keep the soil's humus content stable.
2. The conversion rate of organic matter to organic-C for fallow vegetation is around 40%, but for N-rich prunings from leguminous species 25% is more likely.
3. Hence, to produce 1,450 kg of new humus from hedgerow prunings alone, about 5,800 kg of them must be worked into the soil annually.

(b) Available and not so available phosphorus and potassium

I will assume that part or all of the nutrients, which the crops take up leave the field with the produce, never to return. For crop yields to remain stable, the P and K taken out must be replenished from somewhere in the soil profile, or from fertiliser or manure, otherwise the crop yields must eventually go down. There are large differences in the amount of nutrients tropical soils contain and in the amounts, which are available to the plants (that is not the same thing), so we should also expect large differences in the soil's ability to sustain long-term continuous alley cropping. Soil laboratories use standard tests to measure available nutrients, but the interpretation of the test results is not as straightforward as the word 'available' suggests. And furthermore, the tests usually measure what is available now, not what may become available over a longer period of, say, 20 years, and that is what we need to find out as well. Surprisingly little is known about long-term nutrient release by the soil. Perhaps we should measure the total amount of nutrients stored in different forms in the soil in order to predict what may be expected in the medium to long term? We could, for instance, use aggressive chemicals which will even break open undecomposed minerals, but that is of little use, because some of the nutrients are so tightly bound that it takes decades or even centuries for them to be released naturally. The amounts available over a period of, say, 20 years must be somewhere between what is immediately available and the total stock. Let us see how much we know. You should take the figures I will present with a grain of salt, but they should be in the right order of magnitude.

Phosphorus

First phosphorus. Table A4-2 shows my estimates for different forms of phosphorus in a 'good soil' and a 'poor soil'. The data are based on the handbook *Properties and Management of Soils in the Tropics* by Pedro Sanchez (who got most of them from somewhere else again), but do not blame him for what I did with the figures.

For our purposes the important P-sources are the available and moderately bound forms. Humus-P, which is almost completely located in the topsoil, is in the latter category. Very tightly bound P is not accessible for plants, although a trickle may be released into the other pools. I will ignore that for the budget.[2]

'Available P' means what it says, available to the plants, but scientists have had great difficulty to measure that. They have proposed different extractants to mimic the way plants extract P from the soil, none of them giving reliable results for all types of soils. I will not go into that and just assume that the authors, whose figures we will use, knew what they were doing and that their figures did indeed

---

[2] The fact that the totals for tightly and moderately bound P in the table are the same is not a mistake, it just means that the amounts are expected to be roughly similar.

Table A4-2. Average contents of different forms of phosphorus in West African soils, kg/ha

| Depth (cm) | 'Available' P | Moderately bound P | | | Very tightly bound P |
		Non-humus	Humus	Total	
'Good soil'					
0–30	40	160	240[1]	400	400
30–100	30	300	–	300	300
Total	70	460	240	700	700
'Poor soil'					
0–30	22	80	145[2]	225	223
30–100	23	220	–	220	225
Total	45	305	145	450	450

[1] OC content of 1.63%.
[2] OC content of 1%.

measure what plants can extract. The next question is how easily they can extract it and that depends also on how much there is: the more the easier it is taken up. Soil scientists therefore work with threshold values. Below the threshold the soil is said to be P-deficient. It usually means that it is expected to respond to P-application *when the other nutrients are available in optimal quantities.*[3] Reported thresholds for maize are usually around 10 ppm (parts per million) of available P, that is 10 mg/kg of dry soil.

What about the moderately bound form? In the tropics a large part of that is in the humus and will be slowly released into the available pool as the humus decomposes. Box A4-3 shows how to calculate the amount of P locked up in the humus, the rest is in the form of insoluble inorganic phosphorus compounds. There is a large difference between species in their ability to extract phosphorus from the insoluble form. Maize is particularly poor in that respect, while cassava and cowpeas are more capable. Trees can also extract more P than most crop plants, because of the association of their roots with the mycelium of soil fungi, called mycorrhiza.

---

[3] In Appendix 2 these things were explained in more detail and illustrated with a graph.

Box A4-3. How to calculate the amount of P in the humus

1. The average ratio of carbon and P in soil humus (C/P ratio) is about 200.
2. If we make the reasonable assumption that the humus is concentrated in the top 20 cm, a soil with an average of 1% organic carbon in that layer contains $2 \times 1.47 \times 10^6/100 = 29,400$ kg of C (see note 2 in Appendix 2).
3. The poor soil of Table A4-2, with an organic-C content of 1% therefore contains 29,400/200 = 147 kg of humus-P (rounded to 145).
4. Our model soil with 1.63% organic-C contains 48,000 kg of organic-C and 48,000/200 = 240 kg of organic-P.

Potassium

For K, I used data published by A.S.R. Juo (for many years IITA's soil chemist), and H. Grimme (1980), compiled in the following rough table, which shows different forms of K, for a 'good soil' and a 'poor soil' in West Africa. I am not sure the authors will like this,[4] but again they are not responsible, it is my interpretation. They distinguished three forms of K: (i) very tightly bound, which is locked up in undecomposed minerals and comes available very slowly over a long period; (ii) moderately bound, also called non-exchangeable K (an odd name since the very tightly bound form is even more non-exchangeable), which is in equilibrium with and slowly replenishes the (iii) exchangeable K, adsorbed to the clay–humus complex.

The exchangeable K-contents in the table are close to those reported by Nye and Greenland (Appendix 2). Exchangeable K is in equilibrium with the soil solution and the less there is at the adsorption complex, the lower the concentration in the solution and the more difficult it is for the plants to take it up. Soil analytical laboratories report exchangeable K-content in terms of milli-equivalents (meq) per 100 g of dried soil[5] and soil scientists have established thresholds, below which a soil is considered as K-deficient. For maize, a threshold of 0.15–0.20 meq is often used, but, like the P-threshold, it applies when everything else is in optimal condition.

---

[4] Tony Juo is no longer in a position to object as he passed away in 2005.
[5] one milli-equivalent K is one-thousandth of a gram-equivalent which is the weight of one gram-atom, i.e. 39 grams (the atomic weight of K is 39).

About non-exchangeable K, Juo and Grimme remarked in 1980: "The rate of release is often too low to meet the K demand of vigorously growing crops [...]. Much research is needed regarding the contribution of non-exchangeable K to the growth of cassava and other slow-growing [...] crops", which is the scientist's equivalence of an admission of ignorance. Surely, more than 20 years onward that should have been resolved? It had not. Moritsuka et al. from Kyoto University in Japan stated at a conference in Thailand in 2002 that "the source and the releasing processes of non-exchangeable K from the rhizosphere is not understood well for natural soils...". So, the only thing we can do is some intelligent guessing backed up by the little that is known. Two things are clear. Earlier work by Grimme (quoted in the Juo and Grimme paper) showed that more non-exchangeable K will be extracted as the exchangeable K gets depleted and a crop, which depends on non-exchangeable K alone will yield far below its potential, somewhere between 10% and 25%. And Moritsuka et al. showed that only about 1–4.5% of the K-uptake by maize in a 'natural soil' (with a lot of exchangeable K) came from non-exchangeable sources. Those are the figures I will use for the nutrient balance model.

A4.3 The nutrient budget for a 'poor soil'

We are now ready to predict what will happen over time when each year a crop of maize followed by cowpeas is grown in the alleys on a 'poor soil' in the West African sub-humid forest zone. We start with a high humus- (and thus topsoil P-) content because the land was assumed to have been under shifting cultivation for many generations. Farmers are unlikely to plant hedgerows in such a soil, though – if they were interested at all they would probably wait until fertility had declined. The purpose of the alleys would then be restoration of fertility rather than maintenance, but in this first exercise we will look at fertility maintenance. The calculations should show whether alley cropping can actually maintain humus content of a former shifting cultivation field and whether enough nutrients can be mobilised to keep the system going for a long time. We will look at other, probably more realistic situations as we go along.

Simulating nutrient dynamics in the fairly complex alley cropping system is typically the kind of thing dynamic computer modelling was designed for,[6] but I will use a fairly simple spreadsheet budget instead, which substitutes simplicity for the semblance of sophistication. For both methods, the reliability of the outcome is only as good as the validity of the assumptions, which in the spreadsheet case are visible to the naked eye. The simulation starts with a field, freshly cleared from forest fallow and then cropped annually for 20 years. I will explain, step by step, how that was done, whether I used real data or assumptions, what the model predicts in respect of the evolution of fertility and crop yields and how the predictions compare with field measurements. But before doing that I will first give a broad synopsis of the processes, which are at play in the growth of crops under alley cropping.

A4.3.1 A brief synopsis

In a stable alley cropping system, humus breakdown must be matched by fresh humus build-up from the prunings and the crop residues, otherwise humus content would decline slowly, a little less nitrogen would be released every year and yields would inexorably go down. We will see that the prunings of vigorously growing hedges actually contain more nitrogen than is needed to build up the new humus, so the soil's nitrogen status will be even better than in a system with the same soil organic matter but no alleys. Immediately after planting the hedges, however, there may actually be N-shortage because of the conversion of crop residues into humus. That is because the nitrogen content of the residues, especially maize stover, is lower than that of humus, so their humification will immobilise some free nitrogen causing more acute N-shortage for the next crop.[7] As the amount of biomass from the hedgerows prunings gets larger, however, a nitrogen surplus will build up and the crop yields should actually go up. With P, something similar happens. The decomposing prunings and crop

---

[6] Dynamic modeling simulates processes in fairly great detail in a stepwise fashion, with timesteps of, say, 1 day. How that works is one of the topics of Chapter 9.

[7] That process is also responsible for the poor growth of (unfertilised) maize in a savannah field after incorporation of a grassy vegetation into the soil.

residues release P, part of which is built into new humus, while the surplus remains in the topsoil and can be taken up by the crops along with the P from older decomposing humus.

The hedges themselves obtain their P and K mostly from lower soil levels, because their roots are forced down by tillage, except in the first year or two. Their leaf litter and prunings will be deposited on the soil surface and the nutrients, which are released from them replenish the available P and the exchangeable K at the adsorption complex.

Nutrients do not just stay in the topsoil waiting to be taken up or be converted into humus. Nitrogen (or rather the nitrate ion) is quite mobile and a lot of it is leached into the subsoil if the rains exceed evapotranspiration. Part of it is leached even further down, out of reach of the hedgerow roots, while another part is lost by conversion into gaseous forms, by a process called denitrification. The nitrate ions drag along positively charged ions, in particular K^+, which is therefore also sensitive to leaching. Available P is not very mobile but it may be converted into an insoluble form, at a rate which depends on soil type and soil acidity.

As we have seen in Appendix 2, nitrogen will be the first nutrient limiting the growth of crops like maize and cassava after a long fallow. An organic-C content of 1.63% allows maximum maize yields in the first year of about 2,000 kg. Nitrogen is not the major limiting factor for cowpeas so I have adopted a ceiling yield of 900 kg while the hedgerows are small, going down to 500 kg once the alleys are well established. That is not caused by shortage of plant nutrients, but by shading from the hedges.

Part of the soil nutrients leave the system altogether each year – they are removed with the crop produce by the farmers or they are irretrievably lost by other causes, N by denitrification, and N and K by deep leaching. Everything else flows around within the soil–plant system: from the plant residues and the humus into the topsoil, from the topsoil into the crops, the humus and the subsoil, from the subsoil into the hedges and from the hedgerow prunings back into the topsoil. How much will be where at what time determines how much the crops can take up and how much they will yield. The exciting thing about nutrient budgets is that you can actually calculate these flows, rather than just talk about them in an imprecise way. That is what we are going to do now.

A4.3.2 Initial conditions

The initial conditions are very important. We start with a field with a
high humus content, the cumulative result of many cycles of shifting
cultivation. The vegetation has been cleared and burned in situ, and
the hedgerows and the first maize crop have been planted. It is unlikely
that farmers would be interested in alley cropping under such conditions, but that is another matter, which is discussed in Chapter 7. I
have further postulated the native chemical fertility of a 'poor soil', as
shown in Tables A4-2 and A4-3, except for the N and P associated
with the humus, which is that of a rich soil. Furthermore, the soil is
enriched in the first year by the nutrients from the burned fallow vegetation. The complete initial nutrient contents are shown in Table A4-4.
They are made up of the N and P in the humus, the native P- and
K-stocks from Tables A4-2 and A4-3, and the P and K added by the
ashes. Box A4-4 explains how the table was put together.

Table A4-3. Forms of potassium in West African soils, kg/ha

Depth (cm)	Exchangeable	Moderately bound	Very tightly bound
'Good soil'			
0–30	400	800	8,000
30–100	700	1,400	12,000
Total	1,100	2,200	22,000
'Poor soil'			
0–30	200	380	3,800
30–100	300	420	4,200
Total	400	800	8,000

Table A4-4. Initial nutrient and organic-carbon contents of a poor soil
after long fallow

Phosphorus, kg/ha				Potassium, kg/ha				OC, kg/ha
0–30 cm		30–100 cm		0–30 cm		30–100 cm		
Available	Moderately bound	Available	Moderately bound	Exch.	Non-exch.	Exch.	Non-exch.	
48	80	23	220	450	380	300	420	48,000

> **Box A4-4.** Initial humus, P- and K- contents
>
> 1. The organic-Ccontent is set at a conservative value of 48,000 kg, within the range for soils after many long fallow cycles (see Appendix 2).
> 2. Available P and K in the topsoil are taken from Tables A4-2 and A4-3, augmented by 250 kg of K and 26 kg of P from the burned vegetation (Nye and Greenland's figures).
> 3. Non-humus moderately bound P in the topsoil is that of a poor soil, 80 kg/ha(Table A4-2).
> 4. Moderately bound ('non-exchangeable') K-content of the topsoil and P- and K-contents of the subsoil are those for a 'poor soil' from Tables A4-2 and A4-3.

These figures will form the first row of the spreadsheet program that calculates the nutrient budget.

A4.3.3 Nutrient dynamics in the hedgerow–crop system

I will explain the processes, which take place in an alley cropping field at two levels of detail. The main text explains each process in broad terms, whereby reference is made to the numbered rows in the spreadsheet of which a fragment is shown in Table A4-5. Details of the calculations are explained by the explanatory notes accompanying the table, to be studied by those with a taste for detail and ignored by others. It is a bit of a puzzle, and some of the assumptions will be recognised as rather wild guesses. but at least they are out in the open and can be replaced by better ones, or by real hard information.

(a) Nitrogen, phosphorus and potassium cycling

The calculations start from the initial conditions of Table A4-4, which is also line 01 of the spreadsheet model (Table A4-7). The topsoil is continually replenished with N, P and K from decomposing crop residues and prunings (line 03) and N and P from decomposing humus (line 04). At the same time new humus is formed from hedgerow prunings and crop residues (line 05). According to Nye and Greenland the rate of conversion of forest biomass to organic C averages around 40%, but I used 25% instead, because the bulk consists of young shoots pruned off the leguminous hedges. They also contain more N and P than is needed for the new humus, so some of it remains in the topsoil in available form.

Table A4-5. Detailed model calculations in year 5 for a maize-cowpea crop in a 'poor soil' after a long fallow; intial conditions as in table A4-4

Line		Nitrogen, kg/ha		Phosphorus, kg/ha				Potassium, kg/ha				Yield, t/ha			Soil OC
				0–30 cm		30–100 cm		0–30 cm		30–100 cm					
		0–30 cm	30–100 cm	Available	Bound	Available	Bound	Exch.	Non-exch.	Exch.	Non-exch.	Maize	Cowpeas	Prunings	kg/ha
01	Initial status	0	0	48	80	23	220	450	380	300	420				48,000
	Balance year 1	11	23	35	85	23	220	354	378	325	419	1.95	0.86	0.35	46,560
	Balance year 2	16	28	27	89	21	219	302	377	321	418	1.96	0.48	1.38	45,451
	Balance year 3	13	7	17	93	14	215	244	375	236	411	2.12	0.48	5.28	44,547
02	Balance year 4	13	8	11	94	8	210	242	373	167	399	2.46	0.45	5.33	44,308
	Year 5														
03	Biomass decomposition	184		19				191							
04	Humus decomposition	111		7											
05	New humus	-136		-8											44,612
06	Balance 1	173	8	29	94	8	210	433	373	167	399				
07	Leaching+ denitrification	-86						-48		48					
08	Immobilisation		69	-3	3										
09	Balance 2	86	77	26	97	8	210	385	373	215	399				

10	Uptake maize	-50		-13	0			-52	-1			
11	Balance 3	36	77	13	97	8	210	333	371	215	399	2.80
12	Uptake cowpeas	-11		-2	-1			-10	0			
13	Balance 4	25	77	11	97	8	210	323	371	215	399	0.39
14	Uptake hedgerows	-8		-2	-1	-3	-5	-35	-1	-74	-10	
15	Balance 5	17	28	8	96	5	206	315	288	370	141	
16	Leaching losses		-14								0	
17	Balance year 5	17	14	8	96	5	206	315	288	370	141	4.15

The table gives the details of the model calculations for year 5, when the hedges have attained their full development and the growth processes have become steady. With some effort and the details given below it should be possible to recreate the spreadsheet. Do not be put off by the rather preposterous precision of the figures. They are just as they arise from the calculations, that is how modelling works. What counts are the orders of magnitude and the trends.

Biomass decomposition

The nutrient contents of the prunings, taken from Kang et al. (1981), are N: 3.2%, P: 0.27% and K: 2.9%. The nutrients removed by the crops were taken from table A4-1. The weight of the dry matter in the maize residue is assumed equal to the grain weight. The cowpea hay is taken out by the farmer, as he normally would. The nutrients from the maize residues of maize of year 4 and half the prunings each of year 3 and year 4 are assumed to become available in year 5, and so on. Hence, the nutrients released in year 5 are, for N: $2.46 \times 6 + 0.5 \times (5.33 + 5.28) \times 32 = 184$, for P: $2.46 \times 2 + 0.5 \times (5.33 + 5.28) \times 2.7 = 19$, for K: $2.46 \times 15 + 0.5 \times (5.33 + 5.28) \times 29 = 191$ kg/ha.

(continued)

Table A4-5. (continued)

Humus decomposition

The organic-C content at the beginning of year 1 is set at 1.63% and the model calculates corresponding amount of humus-C as 48,000 kg/ha. At the end of year 4, it has gone down to 44,308 kg. At a decomposition rate of 3%, 1,329 kg humus is broken down in year 5. At a C/N ratio of 12 and a C/P ratio of 200 it releases 111 kg N and 7 kg P in year 5.

New humus

The conversion rate of all residues and prunings is set at 25%, while only 50% of the maize is converted. The rest is decomposed on top of the soil or eaten by termites. The decomposed maize residues from year 4 and the prunings from year 3 and 4 add $0.25 \times (0.5 \times 2.46 + 0.5 \times (5.33 + 5.28)) \times 1000 = 1{,}634$ kg of fresh humus-C, containing 136 kg N and 8 kg P/ha. The balance of humus build-up and breakdown in year 5 is +304 kg.

Leaching and denitrification, immobilisation

About 40% of the N, in balance 1, is leached to the subsoil and 10% is denitrified; 15% of the available P is immobilised; the remainder (balance 2) is available for uptake by the crops.

K-leaching is a function of nitrate leaching and exchangeable K-content. Nye and Greenland put the percentage of the nitrate ions taking along K at 30%. The model postulates a linear trend with exchangeable K-content, such that it becomes zero at 100 kg and 30% at 500 kg exchangeable K per ha.

P-immobilisation is set at 10% per year. It is strongly affected by soil type, so this parameter will be higher in soils exhibiting P-fixation.

Crop yields, nutrient uptake by crops

Potential maize yield is set proportional to the N available in the first season (60% of balance 2). Cowpeas take up all the N they need when available N exceeds 60 kg/ha and proportionally less between 60 and 10 kg. Potential cowpea yield is set at 900 kg/ha in the first year and 500 kg thereafter, for reasons explained in the main text. Actual yield of both crops is calculated by applying a reduction factor depending on available topsoil P or K, whichever is the most limiting. The factor decreases linearly from unity at the threshold (20 kg P or 170 kg K) to zero when all the available nutrient is depleted.

Maize takes up available P only while a minimum of 5% of its K-need comes from non-exchangeable K, increasing to a maximum of 25% as the amount of exchangeable K is smaller.

Cowpeas take up a minimum of 10% of P from bound (non-humus) P, increasing to a maximum of one-third as the amount of available P decreases. For K, the same strategy is used as for maize.

Growth, nutrient uptake and pruning yield by hedges

In the first year, the hedgerows behave as a crop and feed on nutrients from the topsoil. Thereafter uptake shifts from top to subsoil. From year 3, the hedges take 80% from the subsoil.

N-, P- and K-uptake by the hedgerows is calculated in the same way as for cowpeas.

Maximum pruning yield is set at 9 t/ha/year. Below the thresholds for available P and K (45 kg P and 400 kg K, because of the larger subsoil volume) pruning yield is constrained in the same way as crop yield, after discounting uptake by the crops.

Irretrievable leaching

Irretrievable leaching of N and K from the subsoil is treated in the same way as leaching from the topsoil, after discounting uptake by the hedges.

Another group of processes will deplete the available stocks again: leaching of N and K, and denitrification (line 07) and immobilisation of P (line 08). N and K leaching are coupled. The negatively charged nitrate ions are most sensitive to leaching and they drag along positively charged cations, in particular K^+, which is most abundant. The higher the K-concentration in the soil solution, the more of the leached nitrate ions will be accompanied by K. Nye and Greenland estimated that in the first season after a long fallow, 30% would be accompanied by K. Part of the N and K, which is leached into the subsoil will be recaptured by the hedgerow roots (line 14) and returned to the topsoil with the prunings. Ideally, nothing would be lost, but that is unlikely, because the activity of the hedgerow roots is reduced during the cropping phase when there is a net downward movement of moisture. The model treats subsoil leaching of N and K in the same way as that of the topsoil (line 16), after accounting for uptake by the hedges (line 14). That perhaps overestimates subsoil leaching because the downward movement of moisture is less, but the effect will be small anyway.

(b) Nutrient uptake by crops and hedges

Maize growth is primarily restricted by nitrogen, even under the comparatively lavish conditions immediately after a long fallow. The nutrient budget therefore first calculates how much growth the available N would allow and then adjusts it to the soil's P- and K-content if there is not enough to match the available N. The difficult question, however, is what is enough? Obviously, plants cannot just take up all the P and K they need until everything is finished. At some point, uptake will decline because the concentration in the soil solution becomes too low.[8] We have seen that soil scientists have set thresholds for P and K, about 10 ppm and 0.15–0.2 meq/100 g respectively, which hold when everything else is ideal, for instance there is no N-shortage. In that case N-uptake will be high and so is the amount of P and K, which must be taken up to match it. When yield is limited by nitrogen, however, like in our case, the crop needs much less of the other nutrients and they may still be able to take up enough P and K, even when available P and exchangeable K are well below the threshold. That was explained in more detail in Appendix 2. For the

---

[8] That also applies to N of course, but N is much more free-moving than P and K. In fact, I have simply ignored it.

budget, I simply assumed that under the conditions obtaining in our model field, the threshold below which uptake becomes depressed are 4.5 ppm for available P and 0.1 meq for exchangeable K, taken over the 30 cm soil depth from where the crops are expected to take all their nutrients. That is equivalent to 20 kg P and 170 kg K/ha.

Potential yield of the cowpeas is set to 900 kg in the first year and 500 kg/ha later, for reasons explained earlier. Their actual yield depends on available topsoil P or K, whichever is most limiting (line 12). K-uptake by cowpeas is handled by the model in the same way as for maize, while P-uptake is treated in the same way as that of the hedgerows (discussed below), both being leguminous species, which should be able to extract P more easily than non-legumes. When there is a lot of nitrogen in the topsoil, cowpeas take up all the N they need instead of fixing their own, but the uptake decreases in favour of symbiotic nitrogen fixation as there is less of it.

What about the hedges, how do they respond to nutrient availability? In the beginning their seedlings will behave much like crops and feed on the same P- and K- sources, until their roots extend into the subsoil and those in the topsoil are cut off by cultivation. Contrary to crops like maize and cassava, however, growth of leguminous hedges is less restricted by nitrogen. One could therefore assume that the thresholds for P- and K-uptake would be in the order of 10 ppm P and 0.2 meq/100 g K. However, the perennial hedges are better at taking up nutrients at low concentrations, so I will still use the lower thresholds of 4.5 ppm and 0.1 meq/100 g for them as well, the same as for N-limited crops. Since the older hedges take up most of the nutrients over a greater depth, the thresholds are set at 45 kg P and 400 kg K, taken over the soil layer between 30 and 100 cm. Uptake and growth by the hedges are shown in line 14.

So much for the uptake of nutrients in readily accessible form; uptake from moderately bound sources by crops and hedges, which is obviously important to keep the system going, has hardly been studied. I therefore have to pile more assumptions on top of an already worrying lot, but it is better to make reasonable assumptions than give up halfway because of lack of data, I think, so let us continue.[9] Based on scant evidence I have presumed that the hedges

---

[9] Many agronomists do not agree with this principle, though. B.T. Kang, the father of alley cropping for one, commented thus on an earlier draft of this Appendix: "I have some difficulties in reviewing your paper on alley cropping, since you used many assumptions." Fortunately he continues: "In general I agree with your conclusions."

can extract up to one-third of their P from moderately bound (non-humus) P. That is because perennial species are more capable at getting to less available nutrient sources, for example, through symbiosis with soil-borne fungi. The symbiosis is called mycorhiza which contains the Greek words for fungus (mukos) and roots (rhizos). Crops and hedges are assumed to take at most one-fourth of their K from non-exchangeable forms. Finally, uptake of P and K from bound sources is taken to be proportionally less as there is more of it in the available form, to a minimum of 10%.

(c) Nutrient balances and yields

We are now ready to examine the model predictions, which have been brought together in Table A4-6. The table shows the annual nutrient balances and the yields of crops and hedgerow prunings, extending over 20 years. The detailed calculations, an example of which was given in Table A4-5, are not shown. Readers who have made it up to this point may have started wondering what the reality value of the virtual crop–hedgerow system is, and rightly so. But not to worry, after completing the analysis of the model predictions, I will confront them with real data.

Available P in the topsoil is predicted to decrease steadily, by almost a third in the first year and a progressively smaller decline later. Exchangeable K-content of the topsoil decreases by more than half during the first 4 years, but after that it increases, because the hedgerows now pump K from the subsoil, until it declines again later on as growth of the hedgerows slowly diminishes.

What about the yields? The model predicts a slow increase in maize yield until year 5, after which it declines. The trends are similar for cowpeas and hedgerow prunings, but the causes are quite different. The predicted decline in cowpea and pruning yield is due to P-shortage. The maize yield in turn declines because of the progressively smaller amount of hedgerow prunings and hence the decline in N-release. Humus content also decreases steadily by the same cause. Maize only starts to experience P-shortage much later, somewhere around year 15. P becomes a limiting factor for cowpeas as early as year 4, while the maize, being each season's first crop, can still take up as much as it needs. That is because it is replenished by fresh release from the early season flush of humus decomposition. That can be seen in the detailed analysis in Table A4-5 where the P-balance for maize (line 09) is well above the threshold. After the

Table A4-6. Calculated nutrient budget and yields for an alley cropping system with maize and cowpeas, 'poor soil' conditions, after long fallow

	Nitrogen, kg/ha		Phosphorus, kg/ha					Potassium, kg/ha					Yield			
			0–30 cm		30–100 cm			0–30 cm		30–100 cm						
Balance	0–30 cm	30–100 cm	Available	Bound	Available	Bound	Total	Exch.	Non-exch.	Exch.	Non-exch.	Total	Maize	Cowpeas	Prunings	OC
Initial	0	0	48	80	23	220	371	450	380	300	420	1550				48,000
1	11	23	35	85	23	220	362	354	378	325	419	1477	1.95	0.86	0.35	46,560
2	16	28	27	89	21	219	356	302	377	321	418	1418	1.96	0.48	1.38	45,451
3	13	7	17	93	14	215	339	244	375	236	411	1266	2.12	0.48	5.28	44,547
4	13	8	11	94	8	210	324	242	373	167	399	1181	2.46	0.45	5.33	44,308
5	17	14	8	96	5	206	315	288	370	141	389	1188	2.80	0.39	4.15	44,612
6	19	19	7	97	3	202	308	325	368	133	380	1206	2.74	0.33	3.44	44,808
7	20	22	5	97	2	197	302	342	366	134	372	1214	2.61	0.29	2.96	44,754
10	25	27	4	98	0	185	287	345	361	147	353	1206	2.05	0.25	2.39	43,793
20	28	29	4	97	0	149	250	289	347	155	305	1097	1.60	0.24	1.98	39,014

Yields by year and OC content, tons per hectare

	1	2	3	4	5	6	7	10	20
Maize	1.95	1.96	2.12	2.46	2.80	2.61	2.61	2.05	1.60
Cowpeas	0.86	0.48	0.48	0.45	0.39	0.29	0.29	0.25	0.24
Prunings	0.35	1.38	5.28	5.33	4.15	2.96	2.96	2.39	1.98
OC	46.56	45.45	44.55	44.31	44.61	44.75	44.75	43.79	39.01

maize has taken up its share, however, the level of available P falls below the threshold (line11). Available P in the subsoil decreases steadily and so does the amount of P taken up by the hedgerows. As a result, pruning yield never reaches more than about 60% of the assumed maximum of 9 t in this relatively poor soil, in year 4.

Exchangeable K in the topsoil always remains well above the threshold, but not in the subsoil. K does not restrict the growth of the hedgerows, however, because P is the more limiting nutrient. After 20 years, the end of year balance of available P in the topsoil is predicted to be very low, while in the subsoil practically all available P is gone and the hedgerows now have to rely almost entirely on bound P. Exchangeable K in the topsoil never returns to the original level again.

Remember that so far I have been talking about *simulated*, not real results, so let us now see whether there are real data for conditions similar to those in the modelling exercise of Table A4-6. Surprisingly little has been published about soil changes in the first few years under alley cropping. There is only a publication from 1981 by Kang, Wilson and Sipkens I am aware of. There was a clear downward trend in topsoil P- and K-contents after 4 years, as in the model, but the decline was smaller than the model predicted, no doubt because of the fertiliser which was applied in the trial. The first season maize yields were very close to our predicted yields, just above 2 t/ha. I will not speculate further about these model runs, they were really just meant to test the model's inner workings. Instead, as an advanced test I will now look at a well-documented long-term trial, which was carried out under more representative conditions and see whether the model is capable of predicting its results.

A4.4 A long-term alley cropping trial

The trial was carried out between 1982 and 1993 with maize and cowpeas as test crops, indeed a long-term trial for today's standards. A major paper was published about the work in 1999 by Kang et al., with detailed information about the initial conditions, their changes over time and of course the yields of the crops and the biomass produced by the hedges. That was very welcome: if the model cannot be tested against real data the whole thing remains a rather frivolous affair. Before the trial started, organic-C, available P and exchangeable K-content of the topsoil were measured and again 10 years later.

Subsoil nutrient contents were not measured, so I assumed those of a 'good soil' from Tables A4-2 and A4-3, because that is what it was.[10] There was no substantial enrichment of the soil from burned fallow vegetation prior to the trial, because the field had been used the previous 7 years for mechanised crop production. I converted the data into kilogram per hectare (the calculations are shown in Box A4-5) and plugged them into the model with the results shown in Table A4-7.

Several hedgerow species and fertiliser rates were tested in this trial, but I will only look at the treatment with *Leucena* as the hedgerow species and the lowest fertiliser rate of the trial. In the first

Table A4-7. Initial nutrient and organic-carbon contents of a poor soil after long fallow

	Phosphorus, kg/ha			Potassium, kg/ha				OC, kg/ha
0–30 cm		30–100 cm		0–30 cm		30–100 cm		
Available	Moderately bound	Available	Moderately bound	Exch.	Non-exch.	Exch.	Non-exch.	
16	160	30	300	501	800	700	1,400	36,456

Box A4-5. Converting soil analytical data to amounts of nutrients for the long-term maize–cowpea trial

1. Organic-C content was 1.24%. It is taken to apply over the top 20 cm of soil depth, so there would be $1.24 \times 20 \times 1.47 \times 1000 = 36,456$ kg of humus-C.
2. 6.18 ppm of available P was measured in the top 15 cm, which equals about 13.5 kg/ha ($6.18 \times 1.5 \times 1.47$). Most of it will have come from burned fallow. I assumed an additional 20% in the 15–30 cm layer, so the total would be about 16 kg/ha of available P over 30m cm soil depth.
3. Measured exchangeable K was 0.35 meq/100 g, that is $0.35 \times 39 \times 2.2 \times 10 = 300$ kg/ha over 15 cm (for the calculations see footnote 6 in Appendix 2). I assumed that the soil layer of 15–30 cm contained two-third of that amount, hence, total exchangeable K of 501 kg/ha.
4. For the subsoil (30–100 cm) I used the figures for a 'good soil' from Tables A4-2 and A4-3.

---

[10] An *Oxic paleustalf* to be precise.

year no fertiliser was applied at all in this treatment, but this was changed in following years, probably because of the low maize yield. The fertiliser rate in successive years were as follows, in N:P:K:

1982	1983	1984	1985	1986	.....	1993
0:0:0	45:20:20	45:20:20	0:12:25	0:12:25	.....	0:12:25

The spreadsheet model can easily accommodate fertiliser treatments; the amounts are simply plugged in at the beginning of each year. The maize stover and the cowpea hay were left in the field, as was standard practice at the IITA farm.

In the first model run, a major difference showed up between real and predicted P-content of the topsoil, which needed scrutiny: the predicted content after 10 years was more than twice the measured content. Why did the model overestimate P in the topsoil? The change in P-content over the years is strongly influenced by one parameter: the percentage of available P immobilised each year, which was set initially to 10%. By increasing the immobilisation rate to 25% a much better 'prediction' was obtained. I will come back to this fixing procedure later. Table A4-8 shows the results after this adjustment. They were quite interesting. Both the topsoil P- and K-contents were predicted to increase after an initial dip and after 12 years P was not far from and K exceeded the initial values, which is what the actual measurements indeed showed. The predicted maize yields were also close to the measured yields.[11] Average pruning yield predicted by the model was about 20% lower than actual yield. Another important prediction was the complete depletion of available P in the subsoil, forcing the hedgerows to rely practically on bound P alone. The paper did not present data to confirm or refute this prediction.

Even though increasing the P-immobilisation parameter to 25% was probably in order for this type of soil, the reason for doing it was to fix the difference between the model predictions and the measured data. By this intervention we have entered a slippery road. I will suspend the analysis for a moment to explain why that is so.

---

[11] The paper only mentioned the cowpea yield for the penultimate trial year.

Table A4-8. Calculated annual nutrient balances and predicted and measured yields for an alley cropping system with maize and cowpeas, for the conditions of a long term trail (Kang et al., 1999), with P and K fertiliser applied

| | Nitrogen, kg/ha | | Phosphorus, kg/ha | | | | | Potassium, kg/ha | | | | | Yield | | |
| | 0–30 cm | 30–100 cm | 0–30 cm | | 30–100 cm | | | 0–30 cm | | 30–100 cm | | | | | |
Balance			Available	Bound	Available	Bound	Total	Exch.	Non-exch.	Exch.	Non-exch.	Total	Maize	Cowpeas	Prunings
Initial	0	0	16	160	30	300	506	501	800	700	1,400	3,401			
1	12	16	7	163	28	299	497	420	797	686	1,398	3,301	1.24	0.72	1.63
2	11	0	11	170	18	294	493	344	792	591	1,391	3,118	2.54	0.50	6.56
3	11	0	14	180	11	287	492	359	787	504	1,383	3,032	3.06	0.50	6.94
4	13	6	15	191	6	281	493	466	782	450	1,376	3,073	2.76	0.50	5.73
5	14	9	15	201	4	275	495	547	777	424	1,370	3,119	2.77	0.50	5.00
6	14	12	14	211	2	269	497	602	773	411	1,365	3,151	2.65	0.50	4.56
7	14	13	14	221	1	264	500	638	769	406	1,360	3,174	2.58	0.50	4.27
10	15	16	13	248	0	247	508	693	757	419	1,347	3,215	2.50	0.50	3.87
11	15	16	13	257	0	242	511	701	753	428	1,343	3,225	2.49	0.50	3.80
12	15	16	13	265	0	237	514	707	749	439	1,338	3,234	2.48	0.50	3.74

(continued)

Table A4-8. (continued)

Measured and predicted crop yield, t/ha, by year

	Maize										Cowpeas
Year	1	2	3	4	5	6	7	10	11	12	11
Measured	1.20	2.70	3.32	3.40	2.72	2.63	3.20	3.20	2.42	2.34	0.45
Predicted	1.24	2.54	3.06	2.76	2.77	2.65	2.58	2.50	2.49	2.48	0.50

Measured and predicted soil parameters

	OC %	Available P, ppm	Exch. K meq /100 g
Initial, measured	1.24	6.18	0.35
After 10 years, measured	0.97	5.70	0.49
After 10 years, predicted	1.33	4.85	0.54

A4.5 Some tinkering involved

I think the modelling story unfolded quite logically, until I started to manipulate a poorly understood parameter, P-immobilisation, to make the model behave like the real system. That sounds like a mortal scientific sin, and one which is not absolved by its mere confession. But it is actually not as bad as it sounds, as long as it does not stop there. It is even common practice among honourable modellers, whom nobody would want to accuse of cheating. In order to explain that I like to say a few words about the way an alley cropping model (or any model for that matter) should, ideally, be built and tested.

There are various soil and plant processes involved in alley cropping, which together determine how the system as a whole works. We have already seen the most important ones: humus build-up and decomposition, leaching of nitrates and cations into the subsoil, nutrient uptake and biomass production by the crops and the hedgerows. In order to piece all of them together into a predictive model we must account quantitatively for each process as well as for the way the processes will affect each other. Preferably, the data should come from independent-process studies, but if they are from studies with the target system itself, they should not be used again later to test the predictions, otherwise the whole thing becomes circular. That sounds pretty obvious, but it is a principle much violated in modelling practice. Once a preliminary model has been put together its predictions must be compared with the results of field trials. If they are incorrect, something must be wrong with the process data or with the way the model handles their interaction and the processes would have to be studied again, before the model is rerun with better data or better process routines. That is where the problems with much modelling start. Modellers are often engineers with little appetite or patience for meticulous and detailed studies of basic processes, and if they do not find real data they will proceed with the model anyway using a procedure, which is variously called parameterisation, calibration or tuning, all of them nice words for what is essentially tinkering: changing figures here and there in the model, until its results resemble those of the experiments. The modeller then hopes that next time, with a new set of experimental data, the predictions will come out right, but the more tinkering has been done, the less likely that is. That is the punishment for the sin I referred to: the model will degenerate into what

C.T. de Wit, agriculture's head modeller, called a complicated way of curve fitting. Let us see what amount of tinkering was involved in putting together my nutrient budget model for alley cropping.

The soil processes used in the initial version of the model were quite simple: N- and P-release by the humus, nutrient enrichment of the topsoil, first by the ashes of the burned vegetation and later by the decomposed hedgerow prunings, leaching of N and K, P-fixation, nutrient uptake by the crops and the hedges, the latter shifting gradually to the subsoil. The data I used were from Nye and Greenland's essay, from a study on K by Juo and Grimme and from Pedro Sanchez' textbook, while specific alley cropping data were taken mostly from a key publication by Kang and co-workers. In the first run, without tinkering, the model predicted rapid depletion of N and P in the topsoil. Why was that so? New humus incorporates a considerable amount of N and P and if the biomass from which it is formed has low N- and P-content, the balance has to be taken up from the soil. Initially I set the conversion rate of biomass dry matter into new humus to 40%, which was the average of the figures given by Nye and Greenland for forest vegetation. That caused serious depletion of the available N and P in the model and, as a result, it seriously underpredicted maize yield compared to what was usually observed in the alley cropping trials at IITA. I therefore reduced the dry-matter conversion rate from 40% to 25%, which fixed the problem. It also makes sense from a process point of view, because materials with a low C/N quotient, like the prunings of leguminous hedgerow species, will have a considerably lower conversion rate than the litter under an established forest vegetation. After that change, a lot of N was predicted to be leached out of the soil profile, carrying along a large amount of K. That is not likely to happen either, because the hedgerow roots are sprawling in the subsoil and they will take up N from the soil solution if it is easily available, instead of binding atmospheric N. So I added a simple routine, which causes N to be taken up by the hedges at a rate which depends on its concentration.

Even though the changes were reasonable and consistent with experimental data, they were motivated by poor model predictions, so it essentially came down to tinkering with the model machinery. That was still OK, though, as long as it would henceforth correctly predict the results of other experiments, which had in no way been used before. So after these adjustments, I tested the model again, against the results of the long-term maize–cowpea experiment I presented earlier. The model initially predicted a much faster increase in P-content for that trial than was measured in the field, so things were still not in order. The problem could be fixed by simply changing the P-immobilisation rate from 10% to 25%, which resulted in the predictions of Table A4-8. Again, the adjustment was a reasonable one,[12] except that it would have been much better to *measure* P-immobilisation in some way or other instead of adjusting the rate parameter to get the desired results.[13] In any case, tinkering with the immobilisation parameter is an acceptable though second rate solution, provided the adjusted model is then validated again with a new set of experimental data, which has not been used for 'calibrating' the model (and so forth).

A4.6 Maize and cassava

As the next step in what is called the model validation process we are going to look at another long-term trial, and one which was exceptionally well documented, with maize and cassava instead of cowpeas this time, again carried out at the IITA station. The trial ran from 1989 to 2000 in a field, which had been cleared from a long forest fallow in the Institute's remaining forested area. A problem with earlier alleycropping trials at the institute had been the unusual crop sequence of maize and cowpeas, and the use of fertiliser which would prevent or postpone soil exhaustion. Therefore, in 1988 the then Director of the Farming Systems Program, Dunstan Spencer, insisted that a more realistic alley cropping trial was to be set up with maize and cassava and without fertiliser, the way farmers would do it. And so it happened. The results were published in two recent papers, by Tian et al. (2003) and Tian et al. (2005). The papers gave a lot of detailed information on the yields of crops and prunings as well as some less extensive data on the changes in the topsoil, all of which the model should be able to simulate if it was any good. At the same time any remaining weaknesses in the model were likely to show up. Let us see what it predicted for this trial and how that compared with the measurements.

---

[12] P-fixation is low in light, sandy soils and high in strongly weathered, 'old', acid tropical soils with high iron and aluminium content. In SW Nigeria P-fixation would be low to moderate.

[13] Furthermore, it is unlikely that P-immobilisation can be expressed adequately by one simple parameter.

The pre-trial soil fertility data, measured after the vegetation had been cleared and burned, averaged over the entire experimental area were: Organic carbon: 1.25%; available P: 12 ppm; exchangeable K: 0.32 meq/100 g; all for the top 15 cm of soil. These figures are similar to those of the field where the earlier long-term trial was carried out and which I classified as a 'good soil'. In the present case, however, the soil had been enriched with the nutrients from the burned vegetation, so the intrinsic fertility was considerably lower. I therefore put it in the 'poor soil' category. The measured figures combined with the relevant nutrient contents from Tables A4-2 and A4-3 were plugged into the model, after converting them into kilograms of nutrients per hectare in the top 30 cm (Box A4-6), with the results shown in Table A4-9.

Box A4-6. Converting soil analytical data to amounts of nutrients for the maize–cassava alley trial

1. The organic-C content of 1.25% is taken to apply over 20 cm of soil depth, so there would be 36,700 kg of humus-C.
2. 12 ppm of available P in 15 cm equals about 25 kg/ha ($12 \times 1.5 \times 1.47$). Most of that will have come from the burned fallow vegetation. I assumed an additional 20% in the 15–30 cm layer again, so the total would be 30 kg/ha of available P.
3. Exchangeable K in the top 15 cm at 0.32 meq/100g is equivalent to 275 kg/ha (see Box A4-5). I further assumed that the soil layer of 15–30 cm contained two-third of the amount of K of the top 0–15 cm. Hence, total exchangeable K was 450 kg/ha.
4. For the subsoil (30–100 cm) I used the figures for a 'poor soil' from Tables A4-2 and A4-3, because the topsoil, without the additions from the burned vegetation, would also be in the poor soil category.

Table A4-9. Initial nutrient and organic-carbon contents in the long-term maize–cassava alley cropping trial

Phosphorus, kg/ha				Potassium, kg/ha				OC, kg/ha
0–30 cm		30–100 cm		0–30 cm		30–100 cm		
Available	Moderately bound	Available	Moderately bound	Exch.	Non-exch.	Exch.	Non-exch.	
30	80	23	220	450	380	300	420	36,700

The soil texture was similar to that of the long-term trial, so I used the same P-immobilisation of 25%.

The annual cropping pattern being maize intercropped with cassava, the field was occupied practically the whole year around. The cassava was harvested after 12 months, so the hedges would have much less time to recover than with maize and cowpeas. I nevertheless maintained the maximum pruning yield at 9,000 kg/ha. Cassava can extract some moderately bound (non-humus) P and I assumed that capacity to be (numerically) the same as that of cowpeas and hedgerow shrubs (see the notes below Table A4-5 for the details). Finally, 50% of the maize and cassava residues and the hedgerow prunings are taken to be converted into humus at a conversion rate of 25%, as before. The cassava and most of the hedgerow biomass from a particular year will only start decomposing in the following year and the budget therefore allocates half of the nutrients they release to the first year following the crop and half to the second. With that, the model predicted the results shown in Table A4-10, together with the actual measurements.

As expected, the model tells us that crop yield is constrained primarily by nitrogen shortage. A quite steep decline of maize yield was predicted to occur in years 2 and 3, after which, yields would go up again in years 4 and 5. The decline was caused (in the model that is) by conversion of the crop residues from the previous years into humus, which would eat up quite a lot of nitrogen while there is yet little coming from the prunings to make up for that.

Exchangeable K in the topsoil is predicted to remain adequate throughout, but the annual balance of available P would fall below the threshold right at the end of year 1. That does not mean the crops suffered from P-deficiency that early, though. When they needed it, there was enough to match the available nitrogen released from the decomposing prunings and humus and because the cassava can extract some P from less accessible sources. Yield was predicted to get constrained by P-shortage in year 3, but after that further decline was slow, at the expense of subsoil-P being mined and transferred to the topsoil by the hedges. That caused depletion of available subsoil-P and a slow decrease in production of pruning biomass after it reached a peak in year 4.

How does all that compare with the real data? Table A4-10 shows that the measured and predicted cassava and pruning yields were very similar up to year 9 and 8 respectively. Observed maize yield

Table A4-10. Calculated annual nutrient balances and measured and calculated yields and other parameters for an alley cropping system with maize and cassava, under experimental conditions at IITA. (From Tian et al. 2004)

| Balance | Nitrogen, kg/ha | | Phosphorus, kg/ha | | | | | Potassium, kg/ha | | | | |
| | | | 0–30 cm | | 30–100 cm | | | 0–30 cm | | 30–100 cm | | |
	0–30 cm	30–100 cm	Available	Bound	Available	Bound	Total	Exch.	Non-exch.	Exch.	Non-exch.	Total
Initial	0	0	30	80	23	220	353	450	380	300	420	1,550
1	1	18	16	89	23	220	347	355	378	318	419	1,471
2	5	21	10	93	21	219	343	310	377	308	417	1,412
3	10	11	6	96	14	215	330	265	375	230	409	1,279
4	13	12	3	97	8	210	320	256	373	170	399	1,197
5	16	16	3	99	5	206	313	287	370	143	389	1,190
6	17	19	3	100	3	202	307	314	368	129	381	1,191
7	17	21	3	101	2	197	303	324	366	121	374	1,184
8	18	22	3	102	1	193	298	324	364	116	367	1,170
9	18	23	3	102	1	189	294	318	362	112	360	1,152
10	19	23	2	102	0	185	290	308	360	109	353	1,131
11	19	24	2	102	0	181	286	297	358	106	347	1,109
12	19	24	2	103	0	177	282	286	356	103	341	1,086

Comparison of measured and predicted results

		Year											
		1	2	3	4	5	6	7	8	9	10	11	12
maize yield	measured		1.90	2.47	0.71	1.49	1.28	1.02	0.29	0.32	0.47		0.50
	predicted	1.49	1.12	0.90	1.17	1.42	1.30	1.14	1.05	0.94	0.94	0.91	0.89
cassava yield	measured			5.20	4.20	4.60	3.20	7.00	5.20	5.20	2.40		2.60
	predicted	4.89	3.68	4.07	4.99	5.38	5.34	5.22	5.15	5.10	5.06	5.03	5.00
pruning biomass	measured		0.81	3.80	4.80	3.70	3.80	3.00	2.80	1.10	2.50		
	predicted	0.34	1.27	4.62	4.54	3.58	3.04	2.70	2.48	2.33	2.23	2.15	2.10
OC (%)	measured	(1.25)			1.30								
	predicted				1.17								
available P (ppm)	measured	(12)			1.49								
	predicted				1.08								
exch. K (meq/100g)	measured	(0.32)			0.12								
	predicted				0.18								

Soil figures in year 1 (between brackets) were measured before the start of the trial.

was erratic throughout and collapsed completely from year 8 onwards. Model predictions for maize were only approximately correct for the fifth to the seventh year. Predicted soil parameters in year 4 were reasonably close to the actual values.[14] Unfortunately, subsoil nutrient content was not measured, so that the prediction that available P in the subsoil was getting seriously depleted could not be verified. That is surprising in view of the importance of the subsoil as a source of nutrients in alley cropping. The papers did show that the amount of P contained in the prunings declined much more than that of other nutrients and they concluded that P in the subsoil was indeed getting depleted.

Some conspicuous differences between predicted and measured results need special attention, in particular those in respect of maize and to a lesser extent cassava yield:

- The initial decline in maize yields in years 2 and 3 predicted by the model does not seem to have occurred in the trial.
- Maize yields in the trial declined rather suddenly to a very low level in year 8 and remained there afterwards while the model predicts a much slower decline.
- Something similar happened with cassava yield two years later when its yield fell to half of that of the previous years.

The first discrepancy must have something to do with the conversion of crop residues into soil organic matter. Perhaps the immobilisation of nitrogen by the decomposing plant material was less severe than the model assumes, leaving more N for the crops to feed on. Termites may have eaten a larger part of the crop residues, which were left on the soil surface. That means that the residues would not enter directly into the humus cycle but rather through the stomachs of the termites and severe nitrogen depletion due to direct conversion into humus would not have occurred.

The steep yield decline of maize in the trial from year 8 onwards and that of cassava a little later is more serious. It is a sign that the system was breaking down and that permanent cropping in this configuration without external inputs was an illusion. The papers

---

[14] It looks as if there was a considerable difference between measured and observed available P, but the level of both is very low and there is always a degree of uncertainty in P measurements, so the difference was not really significant.

published about the trial suggested that the severe pruning regime might have reduced root growth of the hedges, which led to their decline and would eventually even result in their death. The decline in P content of the prunings mentioned earlier seems to confirm that. The model correctly predicted the decline in growth of the hedges, but it failed to predict the collapse of crop yields in year 8 and 10. Perhaps that collapse was due to lower root activity of the hedges and a reduced ability to unlock the soil's more tightly bound nutrient reserves, which the model does not account for. Or, in view of the sudden nature of the decline, the causes may not even have been directly related to alley cropping.

A4.7 Summing up

A4.7.1 Virtual and actual nutrient dynamics and yields

So, what is the verdict about the alley cropping model? After some tinkering, it did quite a credible job in simulating the maize–cowpea cropping pattern and the yields and soil parameters the model predicted for a long-term trial conducted at IITA were reasonably close to the measured data. The model pointed to P as the element which would be the first (after N) to limit crop yield. After a few more, relatively small, adjustments, the model was applied to a new set of data collected in another long-term trial, this time with maize–cassava. The results were promising again but some new problems showed up. An early-yield depression of maize predicted by the model did not occur in the trial, while a collapse of the system between year 8 and 10 occurred but was not predicted. For the intervening years the simulation results were reasonable and they showed that P-depletion must have occurred in the subsoil and would cause the system to collapse in the long term, if no corrective measures were taken.

The major conclusion drawn in the two papers about this last trial (which involved many more treatments), was that continuous alley cropping with maize–cassava was not possible and that 2 years of fallow were needed for each year of cropping to allow the hedges to recover from the severe pruning regime. Considering the stocks of available and moderately bound P (see Table A4-9), it would take 120 years with that kind of low-intensity land use for all the P in the subsoil to be consumed, unless it was replenished in the meantime from the tightly bound form, which would extend the system's lifetime even further. But eventually the time will come when the hedges no longer find enough P in the subsoil to feed on and only a sterile soil skeleton will remain, perhaps after a few hundred years. But long before that it would no longer be profitable to keep the system going because of declining yields. So alley cropping, in spite of its advantages, remains a form of soil mining, like any other system, which does not return to the soil what it takes out. The papers recommended that at least P-fertiliser should be applied, even in a low-intensity alley cropping system. But since it is the sub-soil that is getting depleted and P-mobility is low, P should actually be applied to the subsoil to feed the roots of the hedges, for instance using a slow-release P-source like ground rock phosphate.

A4.7.2 The prospects for alley cropping (and for modelling it)

After a long fallow, the soil's initial nutrient status is very favourable, so why should farmers bother to start alley cropping in the first place? If at all, they are more likely to wait until the qual-ity of the fallow starts deteriorating and crop yields are going down. By that time the humus level will have declined, there will be a much smaller boost of the soil's P- and K-content from the burned vegetation, all together perhaps enough to tempt farmers into trying some alley cropping. I ran the model again for those conditions, assuming the nutrient contents of a 'poor soil' and an organic-C content of 1%. I will not bother you with another table, but just mention that the trends were very similar to those in the previous case, except that the predicted crop yields were consider-ably lower. Maize yield was predicted to quickly fall below 1 t/ ha and reach 800 kg in the sixth year, while cassava yield declined to around 4 t. Those are pathetic yields considering the efforts needed to realise them. In order to make it worthwhile at all it would be necessary to apply N-fertiliser from the start to get acceptable yields, soon to be followed by P and K.

The most important merits of the alley cropping system are the maintenance of the soil's humus content and the nitrogen surplus

---

[15] If a leguminous species is used, but even some non-leguminous species appear to generate an N surplus, perhaps by symbiosis with free-living N-fixing organisms.

generated by the hedgerows' N-fixation[15] and there is of course the nutrient capturing function of the deep rooting hedges. The price to be paid is subsoil mining of P and K, the way it also happens in fallow systems, but at a much lower extraction rate, and much more work. A fallow system with 2 years cropping and 8 years fallow in a good soil, could go on for centuries, especially when the extraction is compensated by a slow release from tightly bound nutrient stocks. With continuous alley cropping, the extraction intensity is five times higher. In theory, there may be enough P to continue for 40 years, but as the reserves get depleted, the uptake rate will go down and serious problems will occur even earlier, as we have seen in the example of the long-term maize– cassava trial. So, there will be no escape from applying fertiliser, especially phosphorus, as the trial results confirm. Even in a good soil, permanent alley cropping without mineral nutrition is impossible in the medium and long term, because of intensified soil mining. Intermittent fallowing and regular fertilisation, especially with P are needed to maintain the yields at a reasonable level and prevent the system from collapsing. And at least part of the P must be applied to the subsoil, otherwise the hedges cannot function properly. Alley cropping will then minimise losses, maintain organic matter and generally create a stable production environment, be it at the cost of a lot of work.

What I have especially wanted to show by this exercise is that it is not all that difficult to make a fairly realistic model for the nutrient dynamics of alley cropping. Even a simple model can be helpful in indicating the likely long-term prospects of alley cropping or at least show important knowledge gaps. It is surprising that during all those years of alley cropping research it has never been done. Even now I think it is still worth a few years of dedicated work by some smart youngster to come up with a really solid and practical model. If done well, that should easily qualify as a Ph.D. thesis, or several, as these things go.

Appendix 5. More Farmer Technologies

Why should I spend more time on farmers' technologies, if in the end they are bound to disappear anyway? Well, some of them may refuse to go away because they are simply better than the ones science can offer, and those that are not may yet contain elements which are worth keeping.

There have always been westerners who were fascinated by exotic cultures and the way they met the challenge of manipulating nature for their benefit. But they were usually anthropologists and ethnobotanists and perhaps the odd colonial officer or ethnologically interested pharmacist or agronomist. Most agronomists had little patience with farmers' own practices, which they saw as hopelessly outdated. Until the advent of Farming Systems Research, that is when those practices acquired respectability, even in the eyes of people who continued to think that they would stand no chance under the onslaught of the modern times. But in many areas, modern times were slow in coming and meanwhile the farmers' age-old practices continued to serve them quite well. And when they arrived, perhaps those practices could still be transformed into something modern and given a new lease of life.

I have already given some interesting examples of ingenious African cropping systems, but they are not the only thing to show that backwardness is in the eye of the beholder. I would therefore like to spend some time on a wider selection of things African farmers have invented in their struggle for survival. My examples are ordered in three groups: (1) clever land use systems (2) crop manipulation and (3) crop processing. and I will give one or a few examples of each.

A5.1 Land use systems (or how to exploit a difficult environment)

Many examples in this category were given in the main text, and many more pages could be filled with other examples, but I will just give brief summaries here of systems, which I find particularly interesting. If you are interested in more details, it is relatively easy today to find them.

A5.1.1 Terrace farming

In some hilly areas in Africa with very meagre rocky soils, farmers
have developed intensive land use systems, which have allowed them
to practise practically continuous cropping. They all feature meticulous terracing of the land and the use of rocks and boulders to
reinforce the terrace borders. Such systems are found scattered over
Africa and they all have their own story about why the people
settled under such difficult conditions instead of staying in the
plains. Just a few examples, which I am more or less familiar with:

– The Mandara Mountains in northern Cameroun
– The hilly areas in north-eastern Togo and the adjacent areas in
  Atacora, Bénin Republic
– The *Pays Somba* in Mali
– The Konso area in southern Ethiopia

They all share a few essential features: effective stabilisation of the
slopes and prevention of erosion by reinforced terraces, very
labour-intensive cropping practices and careful choice of crops and
rotation practices.

Egbert Westphal in his book on farming systems in Cameroun[1]
describes an intriguing crop rotation practice from the Mandara
mountains whereby *all* the farmers simultaneously rotate sorghum
and millet in their fields. It was not clear at the time what motivated
this practice, but it must have been a pressing problem for the people to respect such a drastic self-imposed rule. Maybe the riddle has
been solved since then, but I have not found any references, which
show it has.

A5.1.2 Intriguing fallow management practices

Just two examples here: one from Sierra Leone and one from
Zambia.

I do not know much about Sierra Leonean agriculture, but I was
intrigued by a fallow management system, which I saw there. Crop
fields in some area (I do not remember exactly which one) are dotted with numerous knee- to waist-high barren tree stumps, giving

---

[1] There is another book by Westphal, on the farming systems of Ethiopia, which
also describes the Konso area.

the field the appearance of a strange graveyard. Tessmann would probably have strongly disapproved of such a messy system, but in fact it is quite sensible. Once the field has been abandoned after a few years of cropping, the vegetation will rapidly recover and establish a closed canopy, which is very important in this humid environment with its heavy downpours. The tree stands are much denser and the stems are much thinner than would be normally the case in a forest, which probably resulted from practising this system over a long period with many cycles. It has definite similarities with the alley cropping system, the way it would have to be practised in a forest area as we have seen in Appendix 4 and Chapter 7.

Yet another special fallow management system is practised in Zambia – *Chitemene*. It somewhat resembles the old European practices whereby the fertility from the 'wasteland' was concentrated on a much smaller area of intensively cropped farm land. In *chitemene,* branches are lopped off the trees in a wide circle around what is going to be the crop field for the coming rainy season. The branches are piled up over the future field and burned, thereby enriching the soil with the nutrients from the ashes. The first crop planted is finger millet, which only thrives in rich soil. The system looks quite destructive, but I am not so sure it is, because the lopped trees will quickly recover and as long as the actual fields are small the tree roots at greater depth may catch the nutrients, which are leached out of the topsoil. It is another example of the many inventive ways in which the African farmer has learned to manage the scarce fertility of the land.

A5.1.3 Wet feet cropping

This example is from inland Bangladesh, from the floodplains of the large river system with their peculiar conditions of annual flooding of a large part of the agricultural land. The traditional form of land use, which was practised there until quite recently, was marvellously adapted to these challenging conditions.

The agricultural year is counted in Bangladesh from April, when the monsoon rains break. First flooding of the low-lying lands usually occurs 2 months into the rainy season and progressively more land goes under water as the season progresses. The land starts to dry again after the end of the rains in November and a period of cool weather follows, which lasts until early February. There were broadly three land use patterns, corresponding with three land elevation classes. In the highest fields, either dry seeded rice (*Aus*) or jute would be planted with the first rains in April, which completed most of their cycle under dry land conditions. Towards the end, however, in August, even the higher lands would get inundated and the crops completed their cycle in standing water. Rice is well adapted to that – it can easily change over from dry to wet conditions (not the other way around), and for jute it was quite convenient, because it tolerated inundation and the cut stems could be left to ret[2] practically in situ. These pre-monsoon crops were followed by the monsoon crop, which was invariably *Aman* paddy, transplanted from late July to early August and harvested in November. And finally, once the land was dry again, a variety of winter crops were planted: wheat, barley, mustard, onions, vegetables and spices.

In the fields of medium elevation (we are talking about differences which may be less than three feet!) it was not possible to grow jute, because by the time it was ready for harvest, the land would already be too deeply flooded to plant *Aman* paddy, the most essential crop for survival. *Aus* paddy was still possible, however, provided it matured early. For the rest, the pattern was the same as the previous one. In the really low-lying fields, no dryland crop would be planted, but the land would instead be readied for deep water rice, to be seeded or transplanted in June or early July. The surplus rainwater flowing to the lowest spots would rapidly inundate the fields and the deep water varieties could keep their heads above the water by extending the straw for as long as the water level went up. The fields would only be accessible again when the water had gone down in November and the paddy was ready to be harvested. Seed of the grasspea (*Lathyrus sativus*) would then be broadcast into the stubble, producing an interesting amount of fodder for the animals before the land had to be prepared again for the next monsoon crop.

In the 1970s, farmers put this whole intricate system on its head by starting to pump water from shallow and deep wells during the dry season, which allowed them to grow high yielding paddy varieties under almost perfect water control, with very high yields. It is not exaggerating to say that this saved the country from the always

---

[2] Retting is a bacterial process taking place in standing water whereby the jute fibres are set free from the woody stalk.

imminent famine – and provided a timely alternative to jute, which was no longer remunerative. The expansion of this so-called *Boro* rice crop drastically changed the land use system – another interesting story, which I must leave untold here.

A5.1.4 Extremely integrated farming

Just a few brief words about the way some Chinese farmers have made integration of different components of farming (and the rest of life) into an art. I have not had the opportunity to observe their artful ways directly, beyond seeing them carry the content of their pit latrines to the field in the early morning, so what I have to say is mainly from hearsay. Every part of their farm is said to be linked to every other one, with the conservation of fertility as the common denominator. That means that the duck pen is built over the fish pond to feed the fish on duck dung, all excrements, including those of the humans, are conserved and used as manure to feed the fish or the crops. And the crop residues are used in the best possible way, as animal fodder or as component of the compost pile. You get the general idea, it is about maximum nutrient cycling and minimum loss. Several books have been published about Chinese integrated farming in the past few decades. It has attracted a lot of interest in circles of ecological farming and some of it has been tried in Africa, with very little lasting success as far as I know.

The first example I have seen in Africa of something resembling the Chinese method was an interesting little missionary project in the southern Guinea savannah in Cameroun in the 1970s. That is to say, interesting to a European visitor, because the local farming community paid no attention. The missionaries produced practically everything they needed for their own upkeep on their small farm where they cropped the land intensively, returning everything they did not consume to the soil or to their animals and using legumes as the main source of nitrogen. I think they even produced biogas from the contents of their latrines. And they were occasionally visited by a soil scientist from Wageningen University who gave them free advice on how to further improve their fertility management. There have been numerous cases in Africa of goodwilling organisations and individuals who tried similar things, to convince farmers, that was the way to go, with as little success as the Camerounian one. Not because the ideas were not good, they probably were, but they required something bordering on a mental revolution for the African farmer and revolutions do not just happen to individuals as we know.

Another problem is that the ideas are sometimes oversold. It is always wise to question claims that a farming model is completely self-sufficient, and examine whether there is not some external fertility flux, which is invisible to the casual observer. I did not carry out such an analysis of the missionary farm in Cameroun to find out, as I should have done, but I will give another example where a group of consultants was fooled into believing they were watching a miracle of fertility cycling, bordering on a *perpetuum mobile*.

At the occasion of a consultancy on regional development in Kenya in 1995, we visited a medium size dairy farm with 12 permanently stabled lactating cows plus young cattle and 2 ha of lush Napier grass. The cows were said to produce an average of 5,000 l of milk per year. The liquid dung from the stables was led directly into the field to fertilise the grass. The owner said that his was a closed system with the cattle feeding on the grass and the grass feeding on the nutrients from the dung, or something to that effect. That impressed some of the visitors very much. Being an agronomist, I had to question this fairy tale, of course, and a simple calculation showed that the nitrogen exported with the milk alone was almost half the amount the Napier grass could have contained. Even if the dung had contained the same quantity, there was far too little to keep the Napier going, so there had to be another source of nitrogen. And indeed, it turned out that the man purchased molasses from the sugar factory and spent grain from the local brewery. It was still an impressive operation, I hasten to say, but not the fairy tale some people liked to see.

A5.2 Manipulating a crop: Enset

Manipulating crops to make them behave the way you want is of the essence of farming. For example, growing seedlings in a nursery and transplanting them to the field once they are strong enough is a very common practice. Other examples are training plants on poles or trellises, grafting or budding of fruit trees, and drastic pruning of tea bushes to rejuvenate them, all of these things were invented by farmers ages ago and sometimes improved by scientists.

Figure A5-1. A young Enset plantation in southern Ethiopia. Photo courtesy of Siseraw Dinku

I will leave those well-known practices alone and talk about an intriguing method from Ethiopia to multiply the *enset* or false banana (*Ensete ventricosum*). *Enset* (Figure A5-1) is only grown as a crop in Ethiopia, in other East African countries it is sometimes planted as an ornamental (I am not sure it is the same species, though).

When the (pseudo-) stem of the *enset* reaches adulthood it flowers, produces a fruit bunch and dies, much like the true banana, only it takes much longer, several years in fact. And it is not the fruits the farmer is after, but rather the stem, which stores considerable amounts of starch. Getting the starch out and processing it into *kotcho* is a laborious affair, which I will leave aside. I have never acquired a taste for *kotcho*, I think you have to live in the country for quite a while before you do. What I like to describe here is the way the *enset* is manipulated to bring it from plantlet to maturity as efficiently as possible. First you must get suckers, which you can separate from the mother plant, to plant in a new plantation (in the same way again as with bananas and plantains). But *enset* does not produce suckers spontaneously while still in the vegetative stage, because of a phenomenon which biologists call 'apical dominance' – hormones from the apex prevent the growth of branches or shoots. So, by destroying the apex you should be able to force it to throw suckers. And indeed, that works. So farmers choose a vigorous plant, bore a hole through its centre and wait for the suckers to appear. And they will come in large numbers. The suckers are dug out and transferred to a field nursery where they will complete the first phase of their cycle. From there, most of them will be transferred to the next phase, leaving space for the remaining ones to develop. This process is repeated once or twice, whereby each time the plant spacing is increased. If the suckers were planted into their final location right from the start that would be very wasteful of land and intercropping with crops like maize or beans is not a good idea for various reasons, so this relay system is really a very elegant solution.

A5.3 Crop processing

I have two examples here, both from Yoruba land in Nigeria (and neighbouring Bénin Republic): oil palm and cassava processing. The reason is that I am most familiar with that area, but equally interesting examples of traditional processing techniques can be found elsewhere, like sago palm processing in Papua, rice parboiling in India and Bangladesh and the preparation of *curare* or arrow poison by South American Indians.

A5.3.1 Oil palm

The oil palm is a really indigenous species of West Africa and its use is intimately interwoven with traditional village life. Palm fronds are used for thatching, stems for building, palm wine is tapped from the trees' terminal growing point or the young inflorescence (which will be male if the tree is intensively tapped). But most important is the red palm oil, extracted from the fruits' fleshy and fibrous mesocarp or fruit pulp. That is done as an extended family operation, because it is a lot of work and the processing plant (Figure A5-2) is best built and operated by a few households together. The first thing is to get the fruit bunches down from the top of the palms, which may be up to 20 m high. There are always

Figure A5-2. Village oil palm processing plant

men in the village who specialise in tree climbing, either for tapping palm wine or to cut the bunches. The bunches drop at the feet of the trunk where they are left for a few days until the fruits detach easily from their peduncles. Not too long, though, because then the oil will start splitting and the acid content increases. The fruits are collected by women and taken to the processing plant in baskets. There they are steamed in big oil drums to deactivate the enzymes responsible for splitting the oil and to set the oil free from the fruit tissue (I do not know what kind of container was used before the arrival of the omnipresent oil drum, a fire-resistant clay pot no doubt). Next, the macerated fruit mass is transferred into a kind of mortar, which is part of a concrete double basin (formerly hewn out of rock, probably), where the fruits are pounded. The mixture of oil and water extruded from the pulp flows into the larger basis, where the slurry is then hand-pressed. The fibrous press cake, called *ogunsho* in Yoruba, is formed into small saucer-like shapes and stuck to the house wall for drying. It still contains enough oil to be an excellent material to kindle a fire. The mixture of water and oil in the basin is then allowed to settle whereby the oil will float on top and can be skimmed off. The product, red palm oil, is called 'unrefined' in the trade, which really is a nonsensical term for those who like the special pungent taste of food prepared with it and find the cheap refined palm oil from Malaysia, bland and without character. There is nothing like cowpea cakes or doughnuts, deep fried in red palm oil for breakfast, or so I think anyway. And so did the Brazilian lady for whom I bought some on a field trip and who recognised them as Brazilian *acarajé*. Apparently, the recipe had been carried across the Atlantic by the Yorubas who travelled there involuntarily and found a way to prepare in their new country what looked like the thing called *akara* at home.

What will development experts do when they see a successful native industry like palm oil production? They will try to improve it. Both the FAO and the World Bank have designed so-called intermediate technology for part or all of the process, like an improved steaming unit and a hydraulic press. That, of course, costs money, there have to be workshops to manufacture the equipment, expert advice to help the village people operate and maintain the equipment, and a non-government organisation to organise groups of processors around the equipment, mediate for a bank loan and perhaps market the oil. You start seeing what will happen. It is unlikely to work, too complicated and especially too dependent on factors over which the villagers have no control. That is a scenario for failure in Africa. And not for lack of potential, but rather because of the amateurism and lack of institutional memory of the so-called development organisations, which keep coming up with the same unworkable schemes. I talked about all that in Chapters 10 and 11.

A5.3.2 Cassava

When asked what is the most typically West African crop, a keen observer, but one with no historical knowledge, would say it is cassava. And yet cassava was introduced from South America, in the seventeenth century and only started spreading widely in the nineteenth century, perhaps under the influence of freed slaves returning from Brazil who brought the processing methods of the Indians with them. Cassava has rooted so deeply in the soils of West Africa and in the souls of its people that life without cassava is now unthinkable in the forest and moist savannah. The most widely spread foodstuff made from cassava is *gari*, which is very similar to the Brazilian *farinha*.[3] Its centre of production as well as the home of the best quality *gari* is south-western Nigeria, not surprisingly,

---

[3] Perhaps *gari* is a corruption of the word *farinha*, but I have not been able to find out.

Figure A5-3. Gari pressing and fermentation

since many 'returnees' from Brazil settled there. They brought with them the recipe for *farinha* in exchange, so to speak, for the *akara* recipe they had taken to Brazil a few generations earlier.

Making *gari* goes like this (Figure A5-3). Freshly dug tubers are washed and peeled and then grated. The resulting mash is entered into gunny or woven plastic bags and put under pressure from heavy stones or a car jack, or yet another contraption, such as the one in the photograph, to press the moisture out while the mash ferments. The process takes from 2 to 4 days and removes the toxic cyanogens, which occur in the roots of many cassava varieties. The mash is then stir-fried in a large flat iron disk over a wood fire until it is dry and has acquired a golden colour. It can be kept for a few weeks and when steeped in hot water becomes a nice, slightly sour paste, or it can even be eaten directly, without any further preparation.

A bottleneck in the processing used to be the grating of the fresh tubers, but in the 1970s a simple motorised grater was developed in western Nigeria, I do not know by whom. The whole assembly could be mounted on a wheeled frame and some small entrepreneurs started to go around the villages to do the grating for a fee. That was a typical African innovation which could not be beaten by the larger machinery, which was tried later by aid organisations – and which invariably failed, in much the same way as the processing equipment for palm oil.

Appendix 6. Papers Presented at the 2005 Symposium of the International Farming Systems Association

1. Diagnosis, policy analysis

– Multi-stakeholder analysis of farming systems development and future policy and institutional challenges for achieving SARD
– Profitability of diversified farming under rainfed rice-eco system of Chhattisgarh state in India
– The changing land use system on Vertisols in Kenya: challenges and opportunities
– Les mutations récentes du système de production oasien dans la vallée de Oued Righ (Algérie)
– Sustainable agriculture and rural development (SARD) analysis
– Complexity in farming systems, livelihood and natural resource management, a case study *in "Bazoft Watershed" in Iran*
– An integrated approach to food security assessment in the context of farming systems in fragile areas
– The role of markets in disaster and recovery to bolster vulnerable and poor farmers
– Forest resource degradation in Ethiopia: major causes, development attempts and future deliberations
– The challenges of farming systems in Bangladesh in the post-globalization period
– Public policies and farming practice changes in French overseas departments – the old times and the modern times
– Developing strategies for decreasing poverty in rural farming areas in Uzbekistan
– An analysis of agriculture–environment interactions and policy options for sustainable agriculture in Eastern Al Ghouta (Syria)
– Traditional native food, biodiversity and culture
– Poultry meat export in the economic dynamics of Iran and the Middle East region
– Market uncertainty and diversification strategies for rubber farmers: a comparison in Indonesia and Cambodia
– Fair trade food systems

- Sustainable, suitable and stable diversified agro-enterprises to augment income of farm families under rice-based production system of Chhattisgarh State in India
- Olive production in Greece
- Stock breeders transformed into agropastors by changing their systems to deal with drought and improving their living
- System characterization in a community of artisan fishermen in the south of Brazil: the case of São Lourenço do Brazil
- Incentives are not enough: could knowledge gaps vis-a-vis natural resource management be constraining rural livelihoods?
- Difference as a resource for sustainable agricultural development: responding to the globalisation of modern agriculture by supporting local agrobiodiversity
- La plasticulture itinérante dans les Ziban (Algérie)
- Plan de développement pastoral participatif en Tunisie centrale
- Identifying strategic development pathways for African agriculture
- Livelihood diversification and other strategies to improve food security, income, diet and health and local capacity of vulnerable populations in Uganda
- Smooth transitions from relief to reconstruction and sustainable agriculture
- Land, water and forest resources degradation in Ethiopia: major causes, development attempts and future deliberations
- To fight against the hunger and poverty by the agropastoral: contributed to the development in DRC: case of the cultures around the Field and Hunting preserve of Bombo Lumene in Kinshasa
- Using ethnographic linear programming to assess natural resource management alternatives among smallholders in the western Amazon
- Multi-use landscapes in the US: developing new synergisms between wildlands and farmlands
- Potentials of tree domestication to improve carbon sequestration and farmer livelihoods in smallholder production systems of the humid forest zone of the Congo basin
- Health hazards associated with occupational exposure to pesticides
- Abandoned pesticide waste sites in Georgia
- Constraints and challenges in the maize-based farming system in southern Africa: experiences from Zambia and Mozambique
- Development strategies, pathways and synergies investing in sustainable diversified agriculture for the smallholder vulnerable but viable farmers

- Revaluing the social domain: using the ORCA model to understand how social factors determine farming systems in Trinidad
- Focus on biodiversity, as this is in decline, and it affects all ecosystems including the human aspects of them such as human well-being and poverty, vulnerable systems and populations
- From pastoral to sedentary farming systems: making a difference to Bedouin Communities in Northwest Coast, Egypt
- The relationship between agroforestryand agroecology vis-a-vis the development of sustainable land use systems
- Small low resource farmers complementary and supplementary farming activities for productivity and livelihood sustainability
- Vulnerability of small farm systems and farmers' coping mechanisms towards land use change and land conversion
- The comprehensive African Agriculture Development Programme (CAADP) NEPAD vision for addressing food and rural development issues in Africa: originality, methodology and way forward in remodeling development policies
- Food security and the sustainability of rural livelihoods: recent trends in Syria
- The essential role of livestock for poverty alleviation in seasonal rainfall environments
- Diversification and other strategies to improve the income, diets and nutritional status of families with vulnerable children in Ghana
- The upland vegetable farmers in northern Philippines: the initial impact of trade liberalization

2. Choice, testing, evaluation of innovations, technology

- More benefit from less land: rice – pulse (as vegetable + fodder) rice is a more profitable cropping pattern for resource – poor farmers in Bangladesh
- Sugarcane-based farming systems research and some developed technologies
- New potato planting technique with rice straw mulching under no-tillage in rice–potato cropping system
- Precision agriculture, best alternative approach for sustainable agricultural development
- Zero-tillage: another revolution in third world agriculture
- Sustainable development of arid lands through appropriate and innovative farming systems and rational use of water resources

- Improving rural livelihoods through efficient on farm water and soil fertility management (in Tanzania)
- Soil and water conservation practices and improved livestock farming systems for sustainable agriculture and food security achievement in the semi-arid region of Burkina Faso
- Mechanised farming? The answer to drastic food security solution for Malawi
- Innovation for sustainable household food security in millet-based farming systems in West Africa
- Coexistence between genetically modified, conventional and organic crops
- A sustainable innovative yield booster in rice farming system in north-western zone of Tamil Nadu, India
- Water harvesting, to abridge the food gap and conservation of resource base in western Sudan (Kordofan Region)
- Animal feeds for smallholder farms in southern Laos
- Corn–livestock integrated farming system in selected corn growing areas in the Philippines
- Small livestock for landless and small farmers: improving farmers' lives through improved goat production practices in India
- Smallholder timber: livelihoods diversification and landscape sustainability
- Butterfly farming: a sustainable micro-enterprise model for biodiversity conservation
- Le système vétiver: une solution pour préserver l' environnement. Pourquoi le système vétiver?
- A congress on organic solutions for world farming
- Poverty reduction in hill farming systems of Nepal through more equitable access to local resources
- A participatory approach for salty soil reforestation in Senegal
- Building on synergies: achieving joint production, conservation and livelihood outcomes at a landscape scale. The case for "ecoagriculture"
- Systèmes de Production Agricole Durable et Lutte contre la Pauvreté: l'expérience de SYSPRO à Sébikotane
- Features of the System of Rice Intensification (SRI) apart from increases in yield
- Préservation Des Ressources Naturelles Et De L'environnement Par Une Valorisation Des Ressources Genetiques Du Rhizobium Autochtone De Quelques Legumineuses A Interet Pastoral Et Fourrager Des Zones Arides Et Semi Arides D'algerie,

– Farming systems (FS) approach to mitigating the effects of HIV/AIDS on rural livelihoods in southern Africa

3. *Extension, knowledge transfer, learning methods, on-farm research*

– Enhancing the role of farming systems research and development in local government development planning and action
– Practices and challenges of learning and experience sharing processes in farming systems; an Indian example
– Participatory processes in the development of livestock farming systems
– Decision support system (DSS) tool for small farm livelihood systems
– Learning processes through participatory on-farm research: experiences of the Lao-Swedish Upland and Forestry Research Program (LSUAFRP)
– Learning and knowledge transfer systems among the people of the Upper West Region of Ghana
– Enhancing capacity of academic institutions to produce gender-sensitive research and development professionals in farming systems
– Sustainet
– Agricultural and rural advice management approach: reference executive of the frame of partly industrialized aviculture of the region outskirts of Dakar, Senegal
– Transforming of agriculture-based agribusiness systems through process re-engineering: a shift from low-value commodities to high-value and price-stable farming systems-based commodity
– Extension workers-farmer-to-farmer learning and organisation in the villages of Papua New Guinea
– Production second: training farmers to move beyond the "big pumpkin fallacy" to customer demand
– Experiential learning on land and water management: a practical field guide for FFS facilitators
– Agricultural innovation system; capacity to address food and nutrition security and poverty
– Turning into photographers: Information and Communication Technologies (ICTs) for the generation of local knowledge in the context of ecological small farmers' agriculture, of the AGRECOL Andes Foundation of Cochabamba

– Institutionalisation of community-based experiential learning and empowerment – mechanisms and methods (including Global FFS Network and Resource Center)
– Decision-making in transition to integrated farming systems in small watersheds in northeast Thailand multiagent systems (MAS) model
– Farmers' knowledge support systems in Myanmar
– Management Advice for Family Farms (MAFF) process in North Cameroon, a framework of mutual learning of extension agent and farmers
– In search of excellence: exemplary forest management in Asia and the Pacific – lessons for agroforestry
– The livestock working group: partnering for improved benefits of livestock centered development
– Decision support system for the economic analysis of smallholder farming systems in South Pacific
– How a community is passing on seed and information? Farmer to farmer dissemination of fodder shrubs in central Kenya
– Institutionalisation of community-based experiential learning and empowerment – mechanisms and strategies
– Research, extension, and education: multi-stakeholder collaboration, organizational development and the future.
– Resolving conflicts to promote prosperous farming systems
– Poverty alleviation in Uganda through farmer empowerment, informal adult education and demand driven advisory services
– Factors contributing to communication fidelity
– Local knowledge and agricultural development: considering multiple dimensions
– Facilitate the learning process by small farmers: an experience using illustration-based educational El Salvador, Central America
– Considerations for effective service delivery: using participatory communication for understanding farmers' realities
– Small-scale farmers as teachers to build rural entrepreneurism and increase family income
– Processes for systemic learning, research and change in complex systems
– Focus on people-centred development
– Indigenous knowledge in natural resource management to improve livelihoods and heal landscapes in pastoral and communities: lessons from eastern Africa

- Improving subsistence farming: educating the farmers simple principles of economics
- A peasant University for the Nordeste of Brasil
- Peer-to-peer learning among farmer groups and service providers that are separated by large distances
- Participatory learning processes in a community-based activity experiences in Laguna and Palawan, Philippines
- Building leadership capacity for sustainable rural development
- Local knowledge in sustainable agriculture : the example of the Sustainable Agriculture Farm Network in Midipyrénées, France
- Shaping integrated pest Management knowledge and practice
- Field experience as a master trainer in the Farmers Field School (FFS) approach as adapted to the situation from its origin in Asia
- Adaptive Collaborative Management (ACM) in practice: lessons from Mafungautsi Forest Reserve in Zimbabwe
- Farmers begin to invent water saving cultivation in northeast Thailand
- From reductionism to farmer innovation systems: implications for multi-stakeholder collaborative learning and client orientation in Uganda
- Strategies beyond peer-to-peer knowing and learning for cocoa quality improvement in PNG
- The evolution of a participatory learning approach for agricultural nutrient management in the north-eastern USA
- Forming a farmer experimental group to develop technology for integrated farming in rainfed northeast Thailand.
- Agro-advisory services based on medium-range weather forecasting in the new alluvial zone of West Bengal, India
- Bonnes pratiques de formation pour un projet de développement rural: L'exemple du Projet Emploi Rural en Algérie
- Fortalecimiento del Programme de desarrollo de las Montanas Cubanas
- Participatory on-farm trials and demonstrations in support of improved food security and agriculture productivity – experiences and lessons from the South Pacific Island Region
- Use of swot analysis in participatory soil conservation planning for smallholder farming systems: a case study
- Agriculture and rural development project in Cambodia with scope for human security-nurturing self-reliant rural communities

4. *Impact studies*

- Impacts of crop–livestock R&D on smallholder farming communities in Bangladesh
- When do smallholder farmer–market linkages increase adoption of improved technology options and lead to increased use of natural resource management strategies?
- Impact of agroforestry: lessons from three sites in Africa and Asia
- Contributions of agricultural extension to rural poverty reduction in Myanmar
- Methodology for assessing impact of farming systems groups (FSGs) in rural communities, Australia
- Processes in the development of livestock farming systems of Northern Ghana
- Evaluating the success of forest conservation efforts by smallholder cacao producers in Southern Bahia, Brazil
- Understanding and sharing of successful local development practices

5. *Socio-economic, institutional, market, policy environment, development platforms*

- Role of supporting organisations for agricultural producers in the Czech Republic,
- The role of cooperatives in improving quality of life and providing sustainable development
- Developing public–private partnerships in agribusiness development: easy to say but a challenge to do
- Helping small farmers through capacity building of government service providers: *lessons from Pacific Island Countries*
- Reducing rural poverty – the farmer–market linkage and the farmer–agro-industry linkage
- Organizational legitimacy as principle for private provision of rural development activities: evidence from Czech Agriculture
- Farmer–private sector partnerships and smallholder producer competitiveness: the nucleus estate experience in Nigeria.
- The role of small retailers and small-scale producers' organisations in enhancing and strengthening markets linkages and local qualification process
- Building social infrastructure (*platforms*) for decentralized management of natural resources

- Harnessing the power of partnerships in the marketplace: using a learning alliance for agro-enterprise integration into agricultural recovery
- Effective partnerships for sustainable rural livelihoods: a critical review and case studies from Africa
- International Partners for Sustainable Agriculture (IPSA)
- The EU novel food regulation – a non-tariff trade barrier for small farmers and trade companies in developing countries
- Environmental decision-making as a framework for farm policy
- A participatory approach in agro-environmental policy development and decision-making
- Globalisation, malcontents, and asymmetric impacts on smallholders
- Supermarkets and small growers
- Importance of improving the conditions of commercialisation for farmers rubber smallholders in the SouthWest province of Cameroon
- Organic agriculture and alternative certification
- Implications of changes in the structure of fertilizer prices in Malawi
- Standards in organic and sustainable agriculture
- The role of middle man
- Collective action by smallholder organic farmers in South Africa
- Improving farmer decision-making and research-extension-private sector linkages for identifying organic and fair trade export opportunities in Uganda
- Guarantee systems for organic production
- Market-orientation of agricultural research in low-income countries: the case of Lake Zone, Tanzania
- Opportunities for agriculture marketing : the case of smallholder farmers in Malawi
- The incomes of the producers within the framework of the production of cocoa resulting from organic the cocoa "bio équitable" in Ecuador
- Agrifood systems
- The organisational culture and the food chain
- Décentralisation a conduit à un conflit de compétence entre les acteurs de gestion des ressources forestières au Mali-Sud
- Networks of NGOs and of governmental organizations – contrasting the experiences of FIDAMERICA and Grupo Chorlavi

- Familiar agriculture from the farmers' organization point of view in the northern Costa Rica
- Multifunctionality, stakeholder participation, R&D, policy making – what has learning got to with these?
- Institutionalization of trans-disciplinary activities for rural livelihood development
- Rôle de l'accès aux ressources naturelles communes dans la réduction de la pauvreté; Cas des parcours collectifs au Moyen Atlas, Maroc
- Building a sustainable community food system in Seattle and King County, Washington, USA: developing a local food policy council
- Productivity enhancement and welfare gains on smallholdings in south-western Kenya: interaction between institutions, technology in transforming farming systems
- A framework to support effective policy making on biodiversity–poverty relations in farming systems in developing countries
- Sustainable Tree Crops Program (STCP): realizing a new development paradigm
- Surprising new partnerships in the journey to a sustainable rural area in Noord-Brabant in the Netherlands. Theory and practice
- Relationship between farmers and service providers
- The power of agroforestry and multisectoral partnership in sustainable upland development: the case of the "Agroforestry Support Program for Empowering Communities Towards Self-reliance (ASPECTS)" project in the Philippines

Appendix 7. Quantification of Cotton Growth and Development

Cotton is an excellent species to demonstrate what is meant by morphogenesis and why a simulation model should account for it if it is to be realistic. First I will briefly describe how a cotton plant establishes its structure and then take a quantitative look at two aspects of growth and development: (i) the link between stem growth and leaf growth and (ii) the relationship between square[1] and flower production on one hand and boll growth and seed cotton production on the other.

A7.1 Establishment of plant structure

The size of the cotyledons and the initial size of the growing point are already determined in the seed. When the seed germinates, the growing point (also called the apex) starts to form leaf primordia and associated axillary buds. Their size is related to that of the apex at the time they are initiated. As the apex increases in size, so do the leaf primordia and the axillary buds that split off from the apex. There will therefore be a regular progression in the size of the main stem leaves and associated branches, unless something happens to reduce the growth of the leaf primordia and the young leaves. For example, the plant may not produce enough carbohydrates to maintain potential growth, or the plant's turgor may be reduced by water stress and the final leaf size will then be smaller than without such stresses. Stress may also affect growth of the apex itself and if that happens, its effect will be felt for some time after, because the leaf primordia will now be smaller than they would have been in the absence of any stress.

From the fifth or sixth main stem node upwards, a sympodium will emerge from each axillary bud, which produces squares at regular intervals. Their total rate of production is slow at first when there are only few sympodia, but increases rapidly as more are formed. As long as there are no calamities, all these buds will produce a

---

[1] 'Square' is the name for a flower bud in cotton.

flower. Whether they go on to produce a boll is another matter, and depends on many things. Among these we shall look in particular at the availability of assimilates. Growing bolls have priority access to assimilates and as more bolls appear there will be less assimilate available for other growth, including new bolls. Under most conditions the increasing boll load will eventually cause growth to come to a full stop, which is called cut-out.

A7.2 Growth of stem girth and weight

This is my favourite in the quantification of cotton growth, taken from the KUTUN simulation model which earned me a Ph.D. When you look down a cotton plant from the top, the girth of the main stem is seen to increase fairly smoothly as it descends, except below the nodes connected to monopodial branches where it suddenly leaps up. I will explain why that happens. Below each new leaf there is an internode which is formed in association with it. Initially, the internode's weight will have a more or less constant relationship with that of its associated leaf, as both originate from the same apex. Hence, the amount of new stem tissue $(\Delta W_S)_1$, directly associated with a new leaf, will be proportional to the weight ΔW_L of that leaf:

$$\frac{(\Delta W_S)_1}{\Delta W_L} = a \tag{1}$$

In addition to that, as the new leaf grows, the stem tissue which is present below it also has to increase a little all the way down in order to accommodate the increased upward and downward water and sap flows. It is reasonable to assume that the amount of new stem tissue $(\Delta W_S)_2$ formed per unit of new leaf along the path from the new leaf to the plant's base (where it emerges from the soil), is proportional to the length of the path, which in turn will be more or less proportional to the weight of the leaves, W_L, present along the path. Therefore:

$$\frac{(\Delta W_S)_2}{\Delta W_L} = b W_L \tag{2}$$

Hence, the total amount of new stem tissue formed in association with new leaf tissue is the sum of (1) and (2):

$$\frac{\Delta W_S}{\Delta W_L} = a + b W_L \tag{3}$$

where W_L is taken *along the path* between the new leaf tissue and the base of the plant. That explains the sudden increase of main stem girth below the point where a monopodial branch is implanted: the main stem section below the monopodial branch has to cater for the leaves on the monopodium, in addition to those on the main stem section higher up. This observation makes it plausible that the postulated mechanism may be correct but it does not prove it, of course, so we are going to look for supporting quantitative data. It would be best if we had data for stem and leaf weight along the path from a new leaf to the plant base, but that is expecting too much. There are data for the whole plant, however, so let us see whether they can be helpful. As long as leaf weight is small, expression (3) should still hold, because there is only a main stem and no branches. From the data plotted in the left-hand part of Figure A7-1, I decided that a linear relationship is reasonable for young plants. The parameters of the linear equation were estimated from the measured data as:

$$\frac{\Delta W_S}{\Delta W_L} = 0.3 + 0.11W_L \qquad (4)$$

What about older plants? More and more of the new leaves will be formed on the sympodia and since expression (3) is postulated to hold along the path from the new leaf to the basis, only leaf weight met along that path counts. And since that is less than total leaf weight, the relationship should depart from the straight line and

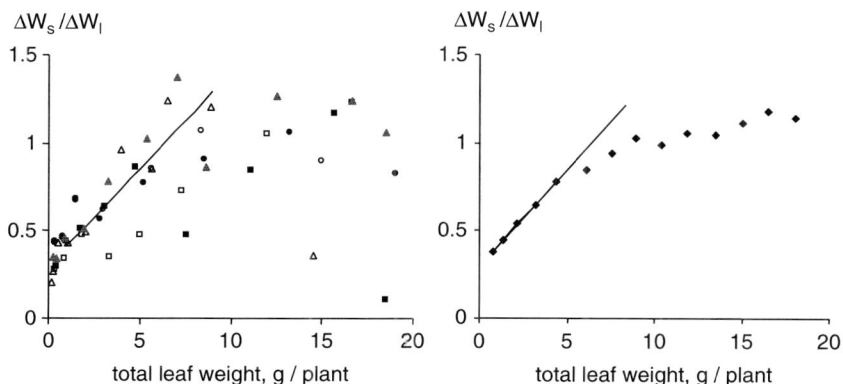

Figure A7-1. Measured (left) and simulated (right) relationship between stem and leaf growth of cotton. (From Mutsaers, 1984.)

gradually level off. That is what the data do indeed show. I formulated a simple spreadsheet program which calculates stem growth associated with each leaf by applying expression (4) along the path between a new leaf and the plant's base. The right-hand part of Figure A7-1 shows that this mechanism results in a pattern which is similar to that observed in the measured data (the flutter of the simulated points to the right is an artefact caused by looking at fully grown leaves only). So, I think we can be reasonably confident that it works like this in the real plant. It is straightforward to build the mechanism into a simulation program for cotton, and avoid the purely empirical relationships which had become standard in crop modelling by the end of the last century.

A7.3 Numbers of squares and bolls

A7.3.1 Build-up of square load

The number of flower buds ('squares') increases in two dimensions. Each sympodium forms new squares at regular intervals, while a new sympodium issues from the axil of each new main stem leaf. That invites mathematical analysis, which is quite simple really. Most of the mathematics involved in crop growth models is essentially simple. How does the (theoretical) number of squares increase with plant size? In order to avoid lengthy phrases, first we define some terminology. The interval between first squares on successive sympodia is called the vertical flowering interval (VFI). It will be practically the same as that between unfolding of the corresponding main stem leaves. The interval between two squares on the same sympodium, called the horizontal flowering interval (HFI), will be considerably longer, because each time several things happen: an axillary bud must start growing, it first forms a prophyll and then a true leaf before finally ending in a floral bud. The ratio between VFI and HFI will be called r. There are varietal differences in the value of r, but it is usually around 0.4.

We count the total number of squares present on all the sympodia together, each time the first square appears on the youngest sympodium. Assume that the first square has just appeared on sympodium n, so there are now n first squares. On sympodium n, there are no additional squares of course, on sympodium $(n-1)$ there are r, but since r is smaller than 1, the next square is not yet visible. On sympodium $(n - 2)$, two main stem intervals have elapsed since the first square, so the number of additional squares is $2r$. And so forth:

Number	$Symp_n$	$Symp_{n-1}$	$Symp_{n-2}$		$Symp_2$	$Symp_1$
n	1	$1 + r$	$1 - 2r$		$1 + (n - 2)r$	$1 + (n - 1)r$

If we ignore the fractional parts of the terms with r (being squares which are not yet visible), the total number of squares is the sum of the terms in the table. The coefficients of r are a simple arithmetic series, 1, 2, 3, 4.... $(n - 1)$, whose sum is $1/2n(n - 1)$. So the total number of squares, including some which are not yet visible, equals $n + 1/2n(n - 1)r$. The number of *visible* squares is found by adding only the integer parts of the terms with r:

$$n + \sum_{i=1}^{a-1} \text{int}(i \cdot r)$$

The graphs for the two functions are shown in Figure A7-2 for an r value of 0.4, whereby the lower curve is the one for visible squares. If a simulation model correctly accounts for the regular addition of main stem leaves and sympodial branches and the associated squares, it will of course automatically generate this square production.

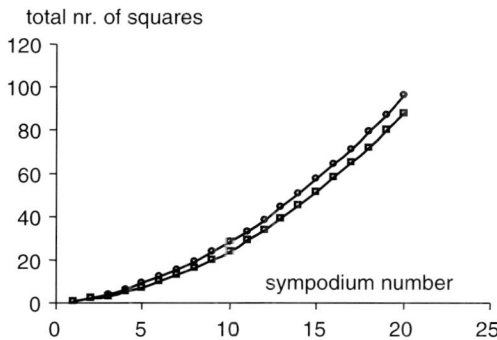

Figure A7-2. Calculated increase of number of squares

A7.3.2 From square to cotton boll

How many of these abundant squares will eventually make it into a mature boll? Let us do some simple calculations for a rough estimate. Consider a field crop with 100,000 plants/ha, a common density for irrigated cotton. The seed cotton (fibre plus seed) from a single cotton boll weighs around 5 g. A very good cotton crop may produce up to 4,000 kg/ha of seed cotton, that is 40 g/plant, for which an average of 8 bolls/plant are needed. According to Figure A7-2, a plant with 15 sympodia could theoretically produce some 50 squares, but at this high planting density, a sympodium will not extend beyond 5 nodes, so the maximum number of squares is more likely to be around 40. That means that even in a very good crop only 20% of the squares make it into a mature boll. The rest must have been shed at one time or other. How and when does that happen?

Squares are rarely shed by the cotton plant – unless they are attacked by insects or there is some other calamity like severe drought, they will develop into a flower. Cotton bolls can also not be stopped, once they are growing vigorously, except again by some calamity. Hence, shedding of a fruiting point must occur in a narrow window of time just after flowering. What triggers this shedding? The so-called nutritional hypothesis claims that it is the availability of growth substrates and that the way the plant partitions them over its organs determines whether a particular flower will actually become a boll. Since squares and growing bolls appear to have absolute priority for growth substrates, it must be substrate shortage just after flowering, which triggers shedding of a fruiting point. Let us see how far you can get with that to explain the fruiting behaviour of cotton.

A7.3.3 Shedding of fruiting points

Sympodia always grow more or less horizontally and a cotton canopy is therefore highly stratified. A fruiting point at the end of a sympodium deep inside the canopy is in the shade of the heavy foliage overhead and far removed from the leaves, which contribute most to carbohydrate production. So you would expect that it has little chance to develop into a boll, and that is how it is: there are often very few bolls beyond the first two sympodial positions, especially in dense crops. Most of the others are shed after flowering. As the number of growing bolls increases, substrate shortages will develop elsewhere as well, more and more fruiting points will drop off, especially at the extremities of the sympodia, leaf growth slows down also and eventually all new growth will be suppressed.

Another interesting thing is that in a very dense and luxurious crop, shedding of fruiting points may be so excessive that the crop just goes on making more leaves and branches. Every time a fruiting point is ready to become a boll, there is already so much new foliage overhead that it is starved of food and drops off and cut-out does not occur, or is much later than usual. The nutritional hypothesis can explain all that, although it is not really proved by it, of course, but simply made plausible.

Appendix 8. Organ Growth and Assimilate Partitioning in Four Modelling Families

A8.1 Introduction

It is not all that difficult to understand how models handle organ growth and assimilate partitioning, although you may think it is, when glancing at the publications and the computer code. That has been one of the problems with most of the modelling literature: people who could have taken advantage of modelling have usually been put off by the unfriendly looks of the modelling texts and the austerity of the computer code. Let us see whether we can get at the essentials of organ growth and assimilate partitioning, which I think are among the most important aspects of plant growth for a model to simulate. First, a few words about why that is important.

In growing plants, assimilate production is distributed over the assimilating leaves and it is somehow linked with a likewise distributed assimilate demand by growing organs or 'sinks' throughout the plant. Furthermore, the growth of different organs is steered by an underlying template while at the same time reflecting the previous growth history. For a crop model to mimic the resilience of a real plant, it must faithfully represent the way the growth of different organs is regulated and the way assimilate supply and demand are coordinated, in other words, how assimilates are partitioned. Computer models were precisely meant to help take care of that kind of complexities.

The production of assimilates and their conversion into plant substances have been handled adequately in most crop models since the early 1970s, but growth of the plant organs and the way assimilates are partitioned among them have not been equally well represented in all but one model family, as I will show in this Appendix. That has made the reliability of the models questionable.

A8.2 Models of the Wageningen school

A8.2.1 ELCROS and BACROS

Some of the plant's vital regulatory mechanisms were well represented in ELCROS, although in a rather rudimentary form. In the
years after ELCROS, not much progress was made in building good
growth physiology and morphogenesis into the Wageningen models, in spite of de Wit's insistence that that should have first priority. But there were some serious attempts. In 1979 Horie et al.
published a long and very difficult series of papers on leaf growth
in cucumber. The idea was to work out in detail how the growth of
successive leaves was linked and whether that could be captured in
a mathematical formalism. Bensink, one of the authors, was a plant
physiologist who had done something similar with lettuce, in a different University department. I have rarely seen the cucumber
papers quoted, perhaps they were too difficult. In any case, after
that de Wit did not return to morphogenesis again, apart from
supervising my Ph.D. thesis, which dealt with the morphogenesis of
cotton and which attracted only slightly more attention than did his
work on cucumber. In the years that followed, members of the
Wageningen group occasionally did publish papers on morphogenesis, but as far as I know the results have not found their way into
the main models.

   ELCROS' successor, BACROS, which was published in 1978, was
stripped of those processes which were less well understood. The
most essential difference with ELCROS was that potential growth
as the demand factor for assimilates was no longer there and the
model had become entirely driven by assimilate supply. The assimilates were now simply allocated to the growing organs according to
fixed partitioning factors. The elimination of ELCROS' demand-
supply mechanism and its replacement by a purely empirical assimilate allocation scheme were a demonstration of the failure of de
Wit's school to realise one of the old master's ambitions: to build
models based on the plants' in-built growth templates as organisational principle.

A8.2.2 BACROS and its successors

After the BACROS model had been published, the Wageningen group did two things.[1] The first was to develop a 'Simple and Universal Crop Growth Simulator', SUCROS, which was based on simplified and partly redesigned versions of the BACROS routines for carbon assimilation and conversion of assimilates into plant substances. SUCROS was meant for practical applications and has indeed been used as a crop growth module in larger programs.

The second line of work was the development of models for particular crops. They also used basic process modules from BACROS augmented with special routines to simulate growth phenomena, which were specific for the target crops. The best known are the models of the ORYZA group, for rice, which I will also describe briefly.

Although these models used improved BACROS modules as their foundation, they were not just assemblies of reworked BACROS parts. They contained two innovative elements, which strengthened the models' physiology content. One was the representation of the plants' phenological development, that is the succession of growth stages – seeding to emergence, emergence to first tillering, tillering to first flower etc. – and the effect of environmental factors on their duration. The second innovation was concerned with the remobilisation of carbohydrate reserves stored in the shoots of cereals. That process was discovered by Spiertz and Ellen in 1978, who found that at the beginning of grain filling in wheat up to 20% of stem weight consisted of reserves, which could be translocated to the growing grain. Without that process the plant would not be able to satisfy the demands made by the grain at the time its growth rate was greatest. Apart from these two innovations, however, the models were all physics and biochemistry; there was no plant. The representation of the crop consisted of little more than loose collections of roots, stems, leaves and storage organs with some empirical parameters to sew the whole thing together.

I will briefly describe representation of organ growth and assimilate partitioning in two examples of the post-BACROS models of the Wageningen group, SUCROS and ORYZA, in order to clarify the points I have made.

---

[1] The genealogy of the Wageningen models was traced in a rather uncritical publication by Bouman et al. in 1996 and again, a little less uncritical in 2003 by van Ittersum et al.

(a) SUCROS

The *Simple and Universal Crop Growth Simulator* was what its name
says: simple, as well as elegant and transparent, at least for other
modellers. Its core was essentially a compact version of BACROS,
the fruit of several decades of simulating well-understood basic
crop growth processes, augmented with some new physiology –
phenology and the storage and remobilisation of stem reserves. The
final version published in 1997 was for spring wheat, but in principle SUCROS could be adapted for any crop for which the necessary
data were available. The flow diagram of Figure A8-1 shows how
the model worked. Canopy photosynthesis (in the upper left hand
corner) produces a flow of assimilates from which maintenance respiration is subtracted first. What remains is split into two parts, one
for the roots and one for the shoots. Which fraction goes to each
depends on the (phenological) development stage and is looked up
in a two-way partitioning table (lower left hand corner). The table
is purely empirical, based on data measured in the field, but allocation can be modified in the case of moisture stress to favour root
growth. The assimilates for shoot growth are further partitioned to

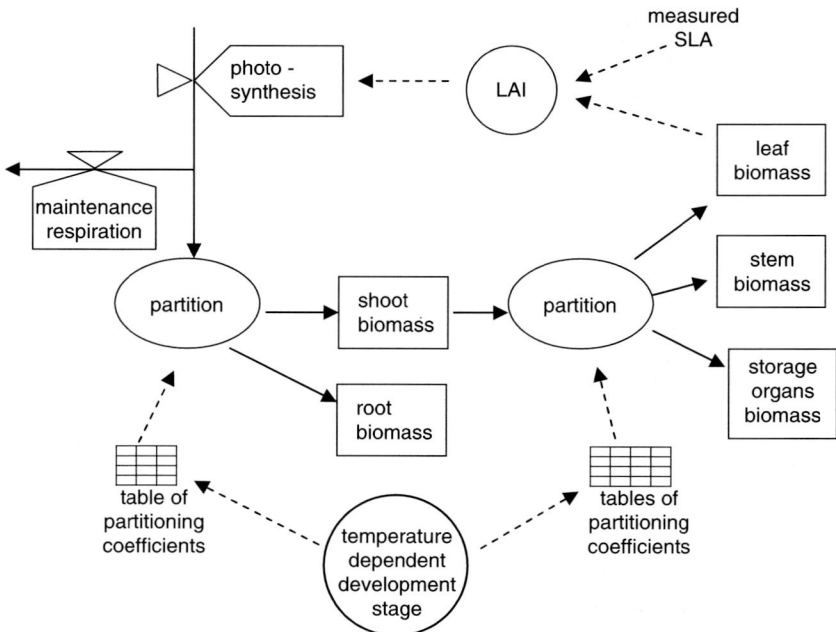

Figure A8-1. Flowchart of the SUCROS simulation model

the leaves, stems and storage organs, using another table with empirical partitioning factors depending on the plants' phenological stage. Obviously, this is all highly artificial. There are no innate partitioning factors in a plant. Partitioning is the end result of the plants' regulatory mechanisms, not something that is fixed a priori. ELCROS stayed much closer to the way real plants work. There, growth partitioning resulted from demand by the growing tissues and a delay related to their distance from the assimilate source, while moisture stress changed partitioning indirectly through a reduction of potential leaf growth, as happens in real life.

One of the linchpins holding a model together is the simulation of the crop's leaf area, because it determines how much assimilates the crop will have available at any time. In the early years, much weight was therefore attached to the development of reliable routines for leaf growth, but without much success. It was one of the things dropped in later models, including in SUCROS, and replaced by a simple calculation of LAI as the product of leaf biomass and a specific leaf area (SLA), the area per unit of leaf weight, measured in the field. That gave the model great stability, at the cost of getting even further removed from the way real plants work.

There was also a version of SUCROS which could simulate growth under water stress. It calculated moisture conditions in the soil, water uptake by the roots and transpiration by the leaves and then calculated the ratio between actual and potential transpiration. If the ratio was smaller than 0.5, growth partitioning to the roots was decreased at the expense of shoot growth.[2] This at least was a sensible mechanism which resulted in relatively more roots when the plants needed them, the way it happens in the real world.

(b) ORYZA

The Wageningen group also developed 'experimental' models for a variety of crops, most of which incorporated BACROS-derived modules, as well as routines for crop-specific growth phenomena, such as a representation of its phenology and empirical growth partitioning factors related to the development stage of the crop. The best known models were those for irrigated rice, collectively named ORYZA. In 2001 a practical implementation was published by Bouman et al. under the name ORYZA2000 for the general user. Well, general user is a little optimistic I think, but with some effort

---

[2] The moisture status also affected the canopy assimilation rate.

non-modellers could at least run it on their PC. It could simulate
rice growth under potential conditions as well as under moisture
and nitrogen stress. ORYZA2000 contained more physiology than
SUCROS, but most of it was also of an empirical nature and implemented in the form of tabulated input, such as phenology-dependent
partitioning factors for assimilates and nitrogen.

A8.3 The CROPGRO and CERES models

CROPGRO was a 'generic' model, developed at the University of
Florida, Gainesville by a group of researchers led by Ken Boote
and Jim Jones, to simulate the growth of dicotyledonous crops. It
was derived from earlier, separate versions for soybeans, groundnuts and beans, which were later merged into one generic model
capable of simulating a wide variety of crop species by feeding it
with species- (and variety-) specific equations and parameters. The
creation of a single crop model with common logic for all (dicotyledonous) species was a significant achievement.

Another influential model, which became the progenitor of a whole
family, was CERES-Maize, developed at Texas A&M University by
a group of modellers around J.T. Ritchie. The CERES models
were designed to simulate the growth of cereal crops and there were
versions for each of the major cereal species. CERES-Maize has been
one of the most successful crop models in the world and it has been
tested, 'parameterised' and applied very widely on all continents.

Both CROPGRO and the CERES group were incorporated in
the DSSAT developed by a consortium of US universities.[3] The
idea was that the DSSAT package would evolve into a practical tool
to be used by agronomists and development workers for crop management and planning purposes. DSSAT has been used a lot by
researchers, mostly in crop production and to a lesser extent in
Integrated Pest Management and economic analysis. The ambition
to make DSSAT part of the toolbox of practical agriculturists,
however, was not quite realised, although there have been a few
serious attempts.[4] In 2006 a consortium of Universities were using

---

[3] A good general text on the implementation of the CERES and CROPGRO models in DSSAT is Jones et al., 2003.
[4] Jim Jones informed me that some of the projects were discontinued because of
institutional changes, not because of problems with the models.

information generated by the DSSAT models in its extension programme on climate risk in the south-eastern USA.

I will describe some key procedures of the CROPGRO and
CERES models, with emphasis on how they handled morphogenesis
and growth partitioning. I have reserved most space to the CERES
logic, because it has been the most widely used group of models in
the tropics.

A8.3.1 CROPGRO

Simulation of basic growth processes by CROPGRO[5] was quite
similar to that of the Wageningen models, with a few major differences. Like SUCROS, CROPGRO was a generic crop simulation
model, but the latter has been implemented for a much wider range
of crop species. Furthermore, growth partitioning was handled
differently by the two models. SUCROS used a two-stage approach
by first applying a partitioning between shoots and roots, followed by
a further partitioning of shoot growth into leaves, stems and ears.
CROPGRO, on the other hand, used a single-stage approach with
much more elaborate partitioning functions for different growth
stages which allowed it to adjust growth flexibly to carbohydrate
supply, moisture and nitrogen availability and more generally to
different environments. The chart in Figure A8-2 shows the 13 vegetative and reproductive stages the model distinguished, each with
its own coefficients and responses to temperature and in some cases
to day length.

(a) Vegetative growth

During vegetative growth, the model used detailed empirical partitioning coefficients to allocate biomass growth to roots, shoots and
leaves. Growth in leaf area was calculated by multiplying leaf biomass
growth by Specific Leaf Area (SLA). Both the partitioning coefficients
and 'normal' SLA were growth-stage-dependent and had to be supplied as species- and variety-specific inputs, obtained from field measurements. They were then dynamically modified by the model in
dependence of C- and N-supply and water stress. In case of (severe)

---

[5] More details can be found in Boote et al., 1998 and Jones et al., 2003.

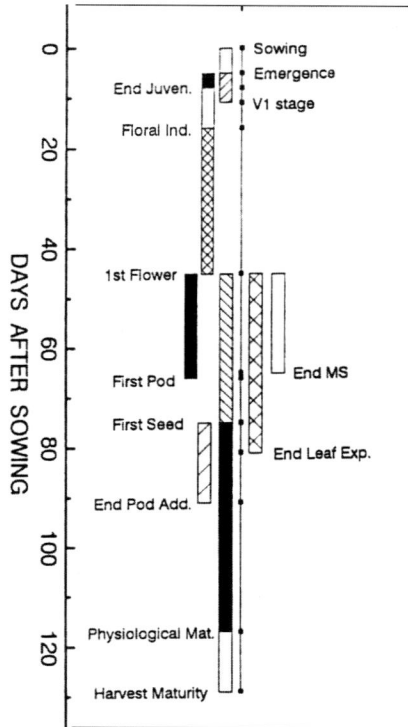

Figure A8-2. Vegetative and reproductive growth stages in CROPGRO (Boote et al.,1998)

nitrogen stress, for instance, the partitioning coefficients were modified to favour root growth. Moisture stress also caused the model to increase partitioning to the roots as well as reducing leaf area growth by lowering SLA.

What has been said earlier about growth partitioning in BACROS and SUCROS applied equally to CROPGRO: in real plants partitioning is the result of the interplay of assimilate supply and demand, rather than a kind of innate or genetic property. Defining standard partitioning coefficients a priori, even if they can be adjusted to substrate supply and environmental conditions, can only be an ad hoc solution, until potential organ growth and the internal balancing process are better understood.

(b) Reproductive growth

I think the simulation by CROPGRO of growth partitioning to 'reproductive tissue' (that is the collective name for flowering buds, flowers, fruits or pods and seeds) came close to the way such processes are likely to take place in real plants. During the reproductive growth stages, the model added daily 'cohorts' of fruiting points, their potential addition rate being a (species- and variety-dependent) input into the model, also obtained by field measurements. Actual addition rate and pod and grain growth were dynamically adjusted to carbohydrate and N-supply. First priority for biomass growth was given to reproductive tissue, up to a 'maximum reproductive partitioning factor'. The factor was 1 for an absolutely determinate species, while the remainder, if any, continued to be invested in vegetative tissue. Indeterminate and semi-determinate plants had a smaller maximum partitioning factor for reproductive growth, which means that there always remained something for vegetative growth, until the 'leaf expansion phase' ended and the model terminated leaf growth.

The model kept track of each of the cohorts of pods and seeds and their potential and actual growth rates. Which part of the fruiting points were retained or shed and how much biomass was accumulated depended on assimilate (and nitrogen) availability. In case there was not enough to allow potential growth, the model would mobilise reserves from the leaves and shoots. And if that was still not enough, actual growth of the cohorts was adjusted by shedding part of them and reducing growth of the rest. This procedure also came close to the supply–demand balancing as it probably occurs in the plants. Finally, the model could account for pest damage by shedding fruiting points.

Using differential maxima for reproductive growth partitioning in determinate and indeterminate species was a nice and simple trick, which was able to generate growth types as different as those of soybeans and groundnuts. But again, the reality value of such fixed maxima is doubtful. Substrate allocation in the plant, resulting from the interplay of supply and demand and its interaction with the environment, is not an intrinsic or genetic property. Something similar can be said about termination of leaf expansion by the model as a programmed event. In reality I think leaf growth only stops because of substrate shortage, not because there is some other signal which tells the plant to stop making leaves. In cotton, for example, once the load of growing fruits declines, vegetative growth will resume if the conditions are right.

A key process taking place in leguminous species is symbiotic N-fixation in the root nodules. CROPGRO had elegant and convincing routines to simulate the balance between N-uptake and N-fixation, but I cannot go into that here. If you are interested read the group's own publications.

A8.3.2 CERES

The best known versions of CERES were those for maize, wheat and sorghum, all using the same simulation logic but different growth equations and parameters. As an example I will look at the way CERES-Maize handled phenology, organ growth and substrate allocation. Most of the details are from Kiniry (1991) and from lecture notes kindly made available by Dr. William Batchelor of Iowa State University, who was leading CERES-Maize development at the time I wrote these paragraphs.

The CERES approach to growth partitioning was quite different from that of CROPGRO. While the latter was entirely supply-driven (except during the earliest vegetative growth phase) using a variable partitioning scheme, the former employed an interesting demand–supply strategy. The model first calculated potential growth of the above-ground parts (the demand or sink), allocated the available assimilates[6] (the supply or source) to them according to their relative demand and sent the balance to the roots. If less than a certain percentage of the available carbohydrates would remain for the roots, then the growth of the above-ground parts was adjusted downwards. The minimum percentage allocated to the roots varied with the growth stage. This scheme was perhaps not as physiologically sound as that of my favourite, ELCROS, but at least it contained an element of internal growth dynamics, which was missing in the purely supply-driven models. Let us see how it worked. It is a good illustration of the many difficulties involved in simulating morphogenesis and substrate demand, even with an essentially empirical approach like that of CERES.

---

[6] The model actually converts the assimilates into 'dry matter' before growth partitioning for reasons of modelling convenience.

(a) Phenology

First we must look at maize phenology, because the CERES growth equations were closely linked with the eight different growth stages CERES-Maize distinguished. The stages are shown in Table A8-1.

The third column contains the 'genetic coefficients' associated with each growth stage, which had to be measured or estimated in some way or other and fed into the model as an input. I will explain some of them as we go along. The length of each stage depends on temperature and on the variety. To account for the temperature

Table A8-1. Growth stages distinguished by CERES-Maize and associated 'genetic coefficients'

Stage	Description	Key processes	Genetic coefficients
S8	Sowing to germination	5 leaf primordia are already laid down in the seed	
S9	Germination to emergence		
S1	Emergence to end of juvenile stage	Vegetative phase, leaf initiation and growth	P_1, duration of stage S1 (GDD[1]) I_L, leaf appearance interval (GDD)
S2	End of juvenile stage to tassel initiation	Flower induction, leaf initiation continues	P_2, photoperiod sensitivity (0–1)
S3	Tassel initiation to silking and end of leaf growth	Apex transforms into tassel, ears are initiated	
S4	Silking to beginning of effective grain filling	Pollination	P_5, duration of S4–S6 (GDD)
S5	Effective grain filling period	Grand (linear) period of grain growth	G_2, potential kernel number G_5, potential kernel growth rate per day
S6	End of effective grain filling to physiological maturity	Completion of grain growth and maturation	

[1]GDD (Growing Degree Days) are explained in the text.

effect, CERES used the 'growing degree day' concept, which I will first explain. It played an important role in modelling in general and in CERES in particular.

Suppose a plant species will only grow when the temperature is between 10° and 34°C. Temperatures below 10°, called the base temperature, and above 34°, the maximum, do not contribute 'growing degrees' while between those limits the contribution is the difference between the actual temperature and the base temperature.[7] So, a temperature of 20° during 1 h contributes $(20 - 10) \times 1/24$, or about 0.42 'growing degree days' (GDD), and temperatures above 34° contribute nothing. Now make a graph of the daily temperature pattern, as in Figure A8-3, add up the hourly growing degrees where the temperature was between 10° and 34° and divide by 24. For the temperatures in the graph the result is 10.8 GDD for that day (It will be a little higher because the temperatures above 34° still contribute something). Since most people only measure daily minimum and maximum temperatures, the CERES model has a routine, which generates a daily pattern from them and then calculates the approximate GDD. It is often found that a specific number of GDD is needed to complete a particular growth stage, which makes it such an attractive parameter. GDD is a rather crude concept, though, and its physiological underpinning is weak, but it often

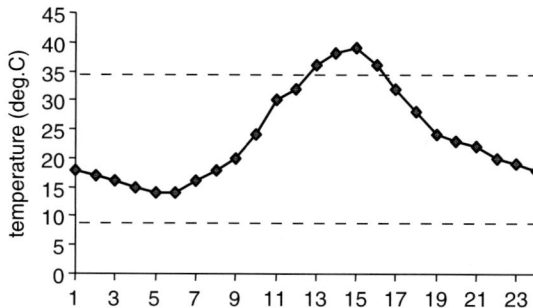

Figure A8-3. Specimen daily temperature pattern

---

[7] That is a little oversimplified. In CERES the absolute maximum is actually 44°C. There is no growth above 44° and temperatures between 34° and 44° contribute progressively less.

works quite well. Practically all crop models use GDD to simulate the length of different growth stages and seed companies often publish the lifetime GGD for their varieties.

(b) Growth of leaf area

The growth habits of species with terminal inflorescences like the cereals are very different from those of dicotyledonous plants. Leaf growth, therefore had to be simulated in more detail than CROP-GRO did, which could get away with treating leaves as "one aggregated or lumped class with no age or positional structure".[8] Simulating leaf growth was quite a complex affair in CERES, which I will try to explain as briefly as possible. In order not to complicate things too much I will first look at potential leaf growth, modified by moisture and nitrogen stress. That is the demand function in the model during vegetative growth. Later I will bring in the carbohydrate supply side and the way growth is adjusted in case of shortages.

The interval between the appearance of successive leaf tips from the leaf whorl had to be given to the model as an input (I_L in the table, in GDD) and was obtained from field measurements. From the actual temperature course, the model could then calculate how many visible leaves there were at any time. Next, the growth rates of the leaves were needed. When looking at a full-grown maize plant, you see a typical size pattern along the stem with small leaves at the bottom, large ones in the middle and small ones again towards the tassel (the beautiful picture of Figure A8-4 is from Galinat, 1979).[9] It looks tempting to try and find some algorithm that can generate the growth of successive leaves, but the creators of CERES-Maize took another road. They derived several empirical equations for the potential growth rate of the leaf area in successive periods from the results of field measurements. The rates were then adjusted by multiplying them by a reduction factor for moisture or nitrogen stress, whichever is more severe. Total daily growth in leaf area was found as the sum of the rates of all the leaves and total leaf area as the cumulative sum of the daily growth rates.

---

[8] Boote et al., 1998.

[9] The only thing that is wrong in the drawing is the number of leaves which is more likely to be around 20 in cultivated maize. Also, in a field crop, maize plants usually have one or at most two ears.

Figure A8-4. Drawing of a maize plant (by Galinat,1979.). Reproduced by permission of Syngenta AG

(c) Total Number of Leaves and Total Leaf Area

There is a limit to the number of leaves a maize plant can initiate, because eventually the apex stops making leaf primordia and converts into a tassel. That happens after a period of tassel induction, which takes place in Stage S2. As you can see from the drawing, the last few leaves below the tassel become progressively smaller and CERES-Maize therefore used a separate growth function for the final leaf growth phase, when the last three leaves are expanding. But how did the model know which leaves were the last three? That is only possible if it knows how many leaves the plant will eventually grow, which can be calculated from three things: (i) the interval between the initiation of two successive leaves (I_p, in GDD); (ii) the total GDD from emergence to tassel initiation (G_T); and (iii) the number of leaf primordia already present at emergence. The number of leaves (N_T) then equals:

$$N_T = \frac{G_T}{I_P} + 5$$

Leaf initiation at the apex cannot be observed, however, so we want an expression based on the appearance of leaf tips instead. By dissecting plants and counting leaf primordia it was found that the interval between leaf tip appearances was about twice as long as the interval between successive primordia, hence the expression that was used by CERES-Maize:

$$N_T = \frac{G_T}{I_L \times 0.5} + 5, \text{ where } I_L \text{ stands for the leaf tip appearance interval}$$

The remaining unknown is G_T, the time needed from emergence to tassel initiation (in GDD), that is the sum of the GDD of the growth stages S9, S1 and S2. For stage S9 it was calculated as a simple function of sowing depth. The authors also devised a method to calculate the duration of S1 and S2, which involved two parameters, P1, the duration of stage S1 and P2, which represents the variety's sensitivity to photoperiod. Both parameters had to be supplied as an input into the model. That would be fine if they could be measured in the field, but that was practically impossible. One problem is that everything happening inside the leaf whorl and at the apex is hidden from view, until the tassel finally emerges. The second is that you cannot actually distinguish the stages S1 and S2, even in dissected plants, which makes it impossible to pinpoint the transition from S1 to S2 from simple inspection. In practice, modellers usually estimated the 'genetic' coefficients P1 and P2 by the infamous procedure of model calibration. That means running the model many times with different values until it correctly 'predicts' the total number of leaves. That has been one of the major weaknesses of the model, which to my knowledge has not been resolved. Anyway, let us assume the model had found the correct number of leaves, then it knew when to start using the last of the leaf area growth functions. Potential growth of leaf area could now be simulated throughout the crop's life. Potential increase in leaf biomass was then calculated by multiplying by the SLA ,which was growth-stagedependent but otherwise fixed.

(d) Biomass production and growth adjustment

So far we have only considered potential leaf growth, modified by moisture and nitrogen stress. Let us now look at the way the model calculated actual biomass growth, for vegetative growth stage S1 only, to keep things simple. During each time step (1 day), the model calculated the amount of solar radiation absorbed by the canopy, using the LAI from the previous iteration, multiplied by a fixed conversion factor to obtain the rate of biomass production. That is justified by the finding that dry matter production is more or less proportional to absorbed radiation. Since there are no stems, tassels and ears, yet biomass growth is partitioned to the leaves and roots only. Leaves are given first access, allowing them to grow at their potential rate, provided at least 25% remains for the roots. Otherwise leaf growth is adjusted downwards. I do not know where the figure of 25% came from, but I presume there must have been some experimental basis for it.

There were many more complex issues involved in growth simulation by CERES-Maize, like calculating the length of the growth stages S3– S6 and simulating cob growth and grain yield. I will not go into that, just a few words about the interesting way the model handled growth partitioning to the grain. First there is an element of forward coupling built into the size of the cobs, or rather the number of initiated grains, by making it dependent on the plants' average photosynthetic rate from the beginning of silking to grain filling. Like in most models growth of the grain had absolute priority and if necessary the production by the canopy was supplemented by translocation from the stem followed by the leaves. In case of remaining shortage, grain growth itself was adjusted.

(e) Model calibration

Like in other models, many ad-hoc assumptions and empirical relations were built into CERES-Maize for lack of real data or real insight in the underlying physiology. I am not particularly familiar with the maize literature, but T. Tollenaar, a long-time maize researcher and modeller wrote, for example, in 2002 on his Internet website: 'Little is known about environmental and genotypic influences on the duration of the various sub-phases of development relative to each other.' And furthermore, apart from the effect of radiation on seed setting, there was no 'memory' in the model, which would, for example, link the size of a leaf to those of the earlier ones through the size of the apex. Without doubt that is the way it happens in real plants. Not surprisingly then, the many journal papers, which have reported on the performance of CERES-Maize outside the ecology where it was developed, demonstrate the model's relatively weak physiology base. In practically every case new calibration had to be done, not just of the so-called genetic coefficients, but also of the other parameters of the growth functions, which were presumed to be fixed.

A8.4 The ARS cotton production model

Cotton modelling in the USA started in the early 1970s and attracted many gifted scientists of different denominations, notably agronomists, plant physiologists, soil scientists and engineers. The papers published in *Crop Science* and other journals in those years convey the enthusiasm with which they scraped together all that was known at the time about how cotton plants grow and put it all, ripe and green, into the comprehensive framework of a simulation model. Meanwhile, a lot of experiments were conducted to fill the knowledge gaps, on leaf and canopy photosynthesis, flowering and fruiting, leaf growth and a host of other things. Several cotton simulation models were built, but the one which in the end carried the day was GOSSYM, put together under the leadership of Don Baker in Mississippi, who was later joined by Basil Acock.

During the 1980s GOSSYM was converted into a practical decision tool for cotton producers, probably too early, because the reliability of its predictions was found wanting. Instead of attacking the problems at their roots, by replacing the deficient plant physiology, the modellers started tinkering with the parameters of half-understood processes, as their colleagues around other crop models did, in order to make the simulated crops look like the real thing in the field. Around 1996, however, the USDA Agricultural Research Service decided it was time to start afresh and bring together the best elements from earlier cotton models as well as some new physiology in a new model, rather than trying to mend GOSSYM. The new model, called the Cotton Production Model (CPM), was developed by a team led by Basil Acock. It underwent some field testing up to 2000, when USDA apparently lost interest and stopped the funding. Since CPM reputedly contained the best possible representation of

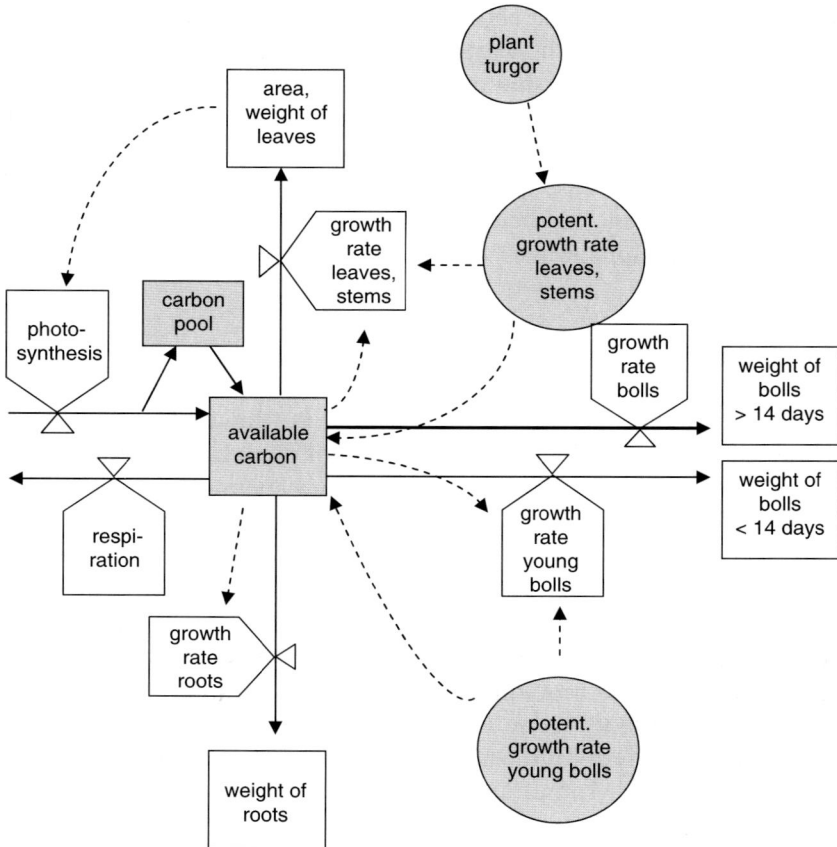

Figure A8-5. Flow diagram of the ARS Cotton Production Model

the knowledge about cotton around the turn of the century, I will describe how it handled organ growth and assimilate partitioning.

CPM was very different from source-driven models like SUCROS and CROPGRO. The simulation hinged on the potential growth of different organs, adjusted if necessary to the availability of substrates, much like Brouwer and de Wit's original ELCROS, my cotton model KUTUN and to a lesser extent CERES. Its flow diagram in Figure A8-5 shows the essential processes of organ growth and assimilate use,[10] which I must briefly explain.

---

[10] I put the diagram together from the CPM documentation available to me; Basil Acock suggested some improvements while confirming its overall correctness.

Potential and actual growth were calculated with time steps of 1 day, while photosynthesis, transpiration, nitrogen uptake and leaf water content were simulated on an hourly basis. My description of the CPM logic was mostly gleaned from the extensive documentation posted on the USDA website,[11] apparently in the desperate hope that somebody might take over the unfinished job.

A8.4.1 The demand side: potential growth

The model's core concept was the separate simulation of potential organ growth and assimilate production, followed by the adjustment of growth to substrate availability. The model kept track of the growth of individual leaves and fruiting points and adjusted boll load to the assimilate status of the crop, as we will see presently.

(a) Potential growth of leaves, stem and branches

If you want to carry out a detailed calculation of potential leaf area growth, four things are needed. First, the interval between successive leaves has to be clearly defined, because it is sensitive to both temperature and to availability of growth substrates. In cotton an unambiguous reference point is the moment when the veins of an unfolding leaf become visible. The reasonable assumption is that this represents the same developmental stage on different leaves. Furthermore you must know the areas of successive leaves at unfolding, their growth duration and their growth pattern. The CPM used a sigmoid pattern with initial size and growth duration as parameters, whereby the initial size of successive leaves was taken to be a fixed function of their location on the plant, as measured on optimally growing plants. Finally, growth of leaf weight was found by multiplying growth in area by SLA, which was treated as a function of temperature and CO_2 concentration. With that, potential leaf growth was completely determined.

Growth in length of stem internodes was linked with growth of the leaves subtending them. Their thickness and weight, however,

---

[11] In 2006 the package could be downloaded via http://www.arsusda.gov/acsl/research/accomp.html

continue to grow long after the leaf has stopped growing and the model increased its weight in proportion to the weight of the plant above it, similar to the way KUTUN handled stem growth.[12]

As leaf turgor pressure fell, shoot expansion was reduced progressively and the potential growth rate of all the organs on the shoot decreased in the same measure.

(b) Potential growth of fruiting points

All fruiting points were initiated with the same weight. After flowering, two growth stages were distinguished: 'young green bolls' and 'green bolls'. The potential growth rate for each stage and their duration were functions of temperature.

A8.4.2 The supply side: canopy assimilation and translocation

(a) Potential Canopy Photosynthesis

The model calculated potential canopy assimilation rate hourly from light interception and the photosynthetic rate of individual leaves. The rate could be reduced by very low air temperature, by plant nitrogen content or by the age of the uppermost, whichever was most limiting. Light respiration and maintenance respiration were subtracted from gross photosynthesis to give potential net canopy photosynthetic rate under well watered conditions. The net rate is negative at night and sometimes in low light.

(b) The shoot carbon pool and translocated carbon

Part of the model's carbohydrate flows passed through a 'carbon pool', which buffered against short-term fluctuations in assimilate production and built up reserves during daytime to be used for growth and respiration at night. The pool could not be missed because of the hourly time steps of the model's assimilation. Part of the assimilates were stored in the stems for a longer time and could be mobilised later to feed growing bolls when their demand was greatest and assimilation was declining due to senescence.

---

[12] The procedure of KUTUN is explained in Appendix 5.

A8.4.3 Actual growth, growth partitioning and functional balance

(a) Actual growth

If there was an insufficient supply of carbohydrates for an organ to grow at its full potential rate, the growth rate was reduced to match the two. At the next time step, new potential growth rates were calculated without taking into account the growth reduction in the previous time step. In other words, there was no 'memory' of the previous growth history. Growth adjustment to the availability of nitrogen was handled in a similar way.

(b) Growth partitioning and functional balance

The partitioning logic was based on Brouwer's concept of functional balance, which favoured growth of the organ closest to the factor, which was in short supply. Except in young plants where the model allocated 50% of the available carbohydrates to the roots to simulate their rapid early growth, decreasing to zero at first square.

CPM's partitioning worked as follows. In the absence of water or nitrogen stress the model gave priority for carbohydrates to the shoot. They were divided as follows. Bolls older than 14 days[13] had absolute priority to allow them to grow at their potential rate (except in the case of water stress as we will see presently). What remained after that first went to bolls of 7–14 days old in the order of their age, and then to leaf and stem tissue. Young fruiting points, up to 7 days after flowering, competed with leaves and internodes on an equal footing. Next the taproot was provided and what remained after that was stored in the carbon pool, up to a maximum of 30% of total shoot weight. When the pool was full the rest went to the feeder roots.

Now suppose the plants experienced gradually increasing water shortage, what would happen? As we have seen, water stress reduced potential growth of the above-ground organs and therefore their demand for resources. Thus, as leaf turgor decreased near midday, every day the amount allocated to the roots would gradually increase, the effect being that the plant mitigates the effect of

---

[13] I presume this means 14 days under some standard temperature conditions.

water stress by increased root development. Until water is applied, full turgor is restored and shoot growth returns to its potential rate.

The mechanism for nitrogen was different. If there was not enough nitrogen to match the carbohydrates for growth of the shoots, the model would direct the surplus carbohydrates to the roots and extract the nitrogen needed for their growth from the vegetative tissue.

What about carbohydrate stress, that is, what happened when the canopy did not produce enough carbohydrate to allow the above-ground organs to grow at their potential rates? For example, because of extreme cloudiness. All the carbohydrates would be shunted to the above-ground organs and if low radiation conditions continued for a long time, the ageing roots would gradually lose their uptake capacity, which must eventually result in loss of turgor and reduced top growth. Surplus carbohydrates would now become available again to the roots and a new balance would be attained, at a higher shoot–root ratio than before, as it should be. As far as I have been able to figure out, the CPM did not have an ageing routine for roots, so this mechanism would not occur. I do not think this incomplete-ness affected the model's predictions under 'normal' conditions, but it could have unexpected consequences. That could only have been found out by running the kind of sensitivity tests, which are part of model development but which were unlikely to take place after the CPM had become an orphan.

(c) Abscission of fruiting points

One of the most important mechanisms, the cotton plant has to adjust growth to resource availability, is abscission of fruiting points. By doing that it avoids the need to curtail the growth of fruits, which are beyond a certain development stage and harm the quality of their lint and seeds. CPM abscised fruiting points if they had not been fed for 2 days and their dry weight as a fraction of potential dry weight had fallen below a threshold value. That threshold decreased with the boll's age. Bolls older than 14 days after flowering were not abscised whatever happened. That looks sound enough, except for one thing. In CPM, like in practically all models, a crop was treated as if it consisted of a large number of identical plants: it was an 'average plant' instead of a population model, as real crops are. That works fine as long as growth adjust-ments occur on a continuous scale from zero to unity. Growth adjustment of fruiting points, however, is discrete: they are either retained or abscised. In a crop at wide spacing the effect of the abscission process, while still discrete, is smoothed by virtue of the many bolls per plant, but when the model aborts a boll in a very dense crop it becomes unbalanced. That was one of CPM's weakest features of which the modellers were well aware (in fact Basil Acock told me about the imbalance) and which was one of the issues to be addressed in the model's further development.[14]

A8.5 Summing up

All the models, which I have reviewed here, were good at simulating basic growth processes like photosynthesis, uptake of water and nutrients, transpiration and conversion of carbohydrates and nutrients into biomass, but all except CPM were really quite weak in the representation of the plant's innate growth template. They all opted for handling the all-important growth partitioning in an empirical way by postulating more or less fixed partitioning factors measured in field trials. That makes them quite stable, but at a high cost: it eliminates the plant with its feedback and feed-forward mechanisms and it becomes highly site-specific because of the empirical parameters.

CPM resembled ELCROS in that it tried to account for the plant's mechanisms by treating potential growth as the demand factor and adjusted actual growth to substrate production through a functional balance approach with a minimum of fixed parameters. That was one of its key features, but it introduced an element of instability which did not occur in the other models because they forced growth into the straight jacket of fixed resource allocation. The CPM model did not go all the way, though. Growth had no memory in that growth potential of a new organ was not affected by the plant's growth history, as in a real plant. Consider a plant which has suffered from a lot of stress and has only attained half the size it would have without stress. Surely, its leaf primordia will also have shared in the growth reduction and the growth potential of the next leaf will be much smaller than the same leaf in an unstressed plant. I did not see that kind of feed-forward in any

---

[14] My cotton model, KUTUN, works with a more realistic 'population' of fruiting points at each fruiting site, so that this imbalance does not occur.

model including ELCROS or CPM, which why is why I said earlier that there was no plant in the models.

Another serious shortcoming of CPM is that it worked with an average plant instead of a population. That turned out to be a real handicap especially when simulating fruiting point abscission as we have seen. For the other models it is less harmful, because they essentially lump together the different tissues into unstructured assemblies of organs, which grow according to the resources allocated to them by practically fixed partitioning schemes, without potential growth as demand. That makes those models more stable, but also less realistic.

Appendix 9. Calculation of Potential Assimilation, Dry Matter Production and Yield

The aim of this Appendix is to show how a rough estimate of potential biomass production and crop yield can be made. The calculations involve three steps. First potential carbon assimilation is estimated using approximate methods due to Goudriaan and van Laar and to Sinclair. Then we need to know how much assimilate is consumed in manufacturing a unit of plant biomass, and finally yield is estimated by using some (rather daring) assumptions about how the canopy develops and how much biomass ends up in the economic product.

A9.1 Potential assimilation

In order to calculate potential assimilation we must know how much photosynthetically active solar radiation (PAR) reaches the earth's surface, how it is distributed inside the canopy, and how assimilation by individual leaves responds to the radiation they receive. There are comprehensive simulation models like those by de Wit and by Duncan and co-workers and their derivatives, which account in detail for the interception and conversion of light energy and carry out separate calculations for direct radiation from the sun and diffuse radiation from all parts of the sky. Those models are very unwieldy and unsuitable for the general user, so approximate or summary models have been developed, which were relatively simple and yet captured the essence of the comprehensive ones. One such model was published by Goudriaan and van Laar in 1978. The reason why I chose this particular one was that it is fairly easily understood and may be used by non-modellers. Next I will look at an even more drastic simplification due to Thomas Sinclair and finally compare the predictions made with both.

A9.1.1 The Goudriaan–van Laar model

The calculations by the Goudriaan-van Laar model were carried out in four steps. First the amount of solar radiation received at the earth's surface on a clear day was calculated for different latitudes

and dates. Then a clever approximate method was used to estimate
daily assimilation under a clear and under an overcast sky. Since the
model was only approximate, so were its predictions. They had to
be 'corrected' to bring them as close as possible to those of a comprehensive benchmark model whose predictions were considered
reliable.[1] The results of the approximate model for different latitudes and dates were therefore regressed on those of the benchmark
model and each time a prediction was made for particular conditions, it had to be corrected by that regression equation. This may
not be immediately clear, but it should become so as we go along.

(a) Solar radiation

How does one calculate the amount of radiation under a clear sky?
That is a purely physical problem, which is treated in any not too
elementary text on climatology. I have given Goudriaan and van
Laar's version of the equations in Box A9-1, in case someone wants
to write a simple program for his own use. Otherwise, skip the box
and go straight to Table A9-1, which was calculated with the same
equations.
    The table shows daily clear sky radiation for different latitudes
and dates. Note that the unit is *mega*-Joule per m^2. That is because
for most people it is easier to memorise 15.4 MJ/m^2 than 1.54×10^7 J/m^2.

(b) Calculating daily gross assimilation

Completely overcast sky

Now comes the clever part. First the authors considered a completely overcast sky. They reasoned that the response of canopy
assimilation to radiation must follow a saturation-type pattern: at
very low radiation the response will be linear, like that of an individual leaf, and at very high radiation it must approach a maximum when all leaves are light-saturated, even though that
situation will never occur in practice. As more leaves reach saturation, the curve very gradually approaches its maximum, as the
example of Figure A9-1 shows. Now let us see whether we can

---

[1] The authors used as a benchmark the assimilation model which was originally
designed by de Wit and refined by various workers from his group, notably
Goudriaan himself.

Box A9-1. Calculating daily total photosynthetically active radiation

---

We start from the top of the atmosphere where a certain amount of radiation arrives per second. Although it varies a little throughout the year, Goudriaan and van Laar assumed it to be constant at 640 J/m^2/s PAR. The extinction of the radiation in its path through the atmosphere is exponential. When the sun is straight overhead the fraction transmitted equals $e^{-0.1}$, so the amount of radiation received at a horizontal surface equals $R = 640e^{-0.1}$. The lower the sun, the longer the distance travelled by the beam through the atmosphere and the higher the extinction. If β is the sun's angle relative to the horizon, the relative distance travelled is approximately $1/\sin \beta$ and the radiation incident on a surface *normal to the sun beam* equals $R_n = 640e^{-0.1/\sin(\beta)}$. We want the radiation incident on a horizontal surface, however, which is $\sin(\beta)$ times that:

$$R = 640 \sin (\beta) \, e^{-0.1/\sin(\beta)} \text{ J/m}^2\text{/sec} \qquad (1)$$

Not surprisingly, amount of radiation is determined by the sun's angle, which is a function of latitude λ, the earth's declination δ and time of day h, in hours. It is calculated as follows (if you want to understand the geometry, consult a meteorological textbook). Call $ssn = \sin \delta \sin \lambda$ and $ccs = \cos \delta \cos \lambda$, then

$$\sin (\beta) = ssn + ccs \times \cos\{2\pi (h + 12)/24\}$$

The only unknown in this expression is the earth's declination δ, which depends on the day of the year, d, according to:

$$\delta = -23.45 \, ccs\{2\pi (d + 10)/365\}$$

Total daily radiation is found by carrying out the calculations repeatedly, say for every 15 minutes, each time multiplying the result by 900 (15 min $\times$ 60 s) and adding it all up. It is not difficult to write a program for the calculations, for example, in BASIC or a spreadsheet.

---

find the parameters of the response curve – the slope of the linear part and the maximum.

The response of an individual leaf at a very low radiation level is called its efficiency ε, in kg CO_2 per J and the efficiency for the entire canopy must be the same, that is the tangent of angle α in Figure A9-1. The maximum leaf rate is A_{max}, so the theoretical maximum for a canopy must be $LAI \times A_{max}$. It will be approached much more slowly than the maximum of an individual leaf. The efficiency and the maximum leaf rate have been measured for many crop species, so we know the start and the end of the curve relating canopy assimilation to incident radiation.

How about the intermediate part of the curve and how rapidly does it approach its maximum, in other words, what is its exact

Table A9-1. Calculated daily PAR incident on a horizontal surface under a clear sky; MJ/m^2/day

North latitude	15 Jan	15 Feb	15 Mar	15 Apr	15 May	15 Jun	15 Jul	15 Aug	15 Sep	15 Oct	15 Nov	15 Dec
0	14.00	14.71	15.16	14.95	14.26	13.77	13.97	14.68	15.17	14.94	14.23	13.77
10	12.17	13.43	14.67	15.43	15.48	15.34	15.41	15.51	15.09	13.95	12.55	11.80
20	9.99	11.72	13.67	15.38	16.22	16.47	16.38	15.84	14.48	12.49	10.50	9.53
30	7.58	9.64	12.20	14.81	16.45	17.11	16.87	15.64	13.37	10.61	8.16	7.05
40	5.05	7.29	10.30	13.73	16.17	17.28	16.87	14.94	11.80	8.40	5.67	4.50
50	2.60	4.78	8.06	12.19	15.43	17.01	16.41	13.75	9.80	5.95	3.18	2.11
60	0.61	2.33	5.56	10.24	14.30	16.43	15.60	12.15	7.47	3.42	1.00	0.32
70	0.00	0.38	2.96	7.97	13.05	16.09	14.85	10.28	4.88	1.10	0.00	0.00
80	0.00	0.00	0.62	5.64	12.85	16.72	15.24	8.82	2.22	0.00	0.00	0.00
90	0.00	0.00	0.00	4.83	13.01	16.99	15.48	8.73	0.19	0.00	0.00	0.00

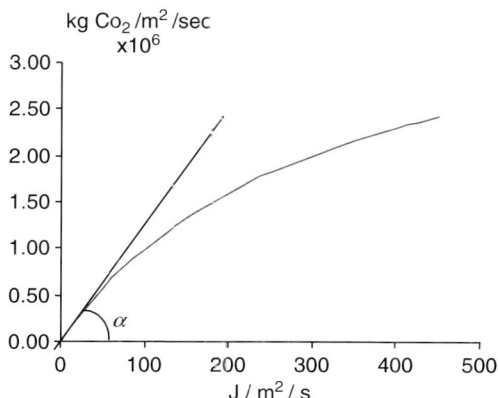

Figure A9-1. Saturation curve for canopy assimilation with incident radiation

shape? The authors chose a so-called rectangular hyperbola, because of its gradual approach to the maximum, as in Figure A9-1. Then they assumed that a similar response pattern will obtain between *daily average* radiation and canopy assimilation, because the relative luminosity of the overcast sky is more or less homogeneous throughout the day. The technical detail about the way the curve's parameters were found is given in Box A9-2, which you may skip if you wish, although it is quite gratifying if you like that kind of thing.

Clear sky

The method used for overcast skies did not work well for clear skies, mainly because light distribution is much less uniform. Therefore, assimilation by directly illuminated and by shaded leaf area were calculated separately and the shape of the canopy response curve was modified. Details are given in Box A9-3.

(c) Adjusting the predictions made by the summary model

The summary model was now complete, except that the results showed a small but systematic difference with those of the benchmark model. That is not surprising with so much content packed into just a few equations. The final step was therefore to correct the predictions made by the summary model by regressing them on those of the comprehensive model, which were assumed to be correct. Box A9-4 explains how that was done.

Box A9-2. Calculating approximate assimilation for an overcast sky

---

A general expression for the hyperbolic response of canopy photosynthesis under an overcast sky (A_{ov}) is:

$$A_{ov} = \frac{X}{X+1} LAI \times A_{max} \qquad (5)$$

What can we say about X? First it must be proportional to average daily radiation under an overcast sky ($X = aR_{ov}$), so that A will be zero when R is zero and approaches $LAI \times A_{max}$ when R is very large. Furthermore, by definition, the initial increase in assimilation with incident radiation is the efficiency ε, therefore:

$$\varepsilon = \lim_{R \to 0} \frac{A_{ov}}{R_{ov}} = \frac{a}{a R_{ov} + 1} LAI \times A_{max} = a LAI \times A_{max}. \text{ Hence,}$$

$$a = \frac{\varepsilon}{LAI \times A_{max}}, \text{ and } X = \frac{\varepsilon R_{ov}}{LAI \times A_{max}}$$

Average radiation equals total daily radiation divided by day length, so we need to know the day length at a particular location and date as well. The astronomical day length (in seconds) is calculated with the following expression:

$$L = 43200 \left\{ \pi + 2 \text{arc sin} (ssn/ccs) \right\} / \pi \text{ (} ssn \text{ and } ccs \text{ were defined in Box A9-1)}$$

The 'effective day length' as defined by the authors is counted after solar elevation exceeds 8°:

$$L_{eff} = 43200 \left\{ \pi + 2 \text{arc sin} [(- \sin (8\pi/180) + ssn)/ccs] \right\} / \pi$$

Finally, radiation under a completely overcast sky was assumed to be 20% of that under a perfectly clear sky, hence average daily radiation equals $R_{ov} = \frac{0.2 R_d}{L_{eff}}$

We can now calculate daily gross assimilation under an overcast sky using the following expressions:

$$R_{ov} = \frac{0.2 R_d}{L_{eff}} \qquad X = \frac{\varepsilon R_{ov}}{LAI \times A_{max}} \qquad A_{ov} = \frac{X}{X+1} LAI \times A_{max} \times L_{eff}$$

(d) Model predictions

We now have the complete approximate model proposed by Goudriaan and van Laar, consisting of summary equations for canopy assimilation plus two regression equation to link its predictions to those of a comprehensive benchmark model. Let us see what predictions they yield for the assimilation by a closed canopy. First we must agree what we mean by a closed canopy. As a canopy gets denser the lowest leaves receive less and less radiation until they become starved for assimilates. Most species will then drop them off so that there is a maximum to the density a canopy can attain, usually

---

Box A9-3. Calculating approximate assimilation for a clear sky

---

For a so-called spherical leaf angle distribution[2] it can be shown that the area of the leaves illuminated directly by the sun in a dense canopy equals $2\sin\beta$, where β is the sun's elevation angle. The authors took the average daily $\sin\beta$ to be roughly half that at noon, when the sun's elevation equals $(90°+\delta - \lambda)$. Hence, the estimated average sunlit (LAI_{sl}) and shaded (LAI_{sh}) leaf areas equal:

$$LAI_{sl} = \sin(90° + \delta - \lambda) = \cos(\delta - \lambda) \text{ and } LAI_{sh} = LAI - LAI_{sl}$$

Next the authors observed that under a clear sky light saturation will be attained more gradually than under an overcast sky and in order to account for that they replaced the variable X in the expression for the saturation curve by $X' = \ln(1 + X)$. Finally, 45% of the incoming PAR was attributed to the sunlit and 55% to the shaded area (these percentages were actually found by trial and error to obtain results as close as possible to those of the benchmark model). So, two sets of equations result:

For sunlit leaf area: *For shaded leaf area:*

$$LAI_{sl} = \cos\{(\delta - \lambda) \times \pi / 180\} \qquad LAI_{sl} = LAI - LAI_{sl}$$

$$R_{cl} = \frac{R_d}{L_{eff}}$$

$$X' = \frac{0.45\varepsilon R_{cl}}{LAI \times A_{max}} \qquad\qquad X = \frac{0.55\varepsilon R_{cl}}{LAI_{sh} \times A_{max}}$$

$$X' = \ln(1 + X) \qquad\qquad X' = \ln(1 + X)$$

$$A_{sl} = \frac{X'}{1 + X'} LAI_{sl} \times A_{max} \times L_{eff} \qquad A_{sh} = \frac{X'}{1 + X'} LAI_{sh} \times A_{max} \times L_{eff}$$

Total predicted gross assimilation for a clear sky now equals: $A_{cl} = A_{sl} + A_{sh}$

---

at a LAI of around 5. A closed canopy is therefore commonly taken to be one with LAI 5, which will intercept more than 95% of incident radiation.

The model then needs to be told the latitude and the date, as well as two properties of the leaves' assimilation curve: its initial slope, i.e. efficiency ε, and the maximum rate A_{max}. Values for different crop species can be found in the literature. They are ugly-looking quantities, which I cannot avoid. For ε I will use 1.3×10^{-8} kg CO_2 /J of incident PAR, which is a much quoted value in the literature. A_{max} ranges from 1–1.25×10^{-6} kg CO_2/m²/s for C_3 plants, such as rice and

---

[2] That is a canopy where the angles and orientations of the leaves have the same distribution as the surface elements of a sphere. It is mathematically convenient and comes close to the real distribution for many species, while modest deviations do not greatly affect canopy assimilation.

Box A9-4. Matching predictions by the comprehensive and the summary model

---

The authors carried out a linear regression of the predictions by the comprehensive model on those of the summary model, which resulted in the following regression equations:

$$\hat{A}_{ov} = 0.9935 A_{ov} + 0.11 \times 10^{-3} \, \text{kg CO}_2/\text{m}^2/\text{day, and}$$

$$\hat{A}_{cl} = 0.95 A_{cl} + 2.05 \times 10^{-3} \, \text{kg CO}_2/\text{m}^2/\text{day}$$

$\hat{A}_{ov}$ and $\hat{A}_{cl}$ are what we are looking for: the prediction of canopy assimilation by the benchmark model, as estimated by the approximate model. In other words, assimilation as calculated by the approximate model is the independent variable in the regression equations whose dependent variables will be as close as possible to the predictions from the benchmark model.

---

cotton, to around 2×10^{-6} for C_4 plants such as maize and sugarcane. Table A9-2 shows daily gross CO_2 assimilation calculated by the approximate model with these input values.

The results for perfectly clear and uniformly overcast skies were tabulated separately, but normal weather is usually somewhere in between, so how do you calculate potential assimilation under actual sky conditions from the table? Solar radiation is routinely measured at many meteorological stations[3] and potential assimilation can be calculated directly from there. Suppose for example that on June 15, at 40° north latitude, PAR measured over the entire day was 12.6×10^6 J/m². The reasoning now goes as follows. According to Table A9-1 the radiation under a clear sky would be 17.28×10^6 J/m². We now assume with Goudriaan and van Laar that radiation under a completely overcast sky is 20% of clear sky radiation and that actual radiation can be handled as if it resulted from spells with fully clear and fully overcast sky. That of course is a gross simplification because all sorts of intermediate skies occur but it has been found to be a reasonable approximation. The fraction x of the day when the sky is clear is then calculated from:

$$17.28 \times x + 0.20 \times 17.28 \times (1 - x) = 12.6$$

---

[3] Meteorological stations measure global radiation, that is radiation over the entire spectrum; about 50% of that is photosynthetically active radiation in the wavelength band from 400 to 700 nm.

Table A9-2. Calculated daily gross assimilation (kg CO_2/ha/day) by closed canopies (LAI 5) with different assimilation parameters

$A_{max} = 1.0 \times 10^{-6}$ kg CO_2/m²/s; $\varepsilon = 1.3 \times 10^{-8}$ kg CO_2/J

Latitude		Jan 15	Feb 15	Mar 15	Apr 15	May 15	Jun 15	Jul 15	Aug 15	Sep 15	Oct 15	Nov 15	Dec 15
0	CL	703	724	737	731	711	696	702	723	737	731	710	696
	OV	306	319	327	324	311	302	305	318	327	323	310	302
10	CL	635	679	721	750	757	755	756	755	736	697	649	623
	OV	270	294	318	333	335	333	334	335	326	304	277	262
20	CL	552	618	690	754	789	802	797	772	720	647	572	533
	OV	226	261	299	333	350	356	354	342	315	276	236	216
30	CL	450	539	642	743	808	837	826	776	688	579	476	426
	OV	175	219	271	324	357	371	365	340	295	239	188	164
40	CL	331	440	574	716	815	861	844	764	637	491	362	302
	OV	120	170	234	304	354	377	368	329	264	193	134	108

$A_{max} = 1.25 \times 10^{-6}$ kg CO_2/m²/s; $\varepsilon = 1.3 \times 10^{-8}$ kg CO_2/J

Latitude		Jan 15	Feb 15	Mar 15	Apr 15	May 15	Jun 15	Jul 15	Aug 15	Sep 15	Oct 15	Nov 15	Dec 15
0	CL	787	812	828	821	796	778	786	811	828	820	795	778
	OV	316	330	339	335	321	311	315	329	339	334	320	311
10	CL	708	759	809	842	849	847	849	848	826	780	723	692
	OV	278	304	329	345	346	344	345	347	337	314	285	270
20	CL	610	688	772	846	886	901	895	867	807	721	633	589
	OV	232	268	309	344	362	368	366	354	326	284	242	222

(continued)

Table A9-2. (continued)

$A_{max} = 1.25 \times 10^{-6}$ kg $CO_2/m^2/s$; $\varepsilon = 1.3 \times 10^{-8}$ kg CO_2/J

Latitude		Jan 15	Feb 15	Mar 15	Apr 15	May 15	Jun 15	Jul 15	Aug 15	Sep 15	Oct 15	Nov 15	Dec 15
30	CL	493	596	714	831	907	939	927	869	767	642	523	466
	OV	179	224	279	334	369	383	378	352	304	245	192	168
40	CL	358	481	634	797	912	965	945	854	706	539	393	326
	OV	122	173	240	313	365	389	380	339	272	198	137	110

$A_{max} = 1.5 \times 10^{-6}$ kg $CO_2/m^2/s$; $\varepsilon = 1.3 \times 10^{-8}$ kg CO_2/J

Latitude		Jan 15	Feb 15	Mar 15	Apr 15	May 15	Jun 15	Jul 15	Aug 15	Sep 15	Oct 15	Nov 15	Dec 15
0	CL	856	886	904	895	867	847	855	884	904	895	866	847
	OV	323	337	347	342	328	318	322	337	347	342	327	318
10	CL	767	826	882	919	927	924	926	925	902	850	785	750
	OV	283	310	336	353	354	352	353	355	345	321	291	275
20	CL	658	745	840	923	967	983	978	946	879	783	684	634
	OV	236	274	316	352	371	377	375	362	333	290	247	226
30	CL	528	642	774	905	989	1,025	1,012	947	833	693	561	498
	OV	182	228	285	341	377	392	386	359	310	250	195	170
40	CL	379	514	684	865	992	1,051	1,029	928	763	578	417	344
	OV	124	176	244	320	373	398	389	346	277	201	138	111

$A_{max} = 2 \times 10^{-6}$ kg $CO_2/m^2/s$; $\varepsilon = 1.3 \times 10^{-8}$ kg CO_2/J

Latitude		Jan 15	Feb 15	Mar 15	Apr 15	May 15	Jun 15	Jul 15	Aug 15	Sep 15	Oct 15	Nov 15	Dec 15
0	CL	967	1,003	1,025	1,015	980	955	965	1,001	1,025	1,014	978	955
	OV	332	347	357	353	338	327	331	347	357	352	337	327
10	CL	862	932	999	1,043	1,050	1,046	1,048	1,049	1,022	960	883	841
	OV	291	319	346	363	365	362	364	365	356	330	299	283
20	CL	733	835	947	1,045	1,097	1,115	1,109	1,073	993	880	763	705
	OV	241	281	324	363	382	388	386	373	343	298	253	231
30	CL	582	713	868	1,022	1,120	1,162	1,146	1,071	938	773	620	548
	OV	186	233	292	351	388	404	398	370	319	256	199	173
40	CL	413	566	761	971	1,120	1,188	1,162	1,044	853	639	456	374
	OV	126	179	249	328	384	409	400	355	284	205	140	113

and x works out at 0.66. Predicted gross assimilation for a crop with an A_{max} of 1.5×10^{-6} kg $CO_2/m^2/s$ (maize perhaps) is then found from (the bottom part of) Table A9-2 as:

$$0.66 \times 1188 + 0.34 \times 409 = 923 \text{ kg } CO_2/ha/day.$$

Finding potential assimilation with the Goudriaan-van Laar method, although fairly straightforward, is a little awkward, unless it is programmed on a computer or scientific calculator. That is not really difficult and it only has to be done once. Alternatively, the figures may be looked up or interpolated in Table A9-2. Note that assimilation is expressed as amount of CO_2, because that is what we observe when measuring assimilation. In the past it was often expressed as glucose, by simply multiplying the weight of CO_2 by 30/44. I have followed the current convention and expressed everything in weight of CO_2.

  If you find the Goudriaan-van Laar equations forbidding, then an even simpler procedure based on a crop's supposedly constant RUE parameter is the way to go. The following describes such a procedure due to Thomas Sinclair.

A9.1.2 Further simplification: radiation-use efficiency (RUE)

In the USA some modellers went one step further in simplifying the simulation of canopy assimilation, for example, Thomas Sinclair, of the University of Florida, who stripped down the process to the barest minimum, with quite amazing results. His argument went like this.

  A large amount of data from a wide range of conditions showed that the amount of biomass produced by a crop is proportional to the amount of radiation it intercepts. The proportionality factor is called the RUE, usually given as grams of biomass produced per MJ of intercepted radiation. We have so far been talking about CO_2 assimilation, but that does not matter, because a more or less fixed conversion factor from CO_2 to biomass may be assumed, at least when looking at a full crop cycle (I will come to that later), so RUE should also be more or less constant for CO_2 assimilation. If that holds true, Sinclair argued (1991), and if the RUE is known for a particular 'easy' case, then CO_2 assimilation can be calculated for any case by simply multiplying intercepted radiation by that RUE.

Sinclair's trick was to calculate an average assimilation rate under some standard conditions for a canopy with LAI 4 by a very simple model and divide that by the amount of radiation intercepted to get a figure for RUE. The RUE thus obtained was then used as a characteristic property for the crop species. For those who are interested, the calculation procedure is shown in Box A9-5. As we will see, the predictions by this simple model for closed canopy assimilation were quite decent over a wide range of conditions.

In order to get a feel for the order of magnitude of RUE and its range, I have compiled a look-up table with values calculated for different combinations of maximum leaf rate and efficiency and a LAI of 4 (Table A9-3). You may want to check some of them using the procedures of Box A9-5.[4] To find the RUE for a particular crop species you need to know its leaf assimilation parameters A_{max} and ε, of course. For completeness' sake I have compiled them for some

---

Box A9-5. Seven steps to calculate RUE by Sinclair's method

1. Take as midday incident PAR 400 J/m²/s and estimate daily average radiation on that day by:
   $I = 400 \sin(45°)$ (45° is the average sun angle)
2. The fraction intercepted radiation equals:
   $F = 1 - e^{-LAI \times .5/\sin(45°)}$
   (5 is the shadow projection of a 'spherical' canopy)
3. The sunlit and shaded LAI equal:
   $LAI_{sl} = F \times \sin(45°)/.5$ and $LAI_{sh} = LAI - LAI_{sl}$
4. Radiation intercepted by sunlit and shaded leaves:
   $R_{sl} = I \times F / LAI_{sl}$ and $R_{sh} = .2 \times I \times F / LAI_{sh}$
5. Assimilation by unit area of sunlit and shaded leaves:
   $A_{sl} = A_{max} (1 - e^{-\varepsilon R_{sl} / A_{max}})$ and
   $A_{sh} = A_{max} (1 - e^{-\varepsilon R_{sh} / A_{max}})$
6. Total assimilation equals:
   $A_{tot} = A_{sl} \times LAI_{sl} + A_{sh} \times LAI_{sh}$
7. And finally, RUE for CO2 equals:
   $RUE = A_{tot} /(I * F)$

---

[4] Note that radiation is taken as PAR again, which is about 50% of total solar radiation at the earth's surface. Sinclair's own calculations were for total radiation, which results in half these values.

Table A9-3. RUE in g CO_2 per MJ of intercepted PAR, for different combinations of A_{max} and ε and LAI 4, calculated by Sinclair's method

Efficiency kg CO_2/J	A_{max}, kg CO_2/m$_2$/s								
	8.0E-07	9.0E-07	1.0E-06	1.1E-06	1.2E-06	1.3E-06	1.5E-06	1.7E-06	2.0E-06
1.0E-08	4.43	4.76	5.07	5.36	5.62	5.87	6.32	6.71	7.22
1.1E-08	4.59	4.94	5.27	5.58	5.86	6.13	6.62	7.06	7.62
1.2E-08	4.74	5.11	5.45	5.78	6.08	6.37	6.90	7.37	7.98
1.3E-08	4.89	5.26	5.62	5.96	6.28	6.59	7.15	7.66	8.32
1.4E-08	5.02	5.41	5.78	6.14	6.47	6.79	7.39	7.92	8.64

important crops, together with their RUE calculated by Sinclair's method (Table A9-4). The assimilation parameters were taken from a 1989 modelling text by Penning de Vries et al. I am not too sure about the quality of some of the maximum leaf rates, though, in particular those of groundnut (too high?) and maize (too low?) but if you have better data it should not be too difficult to recalculate the figures.

With these RUE values and incident radiation, potential canopy assimilation can be calculated for any location and date from:

$$\text{potential assimilation} = \text{incident radiation} \times 0.97 \times \text{RUE}$$

whereby 0.97 stands for the interception by a canopy of LAI 5. If this trick works it would be simple indeed. Let us see whether it does, by comparing the predictions with those from the Goudriaan–van Laar method.

A9.1.3 Comparing the two methods

In order to compare the predictions by the two methods I calculated potential assimilation (at LAI 5) for latitudes ranging from 0 to 40° NL, three A_{max} values, which cover most crop species and a single efficiency ($\varepsilon = 1.2 \times 10^{-8}$ kg CO_2/J). The sky is assumed to be 40% overcast. The results are shown in Table A9-5. The similarity of the predictions by the two methods is remarkable.

Outside this range of conditions the predictions by the Sinclair model get worse. For completely clear skies it overestimates

Table A9-4. Leaf assimilation parameters and calculated RUE for some important crops. (Leaf parameters after Penning de Vries et al.,1989.)

Crop	Leaf assimilation params		RUE g CO_2/MJ
	A_{max} kg CO_2/m^2 /s $\times 10^6$	Efficiency kg CO_2/J $\times 10^8$	
Rice	1.31[1]	1.02	5.95
Maize	1.67	1.02	6.73
Cassava	0.97	1.28	5.74
Soybean	1.11	1.23	5.97
Groundnut	1.39	1.28	6.81
Cotton	1.25	1.02	5.80

[1] Be aware that 1.31 means 1.31×10^{-6}, etc.

Table A9-5. Comparison of potential CO_2 assimilation in kg CO_2 per ha per day, with 40% overcast sky, predicted by the Goudriaan–van Laar and the Sinclair summary models

Latitude	Method	$A_{max} = 1.25 \times 10^{-6}$ kg CO_2/m^2			$A_{max} = 1.5 \times 10^{-6}$ kg CO_2/m^2			$A_{max} = 1.75 \times 10^{-6}$ kg CO_2/m^2		
		Mar 15	June 15	Sep 15	Mar 15	June 15	Sep 15	Mar 15	June 15	Sep 15
0	G-vL	604	565	604	649	605	649	687	639	688
	Sinc	623	566	624	691	627	691	749	680	749
10	G-vL	590	617	603	633	663	648	670	701	685
	Sinc	603	631	620	668	699	687	724	757	745
20	G-vL	560	657	587	600	706	629	634	747	665
	Sinc	562	677	596	623	750	660	675	813	715
30	G-vL	515	685	555	550	736	594	580	779	627
	Sinc	502	704	550	556	780	609	602	845	660
40	G-vL	454	701	507	482	752	541	506	795	569
	Sinc	424	711	485	469	787	537	509	853	582

production by 10–15% and for completely overcast skies the predictions are completely off the mark, but that is to be expected with the equations used. I think it is safe to conclude that for average tropical wet season conditions, Sinclair's very simple model is quite adequate to get a rough estimate for potential canopy assimilation.

A9.2 Potential biomass production

We are now ready for the next step towards our goal of estimating potential crop yield: calculating potential biomass production. We know how to calculate potential assimilation, so the question is now how much biomass can be produced per unit of assimilate. That depends on the chemical composition of the biomass, of course, which in turn differs with the organs you look at. Obviously, the composition of a protein-rich soybean grain is very different from that of the roots. So the question must be broken down in three subsidiary ones:

- How is biomass distributed over the plant parts
- What is their composition and
- How much of each of the major compounds can be produced per unit of assimilate.

To start with the last one, the amount of assimilates needed per unit of compound has been worked out in detail by Penning de Vries et al. (1989), with the following results (Table A9-6):
Note that the figures are for conversion of glucose, not CO_2 this time.

The distribution of biomass over the plants' organs and their chemical composition for particular crop species can be found in the literature (where I am afraid you will find a lot of confusion)

Table A9-6. Conversion factors (g glucose/g compound) for different plant substances (After Penning de Vries et al., 1989)

Compound	Conversion factor	Compound	Conversion factor
Carbohydrates	1.275	Lignin	2.231
Protein; no N-fixation	1.887	Organic acids	0.954
With N-fixation	2.784	Minerals (uptake)	0.120
Fatty compounds	3.189		

or measured directly. I extracted data on organ composition from
Penning de Vries et al. (1989) and added some informed guesses on
biomass distribution on my part, to calculate an average chemical
composition for a number of important species. That is rather
crude, because both dry matter distribution and the composition
of the tissues change over time, but since we are going to look at
potential lifetime biomass production it does not really matter for
our purpose. From the conversion factors of Table A9-6 and the
distribution and composition of biomass an overall, average conversion factor can be calculated for the crop. The amount of crop
biomass that can be produced per unit of CO_2 assimilated can then
be calculated, taking into account that part of the assimilates are
used for maintenance. Box A9-6 shows the calculations for cassava

---

Box A9-6. How much cassava biomass can be produced per gram of
assimilated CO_2?

---

1. We use the following figures for biomass distribution, organ composition
   and overall composition of cassava:

Organ	% of the Biomass	Composition, %					
		CH_2O	Protein	Fat	Lignin	Org. acids	Minerals
Leaves	15	52	25	5	5	5	8
Stems	15	62	10	2	20	2	4
Feeder roots	15	56	10	2	20	2	10
Storage roots	55	87	3	1	3	3	3
Overall		73	8	2	8	3	5

---

2. With the conversion factors of Table A9-4, we calculate that 1 gram of
   cassava biomass will cost: $0.73 \times 1.275 + 0.08 \times 1.887 + 0.02 \times 3.189 + 0.08 \times 2.231 + 0.03 \times 0.954 + 0.05 \times 0.120 = 1.376$ g of glucose, that is 2.019 g of
   CO_2.
3. About 25% of the assimilates was assumed to be used for maintenance,
   leaving 75% for conversion into biomass.
4. Hence, to produce 1 gram of cassava biomass the plant must assimilate
   $2.019/0.75 = 2.69$ g CO_2.
5. Or, from 1 g CO_2 assimilated, 0.37 g cassava biomass can be produced.

as an example. The third column of Table A9-7 shoes the results for various crop species.

The rest of the calculations is straightforward: the species' potential assimilation rate (calculated by Sinclair's or by the Goudriaan-Laar's method) is multiplied by the amount of biomass produced per unit of CO_2 to get its potential biomass production rate. Box A9-7 gives an example, for maize this time.

For the sake of completeness again (and for you to check the answer if you do the calculations yourself) I compiled calculated potential biomass production at different latitudes on July 15 for six crop species in Table A9-7 (hoping I made no mistakes).

Table A9-7. Calculated daily potential biomass production of various crops on July 15 at different latitudes, with 40% cloudiness (potential assimilation was calculated by Sinclair's method)

Crop	RUE g CO_2/MJ	g biomass/ g CO_2	Potential biomass production, kg/ha/day				
			0°	10°	20°	30°	40°
Rice	5.95	0.37	203	224	238	245	245
Maize	6.73	0.35	217	239	255	262	262
Cassava	5.74	0.37	196	216	229	–	–
Soybean	5.97	0.29	160	176	187	193	193
Groundnut	6.81	0.27	169	187	199	205	205
Cotton	5.80	0.31	166	183	194	200	200

Box A9-7. Calculating maximum daily biomass production of maize at 10° NL on July 15

1. Calculate clear sky radiation at 10° NL on July 15 or look it up in Table A9-1. Result: 15.41 MJ/m^2.
2. Radiation with 40% cloudiness equals $0.6 \times 15.41 + 0.40 \times 0.20 \times 15.41 = 10.48$ MJ/m^2.
3. Multiply by the maximum interception (0.97) and by RUE for maize (from Table A9-5: 6.73), to get CO_2 assimilation: 68.4 g CO_2/m^2 = 684 kg/ha.
4. The overall conversion factor of CO_2 to biomass found for maize was 0.35 (Table A9-5), hence, maximum daily biomass production equals: $0.35 \times 684 = 239$ kg/ha.

A9.3 Potential crop yield

The final stage of the exercise is to estimate potential crop yield from potential biomass production, putting in some additional information about the crop in question. There are two sides to this problem: first how much biomass does the crop produce during its lifetime, and second, which part of that ends up in the useful product – the grain, the tuber or the seedcotton. Let us start with lifetime biomass production. We have seen in Chapter 9 how crop simulation models handle it, but we want something simpler here. I am going to define an 'equivalent canopy' for that purpose, which intercepts the same total amount of radiation as the actual crop does, but which is easier to handle numerically. And since biomass production is assumed to be proportional to intercepted radiation, lifetime biomass production by the equivalent canopy should also be the same as that of the real one. The example I am going to use is tropical maize. I must warn you, though, that there are going to be some rather sweeping simplifications, which may be distasteful to purists.

The canopy of a fully developed maize crop growing under optimum conditions intercepts practically all radiation and assimilation takes place at its potential rate. During the crop's juvenile and senescent periods, however, the canopy is not closed and the duration of those periods strongly affects the crop's total assimilation and biomass production. So, in order to calculate potential crop production we must know the lengths of the juvenile and the senescence periods with some accuracy. Graph (a) in Figure A9-2 shows the evolution of a hypothetical maize canopy. In a young maize crop the canopy grows exponentially for a while and then slows down until it attains full development. During the latter part of the crop's life fewer leaves are formed (or none at all in determinate crops like maize) and the ones, which are there, age and drop off. The canopy density therefore decreases rapidly as the crop matures. Graph (b) shows the fraction of incident radiation that is intercepted as a function of LAI.[5] Finally the dotted part of graph (c), which was put together from the first two is what we need: the fraction of the radiation, which is intercepted at different times in the life of the maize crop. With that interception pattern we could,

---

[5] Assuming that interception follows the so-called Beer's law: $I = 1 - e^{-0.8 \times LAI}$.

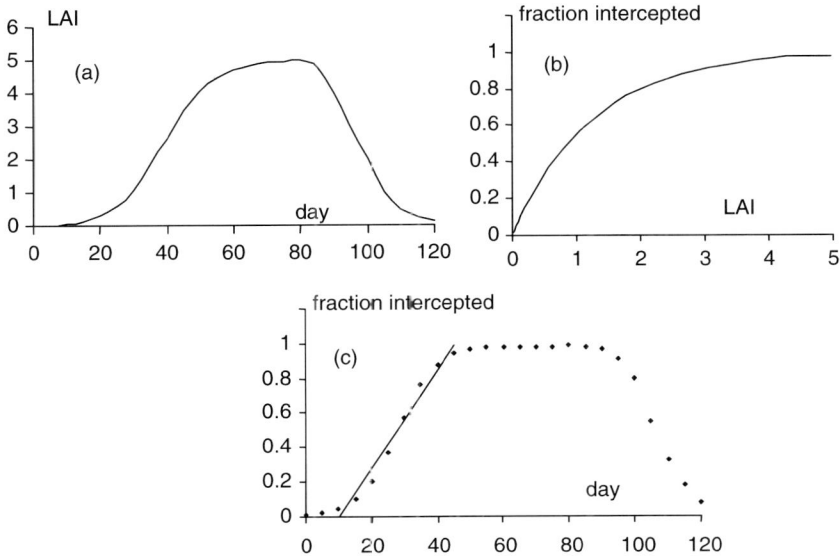

Figure A9-2. Patterns of LAI (a) and radiation interception with LAI (b) and with age (c) by a hypothetical maize crop in the tropics

in principle, calculate the crop's biomass production rate at any time by multiplying incident radiation by the fraction, which is intercepted and by RUE, as explained in section A9.1.2 and then add up the results. The curvilinear ascent and descent of the curve are hard to handle in manual calculations, however, so it would be convenient if we could replace them by straight lines, which trace the same area. I have drawn one for the juvenile part in graph (c), which I think is acceptable. It starts at 10 days after emergence and ends at about 45 days when maximum interception is reached, that is after 35 'effective linear' days. At the end something similar happens in the opposite direction. So there are 70 effective days in all with an average interception and an average biomass production, which is half that of a closed canopy. Or, which is the same thing, 35 days with the same production as a fully closed canopy. Those 35 days are now added to the 45 days during which the crop's interception is maximum (see graph), which gives us the equivalent of 80 days with full interception. In other words, total interception and total biomass production by a maize crop which matures in 120 days would be equivalent to those by a fully closed canopy in 80 days.

Now suppose this crop is grown at 10° NL from May to August and the average cloudiness in that period is about 40%. Table A9-7 tells us that daily growth of biomass around the middle of July would be 239 kg/ha. Radiation varies only little during the growing season at that latitude, so we take it to be more or less constant at the July 15 value and estimate that the crop's total biomass yield will be around $80 \times 239 \approx 19,000$ kg/ha. Part of that will be below ground, in the roots, say 15%, leaving 16,200 in the tops. How much maize yield would that be? Modern maize varieties have a harvest index close to 50%, so this dry matter yield would be equivalent to about 9.5 t of maize grain (8,100 kg dry matter with 14% moisture content). That is indeed the kind of maximum yield that is obtained in that environment.

Literature Cited

Ahn, P.M., 1970. *West African soils*. Oxford University Press, Oxford.

Andrews, D.J., 1972. Intercropping with sorghum in northern Nigeria. *Experimental Agriculture* 8, 139–150.

Baker, E.F.I., 1978, 1979. Mixed cropping in northern Nigeria. 1. Cereals and groundnuts; 2. Cereals and cotton; 3. Mixtures of cereals. *Experimental Agriculture* 14, 293–298; 15, 33–40; 14, 41–48.

Benoît-Cattin, M., 1986. Les Unités Expérimentales du Sénégal. ISRA/CIRAD/FAC. Montpellier.

van Beukering, J.A., 1947. Het ladangvraagstuk, een bedrijfs- en sociaal economisch probleem. Mededeelingen van het Departement van Economische Zaken in Nederlandsch-Indië, No. 9.

Boeke, J.H., 1930. Dualistische economie. Inaugural lecture of Professor Boeke at the University of Leiden. In: W.F. Wertheim (1966) (Ed.) *Indonesian economics: the concept of dualism in theory and practice*. W. van Hoeve, The Hague.

Bonneval, P., M. Kuper and J.-P. Tonneau, 2002. L'Office du Niger, grenier à riz du Mali. CIRAD, Montpellier.

Boote, K.J., J.W. Jones and G. Hoogenboom, 1998. Simulation of crop growth: CROPGRO model. In: R.M. Peart and R.B. Curry (Eds.) *Agricultural systems modeling and simulation*. Marcel Dekker, New York, pp. 651–692.

Boote, K.J., M.J. Kropff and P.S. Bindraban, 2001. Physiology and modelling of traits in crop plants: implications for genetic improvement. *Agricultural Systems* 70, 395–420.

Bouman, A.M., H. van Keulen, H.H. van Laar and R. Rabbinge, 1996. The 'School of de Wit' crop growth simulation models: a pedigree and historical overview. *Agricultural Systems* 52, 171–198.

Bouman, A.M., M.J. Kropff, T.P. Tuong, M.C.S. Wopereis, H.F.M. ten Berge and H.H. van Laar, 2001. Oryza2000: modeling lowland rice. IRRI, Los Baños and WUR, Wageningen, The Netherlands.

Bremer, G., 1928. De cytologie van het suikerriet, 4e bijdrage. Een cytologisch onderzoek der bastaarden tusschen Saccharum officinarum en Saccharum spontaneum. *Mededeelingen van het Proefstation voor de Java-suikerindustrie,* Jaargang 1928, No. 11, 565–795.

Brouwer, R. and C.T. de Wit, 1968. A simulation model for plant growth with special attention to root growth and its consequences. *Proceedings of the fifteenth Easter School in Agricultural Science*, University of Nottingham, 224–244.

Carsky, R.J., M. Becker and S. Hauser, 2001. Mucuna cover crop fallow systems: potential and limitations. In: *Sustaining soil fertility in West Africa*. SSSA Special Publication No. 58.

Carsky, R.J., S. Nokoe, S.T.O. Lagoke and S.K. Kim, 1998. Maize yield determinants in farmer-managed trials in the Nigerian northern Guinea savanna. *Experimental Agriculture* 34, 407–422.

Cobley, L.S., 1957. *The botany of tropical crops*. Longmans, Green & Co., London.

Collinson, M., 1999. *A history of farming systems research.* CABI Publishing, UK.

Donald, C.M., 1968. The breeding of crop ideotypes. *Euphytica* 17, 385–403.

Erickson, R.O. and F.J. Michelini, 1957. The plastochron index. *American Journal of Botany* 44, 297–305.

Evans, G.C., 1972. *The quantitative analysis of plant growth. Studies in ecology,* Vol. I. Blackwell Scientific Publications, Oxford.

Galinat, W.C., 1979. Botany and origin of maize. In: Ernst Häfliger (Ed) Maize. Technical Monograph, Ciba-Geigy Agrochemicals, pp. 6–12.

Geertz, C., 1968. *Agricultural involution: the process of ecological change in Indonesia.* University of California Press, Berkeley, CA.

van der Goot, P., 1948. Twaalf jaar rijstboorderbestrijding door zaaitijdsregeling in West-Brebes (res. Pekalongan) (with summary). *Landbouw* 20, 465–494.

van der Goot, P., 1951. Over levenswijze en bestrijding van sawah-ratten in het laagland van Java. *Landbouw* 23, 123–294.

Goudriaan, J. and H.H. van Laar, 1978. Calculation of daily totals of the gross CO_2 asimilation of leaf canopies. *Netherlands Journal of Agricultural Science* 26, 373–382.

Grubben, G.J.H., 1975. La culture de l'Amarante, légume-feuilles tropical, avec référence spéciale au sud-Dahomey. *Mededelingen Landbouwhogeschool Wageningen 75-6.*

Haar, J. van der, 1993. De geschiedenis van de Landbouwuniversiteit Wageningen. Deel I. PUDOC, Wageningen, The Netherlands.

Hahn, S.K., E.R. Terry and K. Leuschner, 1980. Cassava breeding for resistance to mosaic disease. *Euphytica* 29, 673–683.

Heemskerk, W., N. Lema, D. Guindo, C. Schouten, Z. Semgalawe, H. Verkuijl, B.S. Piters and P. Penninkhof, 2003. A guide to demand-driven agricultural research. *The client-oriented research management approach: rural service delivery for agricultural development.* KIT Publishers, Amsterdam, The Netherlands.

Heinz, D.J. (Ed), 1987. *Sugarcane improvement through breeding. Developments in Crop Science 11.* Elsevier, Amsterdam, The Netherlands.

Hesketh, J.D., D.N. Baker and W.G. Duncan, 1972. Simulation of growth and yield in cotton: II. Environmental control of morphogenesis. *Crop Science* 12, 436–439.

Hildebrand, Peter E., 1999. Don't fence me in. The adventures of a boy from Brush. Unpublished manuscript.

Hildebrand, P.E. and F. Poey, 1985. *On-farm agronomic trials in farming systems research and extension.* Lynne Rienner Publishers, Boulder, CO.

Hilhorst, T. and F. Muchena, 2000. *Nutrients on the move. Soil fertility dynamics in African farming systems.* International Institute for Environment and Development, London.

Horie, T., C.T. de Wit, J. Goudriaan and J. Bensink, 1979. A formal template for the development of cucumber in its vegetative phase. I, II and III. *Proceedings Nederlandse Academie Wetenschappen Serie C* 82, 433–479.

Hijkoop, J., P. van der Poel and B. Kaya, 1991. Une lutte de longue haleine ... Aménagements anti-érosifs et gestion de terroir. *Systèmes de production rurale au Mali,* Vol. 2. IER, Bamako, KIT Publishers, Amsterdam, The Netherlands.

van Ittersum, M.K., P.A. Leffelaar, H. van Keulen, M.J. Kropff, L. Bastiaans and J. Goudriaan, 2003. On approaches and applications of the Wageningen crop models. *European Journal of Agronomy* 18, 201–234.

Jones, J.W., G. Hoogenboom, C.H. Porter, K.J. Boote, W.D. Batchelor, L.A. Hunt, P.W. Wilkens, U. Singh, A.J. Gijsman and J.T. Ritchie, 2003. The DSSAT cropping system model. *European Journal of Agronomy* 18, 235–265.

Juo, A.S.R. and H. Grimme, 1980. Potassium status in major soils of tropical Africa with special reference to potassium availability. Paper presented at the Potassium Workshop, IITA and International Potassium Institute, October 8–10, 1980.

Kang, B.T., G.F. Wilson and L. Sipkens, 1981. Alley cropping maize (Zea *mays* L.) and Leucena (Leucena *leucocephala* Lam) in southern Nigeria. *Plant and Soil* 63, 165–179.

Kang, B.T., F.E. Caveness, G. Thian and G.O. Kolawole, 1999. Long-term alley cropping with four hedgerow species on an Alfisol in southwestern Nigeria – effect on crop performance, soil chemical properties and nematode population. *Nutrient Cycling in Agroecosystems* 54, 145–155.

Kang, B.T., L. Reynolds and A.N. Atta-Krah, 1990. Alley Farming. *Advances in Agronomy* 43, 315–359.

Kassam, A.H., J. Kowal, M. Dagg and M.N. Harrison, 1975. Maize in West Africa and its potential in savanna areas. *World Crops* 27, 73–78.

van Keulen, H., 1977. Nitrogen requirements of rice with special reference to Java. *Contr. Central Research Institute for Agriculture, Bogor, Indonesia*, No. 30.

Kiniry, J.R., 1991. Maize phasic development. In: Hanks, J. and J.T. Ritchie (Eds.) Modeling plant and soil systems. Agronomy Series No. 31. ASA, CSSA, SSA Publishers, Madison, WI.

Koudokpon, V., 1992. Pour une recherche participative. Stratégie et développement d'une approche de recherche avec les paysans de Bénin. DRA-Cotonou et KIT, Amsterdam, The Netherlands.

Kropff, M.J., K.G. Cassman, S. Peng, R.B. Matthews and T.L. Setter, 1994. Quantitative understanding of yield potential. In: Cassman (Ed.) *Breaking the yield barrier. Proceedings of a workshop on rice yield potential in favourable environments.* IRRI, 29 November–4 December 1993. pp. 21–38.

Ley, G., G. Baltissen, W. Veldkamp, A. Nyaki and T. Schrader, 2002. *Towards integrated soil fertility management in Tanzania.* KIT Publishers, Amsterdam, The Netherlands.

Lowe, R.G., 1986. *Agricultural revolution in Africa? Impediments to change and implications for farming, for education and for society.* Macmillan, London.

Meijer, B.J.M., K.J. van Ast and L.C. Zachariasse, 1979. Boer en bedrijf. Resultaat na 8 jaar ontwikkeling. Landbouw-Economisch Instituut.

Monsi, M. and T. Saeki, 1953. Über den Lichtfaktor in den Pflanzengesellschaften und seine Bedeutung für die Stoffproduktion. *Japanese Journal of Botany* 14, 22–52.

Monteith, J.L., 1977. Climate and the efficiency of crop production in Britain. *Philosophical Transactions of Royal Society of London B* 281, 277–294.

Moritsuka, N., J.Yanai and T.Kosaki, 2002. Depletion of non-exchangeable potassium in the maize rhizosphere and its possible releasing processes. *Proceedings of 17th WCSS*, 14–21 August 2002, Thailand.

Mutsaers, H.J.W., 1978. Mixed cropping experiments with maize and groundnuts. *Netherlands Journal of Agricultural Science* 26, 344–353.

Mutsaers, H.J.W., 1984. KUTUN: A morphogenetic model for cotton (Gossypium *hirsutum* L.). *Agricultural Systems* 14, 229–257.

Mutsaers, H.J.W., G.K. Weber, P. Walker and N.M. Fisher, 1997. A field guide for on-farm experimentation. IITA/ISNAR/CTA, Ibadan/The Hague/Wageningen, The Netherland.

Norman, D.W., 1974. Rationalising mixed cropping under indigenous conditions: the example of Northern Nigeria. *The Journal of Development Studies* 11, 1–21.

Norman, J.M. and T.J. Arkebauer, 1991. Predicting canopy photosynthesis and light-use efficiency from leaf characteristics. In: K.J. Boote and R.S. Loomis (Eds.) *Modeling crop photosynthesis – from biochemistry to canopy.* CSSA *Special Publication Number* 19, 75–94.

Nye, P.H. and D.J. Greenland, 1960. The soil under shifting cultivation. Technical Communication No. 51. Commonwealth Bureau of Soils, Harpenden, UK.

Penning de Vries, F.W.T., D.M. Jansen, H.F.M. ten Berge and A. Bakema, 1989. Simulation of ecophysiological processes of growth in several annual crops. Simulation Monographs 29. IRRI Los Baños, PUDOC, Wageningen, The Netherlands.

Piper, Ernest L. and A. Weiss, 1990. Evaluating CERES-Maize for reduction in plant population or leaf area during the growing season. *Agricultural Systems* 33, 199–213.

Purseglove, J.W., 1972. *Tropical crops: monocotyledons* 1. Longman , London.

Radford, P.J., 1968. Growth Analysis formulae – their use and abuse. *Crop Science* 7, 171–175.

Sanchez, Pedro A., 1976. *Properties and management of soils in the tropics.* Wiley, New York.

Simmonds, N.W., 1985. Farming systems research: a review. *World Bank Technical Paper Number* 43. The World Bank, Washington, DC.

Sinclair, T.R., 1991. Canopy carbon assimilation and crop radiation-use efficiency dependence on leaf nitrogen content. In: K.J. Boote and R.S. Loomis (Eds.) *Modeling crop photosynthesis – from biochemistry to canopy.* CSSA *Special Publication Number* 19, 95–107.

Slaats, J., 1995. *Chromolaena odorata* fallow in food cropping systems: an agronomic assessment in South-West Ivory Coast. Ph.D. thesis, Agricultural University, Wageningen, The Netherlands.

Smyth, A.J. and R.F. Montgomery, 1962. *Soils and land use in central western Nigeria.* The Government Printer, Ibadan, Western Nigeria.

Spiertz, J.H.J. and J. Ellen, 1978. Effects of nitrogen on crop development and grain growth of winter wheat in relation to assimilation and utilisation of assim-ilates and nutrients. *Netherlands Journal of Agricultural Sciences* 26, 210–231.

Steenhuijsen-Piters, B. de, 1995. Diversity of fields and farmers. Explaining yield variation in Northern Cameroon. Ph.D. thesis, Wageningen, The Netherlands.

Tessmann, G., 1913. Die Pangwe. Völkerkundliche Monographie eines Westafrikanischen Negerstammes. Ernst Wasmuth, Berlin. Erster Band. 275 pp.

Tian, G., B.T. Kang, G.O. Kolawole, P. Idinoba and F.K. Salako, 2005. Long-term effects of fallow systems on crop production and soil fertility maintenance in West Africa. *Nutrient Cycling in Agroecosystems* 71, 139–150.

Tian, G., B.T. Kang and G.O. Kolawole, 2003. Effect of fallow on pruning biomass and nutrient accumulation in alley cropping on Alfisols of tropical Africa. *Journal of Plant Nutrition* 26, 475–486.

Tripp, R. (Ed.), 1991. *Planned change in farming systems: progress in on-farm research*. Wiley, Chichester, UK.

Versteeg, M.N. and A. Huijsman, 1991. Trial design and analysis for on-farm adaptive research: a 1988 maize trial in the Mono Province of Benin. In: H.J.W. Mutsaers and P. Walker (Eds.) *On-farm research in theory and practice*. IITA, Ibadan, Nigeria, 111–121.

Vine, H., 1954. Is the lack of fertility of tropical African soils exaggerated? *Proceedings of the Second Inter-African Soils Conference*. Leopoldville.

Virk, P.S., G.S. Khush and S. Peng, 2004. Breeding to enhance yield potential of rice at IRRI: the ideotype approach. *International Rice Research Notes* 29.1, 5–9.

Vos, J., 1997. The nitrogen response of potato (*Solanum tuberosum* L.) in the field: nitrogen uptake and yield, harvest index and nitrogen concentration. *Potato Research* 40, 237–248.

Vos, J. and L.F.M. Marcelis (Eds.), 2007. *Proceedings of the Frontis Workshop on Functional-Structural Plant Modelling in Crop Production*, Wageningen, The Netherlands, 5–8 March 2006 (in press).

Westphal, E., 1975. Agricultural Systems in Ethiopia. *Agricultural Research Reports* 826, Centre for Agricultural Publishing and Documentation, Wageningen, The Netherlands.

Westphal, E., 1981. L'agriculture autochtone au Cameroun. Les techniques culturales, les séquences de culture, les plantes alimentaires et leur consommation. *Miscellaneous Papers 20, Landbouwhogeschool,* Wageningen, The Netherlands.

William, A., 1965. The African husbandman. Oliver & Boyd, London.

de Wit, C.T., 1958. Transpiration and crop yield. *Versl. Landbouwk. Onderzoek.* No. 64.6. Pudoc, Wageningen, The Netherlands.

de Wit, C.T., 1960. On competition. *Versl. Landbouwk. Onderzoek.* No. 66.8. Pudoc, Wageningen ,The Netherlands.

de Wit, C.T., 1965. Photosynthesis of leaf canopies. *Agricultural Research Report* No. 663. PUDOC, Wageningen, The Netherlands.

de Wit, C.T., et al. 1978. *Simulation of assimilation, respiration and transpiration of crops. Simulation Monographs,* PUDOC, Wageningen, The Netherlands.

de Wit, C.T., 1992. Resource use efficiency in agriculture. *Agricultural Systems* 40, 125–151.

de Wit, C.T., H.H. van Laar and H. van Keulen, 1979. Physiological potential of crop production. In: J. Sneep and A.J.T. Hendriksen (Eds.) *Plant breeding perspectives*. Pudoc, Wageningen, The Netherlands, pp. 47–82.

de Wit, C.T., R. Brouwer and F.W.T. Penning de Vries, 1970. The simulation of photosynthetic systems. In: *I. Setlik (Ed.) Prediction and measurement of photosynthetic productivity,* 17–23. Pudoc, Wageningen.

Yates, F. and W.G. Cochran, 1938. The analysis of groups of experiments. *Journal of Agricultural Science* 28, 556–580.

Yin, X. and H.H. van Laar, 2005. *Crop systems dynamics: an ecophysiological simulation model for genotype-by-environment interactions*. Wageningen Academic Publishers, Wageningen, The Netherlands.

Zachariasse, L.C., 1974. Boer en bedrijfsresultaat. Analyse van de uiteenlopende rentabiliteit van vergelijkbare akkerbouwbedrijven in de Noord-Oost-Polder. Ph.D. thesis, Wageningen University, The Netherlands.

Zoebl, Dirk, 2002. Crop water requirements revisited: the human dimensions of irrigation science and crop water management with special reference to the FAO approach. *Agriculture and Human Values* 19, 173–187.

Name Index

Subject Index